Impact Engineering and Application

Elsevier Science Internet Homepage - http://www.elsevier.com

Consult the Elsevier homepage for full catalogue information on all books, journals and electronic products and services.

Elsevier Titles of Related Interest

VOYIADJIS & KATTAN
Advances in Damage Mechanics: Metals and Matrix Composites
ISBN: 008-0436013

ZHAO & GRZEBIETA
Structural Failure and Plasticity (IMPLAST 2000)
2 Volume set.
ISBN: 008-043875X

Related Journals

Free specimen copy gladly sent on request. Elsevier Science Ltd, The Boulevard, Langford Lane, Kidlington, Oxford, OX5 1GB, UK

Engineering Fracture Mechanics
European Journal of Mechanics – A/Solids
European Journal of Mechanics – A/Solids with European Journal of Mechanics – B/Fluids (Combined Subscription)
International Journal of Fatigue
International Journal of Impact Engineering
International Journal of Mechanical Sciences
International Journal of Non-Linear Mechanics
International Journal of Pressure Vessels and Piping
International Journal of Solids and Structures
Journal of the Mechanics and Physics of Solids
Mechanics Research Communications
Theoretical and Applied Fracture Mechanics
Tribology International
Wear

To Contact the Publisher

Elsevier Science welcomes enquiries concerning publishing proposals: books, journal special issues, conference proceedings, etc. All formats and media can be considered. Should you have a publishing proposal you wish to discuss, please contact, without obligation, the publisher responsible for Elsevier's engineering mechanics publishing programme:

Ms. Nicola Garvey	Phone:	+44 1865 843879
Publishing Editor	Fax:	+44 1865 843931
Elsevier Science Ltd	E.mail:	n.garvey@elsevier.co.uk
The Boulevard, Langford Lane		
Kidlington, Oxford		
OX5 1GB, UK		

General enquiries, including placing orders, should be directed to Elsevier's Regional Sales Offices – please access the Elsevier homepage for full contact details (homepage details at the top of this page).

Impact Engineering and Application

Proceedings of the 4th International Symposium on Impact Engineering

*16 -18 July 2001,
Kumamoto, Japan*

Edited by

Akira Chiba
Kumamoto University, Japan

Shinji Tanimura
Osaka Prefecture University, Japan

Kazuyuki Hokamoto
Kumamoto University, Japan

Volume I

This publication is supported by the Commemorative Association for the Japan World Exposition (1970)

2001
ELSEVIER
AMSTERDAM – LONDON – NEW YORK – OXFORD – PARIS – SHANNON – TOKYO

ELSEVIER SCIENCE Ltd
The Boulevard, Langford Lane
Kidlington, Oxford OX5 1GB, UK

© 2001 Elsevier Science Ltd. All rights reserved.

This work is protected under copyright by Elsevier Science, and the following terms and conditions apply to its use:

Photocopying
Single photocopies of single chapters may be made for personal use as allowed by national copyright laws. Permission of the Publisher and payment of a fee is required for all other photocopying, including multiple or systematic copying, copying for advertising or promotional purposes, resale, and all forms of document delivery. Special rates are available for educational institutions that wish to make photocopies for non-profit educational classroom use.

Permissions may be sought directly from Elsevier Science Global Rights Department, PO Box 800, Oxford OX5 1DX, UK; phone: (+44) 1865 843830, fax: (+44) 1865 853333, e-mail: permissions@elsevier.co.uk. You may also contact Global Rights directly through Elsevier's home page (http://www.elsevier.com), by selecting 'Obtaining Permissions'.

In the USA, users may clear permissions and make payments through the Copyright Clearance Center, Inc., 222 Rosewood Drive, Danvers, MA 01923, USA; phone: (+1) (978) 7508400, fax: (+1) (978) 7504744, and in the UK through the Copyright Licensing Agency Rapid Clearance Service (CLARCS), 90 Tottenham Court Road, London W1P 0LP, UK; phone: (+44) 207 631 5555; fax: (+44) 207 631 5500. Other countries may have a local reprographic rights agency for payments.

Derivative Works
Tables of contents may be reproduced for internal circulation, but permission of Elsevier Science is required for external resale or distribution of such material.
Permission of the Publisher is required for all other derivative works, including compilations and translations.

Electronic Storage or Usage
Permission of the Publisher is required to store or use electronically any material contained in this work, including any chapter or part of a chapter.

Except as outlined above, no part of this work may be reproduced, stored in a retrieval system or transmitted in any form or by any means, electronic, mechanical, photocopying, recording or otherwise, without prior written permission of the Publisher.
Address permissions requests to: Elsevier Science Global Rights Department, at the mail, fax and e-mail addresses noted above.

Notice
No responsibility is assumed by the Publisher for any injury and/or damage to persons or property as a matter of products liability, negligence or otherwise, or from any use or operation of any methods, products, instructions or ideas contained in the material herein. Because of rapid advances in the medical sciences, in particular, independent verification of diagnoses and drug dosages should be made.

First edition 2001

Library of Congress Cataloging in Publication Data
A catalog record from the Library of Congress has been applied for.

British Library Cataloguing in Publication Data
A catalogue record from the British Library has been applied for.

ISBN: 0-08-043968-3 (2 volume set)

∞ The paper used in this publication meets the requirements of ANSI/NISO Z39.48-1992 (Permanence of Paper).
Printed in The Netherlands.

The papers presented in these proceedings have been reproduced directly from the authors' 'camera ready' manuscripts. As such, the presentation and reproduction quality may vary from paper to paper.

PREFACE

This Proceedings is a compilation of papers presented at the 4th International Symposium on Impact Engineering Kumamoto 2001 held in Kumamoto, Japan, during 16th – 18th July, 2001.

Delegates from fifteen countries participated in the Symposium and represented interests of academic and research institutions. In total 159 papers were presented, one of which was a keynote address by Professor Norman Jones of the University of Liverpool.

This symposium was aimed to unify the researchers and engineers of broad regions of science under the key word of Impact Engineering. And so, the researchers and engineers belonging to the various science regions discussed interdisciplinary themes and foresee new breakthroughs. Kumamoto city, where this symposium was held, is the seat of the Kumamoto Prefectural Office and till the Edo period it flourished as the castle town of the ruling Daimyo. Trace of such a glorious past may still be found in this city. It is also called a city of woods because it is blessed with many green trees and abundant water. In the suburbs of Kumamoto city, the great Mt. Aso, an active volcano, retains vividly its primitive state exposing its rough earth. Mt. Aso forms a caldera, believed to be the largest in the world, creating a monumental spectacle.

We are deeply indebted to Japan Society of Mechanical Engineers, Society of Materials Science, Japan, Japan Society for Technology of Plasticity and Japan Explosives Society for having sponsored the symposium and for enthusiastic interest in impact engineering. We would like to acknowledge with thanks the support rendered by the Centennial Anniversary Association, Faculty of Engineering, Kumamoto University, the Ministry of Education, Culture, Sports, Science and Technology Japan, Kumamoto International Convention Bureau, the Commemorative Association for the Japan World Exposition (1970) and other organizations and private firms.

We trust that this volume will serve as memoirs to these efforts and set a tone for role of these issues in the new millennium.

Finally, we would like to thank the staff and students for their efforts in making this symposium a success.

Akira Chiba
Shinji Tanimura
Kazuyuki Hokamoto

SYMPOSIUM ORGANIZATION

Organizing Committee

Akira Chiba, Kumamoto University, Japan (Chairman)
Shinji Tanimura, Osaka Prefecture University, Japan (Co-Chairman)
Masahiro Fujita, Sojo University, Japan (Public relations)
Kazuyuki Hokamoto, Kumamoto University, Japan (Secretary)
Shigeru Itoh, Kumamoto University, Japan (Treasurer)

Coordination Committee

Goro Eguchi, President, Kumamoto University, Japan
Zenta Iwai, Dean, Faculty of Engineering, Kumamoto University, Japan

International Advisory Panel

Y.-L. Bai, Inst. of Mech., CAS, China
G. Ben-Dor, Ben-Gurion Univ. of the Negev, Israel
S.R. Bodner, Technion-Israel Inst. of Technol., Israel
C.Y. Chiem, Ecole Centrale de Nantes, France
G. Gary, Ecole Polytechnique, France
W. Goldsmith, UC, Berkeley, USA
N.K. Gupta, Indian Institute of Technology, India
Y. Horie, Los Alamos National Laboratory, USA
N. Jones, The University of Liverpool, UK
J.F. Kalthoff, Ruhr-Universitat Bochum, Germany
J.R. Klepaczko, Metz University, France
M.S. Kim, Pusan National University, Korea
C.T. Lim, National Univ. of Singapore, Singapore
K.X. Liu, Peking University, China
L.W. Meyer, Technical Univ. Chemnitz-Zwickau, Germany
M.A. Meyers, UC, San Diego, USA
S. Nemat-Nasser, UC, San Diego, USA
T. Nishioka, Kobe Univ. of Mercantile Marine, Japan
W.K. Nowacki, Inst. of Fund. Technological Res., Poland
A.M. Rajendran, U.S. Army Research Lab., USA
S.R. Reid, UMIST, UK
V.P.W. Shim, National Univ. of Singapore, Singapore
W.J. Stronge, University of Cambridge, UK
Tadashi Shioya, The University of Tokyo, Japan
M. Stuivinga, TNO Laboratory, Netherlands
C.T. Sun, Purdue University, USA
N.N. Thadhani, Georgia Tech, USA
T.X. Yu, The Hong Kong Univ. of Sci. & Technol., HK
J.-L. Yu, Univ. of Sci. & Technol. of China, China

National Organizing Committee

Tatsuhiko Aizawa, University of Tokyo, Japan
Shigeru Aoki, Tokyo Inst. of Technology, Japan
Kazuo Asada, Mitsubishi Heavy Ind., Ltd., Japan
Akiyoshi Chatani, Kanazawa University, Japan
Masashi Daimaruya, Muroran Inst. of Technol., Japan
Hidefumi Date, Tohoku Gakuin University, Japan
Osamu Furukimi, Kawasaki Steel Corporation, Japan
Hiroomi Homma, Toyohashi Univ. of Technol., Japan
Nobutaka Ishikawa, National Defense Academy, Japan
Koichiro Kawashima, Nagoya Inst. of Technol., Japan
Kikuo Kishimoto, Tokyo Inst. of Technol., Japan
Toshiro Kobayashi, Toyohashi Univ. of Technol., Japan
Nobusato Kojima, Sumitomo Metal Ind. Ltd., Japan
Tomoaki Kurokawa, Setsunan University, Japan
Ichiro Maekawa, Kanagawa Inst. of Technol., Japan
Terushige Ogawa, Yokohama National Univ., Japan
Hirohisa Sato, Tohoku Gakuin University, Japan
Akira Sawaoka, President, Daido Inst. of Technol., Japan
Kiyoshi Takahashi, Kyushu University, Japan
Kazuyoshi Takayama, Tohoku University, Japan
Shigeo Takezono, Toyohashi Univ. of Technol., Japan
Masamitsu Tamura, The University of Tokyo, Japan
Sadayuki Ujihashi, Tokyo Inst. of Technol., Japan
Takashi Yokoyama, Okayama Univ. of Sci., Japan

Program Committee

Kiyotaka Dohke, Asahi Chem. Ind. Co. Ltd., Japan
Manabu Gotoh, Gifu University, Japan
Kiyoshi Hashizume, Nippon Kayaku Co. Ltd., Japan
Tsunetoshi Hayakawa, Chugoku Kayaku, Japan
Mitsuaki Iida, NIMC, Japan
Yutaka Imaida, Doshisya University, Japan
Kunihisa Katsuyama, NIRE, Japan
Mitsuo Koshi, The University of Tokyo, Japan
Takehiro Matsunaga, NIMC, Japan
Hidetsugu Nakamura, Kyushu Inst. of Technol., Japan
Motohiro Nakano, Osaka University, Japan
Noriaki Nakashima, Asahi Chem. Ind. Co. Ltd., Japan
Hideaki Negishi, Univ. of Electro-Comm., Japan
Taketoshi Nojima, Kyoto University, Japan
Isamu Oda, Kumamoto University, Japan
Yuji Ogata, NIRE, Japan
Kinya Ogawa, Kyoto University, Japan
Akihiko Sasoh, Tohoku University, Japan
Toshimori Sekine, NIRIM, Japan
Katsuhiko Sugawara, Kumamoto University, Japan
Kunihiko Tanaka, NOF Corporation, Japan
Masatake Yoshida, NIMC, Japan
Wenhui Zhu, Osaka Prefecture University, Japan

Executive Committee

Kazuhito Fujiwara, Kumamoto University, Japan
I. Fukuda, Yatsushiro Nat. College of Technol., Japan
Tetsuyuki Hiroe, Kumamoto University, Japan
Ryuichi Tomoshige, Sojo University, Japan
Toshiyuki Nagaishi, Kyushu Sangyo University, Japan
Kunihito Nagayama, Kyushu University, Japan
Y. Nakamura, Yatsushiro Nat. College of Technol., Japan
Tsutomu Mashimo, Kumamoto University, Japan
Syunichi Yoshinaga, Kyushu Sangyo University, Japan
K. Raghukandan, Annamalai University, India

Host Organization

Shock Wave and Condensed Matter Research Center, Kumamoto University, Japan
Faculty of Engineering, Kumamoto University, Japan

In cooperation with

Japan Society of Mechanical Engineers
Society of Materials Science, Japan
Japan Society for Technology of Plasticity
Japan Explosives Society

Sponsored by

The Commemorative Association for the Japan World Exposition (1970)
Centennial Anniversary Association, Faculty of Engineering, Kumamoto University
The Ministry of Education, Culture, Sports, Science and Technology, Japan
Kumamoto International Convention Bureau

CONTENTS

Volume I

Preface	v
Symposium Organization	vii
Some Recent Developments in the Dynamic Inelastic Response of Structures *N. Jones*	1
The Crushing Strength of Aluminium Alloy Foam at High Rates of Strain *S.R. Reid, P.J. Tan and J.J. Harrigan*	15
On the Behaviour Characterisation of Polymeric Foams over a Large Range of Strain Rates *H. Zhao and G. Gary*	23
Dynamic and Static Compression Behaviour of Paper and Polystyrene Foam Boards *H. Kobayashi, M. Daimaruya and K. Tanaka*	29
Deformation and Failure Mechanism of Aluminum Foams Under Dynamic Loading *Y.-L. Yu, X. Wang, Z.-G. Wei and Y. Pan*	35
Experimental Characterization of Mixed Mode Fracture Toughness of Polymer Matrix Composite Materials Under Impact Loading *T. Kusaka, M. Masuda and N. Horikawa*	43
Increasing the Maximum Strain Measured with Elastic and Viscoelastic Hopkins Bars *R. Othman, M.N. Bussac, P. Collet and G. Gary*	49
Newly Developed Dynamic Testing Methods and Dynamic Strength of Some Structural Materials *S. Tanimura and K. Mimura*	57
Effects of Striker Shape and Attached Position of Strain Gage on Measured Load in Instrumented Charpy Impact Test *T. Kobayashi and S. Morita*	65
A Spectral Method for Wave Dispersion Analysis Application to an Aluminium Rod *R. Othman, R. Blanc, M.N. Bussac, P. Collet and G. Gary*	71
Evaluation of Accuracy in Measurement of Dynamic Load by Using Load Sensing Block Method *S. Tanimura, T. Umeda, W. Zhu and K. Mimura*	77
Dynamic Simple Shear of Sheets at High Strain Rates *W.K. Nowacki*	83
Strength and Failure Characterisation and Behaviour Under Impact Loading *L.W. Meyer, L. Krüger and T. Halle*	91
Temperature and Strain Rate Effects on the Tensile Strength of 6061 Aluminum Alloy *K. Ogawa*	99

Dynamic Elastic-Plastic Buckling of Axially Loaded Circular and Square Tubes *D. Karagiozova and N. Jones*	105
Dynamic Buckling Behavior of Foam-Filled Metal Tube *K. Mimura, T. Umeda, K. Yamashita, T. Wakamori and S. Tanimura*	111
Sterilization of Dry Powder Food by Shocks *K. Fujiwara, T. Hiroe, H. Matsuo and M. Asakawa*	117
Microstructural Evolution and Self-Organization of Shear Bands *M.A. Meyers, Q. Xue, Y. Xu and V.F. Nesterenko*	123
VISAR Interferometry: Simple in Design, Versatile in Application *W.M. Isbell*	131
Emission Spectroscopy of Hypervelocity Impacts *D.H. Ramjaun, M. Shinohara, I. Kato and K. Takeyama*	139
Fragment Creation via Low-Velocity Impact Possible in Space *T. Hanada and T. Yasaka*	145
Radiation Observation of Hypervelocity Shock Waves in Gases *H. Honma, K. Maeno, T. Morioka and H. Shibuya*	151
Development of a New Ballistic Range for the Investigation of Oblique Impacts on Solid and Liquid Surfaces *M. Shinohara, I. Kato, Y. Hamate and K. Takayama*	157
Non-Ideal Detonation Properties of Ammonium Nitrate *A. Miyake*	163
Study on the Treatment of Toxic Wastes by Chemical Explosion *T. Matsunaga, K. Miyamoto, M. Iida, A. Miyake and T. Ogawa*	171
Estimation of Shock Sensitivity Based on Molecular Properties *M. Koshi, S. Ye, J. Widijaja and K. Tonokjura*	175
Deep Penetration of Truncated-Ogive-Nose Projectile into Concrete Target *X.W. Chen and Q.M. Li*	183
Peculiarities of Penetration into an Elastoplastic Perturbed Target *Yu.K. Bivin and I.V. Simonov*	189
Explosive Loading of Dual Hardness Steel Plates *J. Buchar, J. Voldrich and S. Rolc*	195
A Unifying Framework of Hot Spots for Energetic Materials *K. Yano, Y. Horie and D. Greening*	201
Fragment Acceleration by the Detonation of High Explosive Charges without and with Aluminium Content *M. Held*	207

Consideration of the Bumper Shield for Hypervelocity Impact Using Smoothed Particle Hydrodynamics Method *D. Watanabe and Y. Akahoshi*	213
Measurement of the Second Debris Clouds in Hypervelocity Impact Experiment *Y. Akahoshi and K. Kouda*	219
Development of Self-Shielding Bumper Against Space Debris *H. Hata, Y. Akahoshi, T. Tsunetomi and S. Kawakita*	225
Development of Bumper Shield Using Low Density Materials *Y. Akahoshi and M. Tanaka*	231
Impact Energy Absorption Affected by the Cell Size in a Closed-Cell Aluminum Foam *T. Mukai, S. Nakano, T. Miyoshi and K. Higashi*	237
Development of Very Low Deceleration Shock-Absorbers *K. Asada, Y. Tan, K. Ohsono, S. Hode, T. Matsuoka and S. Kuri*	243
Synthesis Behaviour and Hot Shock Compaction of Silicon Carbide-Cromium Boride Composites *R. Tomoshige, H. Kainou and A. Kato*	249
Shock Activation and Reaction Synthesis in the Ti+Si+C and Ti+AlN Systems *J.L. Jordan and N.N. Thadhani*	255
Hypervelocity Impact Consolidation of Mechanically Alloyed Powder in the Ni+Al+B System *K. Ayabe and T. Okabe*	261
Preparation of Co-Cu Metastable Bulk Alloy by MA and Shock Compression *X. Fan, T. Mashimo, Y. Zhang and A. Chiba*	267
Explosive Welding of Bulk Metallic Glass Plate on Crystalline Titanium Plate *Y. Kawamura and A. Chiba*	273
Jet Initiation of High Explosive Charges by Lead Covers in Direct Contact *M. Held*	279
Curling of Square Tube by Electromagnetic Forming *S.B. Zhang and H. Negishi*	285
Perforation of High Strength Fabric System by Varying Shaped Projectiles: Single-Ply System *C.H. Cheong, V.B.C. Tan and C.T. Lim*	291
Perforation of High Strength Fabric System by Varying Shaped Projectiles: Double-Ply System *C.H. Cheong, C.T. Lim and V.B.C. Tan*	297
Evaluation of Impact Perforation Characteristics of Laminated Composites Using a Punch-Loading Hopkinson Bar *T. Yokohama*	303
Deformation Mechanism of WHA Composite Block Impacting on Rigid Anvil *Z.G. Wei, Y.C. Li, J.R. Li, X.Z. Hu and S.S. Hu*	309

A Study on Impact Fatigue Behavior of Structure Steels Under Repeated Tensile Loads – Investigation for a High Strength Steel   315
   K. Kitaura and H. Okada

Correlation of Dynamic Behavior with Microstructural-Bias in Two-Phase TiB_2-AL_2O_3 Ceramics   321
   A. Keller, G. Kennedy, L. Ferranti, M. Zhou and N. Thadhani

Computational Analysis of Oblique Projectile Impact on High Strength Fabric   327
   V.B.C. Tan, V.P.W. Shim and N.K. Lee

Analysis of Dynamic Thermal Stresses in Moderately Thick Shells of Revolution of Functionally Graded Material with Temperature-Dependent Properties   333
   E. Inamura, S. Takezono, K. Tao and T. Kawasaki

Impact Erosion on Interface Between Solid and Liquid Metals   339
   M. Futakawa, H. Kogawa, Y. Midorikawa, R. Hino, H. Date and H. Takeishi

Fatigue Strength of Carbon Steels Under Repeated Impact Tension   345
   A. Chatani, A. Hojo, H. Tachiya, N. Yamamoto and S. Ishikawa

Experimental and Numerical Study on High Velocity Impact Against Steel Plate   351
   S. Yoshie and T. Usui

Deformation and Failure Mechanism of Thin-Plate Under Combined Laser Heating and Pre-Stress   357
   Z.G. Wei, Y.C. Li and Z.P. Tang

High Strain Rate Deformation Behavior of Al-Mg Alloys   363
   T. Masuda, T. Kobayashi and H. Toda

A Visco-Plastic Anisotropic Model for the Impact Deformation of Crushable Foam   369
   V.P.W. Shim and L.M. Yang

Deformation and Perforation of Water-Filled and Empty Aluminum Tubes by a Spherical Steel Projectile: Experimental Study   375
   M. Nishida, K. Tanaka and M. Ito

Generation-Phase Simulation of Dynamic Interfacial Fracture Under Impact Loading   381
   T. Nishioka, Q.H. Hu and T. Fujimoto

Dynamic Elasto Visco-Plastic Crack Propagation Analysis Using Moving Finite Element Method   389
   T. Fujimoto and T. Nishioka

Deformation and Fracture of CFRP Under Hyper-Velocity Impact Using Laser-Accelerated Flyer   395
   Y. Yamauchi, M. Nakano, K. Kishida, N. Ozaki, T. Kasai, Y. Sasatani, H. Amaki, K. Kadono, S. Ikai, K. Nishigaki and K.A. Tanaka

Crack-Tip Stress Field Measured by Infrared Thermography and Fracture Strength in Explosion Clad Plate   401
   I. Oda, Y. Tanaka, A. Masuki and T. Izuma

Deformation Near a Crack Close to Interface in Explosion Copper-Clad Mild Steel Plate *I. Oda, K. Shiraishi and M. Yamamoto*	407
A Study on the Dynamic Brittle Fracture Simulation *D.-T. Chung, C. Hwang, S.I. Oh and Y.H. Yoo*	413
Dynamic Fracture Test of Rock Utilizing Underwater Shock Wave *S. Kubota, Y. Ogata, R. Takahira, H. Shimada, K. Matsui and M. Seto*	419
Impact Resistance and Thermal Couplings During Failure and Fracture of Engineering Materials *J.R. Klepaczko*	425
The Effect of Pre-Fatigue on Dynamic Behavior of Some Steels and Aluminum Alloys *M. Itabashi and H. Fukuda*	433
Compression Flow Stress of Ultra Low Carbon Mild Steel at High Strain Rate *N. Kojima, Y. Nakazawa and N. Mizui*	439
Spall Damage Growth Under Repeated Plate Impact Tests *N. Nishimura, K. Kawashima, T. Yamakawa and M. Kondo*	445
Impact Behavior of Mercury Droplet *H. Date, M. Futakawa and S. Ishikura*	451
Stress-Strain Response of S15C Carbon Steel Friction Welded Butt Joints Under Impact Tensile Loading *T. Yokoyama*	457
Evaluation of Impact Tensile Strength for PMMA/Al Butt Adhesive Joints *H. Wada, K. Suzuki, K. Murase and T.C. Kennedy*	463
Finite Element Stress Response Analysis of Butt Adhesive Joints of Hollow Cylinder Subjected to Impact Tensile Loads *T. Sawa, Y. Suzuki and I. Higuchi*	469
Stress Analysis of Lap Adhesive Joint of Hollow Cylinders Subjected to Static and Impact Loads *T. Sawa and H. Nakagawa*	475
Impact Energy Absorption of CFRP-Aluminium Alloy Hybrid Members Bonded Adhesively *M. Wasaki, T. Suwa, T. Karaki and C. Sato*	481
High Pressure Generation Systems for Materials Processing and Evaluation *T. Aizawa*	487

Volume II

Preface	v
Symposium Organization	vii
Future for Explosive Materials Processing *E.P. Carton and M. Stuivinga*	495

Title	Page
Detonation Velocity and Pressure of Non-Ideal Explosive ANFO A. Miyake, K. Takahara, T. Ogawa, H. Arai and Y. Wada	503
Recalibration of Shock Attenuation in PMMA Gap by Direct Pressure Measurement S. Mori, A. Miyake, K. Takahara, T. Ogawa, Y. Ogata and H. Arai	509
Initiation of PETN High Explosive by Pulse Laser-Induced High-Temperature Plasma K. Nagayama, K. Inou and M. Nakahara	515
Numerical Study on Initiation of Thin Explosive S. Kubota, K. Nagayama, H. Shimada and K. Matsui	521
Science of Explosive Welding: State of Art A. Deribas	527
Influence of Specimen Geometry on Adiabatic Shear Sensitivity of Tungsten Heavy Alloys: A Numerical Simulation J.-R. Li, J.-L. Yu and Z.-G. Wei	535
Effect of Adiabatic Heating in Some Processes of Plastic Deformation A. Rusinek and J.R. Klepaczko	541
Mechanical Properties in Structural Magnesium Alloys Under Dynamic Tensile Loading T. Mukai, K. Ishikawa and K. Higashi	547
Effects of Heat Generation in Dynamic Material Testing C.T. Lim, C.L. Lim and V.B.C. Tan	553
Dynamic Splitting Fracture of Tube T. Kurokawa and M. Taira	559
Effects of Grain Size and Density on High Strain Rate Behavior of P/M Aluminum W.Y. Lee, W. Cheng, J. Ma, G.E.B. Tan and C.T. Lim	565
Temperature Dependence of Dynamic Behavior of Titanium by the High Strain-Rate Compression Test Y. Hyunmo, M. Oakkey and P. Kyoungjoon	571
Influence of Strain Rate on the Tensile Mechanical Behavior in $Pd_{40}Ni_{40}P_{20}$ Bulk Metallic Glass T. Mukai, Y. Kawamura, A. Inoue, T.G. Nieh and K. Higashi	577
Equations for Evaluation of Effects of Strain Rate on Concrete Mechanical Properties K. Shirai, T. Yagishita and C. Ito	583
A Numerical Analysis for Stress Wave Propagation Problem of Anisotropic Solids by Discrete Element Method K. Liu, W. Zeng, L. Gao and S. Tanimura	589
Rayleigh Wave in Transversely Isotropic Fluid-Saturated Poroelastic Media K. Liu and Y. Liu	595
Three-Dimensional Stress Wave Propagation in Orthotropic Media K. Liu	601

A New Method for Non-Destructive Evaluation of Concrete Structures Using a Shock Wave and Laser Vibrometers — 607
K. Mori, A. Spagnoli, I. Torigoe, Y. Murakami and S. Itoh

Numerical Simulations of the Response of Reinforced Concrete Beams Subjected to Heavy Drop Tests — 613
M. Unosson

Analytical Study on the Impact Response of Flexible Rockfall Fence — 619
Y. Sonoda, S. Ishii and H. Hikosaka

Experimental and Numerical Simulations of Aluminum Foam Behavior Under Short Duration Dynamic Loads — 625
I. Anteby, O. Haham, A. Schenker, O. Sadot, E. Nizri and G. Ben-Dor

Dynamic Damage in Metal: Porosity as a Test for Damage Models — 633
J. Bontaz-Carion, M. Nicollet, P. Manczur, Y.-P. Pellegrini, E. Boller and J. Baruchel

An Elasto-Plastic Impact of Two Balls — 639
S. Takezono, H. Minamoto, M. Mitsuyama and M. Nagai

Structural Impact of Bamboo/Aluminum Laminate — 645
J.-Y. Zhang, J.-L. Yu and T.X. Yu

Impact of Thin Plates with Losses of Contact — 651
V. Grolleau, G. Rio and S. Maillard

Initial Transient Behavior of a Tall-Building Model Due to a Low Speed Impact — 657
S. Tanimura, T. Sato, T. Saito, T. Umeda, K. Mimura and N. Ogawa

Approximate Analysis of DSIF of Single-Edge Notched Specimen — 663
Luo Jingrun, Chen Yuze, He Yingbo and Zhang Erheng

Interference Effect of Dynamic Stress Concentration Between Two Parallel Notches — 669
I. Maekawa, H. Ueguri and K. Uda

Analytical Methods for Multi-Body Impact with Dissipative Compliant Contacts — 675
W.J. Stronge

Study on Impact Degradation of Materials by High Energy Atomic Oxygen — 683
T. Shioya, K. Satoh and K. Fujita

High-Speed Shape Recovery of SMA — 691
J. Tani, J. Qiu and Y. Urushiyama

Shape Optimization in Impact Problems – An Application to Optimal Design of Golf Club Head — 699
Z. Wu, Y. Sogabe, K. Nakai and Y. Arimitsu

Mechanism of GFRP Damage Caused by Repeated Raindrop Impact — 705
H. Homma, T. Washio and F. Gunawan

Micro Damage Evolution in Penetration Process of Mixed Hardening Material — 713
T. Tsuta, Y. Yin and T. Iwamoto

A Study of Dynamic Deformation Modes of Intact and Damaged Aluminium Honeycombs D. Ruan, G. Lu and B. Wang	719
The Effects of Back Plate Materials on Cone Crack Formation of Glass Plate by Impact with Small Spheres M.S. Kim, H.S. Shin and H.C. Lee	725
A Numerical Study on the Remained Lifetime of the Damaged Steel Member by Impact Load J. Nagahiro, Y. Sonoda and H. Hikosaka	731
Numerical Study on the Local Damage of Reinforced Concrete Columns Caused by a Vertical Impact Load S. Tamura, M. Katayama, M. Itoh, S. Mitake, K. Harada, M. Beppu and N. Ishikawa	737
Studies on Rate-Dependent Evolution of Damage and Its Effects on Dynamic Constitutive Response by Using a Random Fuse Network Model Huang Dejin, Shi Shaoqiu and Li-Lih Wang (Wang Lili)	743
Determination of Damage Thresholds of Fiber Reinforced Composite Laminated Plates by Transverse Impact D. Jiang and W. Shen	749
An Experimental Investigation on the Failure of Rectangular Plate Under Wedge Impact W.Q. Shen, P.S. Wong, H.C. Lim and Y.K. Liew	755
Dynamic Local Buckling of Steel Tube Under Impulsive Vertical Loading from the Bottom N. Ishikawa, M. Mori, K. Suzuki and N. Masuda	761
A Viscoelastic Split Hopkinson Pressure Bar Technique and Its Applications to Several Materials Y. Sogabe, T. Yokoyama, M. Nakano, Z. Wu and Y. Arimitsu	767
A Calibration Method of Pressure Transducer (Hydrophone) Using Collision Between Two Elastic Bars S. Nagata, K. Fukuda and K. Isuzugawa	773
Test on Negative Pressure-Wave Generation in Mercury Target H. Kogawa, M. Ohba, M. Futakawa, S. Ishikura, R. Hino and M. Akiyama	779
Deformation of AD995 Alumina Under Repeated Plane Shock Wave Loading D.P. Dandekar	785
Dynamic Stress and Deformation of Multi-Layered Poroelastic Shells of Evolution Saturated in Viscous Fluid S. Takezono, K. Tao and T. Gonda	791
Dynamic Analysis of Coil Spring Under Impact Load A. Hojo, Y. Wang, A. Chatani, H. Tachiya and J. Shen	797
Experimental Studies on the Dynamic Properties of High Strength Fibers Huang Cheunguan, Zhang Huaqiang and Duan Zhuping	803
The Influence of Initial Shape Irregularity on Crush Strength of Automobile Reinforcing Members Y. Sawairi, M. Gotoh and M. Yamashita	809

Energy Absorption Control of Vehicular-Structure Members *S.K. Kim, K.H. Im, I.Y. Yang and T. Adachi*	815
Dynamic Tensile Behavior of Welded Parts in Automobiles *K. Mimura, K. Hirai and S. Tanimura*	821
Impulse Magnetic Pressure Seam Welding of Aluminum Sheets *T. Aizawa, K. Okagawa, M. Yoshizawa and N. Henmi*	827
High-Strain-Rate Properties of Polymethyl Methacrylate *Y. Sato and W. Kikuchi*	833
Laser Impact in Aluminum and Tantalum *J. Bontaz-Carion, J.C. Protat, P. Manczur, L. Berthe, P. Peyre and E. Bartnicki*	839
Development of Desktop Two-Stage Lightgas Gun *Y. Akahoshi and Y. Sato*	845
Explosively Produced Spallation in Metals and Its Loading Effects *T. Hiroe, K. Fujiwara, H. Matsuo and N.N. Thadhani*	851
Numerical Analysis of Thermal Shock Wave Applied to Metal Alloying Process *A. Abe, R. Sano and Y. Sano*	857
Estimate of Shock Front Profiles Using Thermo-Elastic Stresses *A. Abe, K. Kohara and Y. Sano*	863
Control of Detonation Velocity by Discrete Charges of Explosives *K. Fujiwara, H. Matsuo, T. Hiroe and Y. Hayashi*	869
Energy Absorption Characteristics of CFRP Composite Tubes *K.H. Im, Y.N. Kim, J.W. Park, J.K. Sim and I.Y. Yang*	875
Measurement and Estimation of Falling Rock Speeds Using a Middle-Size Slope Model *S. Murata, T. Tashiro, H. Kawakubo and H. Shibuya*	881
Influence of Geometrical Initial Imperfection on Deformation of Axially Compressed Square Tube *M. Miyazaki and H. Negishi*	887
Initiation and Parameters of Trinitrotoluol Explosion at Impact by Fragment *G.V. Belov, S.M. Bakhrakh, S.P. Egroshin, Ju.G. Fedorova, A.V. Petruchin, S.A. Shavredov and E.L. Jamolkin*	893
Fabrication of Titanium Implants with Porous Surface by Cylindrical Shock Compaction Method *T. Watanabe, A. Chiba and Y. Morizono*	899
Fabrication of Titanium/Hydroxyapatite Functionally Graded Bio-Materials by Underwater Shock Compaction *S. Kimura, A. Chiba and Y. Morizono*	905
Effect of Mechanical Property on the Punching Process Using Impulsive Pressure *Y. Nakiyama, M. Mochihara, A. Kira, M. Fujita and K. Hokamoto*	911

A Trial for an Explosive Forming of a Plate Using a Paper as Die Material *A. Kira, M. Sassa, M. Fujita and S. Itoh*	917
A New Explosive Welding Technique Method Using Reflected Underwater Shock Wave *K. Shiramoto, A. Kira, M. Fujita, K. Hokamoto, S. Itoh and Y. Ujimoto*	923
Generation of Extremely High Shock Pressure by Convergent Collision of High Speed Jets *R. Tomoshige, M. Fujita, A. Kato, K. Hokamoto, S. Itoh and Y. Ujimoto*	927
Numerical Simulation on Mach Detonation Formed by High Explosives *Y. Nakamura, S. Itoh and Y. Kato*	933
Important Parameters to be Controlled for Obtaining Bulk Materials during Shock Consolidation of Powders and Shock Synthesis of Intermetallics *K. Hokamoto, J.S. Lee, K. Siva Kumar, T. Balakrishna Bhat, K. Raghukandan, A. Chiba, S. Itoh and M. Fujita*	939
Impact Fracture Behavior in Alumina Plates *H.S. Shin, S.Y. Oh, Y.Z. Bae, B.S. Jeon and S.N. Chang*	945
Development of Explosive Device for Separation of Stone Slabs from Rock Massive (Granite, Basalt, Marble, etc.) and Further Fragmentation of Them to Pieces of Specified Sizes *E.B. Moiseev, A.K. Botvinkin, N.V. Bryukhanov, A.S. Ermenko, S.A. Novikov and V.N. Khvorostin*	951
Numerical Simulation of Multi Layer Explosive Welding Using Underwater Shock Wave *H. Iyama, A. Kira, M. Fujita, K. Hokamoto and S. Itoh*	957
The Studies on Design and Experiment Technology of Explosively Formed Projectiles with Star Shaped Tail *Yu Chuan, Dong Qingdong, Sun Chenwei, Tong Yanjin, Yan Chengli, Li Fabo, Gui Yulin, Xie Panhai, Li Bin and Yang Jinyan*	963
Buckling Behavior Analysis of a Rectangular Tube Structure by Lateral Impact Load *K.H. Yoon, K.N. Song and Y.H. Jung*	969
Global-Local Finite Element Method for the Drop/Impact Analysis on the Electronic Products with Small Components *X. Zhang and K.H. Low*	975
Shock Analysis on Suspending Structures in the Drop/Impact Simulation of Electronic Products *X. Zhang and K.H. Low*	981
Author Index	I1
Keyword Index	I5

SOME RECENT DEVELOPMENTS IN THE DYNAMIC INELASTIC RESPONSE OF STRUCTURES

Norman Jones

Department of Engineering (Mechanical Engineering)
The University of Liverpool, Liverpool L69 3GH, U.K.

ABSTRACT

This article examines some recent developments in the dynamic inelastic behaviour of ductile structures which are important for the structural crashworthiness field and for energy absorbing systems and other industrial safety problems. The pressures on the performance of structural systems is increasing due to higher transportation speeds, the quest for lighter structures and environmental concerns, as well as an increased public awareness of safety issues. Thus, in order to address these factors, theoretical methods of analysis, numerical finite-element methods and other numerical schemes require more complete information on the dynamic inelastic properties of materials. In particular, the strain rate sensitive behaviour of materials at large strains, the influence of stress triaxiality on the dynamic rupture strain and the criterion for dynamic inelastic failure, are discussed briefly. Some comments are given on the importance of potential transitions to less effective energy absorbing mechanisms that might develop in structures. It is observed that these mode changes are related in a complex manner to increased impact velocities, changes of geometrical parameters and material characteristics and axial inertia, or stress wave propagation. It is vital to identify any potential mode changes, or transitions, during the preliminary design process.

KEYWORDS

Dynamic, impact, inelastic, rupture, failure, structures, crashworthiness, safety, hazard assessments, stress waves, strain rate sensitivity.

INTRODUCTION

Theoretical analyses of the static behaviour of ductile structural members developed rapidly after the limit theorems for a perfectly plastic material were proved about 50 years ago, so that it is now a mature subject, which, for some time, has been used for design purposes. It turns out that the theoretical apparatus developed during this work for static loads can be used to obtain the dynamic inelastic response of ductile structures, the only additional effect being the inertia, or time-dependent effects. It transpires that these so-called rigid-plastic methods of analysis can provide useful estimates for the behaviour of ductile structures which are subjected to large dynamic loads producing inelastic strains much larger than the corresponding yield strain for the material. These methods have been developed further in recent years and employed to design energy absorbing systems and to examine a wide range of structural crashworthiness problems. They have been used to predict the structural response for safety calculations and hazard assessments of structures and components when subjected to extreme dynamic events in industrial design applications.

However, the quest for lightweight structures continues unabated in the search for optimum designs with smaller factors of safety. Thus, it becomes necessary to cater for various effects more accurately as structural designs move nearer to the failure boundaries. For example, the consideration of material strain rate effects should recognise the important variation of properties with the increase of strain, as well as strain rate. This is an important factor in the analysis of many structures because of the large strains produced by the extreme events which are examined in industry for hazard assessments. The greater speed of train travel and the broader survivable envelope for aircraft landing and take-off accidents requires a consideration of stress wave propagation, as well as the structural mode changes caused by lateral inertia effects and other factors.

Rigid-plastic theory assumes that a material has an unlimited ductility, whereas actual materials have a finite strength which could be exhausted during an extreme dynamic event. Some recent studies have focused on this area, but no reliable universal failure criterion for design, or for incorporation into numerical schemes, is yet available.

Some comments on the current status of the above topics are offered in the following sections.

DYNAMIC MATERIAL INELASTIC PROPERTIES

The influence of strain rate on the dynamic flow stress (σ_d) of a ductile metal may be expressed by the Cowper-Symonds equation (Jones (1997))

$$\sigma_d / \sigma_s = 1 + (\dot{\varepsilon}/D)^{1/q}, \tag{1}$$

where σ_s is the static yield stress and D and q are constants obtained from dynamic tests on a given material. Equation (1) is written for dynamic uniaxial tension or compression, but it may be generalised by the usual procedure in solid mechanics for any arbitrary three-dimensional stress state. It is important that the constants D and q are generated over the range of strain rates expected in a particular design and for the actual material of interest. For example, the values D = 6500 sec^{-1} and q = 4 are used frequently for any aluminium alloy. However, these particular values were obtained from experimental data obtained on "duralumin" over 50 years ago, but Figure 1 shows how they exaggerate the strain rate sensitive properties which themselves depend on the type of aluminium alloy. Moreover, even for a given material, aluminium alloy 6061 T6, it was shown in by Jones (1974) that the experimental results for material strain rate sensitivity are contradictory. Nevertheless, the results in Figure 1 indicate that the strain rate sensitivity of an aluminium 6061 T6 is very small, so that differences between authors are due possibly to experimental errors which, in some cases, mask entirely any strain rate sensitive effects.

It is important to obtain the entire stress-strain curve because the strain hardening characteristics of a material might change with strain rate. In other words, the values of D and q in equation (1) could

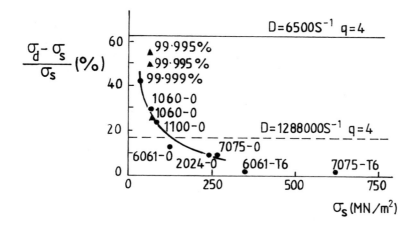

Figure 1: Percentage increase in the plastic flow stress at $\varepsilon = 0.06$ between $\dot{\varepsilon} = 10^{-3}$ s^{-1} and $\dot{\varepsilon} = 10^3$ s^{-1} for aluminium alloys. σ_s and σ_d are the static and dynamic flow stresses at $\dot{\varepsilon} = 10^{-3}$ s^{-1} and $\dot{\varepsilon} = 10^3$ s^{-1}, respectively. (see Jones (1997)).

change with strain. It was shown by Jones (1989a) that this phenomenon could be catered for by the modified Cowper-Symonds equation

$$\sigma_d / \sigma_s = 1 + \{\dot{\varepsilon}/(B+C\varepsilon)\}^{1/q}, \qquad (2)$$

where $\qquad B = (\varepsilon_u D_y - \varepsilon_y D_u)/(\varepsilon_u - \varepsilon_y) \qquad$ and $\qquad C = (D_u - D_y)/(\varepsilon_u - \varepsilon_y)$.

The subscripts y and u are associated with the yield (or proof) and ultimate tensile stresses, respectively. Equation (2), therefore, requires dynamic experimental data to be recorded for the behaviour of the material at both the yield and ultimate tensile stresses. Equation (2) for the specific values $\varepsilon = \varepsilon_y$ and $\varepsilon = \varepsilon_u$ takes the same form as equation (1) but with D replaced by D_y and D_u, respectively. The corresponding dynamic stresses are associated with the yield (proof) and ultimate tensile stresses.

The uniaxial rupture strain of a ductile material might change with strain rate and become less ductile, although some aluminium alloys and other materials become more ductile with increase in strain rate. It was shown by Jones (1989b) that the dynamic uniaxial rupture strain (ε_{dr}) is given by the expression

$$\varepsilon_{dr} = \varepsilon_{sr} \{1 + (\dot{\varepsilon}/D)^{1/q}\}^{-1} \xi, \qquad (3)$$

where ε_{sr} is the static uniaxial rupture strain and ξ is the ratio of the total energies to rupture the material for dynamic and static loads. If the energies consumed by the material up to rupture is invariant for static and dynamic loadings, then $\xi = 1$, and

$$\varepsilon_{dr} = \varepsilon_{sr} \{1 + (\dot{\varepsilon}/D)^{1/q}\}^{-1}. \qquad (4)$$

Again the constants D and q should be obtained from dynamic tensile tests on the actual material, and equations (3) and (4) may be generalised for three-dimensional strain states when assuming no differences between tension and compression.

In addition to the effects of strain rate on the magnitude and shape of the stress-strain curve and the rupture strain value, there are, of course, many other effects of potential importance in the appropriate circumstances (e.g., temperature, radiation dose, etc.). One of these factors is the influence of hydrostatic stress, or triaxiality, an increase of which is taken generally to reduce the rupture strain with material failure initiating at the location of the highest triaxiality, although this behaviour is not always found as shown in Figure 2. A recent study (Alves and Jones (1999)) has observed that some of this confusion occurs because the stresses are often estimated using Bridgman's formula rather than an accurate finite-element analysis (see also Alves et al. (2000)) which reveals that the actual value of triaxiality in a notched specimen is a material and geometric dependent parameter. The experimental tests on mild steel notched specimens reveals that the stress triaxiality is not the only fundamental parameter for initiating failure; the plastic strain appears to be important. Thus, a relationship between the stress triaxiality and the plastic strain might control the actual location where failure commences. Clearly, more experimental studies are required in order to clarify the role of stress triaxiality on the dynamic inelastic rupture of ductile metals.

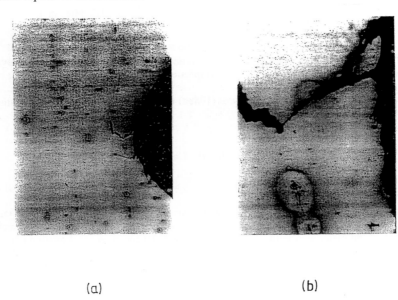

(a)   (b)

Figure 2 (a): Mild steel tensile specimen with a crack initiating at the notch root, whereas maximum triaxiality occurs on the vertical axis of symmetry, (b) detail at the notch root (see Alves and Jones (1999))

DYNAMIC INELASTIC FAILURE OF STRUCTURES

As noted in the Introduction, rigid-plastic methods of analysis assume an unlimited ductility, whereas actual materials do fail for sufficiently severe dynamic loads. The first systematic studies on the dynamic rupture of structures were undertaken experimentally by Menkes and Opat (1973) for blast loaded aluminium alloy 6061 T6 beams, and theoretically by Jones (1976) using a modified rigid plastic method of analysis. Three modes of failure were identified for the beams:

mode 1;   large permanent transverse deflections, W_{max}, or damage without any material rupture, mode 2;   tensile rupture at supports, which occurs when the maximum strain

$$\varepsilon_{max} = \varepsilon_{dr}, \qquad (5)$$

mode 3;   transverse shear failure at supports, which occurs when the maximum shear displacement

$$W_s = kH, \qquad (6)$$

where k is a constant with $0 \le k \le 1$ and H is the beam thickness.

A mode 1 behaviour was predicted by Jones (1976, 1997), using a rigid plastic analysis for the dynamic response of beams which includes the influence of large ductile deformations, or geometry changes. The limit of mode 1 behaviour was taken as the onset of a mode 2 failure when the maximum strain at the supports reached the uniaxial rupture strain. Mode 3 transverse shear failures, for the highest blast loadings, were predicted by assuming that a beam is severed from the supports when the difference between the transverse displacements across both sides of a shear hinge reaches a critical value. Jouri and Jones (1988) presented some experimental data on the threshold conditions and values of k in equation (6) for the transverse shear failure of aluminium alloy beams subjected to mass impact loadings. Figure 3 presents some experimental results for the dimensionless kinetic energy (λ_s) to fail a mild steel beam fully clamped across the span (2L) which is struck by a mass at several distances, L_1, from a support. The energy is greatest for impacts at the mid-span $L_1/2L = 0.5$ when membrane effects are dominant, but is markedly reduced for impacts at other locations especially near to a support, where transverse shear forces are significant. Yu and Chen (2000) have examined the interaction between the bending moment and a transverse shear force at a plastic hinge in impulsively loaded fully clamped beams.

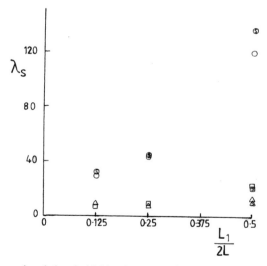

Figure 3: Non-dimensional threshold kinetic energy for severance (λ_s) of beams struck by blunt-ended impactors (see Jones and Jones (2001)). Experimental results: O, □, Δ for 1.6, 3 and 6 mm thick beams, respectively. S indicates failure of beams when loaded statically.

Teeling-Smith and Nurick (1991) subjected fully clamped circular plates to impulsive loadings and obtained the same three failure modes noted above for beams. However, the blast loading of square plates (Nurick and Shave (1996)) and non-axisymmetric impact loadings of circular plates (Jones and Jones (2001)) gave rise to additional modes of failure. The circular plates failed underneath the impactor, at an adjacent support, or at a combination of both sites.

Recently, it has been shown by Li and Jones (1999) when retaining temperature effects in the basic equations, that an adiabatic shear failure might develop for even higher impulses than those associated with a mode 3 failure of aluminium alloy 6061 T6 beams. A similar response could occur for other materials and in other structural members, such as plates and shells.

Li and Jones (2000) used dimensional analysis and a finite-element simulation to study the formation of a transverse shear localisation (shear hinge) in a beam element made from a ductile material and subjected to transverse dynamic loads. It transpires that the beam response under a transverse dynamic load is determined by two dimensionless parameters, represented by the relative values of loading velocity and loading time. When the actual impact or impulsive loading velocity is of the same order as the characteristic loading velocity, or plastic shear wave speed, then the concept of a shear hinge is relevant. In this case, a quasi-static method of analysis can be used to obtain the length of a shear hinge, and the shear strain within the shear hinge may be determined from the shear sliding displacement obtained using a rigid, perfectly plastic analysis. When the impact velocity is much smaller than the plastic shear speed, then bending dominates the beam response. On the other hand, in-plane plastic shear wave propagation is important for impact velocities larger than the plastic shear speed.

The theoretical considerations on failure in this section have been developed for simple structures and are valuable for understanding the various failure phenomena. Sometimes, these elementary methods are suitable for design purposes when bearing in mind the paucity of dynamic material properties and the uncertainties in the dynamic loading characteristics. However, these methods are difficult to use for more complex structural systems and, therefore, it is necessary to seek a more general structural failure criterion.

A series of careful experimental tests were conducted by Yu and Jones (1991) to obtain the precise threshold conditions for the various failure modes of aluminium alloy and mild steel beams subjected to impact loadings. The static and dynamic material properties were obtained using specimens cut from the same block of material which was used to manufacture the beams. The impact tests were repeated with increasing impact energies on virgin beams until the threshold of failure was reached. A calibrated finite-element numerical programme (Yu and Jones ((1997)) was then used to study the beams for the experimental threshold conditions associated with the first appearance of a crack, but without any failure criterion being activated in the programme. This allowed various failure criteria to be examined without prejudice to any particular one, but it transpired that the maximum membrane force in a beam cross-section appears to be the most promising one for predicting a tensile failure. The maximum Tresca stress, or von Mises stress, and the maximum plastic strain energy density, are worthy of further study for predicting a shear failure. Some promising expressions were found for a global failure criterion (i.e., interaction between bending moment, axial membrane force and transverse shear force), but further experimental results are required to confirm the reliability of these criteria.

The experimental studies together with the associated numerical calculations reported above were undertaken to obtain a criterion which governs the failure of beams subjected to impact loads which produce large inelastic strains and material rupture. A beam might be considered as a one-dimensional structure from a global, or design, perspective, although local three-dimensional affects are important at the failure site. The impact failure of plates, which are nominally two-dimensional structures, is examined by Jones and Jones (2001).

An extensive literature has been published on the behaviour of plates struck by missiles (e.g., Corbett et al. (1996)) and a considerable body of theoretical work and several empirical equations are now available for use by a designer. However, the problem is by no means completely understood and the field remains active with articles being published currently on high velocity perforations concerned with adiabatic shearing effects, etc., and on low velocity impacts which produce large transverse displacements which induce membrane forces in a plate before failure (e.g., Wen and Jones (1993), Langseth and Larsen (1994)). The material fails in all of these studies immediately underneath the striking mass. Quite clearly, it is difficult to develop a universal failure criterion on the basis of the behaviour in this highly localised region, even for a low velocity impact (Jones et al. (1997)).

DYNAMIC AXIAL CRUSHING OF THIN-WALLED MEMBERS

The axial loading of thin-walled tubes with circular and square cross-sections can produce progressive crushing for sufficiently large forces. This inelastic phenomenon has been studied exhaustively for static axial forces because of the efficient energy absorbing characteristics (Jones (1997), Gupta and Velmurugan (1997), etc.) as well as the simplicity of the structural members, their ready availability and relatively low cost. Fewer studies have been reported for dynamic axial loading and most have assumed that the response is quasi-static.

Quasi-static analyses have been shown to be accurate for some dynamic inelastic structural problems provided the lateral inertia effects remain small. This situation often occurs when the striking mass is large compared to the structural mass and the impact velocity is low (say less than 10-15 m/s). The dynamic or quasi-static response mode is assumed to be identical to the corresponding static deformation mode, except that it is necessary to cater for any strain rate sensitive properties of the material. In fact, it is not straightforward to incorporate strain rate effects into a quasi-static analysis because the strain rate associated with dynamic loads will vary both spatially and temporally. Fortunately, the phenomenon of material strain rate sensitivity is highly non-linear for many materials. The dynamic flow stress of mild steel, for example, doubles for an increase of strain rate from static testing speeds ($\cong 10^{-4}$ sec^{-1}) to 40 sec^{-1}, approximately, (i.e., an increase of 5-6 orders of magnitude according to equation (1) with $D = 40$ sec^{-1} and $q = 5$). Thus, a reasonable estimate for the dynamic flow strength can then be made by calculating an average value for the strain rate. The associated enhanced dynamic flow stress can then be used in quasi-static calculations to provide estimates for the dynamic behaviour which are found to agree reasonably with experimental results.

Theoretical quasi-static methods have been developed for the dynamic axial crushing of circular and square tubes and for top-hat and double-hat sections, which are of particular interest for automobile crashworthiness calculations. Quite simple and fairly reliable design formulae have been obtained by many authors using rigid-plastic methods of analysis, as discussed by Jones (1999) and illustrated in Figure 4 for an axially impacted top-hat section. These analyses assume that the wrinkles, or lobes, form sequentially until they cover the entire length of a section. The final length of a thin-walled member is typically around one-quarter of the initial length, a phenomenon known as bottoming-out. It has been observed that these methods are suitable for preliminary design purposes, and, in many cases, are adequate for final designs when acknowledging the often incomplete information about the actual design scenario.

A thin-walled section subjected to an axial impact force might respond with global bending, or in an Euler-type buckling mode, if the initial axial length is sufficiently long. This is a less efficient way of absorbing impact energy when compared with a dynamic progressive buckling mode discussed above. However, little information is available on the conditions necessary for a transition between dynamic progressive buckling and global bending. In fact, apart from a study on the static behaviour of aluminium circular tubes by Andrews et al. (1983), no design guidelines appear to be available on the

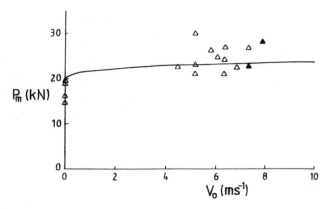

Figure 4: Mean dynamic crushing force (P_m) of 60 x 60 mm mild steel top-hat sections having a 1.2 mm wall thickness and a flange width of 25 mm. V_o is the axial impact velocity. $\triangle$ and $\blacktriangle$: experimental data for regular and irregular collapse profiles, respectively. ———: theoretical rigid-plastic analysis with D = 28,104 s^{-1} and q = 4.61 (see White and Jones (1999) for further details).

effect of the initial length of a tube on the subsequent mode of behaviour. Thus, a study was undertaken and reported by Abramowicz and Jones (1997) on the dynamic response of circular and square mild steel tubes having various axial lengths and struck axially by masses travelling with initial velocities up to 12 m/s. All of these specimens entered the plastic range in an initial straight configuration and buckled plastically in the Euler or global bending mode in the sense that the lateral displacements commenced to grow with a further increase of the axial compressive load. However, many of the tubular specimens switch to a progressive buckling mode and a visual inspection of most of these specimens after a test would lead one to conclude erroneously that progressive buckling had occurred throughout the entire deformation process. Nevertheless, other specimens continued to bend and collapsed with an Euler or global bending mode. The effect of lateral inertia in the dynamic crushing case acts so as to promote, or at least favour, a dynamic progressive buckling mode which may be maintained until near to the end of motion when the resisting inertia force in some cases might be insufficient to prevent development of a global bending mode.

No systematic studies appear to have been reported on the threshold conditions for a transition between progressive buckling and global bending of thin-walled sections having top-hat or double-hat cross-sections and subjected to axial dynamic loads. It is evident that further studies are required on thin-walled tubes having a range of cross-sections and subjected to axial impact forces in order to seek these transitions and to explore the characteristics for thin-walled sections made from aluminium alloys and high strength steels.

INERTIA EFFECTS IN AXIAL IMPACT OF THIN WALLED TUBES

The theoretical analyses, which are reported in the previous section on the dynamic axial crushing of various thin-walled sections, were obtained using the quasi-static assumption, but retaining the dynamic properties of the material. However, these quasi-static methods of analysis are only likely to remain valid for a restricted range of impact velocities up to about 15m/s for large mass ratios, as noted earlier. The influence of transverse inertia effects becomes important for higher impact velocities which lie within the practical range of many transportation accident scenarios.

The wrinkles, or lobes, form sequentially, or progressively, for dynamic progressive buckling, whereas the larger lateral or radial inertia of a circular tube, for example, when struck by a mass travelling with a high velocity, prevents the formation of large lobes and favours the simultaneous growth of smaller wrinkles along the entire length of a tube. This phenomenon is known as dynamic plastic buckling and has higher associated decelerations than dynamic progressive buckling, or a quasi-static response, which would give rise to larger forces being transmitted to adjacent structural members, and, more particularly, make it more difficult to satisfy human injury criteria in transportation systems. The actual conditions for the transition between a dynamic progressive buckling behaviour and a dynamic plastic buckling response are not clear. Nevertheless, it is important to have a reliable criterion since modern structural crashworthiness designs are exploring the advantages of different materials and at ever increasing impact velocities.

Different governing equations and approximations are introduced in the analyses for the two phenomena so that the transition thresholds between them are difficult to obtain. It was suggested by Jones (1997) that a transition from dynamic progressive buckling to dynamic plastic buckling occurred when the axial compressive load equalled the plastic squash load when considering any enhancement of the axial stress for a tube made from a strain rate sensitive material. This guideline was suggested because a perturbation analysis, which is used to study dynamic axial plastic buckling, requires a uniform dominant axial plastic flow. More recently, the role of elastic and plastic stress waves, which are generated by an axial impact on the proximal end of a tube, has been examined. These dynamic analyses retain both axial and radial inertia and, therefore, are complex not least because the time-dependent response involves elastic and plastic behaviour with attendant history effects in the presence of large deformations.

The axisymmetric buckling of elastic-plastic circular cylindrical shells was studied numerically by Karagiozova and Jones (2000) using a discrete model for a shell. The elastic-plastic material was strain rate insensitive with linear strain hardening properties and the Bauschinger effect. Stress wave propagation effects were studied for both moving and stationary shells with different loading conditions given by various combinations of the striking mass and the initial velocity. Good agreement was obtained with some experimental data on aluminium alloy tubes.

It is found that the dynamic buckling process is governed by stress wave propagation effects and, in general, the entire length of the shell is involved in the deformation process for a high velocity impact. This phenomenon is known as dynamic plastic buckling. However, the final buckling shape depends strongly on the inertia properties of the striker and the geometry of the shell. Regular buckling shapes for a high velocity impact occur in relatively thick shells when buckling develops within a sustained axial compressive plastic flow. A localisation of buckling can develop in thinner shells when the buckling process involves a partial unloading of some cross-sections of a shell, thereby interrupting any further axial stress wave propagation. A low mass-high velocity impact causes a large axial compression to develop near to the impacted end, while large bending deformations occur near the stationary end, and a considerable portion of the initial kinetic energy is absorbed in compression during the initial deformation phase. Larger mass-lower initial velocity impacts tend to cause large bending deformations near the impacted end which leads to a progressive folding.

In view of the importance of radial and axial inertia effects on the response of axially impacted elastic-plastic cylindrical shells, which were observed in the above study using a discrete model of a shell, a numerical simulation was undertaken by Karagiozova et al. (2000) with the aid of the ABAQUS finite-element programme.

This study reveals that the dynamic behaviour of cylindrical shells subjected to an axial impact are both velocity and mass sensitive. A larger kinetic energy can be absorbed for higher-velocity impacts when decreasing the striking mass, because larger strains and stresses accumulate in a shell during an initial compression phase. The compression of a shell develops intermittently throughout the entire response for a low-velocity impact which causes a significant shortening of the shell. The shortening of the shell due to compression for a high-velocity impact develops mainly during the initial phase of deformation before buckling commences. This shortening of a shell, for both high- and low-velocity impacts, causes a smaller number of wrinkles to develop in comparison with existing design formulae for progressive buckling, which do not take into account any reduction of the shell length.

Regions in the load parameter space, where different buckling phenomena occur, are shown in Figure 5 for two particular shells. It is evident that the inertia characteristics of a shell, together with the material properties, determine particular stress wave propagation patterns, which cause either dynamic plastic buckling or dynamic progressive buckling to develop during the initial shell instability phase. Thicker shells made of materials exhibition large strain hardening characteristics tend to buckle in a dynamic plastic buckling pattern in contrast to the thinner shells, or shells made of materials with smaller strain hardening characteristics, which buckle progressively, even for high velocity impacts. Harrigan et al. (1999) have explored some aspects of inertial effects in the response of impact energy absorbers, namely tube inversion and aluminium honeycombs.

DISCUSSION

This article has reported on some recent studies on the dynamic inelastic response of structures. This particular area has widespread industrial importance because of the increasing demands to achieve optimum designs and the requirements to estimate more reliably the integrity of safety systems. It is necessary for all these applications to have accurate dynamic material properties since theoretical predictions and numerical calculations are only as reliable as the input data. Unfortunately, there is a paucity of information on the dynamic properties of materials with large strains, partly because of experimental difficulties. For example, it is difficult to generate the dynamic uniaxial stress-strain curves up to rupture at a constant strain rate. Alves and Jones (1999) observed that significant variations in strain rate can develop during some tests up to large strains, apart from other considerations related to instrumentation, filtering and stress wave effects (Karagiozova (2000)). A large amount of data is available for many materials over a wide range of strain rates, but it is often restricted to small strains, whereas many designs require data for large strains up to rupture. Thus, a relatively sparse amount of data is available which gives the influence of strain rate on the dynamic rupture strain of materials and, indeed, even the role of stress triaxiality on dynamic rupture is uncertain. This situation effects also the output from sophisticated numerical codes for problems involving large plastic strains and rupture.

It has been suggested that it is necessary to use micromechanics in order to predict the dynamic inelastic failure of structures. The micromechanics approach is exemplified by Bammann et al (1993) who used finite-element codes to examine the failure of several dynamic problems. This method appears to be attractive and will no doubt become more so as computers become even more powerful. However, it is important to remark that these methods require several material constants which are difficult to obtain experimentally. For example, a Taylor-Quinney coefficient of 0.9 is used in the basic equations by Bammann et al. (1993) which implies that only 90 per cent of the plastic work is dissipated as heat. However, this factor is dependent on the energy absorbed even in a static tensile test and is, therefore, time-dependent. These numerical schemes often use a critical void volume fraction of 0.12 and employ other parameters in the plastic flow potential which were obtained from similar quasi-static studies. No doubt, with the passage of time, the various assumptions required currently in the use of micro-mechanics, for the dynamic inelastic failure of structures, will decrease as a greater body of experimental knowledge becomes available. Currently, it is unrealistic to use the micromechanics approach for most civilian problems associated with transportation and industrial

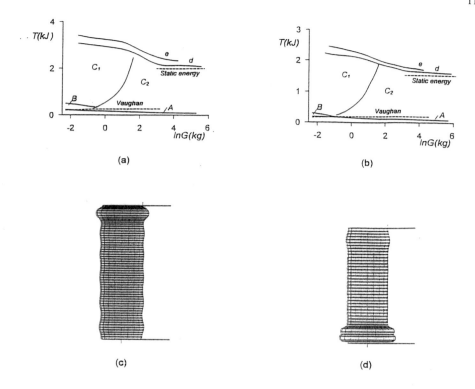

Figure 5: Regions for the load parameters where different buckling phenomena can occur for the axial impact of elastic-plastic, circular cylindrical shells (see Karagiozova et al. (2000) for further details.) **A** – uniform axial compression, **B** – dynamic plastic buckling, **C_1** – initial dynamic plastic buckling and a subsequent progressive buckling, **C_2** – progressive buckling, **d** – maximum energy which can be absorbed by a shell made of strain rate insensitive material, **e** – maximum energy which can be absorbed by a shell made of strain rate sensitive material. Variation of kinetic energy (T) with striker mass (G) for cylindrical shells having a mean radius of 11.875 mm, wall thickness of 1.65 mm and axial lengths (a) 106.68 mm (shell S1), (b) 76.2 mm (shell S2). (c) dynamic plastic buckling of shell S1 (G = 0.4167 kg, V_o = 120 m/s, t = 155 µs), (d) dynamic progressive buckling of shell S1 (G = 262.5 kg, V_o = 4 m/s, t = 8.95 ms).

safety, although it is used for military applications where the time scale for designs are much longer and financial considerations play a much smaller role.

Another method for predicting failure uses the principles of structural mechanics which have been used successfully throughout engineering and are emphasised in this article. Of course, it is evident that many failures are initiated locally, whereas a structural mechanics approach uses global quantities such as generalised stresses (moments, forces) and generalised strains (curvatures, membrane strains). This appears to be a fundamental contradiction. Nevertheless, these methods do allow for some local behaviour in a structural member because, typically, the plane sections assumption gives rise to maximum stresses and strains at a discrete location on the surface of a structural member. In fact, simple methods of analysis are often adequate for preliminary scoping designs and are satisfactory for some final designs when acknowledging the inadequate data available on the dynamic properties of materials and any uncertainties in the projected dynamic loading characteristics.

Recent studies (Shen and Jones (1992, 1993), Yu and Jones (1997)) have revealed that the strain energy density failure criterion provides promising predictions for the dynamic inelastic failure of beams and plates. However, it is quite clear from an earlier section that a considerable amount of additional work is required before this, or any other failure criterion, gains widespread support within the engineering design community. Nevertheless, a failure criterion is required urgently to allow designers to compare rationally the suitability of different materials for a given design scenario and to select the lightest and most robust design. It should be emphasised that these remarks apply to designs obtained using finite-element methods, or other numerical schemes, as well as simple methods of analysis.

It is evident that transportation speeds will continue to increase and that the public now has a considerable awareness of safety issues in transportation systems as well as knowledge of potential industrial hazards. Thus, it is anticipated that many future designs must satisfy more demanding requirements which will emerge from regulatory bodies. At the same time, environmental, economic and competitive pressures will drive designs to become lighter with smaller factors of safety against failure. Clearly, this situation requires more knowledge and a better understanding of the basic mechanics of energy absorbing systems and of the methods used for estimating the response of structures under extreme dynamic loadings. Thus, it is important to examine potential structural and component designs to ensure that a transition to a less favourable mode of behaviour does not occur, as, for example, when global bending intervenes during the response of an axially loaded thin-walled tube, or energy absorber, rather than the much more efficient dynamic progressive buckling behaviour. Moreover, the quasi-static methods of analysis, which are often used currently for structural crashworthiness designs, will become increasingly inadequate as higher impact speeds give rise to stress wave (axial inertia) effects in thin-walled tubes. Stress waves may produce significant mode changes and higher axial decelerations making human impact injury criteria more difficult to satisfy for transportation systems, as well as influencing the energy absorbed in a given design.

CONCLUSIONS

This article reveals that existing design methods, often quasi-static, which are used for the dynamic inelastic response of structures, both numerical and theoretical, require important improvements to cater for the pressures arising from faster speeds of transportation, lighter structures and new materials. More material data is required, particularly for the dynamic behaviour at large inelastic strains, as well as more complete numerical codes and theoretical design methods which cater for the new structural phenomena including those brought about by stress wave propagation effects.

ACKNOWLEDGMENTS

The author is grateful to EPSRC for their partial support of this study under grant number GR/M75044, to Mrs. M. White for her secretarial assistance and to Mrs. I. Arnot for her assistance with some of the figures.

REFERENCES

Abramowicz, W. and Jones, N. (1997). Transition from Initial Global Bending to Progressive Buckling of Tubes Loaded Statically and Dynamically, *Int. J. Impact Engineering*, **19:5/6**, 415-437.

Alves, M., Yu, J. and Jones, N. (2000). On the Elastic Modulus Degradation in Continuum Damage Mechanics. *Computers and Structures*, **76:6**, 703-712.

Alves, M. and Jones, N. (1999). Influence of Hydrostatic Stress on Failure of Axisymmetric Notched Specimens. *J. of the Mechanics and Physics of Solids*, **47**, 643-667.

Andrews, K. R. F., England, G. L. and Ghani, E. (1983). Classification of the Axial Collapse of Cylindrical Tubes under Quasi-Static Loading, *Int. J. Mechanical Sciences*, **25:9/10**, 687-696.

Bammann, D. J., Chiesa, M. L., Horstemeyer, M. F. and Weingarten, L. I (1993). Failure in Ductile Materials Using Finite Element Methods, *Structural Crashworthiness and Failure*, Eds. N. Jones and T. Wierzbicki, Elsevier Applied Science, 1-54.

Corbett, G. G., Reid, S. R. and Johnson, W. (1996). Impact Loading of Plates and Shells by Free-Flying Projectiles: A Review, *Int. J. Impact Engineering*, **18:2**, 141-230.

Gupta, N. K. and Velmurugan, R. (1997). Consideration of Internal Folding and Non-Symmetric Fold Formation in Axisymmetric Axial Collapse of Round Tubes, *Int. J. Solids and Structures*, **34:20**, 2611-2630.

Harrigan, J. J., Reid, S. R. and Peng, C. (1999). Inertia Effects in Impact Energy Absorbing Materials and Structures, *Int. J. Impact Engineering*, **22:9/10**, 955-979.

Jones, N. and Jones C. (2001). Inelastic Failure of Fully Clamped Beams and Circular Plates Under Impact Loading. Impact Research Centre Report No. IRC/185/2001, Department of Engineering, The University of Liverpool.

Jones, N. (1999). Some Phenomena in the Structural Crashworthiness Field. *Int. J. of Crashworthiness*, **4:4**, 335-350.

Jones, N. (1997). *Structural Impact*, Cambridge University Press.

Jones, N. (1989a). Some Comments on the Modelling of Material Properties for Dynamic Structural Plasticity, International Conference on the Mechanical Properties of Materials at High Rates of Strain, Ed. J. Harding, Oxford. *Institute of Physics Conference Series No. 102*, 435-445.

Jones, N. (1989b). On the Dynamic Inelastic Failure of Beams, *Structural Failure*, Eds. T. Wierzbicki and N. Jones, John Wiley, 133-159.

Jones, N. (1976). Plastic Failure of Ductile Beams Loaded Dynamically, *Trans. A.S.M.E., Journal of Engineering for Industry*, **98:B1**, 131-136.

Jones, N. (1974). Some Remarks on the Strain-Rate Sensitive Behavior of Shells, *Problems of Plasticity*, Vol 2, Ed. A. Sawczuk, Noordhoff, 403-407.

Jones, N., Kim, S. B. and Li, Q. M. (1997). Response and Failure Analysis of Ductile Circular Plates Struck by a Mass, *Trans. ASME, J. Pressure Vessel Technology*, **119:3**, 332-342.

Jouri, W. S. and Jones, N. (1988). The Impact Behaviour of Aluminium Alloy and Mild Steel Double-Shear Specimens, *Int. J. Mechanical Sciences*, **30:3/4**, 153-172.

Karagiozova, D. (2000). Inertia Effects on Some Crashworthiness Parameters for Cylindrical Shells under Axial Impact, Int. J. Crashworthiness Conference, published by The Bolton Institute, Eds. E. C. Chirwa and D. Otte, 507-517.

Karagiozova, D. and Jones, N. (2000). Dynamic Elastic-Plastic Buckling of Circular Cylindrical Shells under Axial Impact, *Int. J. Solids and Structures*, **37:14**, 2005-2034.

Karagiozova, D., Alves, M. and Jones, N. (2000). Inertia Effects in Axisymmetrically Deformed Cylindrical Shells under Axial Impact, *Int. J. Impact Engineering*, **24:10**, 1083-1115.

Langseth, M. and Larsen, P. K., (1994). Dropped Objects' Plugging Capacity of Aluminium Alloy Plates, *Int. J. Impact Engineering*, **15:3**, 225-241.

Li, Q. M. and Jones, N. (2000), Formation of a Shear Localisation in Structural Elements under Transverse Dynamic Loads, *Int. J. Solids and Structures*, **37:45**, 6683-6704.

Li, Q. M. and Jones, N. (1999). Shear and Adiabatic Shear Failures in an Impulsively Loaded Fully Clamped Beam, *Int. J. Impact Engineering*, **22:6**, 589-607.

Menkes, S. B. and Opat, H. (1973). Broken Beams, *Experimental Mechanics*, **13**, 480-486.

Nurick, G. N. and Shave, G. C. (1996). The Deformation and Tearing of Thin Square Plates Subjected to Impulsive Loads – An Experimental Study. *Int. J. Impact Engineering*, **18:1**, 99-116.

Shen, W. Q. and Jones, N. (1993). Dynamic Response and Failure of Fully Clamped Circular Plates Under Impulsive Loading, *Int. J. Impact Engineering*, **13:2**, 259-278, 1993.

Shen, W. Q. and Jones, N. (1992). A Failure Criterion for Beams Under Impulsive Loading, *Int. J. Impact Engineering*, **12:1**, 101-121 and **12:2**, 329 (1992).

Teeling-Smith, R. G. and Nurick, G. N. (1991). The Deformation and Tearing of Thin Circular Plates Subjected to Impulsive Loads. *Int. J. Impact Engineering*, **11:1**, 77-91.

Wen, H. M. and Jones, N. (1993). Experimental Investigation of the Scaling Laws for Metal Plates Struck by Large Masses, *Int. J. Impact Engineering*, **13:3**, 485-505.

White, M. D. and Jones, N. (1999). A Theoretical Analysis for the Dynamic Axial Crushing of Top-Hat and Double-Hat Thin-Walled Sections, *Proc. Institution of Mechanical Engineers*, **213:Part D**, 307-325.

Yu, J. and Jones, N. (1997). Numerical Simulation of Impact Loaded Steel Beams and the Failure Criteria, *Int. J. Solids and Structures*, **34:30**, 3977-4004.

Yu, J. and Jones, N. (1991). Further Experimental Investigations on the Failure of Clamped Beams under Impact Loads, *Int. J. Solids and Structures*, **27:9**, 1113-1137.

Yu, T. X. and Chen, F. L. (2000). A Further Study of Plastic Shear Failure of Impulsively Loaded Clamped Beams, *Int. J. Impact Engineering*, **24:6/7**, 613-629.

THE CRUSHING STRENGTH OF ALUMINIUM ALLOY FOAM AT HIGH RATES OF STRAIN

S.R. Reid, P.J. Tan and J.J. Harrigan

Department of Mechanical, Aerospace and Manufacturing Engineering,
UMIST, P.O. Box 88, Sackville Street, Manchester M60 1QD, U.K.

ABSTRACT

Metallic foams can be compressed plastically over a long stroke at a nearly constant load. Such material behaviour is ideal for many packaging and impact/blast mitigation systems. However, the dynamic crushing behaviour of metallic foams has proved difficult to quantify, with experimental investigations on the effects of loading rate producing apparently contradictory results. In order to clarify the effect of strain rate and impact velocity, the authors have conducted an extensive experimental study using a direct impact technique. Closed-cell aluminium alloy foam (Hydro) specimens in two average cell sizes and with a diameter of 45mm were fired directly at a Hopkinson pressure bar load cell at impact velocities up to 200ms^{-1}, corresponding to a nominal strain rate of 4500s^{-1}. Experimental data on the initial dynamic crushing strengths and the plateau stresses of cylindrical specimens are compared for two average cell sizes of approximately 4mm and 14mm. The results reveal significant differences in the dynamic enhancement of the initial crushing strengths and plateau stresses depending upon the impact velocity, the average cell size and the direction of impact relative to the casting directions. The measured force pulses were processed by deconvolution in order to improve their accuracy.

KEYWORDS

Aluminium foam, impact energy absorption, dynamic crushing strength, deconvolution, cell size, morphology, strain rate, inertia, casting direction, closed cell.

INTRODUCTION

Successful implementation of aluminium foams requires that their mechanical behaviour be firmly established over a wide range of loading rates. In the present study, the high rate compressive behaviour of the closed-cell Norsk Hydro aluminium foams (Hydro foams) is investigated.

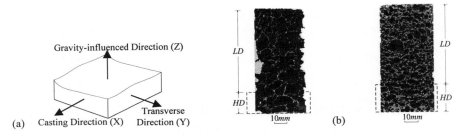

Figure 1: (a) Reference system of a foam slab (b) Example of $80mm$ LC and SC specimens along the Z-direction.

The foams have two average cell sizes of $14mm$ (LC) and $4mm$ (SC). The base alloy composition is Al-Si8-Mg with a density and yield strength of $2730 kgm^{-3}$ and $185 MPa$ respectively. Unless otherwise stated, the circular cylindrical specimens were of equal length and diameter ($\approx 45mm$) with their length axes orientated along the *transverse* (Y) or *gravity-influenced* (Z) directions (Fig. 1a). The latter have thicker cell walls/edges near to their base due to gravitational drainage (Fig. 1b). The density in the Z-direction varies approximately as a step function. Across the step the mean densities typically differ by 40% in an SC foam and up to 400% in an LC foam. No preferential orientation of the cells relative to the principal casting directions was observed.

DEFINITIONS OF QUASI-STATIC MATERIAL PROPERTIES

Both quasi-static and dynamic specimens were tested within a $45mm$ diameter barrel to prevent specimen break-up. Uniaxial quasi-static compression tests were carried out on an Instron testing machine. Typical nominal stress-strain curves of foams compressed along the Y and Z directions for the two average cell sizes are shown in Fig. 2a. The localised nature of cell collapse implies that these curves are non-unique.

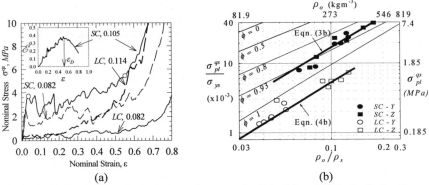

Figure 2: (a) Typical nominal stress-strain curves for Y (solid) and Z (dashed) directions showing relative density values (b) Variation of the normalised plateau stress with relative density

Various *ad hoc* criteria have been proposed to identify key material parameters such as plateau stress σ_{pl}^{qs} and densification strain ε_D (see Deshpande & Fleck (2000), Hanssen et al (2000), Reid & Peng

(1997)). A consistent definition for the parameters is now proposed (Tan & Reid (2001), Langseth (2001), Avalle(2001)). The initial crushing stress σ_{cr}^{qs} is the first peak stress prior to the onset of strain softening. The energy absorption efficiency $\mathcal{E}$ is defined as $\mathcal{E}(\varepsilon_a) = \frac{1}{\sigma(\varepsilon)|_{\varepsilon=\varepsilon_a}} \int_0^{\varepsilon_a} \sigma(\varepsilon) d\varepsilon$. The densification strain ε_D is the strain at which the efficiency is a maximum in the efficiency-strain curve (inset of Fig. 2a). A fit of the experimental data on foams in the Y-direction gives

$$\varepsilon_D = 0.76\left[1 - 3.167(\rho_o/\rho_s) + 2.167(\rho_o/\rho_s)^3\right] \tag{1}$$

Once σ_{cr}^{qs} and ε_D are identified, the plateau stress σ_{pl}^{qs} is defined by the equation

$$\sigma_{pl}^{qs} = \frac{1}{(\varepsilon_D - \varepsilon_{cr})} \int_{\varepsilon_{cr}}^{\varepsilon_D} \sigma(\varepsilon) d\varepsilon . \tag{2}$$

Values of experimental plateau stresses are plotted against relative density in Fig. 2b. Predictions of the scaling law proposed by Gibson & Ashby (1997) are plotted in the same figure for different volume fractions ϕ of the solid contained within the cell edges. The experimental data for both foams (LC and SC) lie close to $\phi = 1$ suggesting plastic collapse is predominantly by cell edge bending. Due to the ratio of the specimen size to cell size (or 'size effect') and morphological defects, the LC foam data falls beyond the limiting value of $\phi = 1$. The 'size effect' is thought to be responsible for this anomaly. The superimposed density dependence dictates that the quasi-static properties of each foam are best described by separate characteristic curves. The scaling law by Gibson and Ashby (1997) was used to fit the experimental data, which gives (by setting $\phi = 1$)

Small cell foam :
$$\sigma_{cr}^{qs}/\sigma_{ys} = 0.522(\rho_o/\rho_s)^{3/2} \quad ; \quad \sigma_{pl}^{qs}/\sigma_{ys} = 0.541(\rho_o/\rho_s)^{3/2} \tag{3a \& b}$$
Large cell foam :
$$\sigma_{cr}^{qs}/\sigma_{ys} = 0.090(\rho_o/\rho_s)^{3/2} \quad ; \quad \sigma_{pl}^{qs}/\sigma_{ys} = 0.166(\rho_o/\rho_s)^{3/2} \tag{4a \& b}$$

For each foam, the plateau strengths along the two principal directions are comparable. This is because the plateau regime of a foam compressed along the Z-direction is dominated by cell collapse within its lower density zone. This density is comparable to that of a specimen aligned in the Y-direction. The compression of entrapped gas does not contribute significantly to the strength of the foams (Tan & Reid (2001)).

DIRECT-IMPACT TESTS

The dynamic crushing behaviour of metallic foams has proved difficult to quantify, with experimental investigations on the effects of loading rate producing apparently contradictory results (Deshpande & Fleck (2000), Mukai et al (1997)). In order to clarify the effect of strain rate and impact velocity, the authors have conducted an extensive experimental study using a direct impact technique. Aluminium foam specimens with attached backing masses were fired directly at a modified Hopkinson pressure bar load cell at impact velocities up to 200ms^{-1}, corresponding to a nominal strain rate of 4500s^{-1}. Details of the experimental set-up (Fig. 3) are described by Harrigan et al. (1998).

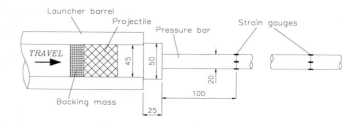

Figure 3: Set-up for direct impact tests on aluminium foam projectiles

The common version of the Hopkinson pressure bar load cell consists of a long, solid, steel cylinder. When an impulsive force is applied to one end of the bar it causes a stress wave to propagate along the axis which can be measured using strain gauges. The aluminium foam deforms at stress levels that are too low to produce measurable strains in a common steel pressure bar. A modified version of the pressure bar load cell was employed to give a greater sensitivity. The bar used was 20 mm in diameter and 2.4m long. A steel ring with a 50mm outer diameter was shrink fitted to the end of the bar before this end was faced-off. Measured force pulses were processed by deconvolution as a one-dimensional stress wave could not be assumed. The transfer function was obtained using the ABAQUS finite element code (Fig. 4) rather than an experimental calibration. An input force pulse $f(t)$ was applied to the model and the strain response $e(t)$ was monitored at a location along the bar. The deconvolution operation is highly sensitive to background noise in both the calibration step and the test results (Inoue et al. 2001). This calibration method had the advantage of removing random noise from the physical surrounding.

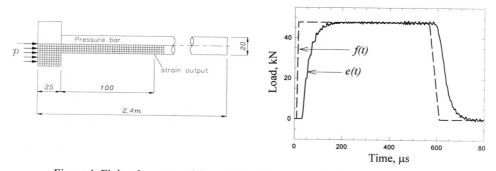

Figure 4: Finite element model and calibration curves for the modified pressure bar

Typical force pulses for *LC* and *SC* foams are shown in Fig. 5. The dynamic initial crushing and plateau stresses are defined from these pulses in a manner similar to that employed for quasi-static results. The *large* and *small* cell foams have a quasi-static scatter in strength of the order of 40% and 20% respectively. Hence the dynamic stress is said to be enhanced only if it exceeds this scatter (Deshpande & Fleck (2000)), i.e. $\sigma^d > 1.4\sigma^{qs}$ (for *LC* foam) and $\sigma^d > 1.2\sigma^{qs}$ (for *SC* foam).

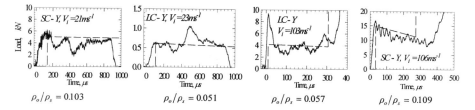

Figure 5: Experimental force traces (heavy line) and theoretical predictions (dashed line)

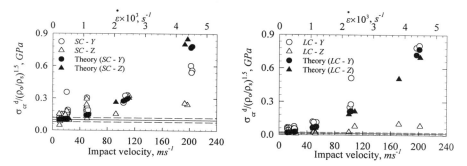

Figure 6: Variation of the normalised initial crushing stress with impact velocity.
Quasi-static scatter in loads shown by dashed lines.

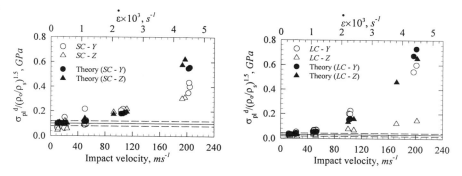

Figure 7: Variation of the normalised plateau stress with impact velocity.
Quasi-static scatter in loads shown by dashed lines.

Effect of Impact Velocity on the Initial Crushing and Plateau Stresses

The variation of the initial crushing stress with impact velocity is shown in Fig.6 for both foams. At sub-critical impact velocities ($V_i < 100 ms^{-1}$) the initial crushing strength of the *SC* foam is significantly larger than that for the *LC* foam. This is due to the 'size effect' as is the case with quasi-static compression. At super-critical velocities ($V_i > 100 ms^{-1}$) the initial crushing strength of the *LC* foam exceeds the *SC* foam suggesting that the stress enhancement mechanism has switched to one independent of this 'size effect'. The initial crushing stress is sensitive to heterogeneity effects, such as local density variations, on the proximal surface. At sub-critical velocities, the strength enhancement is attributed to (i) rate sensitivity of the cell wall material (ii) compression of entrapped gas and (iii) microinertia of cell wall/edge material. At super-critical velocities, further local enhancements in stress are primarily the result of inertia effects associated with the dynamic localisation of crushing (i.e. shock-like characteristics).

The data shown in Fig. 7 indicate significant plateau stress enhancements at $V_i \geq 50 ms^{-1}$ and $V_i \geq 100 ms^{-1}$ for the LC and SC foams respectively. The plateau stress of the LC foam exceeds its SC counterpart at $V_i > 100 ms^{-1}$. This is in agreement with the results of the initial crushing strength enhancement and therefore supports the contention that a switch in the stress enhancement mechanism has occurred. In general, the initial crushing stress is higher than the plateau stress although what controls their differences remains to be clarified. Whenever the dynamic plateau strength is not enhanced ($V_i < 50 ms^{-1}$ for LC foams and $V_i < 100 ms^{-1}$ for SC foams) cursory examination of crushed samples reveals that cell collapse is through discrete crush band multiplication. This implies that there exists a specimen length scale below which a simple continuum model is not applicable, as is also the case for quasi-static compression.

The data in Figs. 6 and 7 were normalised to minimise the effect of differences in relative density. However, this normalisation is inappropriate when a significant density gradient exists. This is the reason for the relatively low level of dynamic strength enhancement for the LC foams along the Z-direction.

Effect of Density Gradient on Initial Crushing Stress

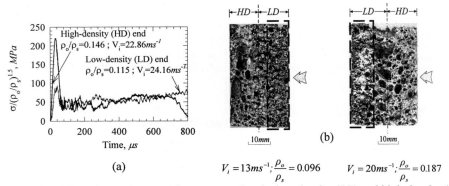

Figure 8: (a) Experimental stress pulses measured at the low-density (LD) and high-density (HD) ends for specimens impacted along the gravity-influenced direction. (b) Crushed samples (impact surface indicated by arrows)

Stress pulses for impacts at the high density (HD) and low density (LD) ends are compared in Fig. 8a. The plateau stresses were less than the peak values by factors of between 2 and 15 depending upon the density variation through the specimens and the impact velocity. The plateau stress is concerned with cell collapse in the LD zone, irrespective of the impact end. The comparable plateau stresses observed in Fig. 8a indicates that the HD material behaves like a rigid restraint by confining cell deformation to the LD zone. This is supported by the photographs of two crushed specimens in Fig. 8b where the boxes highlight crushed regions.

COMPARISON WITH SHOCK MODEL

A simple shock model (Reid & Peng (1997)) based on a rate-independent, rigid, perfectly-plastic, locking (r-p-p-l) idealisation of the nominal stress-strain curve has been used to provide a first order understanding of the dynamic response of foams (Tan & Reid (2001)). In general, the predictions of the shock model compare well with the experimental force-time pulses, as evident in Fig. 5. However, the onset of locking cannot be reliably predicted because the real material is not perfectly rigid at the densification strain. Figs. 6 and 7 show that the predicted dynamic initial crushing and plateau stresses agree reasonably well with the experimental values. The form of their variation with impact velocity is well represented by the theory especially at super-critical velocities $V_i \geq 100 ms^{-1}$, see Figs. 6 and 7. The dependence of plateau stresses on impact velocity rather than strain rate is emphasised by the correlation between the shock model and experimental results shown in Fig. 9. These results were obtained for specimens with different lengths to provide a range of nominal strain rates for the same impact velocity. The quasi-static results show clear scale effects when the specimen length (l_o) is less than four times the cell size (l_c). This suggests that the dynamic strength properties at the super-critical velocity regime are governed primarily by the impact velocity whilst 'size effect' and morphological defects are secondary. Whilst giving reasonable first order estimates, the shock model is not generally applicable in the sub-critical velocity regime due to the non-local nature of the foams' deformation response. A non-classical wave theory capable of taking into account the length scale characteristics of the cellular geometry and its softening characteristics must be formulated.

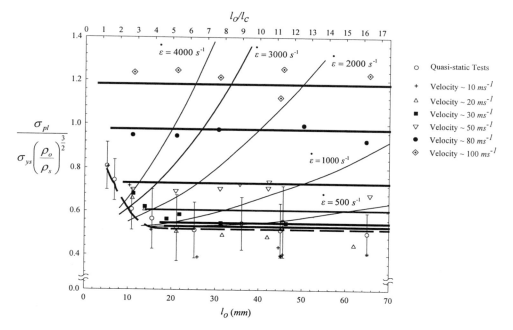

Figure 9: Effect of impact velocity and specimen length on the normalised plateau stress of small cell foam in the transverse direction. The dashed line shows the quasi-static stress. Thick horizontal solid lines are predictions by the shock model and thin solid lines give a family of intersecting contours of constant nominal strain rate.

CONCLUSIONS

A consistent definition for the plateau stress and densification strain is proposed. Inertia effects associated with the localisation of crushing are responsible for the enhancement of the dynamic strength properties at the super-critical velocity regime. The predictions of the shock model compare well with experiment in this velocity regime whilst 'size effect' and morphological defects are insignificant. Local density variation inherent to the manufacturing process has a significant effect on the dynamic peak forces measured. Accurate lower velocity modelling awaits the formulation of a suitable generalised wave theory. Dynamic tests of these types of materials would appear to be best approached using inverse modelling (see Hanssev et al. (2002)). This involves formulating a mathematical model, comparing the predictions with accurate test results (in this case using the direct impact test) and deriving values for parameters in the theoretical formulation. Direct measurement of constitutive behaviour (e.g. using the split Hopkinson pressure bar) would appear to be questionable.

Acknowledgement

The authors are grateful to Prof. M. Langseth, Dr. T.Y. Reddy and Dr. A.G. Hanssen for helpful comments. The financial support and supply of foam specimens by Hydro Aluminium, a.s. (in collaboration with NTNU, Trondheim, Norway) are gratefully acknowledged.

References

Avalle M., Belingardi G. and Montanini R. (2001). Characterisation of polymeric structural foams under compressive impact loading by means of energy-absorption diagram. *Int. J. Impact Engng.*, **25**, 455-472.

Deshpande V.S. and Fleck N.A. (2000). High strain rate compressive behaviour of aluminium alloy foams. *Int. J. Impact Engng.* **24**, 277-298.

Gibson L.J. and Ashby M.F. (1997), *Cellular Solids, Structure and Properties*, 2nd ed. Cambridge University Press, Cambridge

Hanssen A.G., Langseth M. and Hopperstad O.S. (2000). Static and dynamic crushing of square aluminium extrusions with aluminium foam filler. *Int. J. Impact Engng.* **24:4**, 347-383

Hanssen A.G., Hopperstad O.S., Langseth M. and Ilstad H. (2002). Validation of constitutive models applicable to aluminium foams. To appear in *Int. J. Mech. Sci.*

Harrigan J.J., Reid S.R. and Reddy T.Y. (1998). Accurate measurement of impact force pulses in deforming structural components, *Exp. Mech.* (ed. I.M. Allison) 149-154.

Inoue H., Harrigan J.J. and Reid S.R. (accepted for publication 2001). A review on inverse analysis for indirect measurement of impact force *Appl. Mech. Rev.*

Langseth M. (2001). Private Communication, UMIST

Mukai T., Kanahashi H., Higashi K., Yamada Y., Shimojima K., Mabuchi M., Miyoshi T. and Nieh T.G. (1999). Energy absorption of light-weight metallic foams under dynamic loading, in *Metal Foams and Porous Metal Structures* (Banhart, J., Ashby, M.F. and Fleck, N.A. Eds.), Verlag MIT Publishing, Bremen (Germany), 353-358.

Reid S.R. and Peng C. (1997). Dynamic uniaxial crushing of wood. *Int. J. Impact Engng.* **19:5-6**, 531-570.

Tan P.J. and Reid S.R. (2001). Dynamic characterisation of 'Hydro' aluminium foams subjected to high strain rate compression. *Final Technical Report* (Ref.: *UMIST Report ME/AM/6.01/NTNU3*), Appl. Mech. Div., Dept. of Mech. Engng., UMIST

ON THE BEHAVIOUR CHARACTERISATION OF POLYMERIC FOAMS OVER A LARGE RANGE OF STRAIN RATES

Han ZHAO[1] and Gérard GARY[2]

[1]Laboratoire de Mécanique et Technologie-Cachan, Université Pierre et Marie Curie (Paris 6), 61, Avenue du président Wilson, 94235 Cachan cedex, France.

[2]Laboratoire de Mécanique des Solides, Ecole Polytechnique, 91128 Palaiseau, France.

ABSTRACT

The testing and modelling of the mechanical behaviour of some polymeric foams involved in the automotive industry are presented. A variety of current experimental arrangements over a large range of strain rates has been reviewed. Recent improvements of a particularly useful technique for impact loading - the split Hopkinson bar, are presented. A phenomenological model is developed to describe experimental data. Difficulties for three-dimensional testing and modelling are also discussed.

KEYWORDS

Polymeric foams, Experiments, Kolsky bar, Strain rates, Constitutive law, Impact.

INTRODUCTION

Polymeric foams are quite widely used for many industrial reasons. In the automotive industry, they are used for the damping of noises, for the body protection in case of collisions, or for the manufacture of anthropomorphic dummies used in crash tests.

The modelling of the mechanical properties of polymeric foams over a large range of strain rates is then needed. For example, in the study of the crashworthiness, anthropomorphic dummies have been used to investigate the response of the human body and to analyse the interaction with the inhabitant compartment. In order to evaluate the occupant protection by numerical crash simulations in the early phases of car design, the characterisation of the mechanical behaviour of polymeric foams and their strain rate sensitivity is indispensable.

Experimental and theoretical works on mechanical properties of polymeric foams have been reported in the open literature, especially under quasi-static loading. Indeed, the quasi-static behaviours of various polymeric foams have been obtained with ordinary testing machines. Many constitutive foam models have been also proposed, which are based on a micromechanical analysis allowing for the determination of global cellular structure response from the study of a single cell. An extensive review can be found in the book of Gibson and Ashby (1988).

The behaviour of polymeric foams at relatively high strain rates has been an interest of investigators since the 1960s. Experimental results using different devices such as the falling weight or impacting mass technique (Faruque et al., 1997; Lacey,1965; Schreyer et al.,1994; Traegar,1967; Zhang et al. 1997), rapid hydraulic testing machine (Chang et al., 1998; Rehkopf et al., 1996; Wagner et al., 1997), split Hopkinson bar have been reported (Rinde and Hoge, 1971; 1972). Phenomenological constitutive models at high strain rates (Chang et al., 1998; Faruque et al., 1997; Rehkopf et al., 1996; Schreyer et al.,1994; Wagner et al., 1997; Zhang et al. 1997) and those based on the micromechanical analysis are also developed (Mills, 1997; Shim et al., 1992).

In this paper, different testing arrangements are reviewed and commented. The recent improvements of split Hopkinson bars technique are presented. A simple phenomenological model will be also proposed. Finally, difficulties in actual three-dimensional testing and modelling are discussed.

ONE-DIMENSIONAL EXPERIMENTAL ARRANGEMENTS

The quasi-static behaviour of polymeric foams can be obtained with ordinary testing machines. It is a well-controlled technique, provided that the suitable device is used for the force measurement when very weak foams are tested. Experimental techniques differ when the strain rate is more or less important. Rapid hydraulic testing machine or falling weight arrangements are techniques that are often reported in the literature. However, even if those techniques can provide a suitable mechanical loading over a large range of strain rates, the quality of associated measurements decreases with the increase of the strain rate. The main reason is that the assumption of an instantaneous equilibrium in the loading frame as well as the measuring device, which neglect any wave propagation effect, is not valid. Consequently those measuring techniques begin to lose the accuracy at medium strain rates ($\dot{\varepsilon} > 10/s$) and special short loads cells (Holzer,1978) have to be used. For high strain rates ($\dot{\varepsilon} > 200/s$), it is very difficult to obtain reliable results using those devices.

For high strain rates, the Split Hopkinson Pressure Bar (SHPB) is a widely used experimental technique to study constitutive relationship of materials (Hopkinson,1914;Kolsky,1949). Rinde and Hoge (1971, 1972) have performed tests on polymeric foams at high strain rates with SHPB. A typical SHPB setup is composed of long input and output bars with a short specimen placed between them. The impact of the projectile at the free end of the input bar develops a compressive longitudinal incident wave (Fig.1a). Once it arrives at the bar specimen interface, a reflected wave is developed in the input bar, whereas a transmitted wave is induced in the output bar. From those basic experimental data (incident, reflected and transmitted waves), forces and velocities at both faces of the specimen can be calculated.

However, using the SHPB device for foam testing raises two technical difficulties. One consists of a too low impedance ratio between the specimen and metallic split bars. This leads to imprecise measurements of the input force, output force and output velocity. For example, Rinde and Hoge (1971,1972) have to use quartz crystals for the measurement of the specimen stress. These devices are not sufficiently accurate at the early stage of loading so that it was not possible to measure the apparent Young modulus and the yield stress.

The use of low impedance bars which are generally viscoelastic is then proposed. Two 3m long Nylon bars of 40 mm diameter (density 1200 kg/m^3 and speed 1700 m/s) are used. It provides an improvement of about 20 times the impedance ratio, by comparison with a classical steel bar (density being 7850 kg/m^3 and wave speed 5000 m/s). The use of viscoelastic bars in a SHPB setup introduces however complications such as the wave dispersion in a viscoelastic bar (Zhao and Gary, 1995), the calculation of stress and particle velocity from the measured strain and the projectile length limitations (Zhao and Gary, 1997).

Another particular feature of foam testing is the need to achieve a large maximum strain (up to 80%) for the study of the densification, associated with a significant increase of the stress. Ordinary SHPB arrangements cannot measure up to such strains, even if large strains are easily reached because of the very low resistance of foams. Indeed, measuring technique using bars is based on the superposition principle which implies that the stress, the strain and the particle velocity at any cross-section can be considered as the algebraic sum of those values associated with the two elementary waves propagating in opposite directions at this cross-section. Such a technique requires then a separate recording of each single wave propagating in the bar. This is classically realised by using a long bar and a short pulse to insure the existence of a particular cross-section where those waves are not superimposed. The measuring duration of a given SHPB is then limited. To overcome this measuring limitation, one has to investigate the multiple reflections in bars (Campbell and Duby, 1956). A two-gauge method has been reported to separate the two waves in elastic bars (Lundberg and Henchoz, 1977) and viscoelastic bar (Zhao and Gary, 1997). This two gauges method provides an unlimited measuring duration and gives consequently sufficient maximum strain. This method, that sometimes exhibits instability problems has been recently improved by using more than two strain or speed sensors (Bussac et al., 2001).

The use of such an unlimited measuring technique can be also applied to perform tests at medium strain rates (from 10 to 200 s^{-1}), provided that a suitable loading device is used. For example, fig. 1b shows an experimental arrangement called "slow bar" (Zhao and Gary, 1997) which uses the bars as measuring device and a high speed oil jack as loading device.

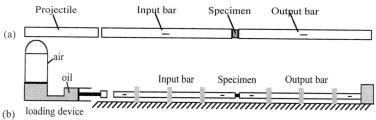

Figure 1. (a) Classical SHPB arrangement (b) "Slow bar" arrangement

SOME EXPERIMENTAL RESULTS AND PHENOMENOLOGICAL MODELS

With suitable techniques, the experimental data over a large range of strain rate can be then obtained. However, all those techniques just provide measurements of force, velocity or displacement at both faces of specimen (SHPB test) or sometimes only at one face of specimen (Standard machine, falling weight). One has to assume that stress and strain fields in the specimen are homogeneous in order to deduce an average stress-strain relation.

Under this assumption, some general features of testing results are presented in the following. Most polymeric foams exhibit strong rate sensitivities. Their mechanical properties depend also on the temperature and moisture. The specimen is a 40mm diameter and 40mm length cylinder. Quasi-static test has been performed at a constant speed of loading. Results at medium strain rates are obtained using the "slowbar" technique. Finally, a Nylon Hopkinson bar provides results at high strain rates. At a strain rate of 250/s, Standard Hopkinson bar gives 27% of strain. With the two-measurement method, the measured strain is increased up to 70%. The cyclical loading of the SHPB test is due to the round trip of wave in incident bar.

A set of stress-strain responses of foams over a large range of strain rates makes their behaviour modelling possible. In the numerical simulation, one uses in general phenomenological models rather than physical ones deduced from micro-mechanical considerations (Gary et al., 1996; Zhao, 1997). It is probably because of the simplicity of phenomenological models (Zhao and Gary, 1996). For instance, a simple model involving a few coefficients but giving a satisfactory curve fitting is presented in the following.

Experimental observations give two indications. The absence of significant permanent deformation suggests a hyper elastic behaviour. The weak influence of the density on the rate sensitivity shows that the viscosity of foams is derived from that of the basic polymeric constitutive materials. The behaviour of foam is considered as the sum of an elastic part describing the progressive densification of the hollow polymeric structure and a viscoelastic part due to the viscosity of the elementary polymeric materials of which foams are made.

$$\sigma = \sigma_{élas} + \sigma_{vis} \qquad (1)$$

The elastic part is modelled by a hyperbola where the parameter h is obviously the asymptotic strain related to the initial foam porosity and k is a parameter having the same dimension as a stress (Eq. 3).

$$\sigma_{éla} = \frac{k}{h-\varepsilon} \qquad (2)$$

The viscoelastic part is chosen as an usual power law, but multiplied by a term due to progressive densification of the hollow polymeric structure. The parameter h is the same asymptotic strain and η is a parameter having a dimension of the stress (Eq.4). $\dot{\varepsilon}_0$ is the reference strain rate.

$$\sigma_{vis} = \frac{\eta}{h-\varepsilon}\left(\frac{\dot{\varepsilon}}{\dot{\varepsilon}_0}\right)^n \qquad (3)$$

As it can be observed in Fig. 2, this four-parameter model gives good results over a large range of strain-rates (where h=0.96, k=0.7 MPa, η=0.3 MPa, n=0.21, $\dot{\varepsilon}_0$=0.0016/s).

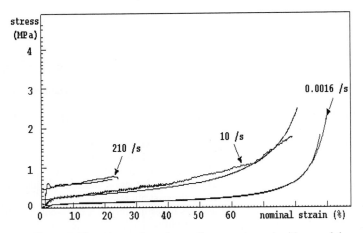

Figure 2. Experimental results on foams compared with a model

4. DISCUSSIONS

The testing and modelling of foams discussed above concerns only the one-dimensional case whereas 3D models are needed for numerical calculations. A rather easy way to generalise a one-dimensional model consists in using the Von-Mises equivalent stress-strain like in the classical plasticity theory. More theoretically pertinent models can be used, provided that complementary three-dimensional testing data is available.

However, the most difficult tasks for behaviour characterisation of foams lie in the three-dimensional testing, especially at relative high strain rates. The usual testing methods involving a three-dimensional effect can not be easily applied for foams. A dynamic test with quasi-static lateral confinement is also difficult as the pressure (with air or oil) cannot be directly applied to the specimen. Using a rigid ring confinement is also difficult as lateral friction forces could be in the range of compression force that has to be measured. More generally speaking, it is very difficult to control and measure loading and boundary conditions. In addition, for high strain rates, the wave propagation effects in the sample make the stress and strain fields non-homogeneous.

Most reported investigations use an inverse identification technique. They try to determine parameters of a given model, using a sophisticated three-dimensional test. It would be better, in author's opinion, to perform simpler 3D tests in which the measuring precision can be well controlled, because identifications by an inverse method are very sensitive to the noise of the measurements.

For this purpose, a double-side indentation of a 60mm diameter foam specimen (length 36mm) is performed with a SHPB arrangement of 40mm diameter. (Fig.3). The advantage of such tests lies in their simplicity so that no supplementary interface is needed. It is consequently certain that the difference between this test and the pure compression one is only due to foam properties.

Figure 3 shows the comparison between stress-strain curves obtained from indentation tests and compression ones. A difference of about 30% of the stress level is found at the same testing condition. In addition, differences between those two kinds of tests vary with the strain rate, which indicate that the three-dimensional feature is not independent of the strain rate effect.

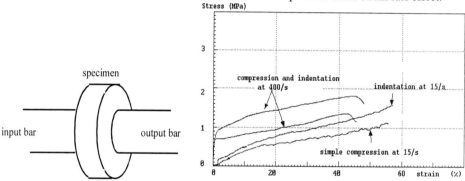

Figure 3. Comparison between indentation and compression tests

5. CONCLUSION

In this paper, different experimental arrangements for the testing of polymeric foams over a large range of strain rates have been discussed. It is shown that the Split Hopkinson bar technique gives accurate results, provided that a special processing is applied for the signals recorded at two points of each bar. With a special "slow bar" arrangement, it is also possible to test foams at medium strain-rates (from 10 to 200 s^{-1}). The knowledge of the response of foams over a large range of strain-rates makes its constitutive modelling possible. A rather simple phenomenological constitutive model correctly describes their one-dimensional behaviour. An indentation test using SHPB bar techniques is proposed to investigate three-dimensional features.

REFERENCES

Bussac M.N., Collet P.C, Gary G., Othman R.(2001), An optimisation method for separating and rebuilding one-dimensional dispersive waves from multi-point measurements. Application to elastic or viscoelastic bars, *accepted for publication in J.Mech.Phys.Solids*.

Campbell J.D. and Duby J., The yield behaviour of mild steel in dynamic compression, *Proc. R. Soc. Lond.,* **A 236** (1956), 24-40.

Chang F.S., Song Y., Lu D.X. and Desilva C. N. (1998), Unified constitutive equations of foams materials, *ASME J. Engng. Mater. Tech.,* **120,** 212-217.

Faruque O., Liu N. and Chou C. (1997), Strain rate dependent foam – constitutive modeling and applications. S.A.E Transactions. *J. Materials & Manufacture,* **106,** 904-912.

Gary G., Rota L., Thomas J.J. and Zhao H (1996), Testing and modelling the dynamic behavior of polymeric foams used in automotive industry dummies, *In: Proc. 9th Dymat technical conference,* (Ed. J.Najar), Munich, Germany.

Gibson L.J. and Ashby M.F.(1988), *Cellular Solids,* Pergamon Press.

Holzer A.J. (1978), A technique for obtaining compressive strength at high strain rates using short load cells, *Int. J. Mech. Sci.,* **20,** 553-560.

Hopkinson B. (1914), A method of measuring the pressure in the deformation of high explosives or by the impact of bullets, *Phil. Trans. Roy. Soc.,* **A213,** 437-452.

Kolsky H. (1949), An investigation of the mechanical properties of materials at very high rates of loading, *Proc. Phys. Soc.,* **B62,** 676-700.

Lacey, R.M.(1965), Response of several materials at intermediate strain rates, *Proc. Fifth Int. Symp. on high speed testing,* Boston.

Lundberg B. and Henchoz A. (1977), Analysis of elastic waves from two-point strain measurement, Exper. Mech., 17, 213-218.

Mills N.J. (1997), Time dependence of the compressive response of polypropylene bead foam, *Cellular polymers,* **16,** 194-215.

Rehkopf J.D., Mcneice G.M. and Borland G.W. (1996), Fluid and Matrix components of polyurethane foam behaviour under cyclic compression, *ASME J. Engng. Mater. Tech.,* **118,** 58-62.

Rinde J.A. and Hoge, K.G. (1971), Time and temperature dependence of the mechanical properties of polystyrene bead foam, *J. Apppl. Polymer Sci.,* **15,** 1377-1395.

Rinde J.A. and Hoge, K.G. (1972), Dynamic shear modulus of polystyrene bead foams, *J. Apppl. Polymer Sci.,* **16,** 1409-1415.

Schreyer, H.L., Zuo, Q.H. and Maji, A.K. (1994), Anisotropic plasticity model for foams and honeycombs, *ASCE J. Engng. Mech.,* **120,** 1913-1930

Shim, V.P.M., Yap K.Y. and Stronge W.J. (1992), Effects of nonhomogeneity, cell damage and strain rate on impact crushing of a strain-softening cellular chain, *Int. J. Impact Engng,* **12,** 585-602.

Traegar R.K. (1967), Physical properties of rigid polyurethane foams, *J. Cellular Plastics,* 405-418.

Wagner D.A., Gur Y., Ward S.M. and Samus M.A. (1997), Modelling foam damping materials in automotive structure, *ASME J. Engng. Mater. Tech.,* **119,** 279-283.

Zhang J., Kikuchi N., Li V., Yee A. and Nusholz G. (1997), Constitutive modelling of polymeric foam material subjected to dynamic crash loading, *Int. J. Impact Engng.,* **21,** 369-386.

Zhao H. and Gary G. (1995), A three dimensional analytical solution of longitudinal wave propagation in an infinite linear viscoelastic cylindrical bar. Application to experimental techniques, *J. Mech. Phys. Solids.* **43,** 1335-1348.

Zhao H., Gary G. (1996), The testing and behavior modelling of sheet metals at strain rate from 10^{-4} to 10^{4}/s, *Mater. Sci. & Engng.* **A207,** 46-50.

Zhao H., Gary G., and Klépaczko J.R. (1997), On the use of a viscoelastic split Hopkinson pressure bar, *Int. J. Impact Engng,* **19,** 319-330.

Zhao H. and Gary G. (1997), A new method for the separation of waves. Application to the SHPB technique for an unlimited measuring duration, *J. Mech. Phys. Solids,* **45,** 1185-1202.

Zhao H., A constitutive model for metals over a large range of strain rates. Identification for mild-steel and aluminium sheets. *Mater. Sci. & Engng.* **A320** (1997) 95-99.

DYNAMIC AND STATIC COMPRESSION BEHAVIOUR OF PAPER AND POLYSTYRENE FOAM BOARDS

Hidetoshi Kobayashi, Masashi Daimaruya and Ken-ichi Tanaka

Department of Mechanical Systems Engineering, Muroran Institute of Technology,
27-1, Mizumoto, Muroran, 050-8585, Japan

ABSTRACT

In this study, the effect of loading rate on the strength and the absorbed energy of paper foam board and polystyrene foam board was examined by a series of dynamic and quasi-static compression tests. In dynamic compression tests, a polyvinyl chloride (PVC) tube instead of an ordinary metal bar was used for the measurements of the load applied to the specimens, in order to relieve the impedance mismatch between stress bars and specimens. The absorbed energy up to a strain of 60 % in compression tests of paper foam board is about 1.3 times greater than that of polystyrene foam board. It was also found that the absorbed energy obtained from dynamic tests for paper and polystyrene foam boards was greater than that obtained at quasi-static rates.

KEY WORDS

Paper Foam Board, Polystyrene Foam Board, Dynamic Compression Test, Stress Wave Propagation Test, Strain Rate-Dependence, Absorbed Energy

INTRODUCTION

In April 1997, Containers and Packaging Recycling Law was partially come into force for glass bottles and PET bottles in Japan. In April 2000, the law was fully become effective, i.e. its target was extended to paper and plastic containers and packaging. To change packaging and buffer materials from polystyrene to paper becomes a clear recent trend. Therefore, it seems to be important to reveal the dynamic behaviour of paper in compression or in tension.

A number of researches concerning compression of paper are reported. In 1980s, dynamic compression behaviour of saturated cellulose mats was investigated by Hoering et al. (1983) and Ellis et al. (1984) to improve the dehydration ability of paper milling machine. Feller & Braenge (1993) examined the effect of water absorption on the compressive strength of papers such as kraftpaper. It was reported by Sakurai et al. (1993) that a kind of paper was newly developed using polyethylene laminated waste paper with the aim of use for shock absorber or reinforced materials. In 1999, paper foam board including micro-bubbles among pulp fibers was developed to use for exhibition panel boards and shock absorber.

In this study, we performed static tensile tests to examine the fundamental mechanical properties of paper foam board and polystyrene foam board. Furthermore, by means of dynamic and quasi-static compression tests, we observed their compression behaviour of both foam boards and tried to reveal the effect of strain rate on their strength and absorbed energy.

EXPERIMENTS

Experimental Materials

Paper foam and polystyrene foam boards used in this study have the thickness of about 2 mm. SEM photographs of cross section in both foam boards are shown in Fig.1. Paper foam board was made by putting acrylic micro-bubbles with diameter of about 10 μm into pulp fibers and then the micro-bubbles were expanded. In Fig.1(a), a lot of spherical foams with diameter of about 50 μm expanded by isobutene gas were involved among complex paper fibers. On the other hand, polystyrene foam board is composed of oval cells (length of about 400 μm, width of about 150 μm), which may cause anisotropy. Therefore, in order to examine the effect of anisotropy of the both boards, L-direction and W-direction were introduced. The L-direction is the longitudinal direction of an oval cell on the polystyrene foam board and the direction of pulp fibers in the paper foam board, respectively. The W-direction is the direction perpendicular to the L-direction in plane.

(a) Cross section of paper foam board

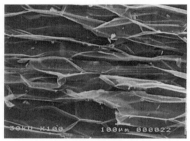
(b) Cross section of polystyrene foam board

Fig.1   SEM photographs of cross section in both foam boards.

Tensile Tests

In order to investigate the effect of anisotropy on the strength of the paper foam and polystyrene foam boards, a series of tensile tests was carried out at 5×10^{-4} sec^{-1} using an Instron universal testing machine (mode 5566). The tensile specimen used is a rectangular type specimen with the gauge length of 20 mm and the width of 10 mm (Fig.2(a)). For the direction of the thickness (T-direction), a disc-shaped specimen with a diameter of 23 mm was used (Fig.2(b)). Tensile tests of L- and W-directions were carried out by chucking the shaded portions indicated in Fig.2(a). The tests of T-direction were performed using the disc specimen (Fig.2(b)) glued onto aluminum jigs with the same diameter as the specimen. The load was measured by the machine load-cell and the elongation was estimated by the movement of the cross-head of the testing machine.

Compression Tests

Quasi-static compression tests were also performed in the T-direction by Instron testing machine at room temperature. The strain rates adopted for the tests were three strain rates of 5×10^{-4}, 7×10^{-3}, 1×10^{-1} sec^{-1}. The load and displacement were obtained from the testing machine as well as tensile tests.

The arrangement for dynamic compression tests is shown in Fig.2(c). In this arrangement, an impact bar with a flange (aluminum circular plate, 75 mm in diameter, 10 mm thick) accelerated by compressed air collides with a specimen attached on one end of the stationary output bar. The impact bar is an aluminum pipe (A6061) with the outer diameter of 19 mm, wall thickness of 1.5 mm and length of 1515 mm. The output bar is a long circular tube of polyvinyl chloride (PVC) instead of an ordinary metal stress bar. To measure a dynamic load, a number of other load-cell types (Yoshino et al (1996), Ujihashi et al (1993)) were used in dynamic compression tests for hollow circular or square cylinders. However, the stress bar is very simple if the stress wave can be measured accurately. In addition, it seems to be quite difficult to measure the stress wave transmitting in metal bars, because the strength of the both foam boards is much smaller than that of metals, i.e. the impedance mismatch between paper specimens and metal bars is supposed to be large. These are the reasons why the PVC output bar was adopted in this study. The outer diameter, wall thickness and length of this PVC tube were 26 mm, 3 mm and 3500 mm, respectively. Of course, a PVC circular plate with the same diameter was attached at one end of the tube. Two semiconductor strain gauges were glued axi-symmetrically on the surface of the tube at the location of 300 mm apart from an end. The output signal from the gauges was recorded by a digital oscilloscope (Nicolet model 400) after passing through a bridge box and an amplifier. The deformation of specimen was measured using markers and an optical non-contact extensometer (Zimmer model 200X), as shown in Fig.2(c).

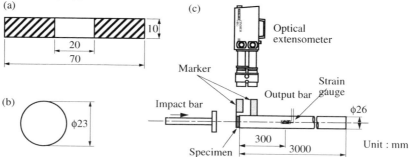

Fig.2 Specimens and arrangement for impact compression tests.

Stress Wave Propagation Tests in PVC Tube

When a stress wave propagates through a PVC tube, the wave is generally attenuated because of the viscosity of PVC. To clarify the attenuation property of PVC, therefore, stress wave propagation tests were required. For the tests, another set of strain gauges was glued at a point of 900 mm apart from the end. For further details about this propagation test, refer to our previous study (Kobayashi et al (1998)). From the wave propagation tests, it was found that the peak value of traveling waves decreased with 12%/m and the duration at the 30% level of the peak value of each wave increased with 10%/m, respectively. The velocity of stress wave, C_0, was almost constant, and the average of wave velocity was $C_0 = 1582$ m/s. The Young's modulus and density of this PVC tube were measured to be $E_{pvc}=3.48$ GPa and $\rho_{pvc}=1390$ kg/m^3, respectively. Since stress waves are measured at the location of 300 mm from the specimen in dynamic tests, a modified load-time curve can be obtained by multiplying the amplitude and the time of a measured curve by 1.04 and 0.97, respectively.

EXPERIMENTAL RESULTS AND DISCUSSIONS

Results of Tensile Tests

The results of tensile tests are listed in Table 1, where E is the Young's modulus, σ_T is the tensile strength, ε_{pk} is the uniform elongation, ρ is the density and σ_T/ρ is the specific strength. Paper foam board has 4 or 5 times greater tensile strength than polystyrene foam board. Although the density of paper foam board is about 5 times greater than that of polystyrene foam board, there is not large difference in the specific strength of both foam boards. In tensile strength, the anisotropy of paper foam board between L- and W-direction is small, about twice, which is similar to that of polystyrene foam board. However, the anisotropy of paper foam board between T-direction and other two directions is extremely large in the rigidity or in the strength.

TABLE 1
MECHANICAL PROPERTIES AND DENSITY OF PAPER AND POLYSTYRENE FOAM BOARDS

		E [MPa]	σ_T [MPa]	ε_{pk} [%]	ρ [kg/m^3]	σ_T/ρ [MNm/kg]
Paper foam	L-direction	480	6.09	2.8		0.033
	W-direction	200	3.58	4.6	185	0.019
	T-direction	1.07	0.075	8.6		
Polystyrene foam	L-direction	70.0	1.56	11.8	38.4	0.041
	W-direction	11.0	0.74	15.5		0.019

Results of Compression Tests

A typical record of the stress wave obtained from the PVC output tube in an impact compression test is shown in Fig.3. The important part in the stress wave is the part in circle A, which is magnified in Fig.3(b). By modifying this load-time curve as mentioned before, we can get an applied load to the specimen. The displacement-time curve is also shown by a broken line in Fig.3(a) on right-hand side scale. From the slope of the curve, the strain rate in the impact compression test can be given. The range of strain rate tested was from about 4800 to 6200 sec^{-1}. From these curves, load-time curve and displacement-time curve, a nominal stress-strain curve of the specimen in impact compression tests was finally obtained.

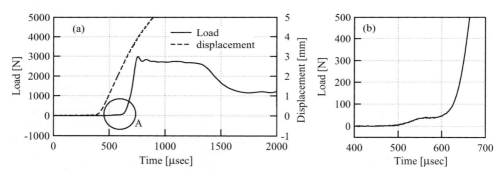

Fig.3 Typical oscilloscope record obtained from impact compression tests.

Effect of Strain-Rate Dependence

Stress-strain curves of paper foam and polystyrene foam boards obtained from quasi-static and dynamic compression tests are shown in Fig.4. Young's modulus observed in the compression tests for paper foam board is $E_c = 0.95$ MPa, which is a little smaller than that of tensile tests. In the curve of polystyrene foam board, it was mentioned by Gibson & Ashby (1988) that the plateau region appears after the linear elasticity region, which is caused by the continuous collapse of the cell walls due to non-linear elastic buckling. In the curve obtained from this study, however, the stress always increases monotonously and a clear plateau did not observed. The strength of the both foam boards was affected by strain rates, i.e. the strength increases with the increase of strain rate, even though the strain rate is within the range of quasi-static rates.

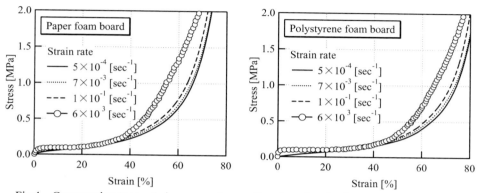

Fig.4 Compressive stress-strain curves of paper foam board and polystyrene foam board.

From these figures, the stresses at strains of ε = 40 and 60% of both foam boards are taken and plotted against the strain rate, $\dot{\varepsilon}$ (Fig.5). Although the increase of the stresses of paper foam board (Fig.5(a)) at strain of ε = 40% is small, the stresses at ε = 60% obtained from dynamic tests are about twice as large as those obtained at $\dot{\varepsilon} = 5 \times 10^{-4}$ sec^{-1}. Similar tendency can be found in the result of polystyrene foam board (see Fig.5(b)). The stress at ε = 60% appears to be the stress close to the start point of the densification of these foam materials. Therefore, we may say that the latter half of the compressive deformation of these foam boards shows a relatively large positive rate-dependence.

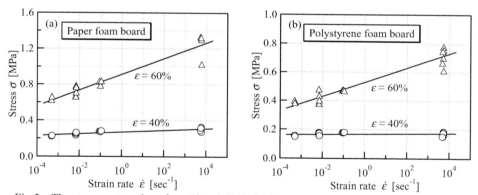

Fig.5 The stresses at strains of ε = 40 and 60% obtained at dynamic and quasi-static rates.

Absorbed Energy

The absorbed energy is often defined by the total area under a stress-strain curve until the densification starts per an unit volume of specimen. In this study, however, the absorbed energy is defined as the amount of energy absorbed up to a strain of 60% because the start point of the densification is not clear in the stress-strain curves. The absorbed energy per an unit volume of specimen, w^*, is shown in Fig.6 with taking the strain rate in the horizontal axis. The absorbed energy obtained from dynamic tests for the both foam boards was about 1.6 times greater than that obtained at quasi-static rates. Furthermore, the absorbed energy of paper foam board is about 1.3 ~ 1.6 times greater than that of polystyrene foam board. Therefore, the paper foam boards may be suitable enough to be used as an ecological shock absorber instead of polystyrene foam boards.

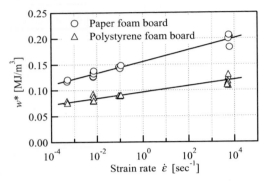

Fig.6　Absorbed energy per an unit volume, w^*, obtained from compression tests.

CONCLUSION

In this study, a series of dynamic and static compression tests for paper foam board and polystyrene foam board was carried out. The principal results obtained are as follows : (1) Although the density of paper foam board was about 5 times greater than that of polystyrene foam board, the specific strength is almost similar to that of polystyrene foam board. (2) In compression tests, there was a little rate-dependence on the stresses at $\varepsilon = 40\%$, however, the stresses at $\varepsilon = 60\%$ showed a relatively large positive rate-dependence. (3) The absorbed energy of the both foam boards obtained from dynamic tests was about 1.6 times greater than that obtained at quasi-static rates. The absorbed energy per an unit volume up to a strain of 60 % of paper foam board was about 1.3 ~ 1.6 times greater than that of polystyrene foam board. As a substitute for polystyrene foam board, therefore, paper foam board may have enough mechanical properties.

REFERENCES

Ellis E.R., Jewett K.B., Ceckler W.H. and Thompson E.V. (1984). Dynamic Compression of Paper. III. Compression Equation for Cellulose Mats. *AIChE Symp. Ser.* **80:232**, 1-7.
Fellers C. and Braenge A. (1985). The Impact of Water Sorption on The Compression Strength of Paper. *Pap. Raw. Mater.* **2**, 529-539.
Gibson L.J. and Ashby M.F. (1988). *Cellular Solids – Structure & Properties*. Pergamon Press.
Hoering J.F., Ellis E.R., Jewett K.B., Ceckler W.H. and Thompson E.V. (1983). Dynamic Compression of Paper. II. Compression Behavior of Saturated Cellulose Mats. *J. Pulp Pap. Sci.* **9:5**, 140-145.
Kobayashi H., Daimaruya M. and Kobayashi T. (1998). Dynamic and Static Compression Tests for Paper Honeycomb Cores and Absorbed Energy. *JSME International Journal* **41:3, A**, 338-344.
Sakurai H., Ito A., Watanabe I. and Yokoyama T. (1993). Development of A Paper Shock Absorber or A Paper Reinforced Material. Conversion of PE-Laminated Waste Paper by Microwave Heating. *The reports of Fuji Industrial Research Institute of Shizuoka Prefecture.* **3**, 23-38.
Ujihashi S., Sogo T., Matsumoto H. and Adachi T. (1993). Energy Absorption Ability of Thin-Walled Members by Crushing under Impact Loading. *J. Soc. Mat. Sci. Jpn.* **42:438**, 1427-1431.
Yoshino H., Ishikawa Y., Mimura K. and Tanimura S. (1996). Dynamic Buckling Behavior of Thin-Walled Aluminum Circular Pipe. *Proc. 1996 Ann. Meeting of JSME/MMD* **96-10, A**, 351-352.

DEFORMATION AND FAILURE MECHANISM OF ALUMINUM FOAMS UNDER DYNAMIC LOADING

J. -L. Yu, X. Wang, Z. -G.. Wei and Y. Pan

Department of Modern Mechanics, University of Science and Technology of China,
Hefei, Anhui 230027, P. R. China

ABSTRACT

In this paper the dynamic behavior of an open-cell aluminum foam is investigated experimentally. The dynamic compressive stress-strain curves are obtained by SHPB technique. No strain rate sensitivity is found. Quasi-static and dynamic bending tests are carried out for sandwich beams made of aluminum skins and the aluminum foam core under studying. The deformation and failure mechanism are revealed and discussed by 'frozen' test using stop blocks. It is found that due to large local indentation and damage the energy absorbing capacity of beams in dynamic cases is lower than that in quasi-static cases.

KEYWORDS

aluminum foam, foam core sandwich beam, dynamic response, impact failure

INTRODUCTION

In the last two decades metallic foams have been developed and are growing in use as new engineering materials. These ultra-light metal materials possess unique mechanical properties, such as high rigidity and high impact energy absorption at low weight, equal properties in all directions giving tolerance to varying direction of loading, stable deformation mode and adaptation to loading condition during deformation, etc. Their applications include energy absorbers in the automotive industry and other equipment for transportation, packaging (protection from shock for heavy components that are sensitive to impact), core material in sandwich structures with special requirements, and core material in hollow structures to prevent buckling.

Quasi-static Mechanical Behavior

When a block of foam is compressed, the stress-strain curve shows three regions. At low strains, the foam deforms in a linear-elastic way, then a plateau of deformation at almost constant stress occurs, and finally there is a region of densification as the cell walls crush together. The extent of each region depends on the relative density ρ/ρ_s. Elastic foams, plastic foams, and even brittle foams all have three-part stress-strain curves like this, though the mechanism is different in each case.

The Young's modulus and compressive strength of metallic foams have been measured by a number of researchers (Prakash *et al*, 1995; Beals and Thompson, 1997; Suginura *et al*, 1997; Andrews *et al*, 1999). However, most commercially available cellular metals, unlike some of their polymer counterparts, do not achieve the properties prediction by theoretical models according to the properties of the cell wall material and the relative density of the foam (Gibson and Ashby, 1997). Various hypotheses have been made regarding the 'defect' that diminishes the properties (Simone and Gibson, 1998a,b; Grenestedt and Bassinet, 2000).

Dynamic Mechanical Behavior

In order to evaluate the capacity of impact energy absorption, strain-rate sensitivity of the foam material must be characterized. SHPB method has been used for measuring the dynamic compressive response of cellular materials, including polymers and metals.

Only limited data are available for the strain rate dependence of the compression strength of cellular materials. Lankford and Dannemann (1998) reported that the strain rate dependence was negligible for a low-density open-cell 6101 Al foam. Recently, Deshpande and Fleck (2000) investigated the high strain rate compressive behavior of a closed cell aluminum alloy foam Alulight and an open cell aluminum alloy foam Duocel for strain rates up to 5000 s^{-1} using SHPB and direct impact tests. It is found that the dynamic behavior of these foams is very similar to their quasi-static behavior. On the other hand, Mukai *et al* (1999a, b) and Kanahashi *et al* (2000) reported that an open cell foam AZ91, an open cell aluminum foam SG91A and a closed cell aluminum foam ALPORAS all exhibited high strain rate sensitivity of the plateau stress. They also found that the absorption energy normalized by the relative density at dynamic strain rates is about 60% higher than that at quasi-static strain rates. Paul and Ramamurty (2000) investigated the strain rate sensitivity of a closed cell aluminum foam under nominal strain rates from 3.33×10^{-5} to 1.6×10^{-1} s^{-1}. Within this range, they found that the plastic strength and the energy absorbed increase by 31 and 52.5% respectively with increasing strain rate.

Foam Core Structures

Metal foams are frequently used in the form of foam core structures. In engineering applications, energy absorbing structures work at the post-failure stage; some sandwich structures must be capable of sustaining large overload, e.g. foreign object impact. In these cases, the non-linear (failure) behavior of the structure is important and the parameters of interest are the integrity of the structure and the energy absorbing capabilities of the structure before total collapse.

Only a little work has been published. Harte *et al* (2000) explored failure modes of aluminum skin-Alporas foam core sandwich panels and constructed maps dependent on the sandwich panel geometry. Hanssen *et al* (2000a, b) investigated static and dynamic crushing of square and circular aluminum extrusions with aluminum foam filler. Santosa and Wierzbicki (1999) and Santosa *et al* (2000) studied the planar bending response of thin-walled beams with low-density metal filler and found that the bending resistance was improved dramatically by aluminum foam filler. The presence of the foam filler changes the crushing mode of the thin walled beam and prevents the drop in load carrying capacity.

In the present paper, the responses of an open-cell aluminum foam and aluminum skin-aluminum foam core beams under dynamic compression and bending, respectively, are investigated and the deformation and failure mechanism are compared with the quasi-static behavior. It is found that the foam tested exhibits no strain-rate sensitivity. Under impact loading the upper skin of sandwich beams wrinkles first, accompanied with local debonding between the upper skin and the core. The final failure occurs when the lower skin is broken. This is similar to one of the failure modes found under quasi-static loading.

DYNAMIC COMPRESSION OF ALUMINUM FOAM

Description of Experiment Techniques

Dynamic compression tests were preformed on open-cell aluminum foam using a split Hopkinson pressure bar. The foam is made by grain-casting method with a relative density of 0.37 and an average cell size of 1.5 mm. The components of the cell wall material are listed in Table 1. The specimens were electric spark machined to circular cylinders of diameter 30mm. Two different values of thickness, 10 and 20 mm, are used in the tests.

TABLE 1
THE COMPONENTS OF THE CELL WALL MATERIAL

Component	Al	Si	Fe	Cu	Impurity
Content (%)	≥ 98%	≤ 1.8%	≤ 1%	≤ 0.05%	≤ 2.0%

Experimental Results

No significant difference in the dynamic response between specimens of 10 mm thickness and those of 20 mm thickness was found. A comparison of quasi-static ($\dot{\varepsilon}=10^{-3}$ s^{-1}) and dynamic ($\dot{\varepsilon}=1000\sim1750$ s^{-1}) stress-strain curves of the aluminum foam specimens is shown in Figure 1. The dynamic response agrees well with the quasi-static response. There is no consistent trend with increasing strain rate. So, the foam material under studying is strain-rate insensitive.

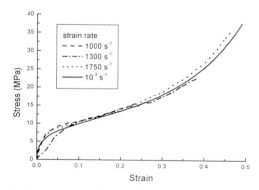

Figure 1: Comparison of the quasi-static and dynamic stress-strain curves of the aluminum foam under compression

Micro structure observation

Figure 2 is the SEM photographs of the aluminum foam specimens. It is evidence that the aluminum foam crushed in a spatially uniform manner both at quasi-static rate and in the dynamic split Hopkinson bar tests. This indicates that the foam deformation is uniform with no crush band formation. We conclude that, for the foam tested, the mode of dynamic collapse ($\dot{\varepsilon}<1800$ s^{-1}) is qualitatively the same as that under quasi-static loading.

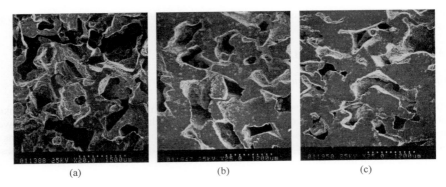

Figure 2: Mid-plane SEM photographs of initial and dynamically deformed open-cell aluminum foam specimens: (a) $\varepsilon = 0$, (b) $\varepsilon = 0.18$, and (c) $\varepsilon = 0.26$

STATIC AND DYNAMIC TESTS OF ALUMINUM FOAM CORE SANDWICH BEAMS

Experiments and Results

Sandwich beams are made up of two thin aluminum skins adhered to an open cell aluminum foam core by an acrylate agglutinant. The core material is the same as that used in the compression tests. Three different values of the nominal core thickness, i.e. 10 mm, 15 mm and 20 mm, are chosen. The thickness of the aluminum skin is 0.5 mm. The length of the beam specimens is 300 mm with a span of 250 mm in tests.

Static tests

Static tests were conducted on an MTS810 testing system using a three-point bending rig at a crosshead speed of 0.025 mm/s. Table 2 gives the details of the tests, where c is the core thickness, B the beam width, I the initial bending stiffness, D the total energy absorbed, P_b and δ_b are force and deflection when the beam breaks off, respectively. The measured load-deflection curves of the static three-point bending of the sandwich beams with different core thickness are given in Figure 3.

TABLE 2
STATIC THREE-POINT BENDING TESTS

Specimen number	c (mm)	B (mm)	I (N·m²)	Failure mode	P_b (KN)	δ_b (mm)	D (J)
A10-1	9.60	35.25	72.31	I	0.46	16.1	8.78
A10-2	9.57	35.03	44.43	II	0.39	22.9	9.74
A10-3	9.51	34.99	61.91	II	0.43	22.9	10.82
A15-1	16.74	34.81	140.53	I	0.58	11.3	8.41
A15-2	15.85	35.07	137.36	II	0.45	20.9	12.05
A15-3	15.83	34.79	139.26	I	0.64	13.3	10.20
A15-4	15.74	35.00	95.821	II	0.50	26.1	14.78
A20-1	20.10	35.19	203.92	I	0.79	9.9	10.10
A20-2	20.31	34.87	199.25	II	0.65	17.1	13.10

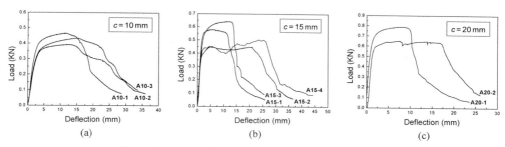

Figure 3: Load-deflection curves of the sandwich beams

Impact tests

Impact tests are conducted on a drop weight machine. The mass of the hammer is 2.58 kg and the drop height is 0.8 m. This holds an initial impact energy of about 20 J. In order to get information of the deformation and failure process, different "stop blocks" are used to limit the final deflection δ. Details of the impact tests are given in Table 3.

TABLE 3
THE GEOMETRY OF THE BEAM AND EXPERIMENT CONDITION OF IMPACT

Specimen number	c (mm)	B (mm)	δ (mm)	Lower Skin Failure
B10-1	9.86	34.96	12.5	No
B10-2	9.84	35.06	32.5	No
B10-3	9.84	34.90	47.0	Yes
B15-1	16.50	34.90	12.0	No
B15-2	16.02	34.72	15.0	No
B15-3	16.20	35.00	19.0	Yes
B20-1	20.13	34.84	13.0	No
B20-2	20.15	34.96	22.0	Yes

An accelerometer is embedded inside the hammer in order to get the velocity and displacement history. A typical record of the accelerometer is shown in Figure 4. The large deceleration after 6 ms in Figure 4 is due to the impact of the hammer with the stop block. The histories of the hammer velocity and the beam deflection are then obtained by integral of the accelerometer signal. The results of three tests, for which final failure occurs, are shown in Figures 5 and 6 for the hammer velocity and beam deflection, respectively. It should be noted that the negative signal of deceleration is due to stress wave effect, for the accelerometer is located 20 cm from the impact point and attached on a surface inside the hammer. Hence the accelerometer signal is related to the deceleration of the mass only in the average sense. The oscillation of the hammer velocity at the early stage is due to the stress wave effect and is not true.

In considering the stress wave effect mentioned above, the impact force can not be direct obtained from the accelerometer record by multiplying the mass. Rather, a six order polynomial fitting is applied to the velocity history and an approximate force history is then obtained by differentiation of the polynomial. The force-deflection curve for specimen B15-3 by this way is shown in Figure 7, in comparison with the corresponding quasi-static tests, specimen A15-2 and A15-4.

It can be seen from Figure 7 that the deflection at failure in dynamic bending is significantly less than that in quasi-static tests. This is presumably due to impact damage of the core material.

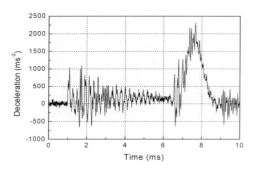
Figure 4: A typical accelerometer record

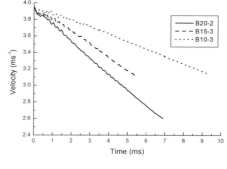
Figure 5: Velocity vs. time curves of three tests

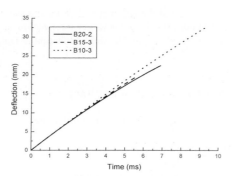
Figure 6: Deflection histories of three tests

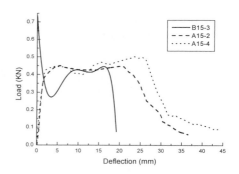
Figure 7: A comparison of quasi-static and dynamic force-deflection curves

Failure Mechanism

Final failure of beams under quasi-static bending occurs when the lower skin is broken. Two failure modes can be distinguished. Mode I is without a wrinkle of the upper skin while Mode II with a wrinkle of the upper skin, as shown in Figure 8. In both modes, the upper part of the core was compressed while a tensile crack was found in the lower part of the core. However, the associated load-deflection curves exhibit different characters. For mode I failure, the maximum load is higher but the tensile failure of the lower skin occurs earlier. For mode II failure, on the other hand, the beam can bear a lower, nearly constant load with a larger deflection at failure, thus absorbs more energy. This phenomenon is more distinct for thicker beams. Only a small wrinkle was found for specimens A10-2 and A10-3. The maximum load increases with the thickness but thin beams can also absorb large energy because of the large deflection, as can be seen in Figure 8 and Table 2.

The failure process of beams under impact loading can be revealed from a set of tests with different final deflections. The side views of specimens B15-1, B15-2 and B15-3 near the impact point are shown in Figure 9. It can be seen from these pictures that, when a sandwich beam is impacted by a mass, a wrinkle will first form near the impact point of the upper skin. This phenomenon is accompanied with local debonding between the upper skin and the core, as well as the compression of the upper core and tension of the lower core, Figure 9(a). Then a local stable squash may take place in the upper core, Figure 9(b). The final failure is the lower core and lower skin tensile failure, Figure

9(c). This failure mode is similar to Mode II of the quasi-static beam tests.

Figure 8: Failure modes of beams under static bending, (a) failure mode I (specimen A15-3); (b) failure mode II (specimen A15-2)

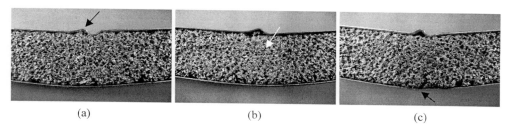

Figure 9: Side view of specimens (a) B15-1 ($\delta = 12$mm), (b) B15-2 ($\delta = 15$mm) and (c) B15-3 ($\delta = 19$mm) after impact

As shown in Figure 7, a peak load is recorded at the beginning of the impact but the dynamic load at the plateau stage is a little bit lower then the quasi-static ones. This could be attributed to the severe deformation and damage near the impact region. When the thin skin on a soft material is impacted, a wrinkle always occurs which will certainly lower down the load carrying capacity of the beam. Hence mode I failure is excluded in dynamic tests. On the other hand, aluminum is strain rate sensitive. So the skin is stronger in dynamic tests. However, a strain gauge record shows that the maximum strain rate of the lower skin in the dynamic tests is less than 10 s^{-1}. So the rate effect is negligible.

CONCLUSION

The aluminum foam under investigation is strain-rate insensitive within the strain range tested. It is shown that the deformation of aluminum foam specimens in dynamic compression is macroscopically uniform in the SHPB tests.

For the aluminum/aluminum foam/aluminum sandwich beams, final failure occurs when the lower skin and the lower part of the core cracked due to tension. Quasi-static bending tests show two failure modes. In Mode I failure no upper skin wrinkle occurs and the beam may carry a high load but is broken early. In Mode II failure an upper skin wrinkle reduces the maximum load but the final deflection of the beam is larger than that in Mode I, leading to a higher energy absorbing capacity. Only Mode II failure is found in dynamic bending tests due to local indentation by impact. The wrinkle, the local damage and the associated skin-core interface debonding are responsible for the degradation of the energy absorbing capacity.

ACKNOWLEDGEMENT

This work is supported by the National Natural Science Foundation of China (projects No. 19672060 and No.10002017).

REFERENCES

Andrews E., Sanders W. and Gibson L.J. (1999). Compressive and tensile behavior of aluminum foams. *Mater. Sci. Eng.* **A270**, 113-124.
Beals J.T. and Thompson M.S. (1997). Density gradient effects on aluminum foam compression behavior. *J. Mater. Sci.* **32**, 3595-3600.
Deshpande V.S. and Fleck N.A. (2000). High strain rate compressive behaviour of aluminium alloy foams. *Int J Impact Engng*, **24**, 277-298.
Gibson L.J. and Ashby M.F. (1997). *Cellular Solids: Structure and Properties*, Pergamon, Oxford, UK
Grenestedt J.L. and Bassinet F. (2000). Influence of cell wall thickness variations on elastic stiffness of closed-cell cellular solids. *Int J Mech Sci*, **42**, 1327-1338.
Hanssen A.G., Langseth M. and Hopperstad O.S. (2000a). Static and dynamic crushing of square aluminium extrusions with aluminium foam filler. *Int. J. Impact Eng.*, **24**, 347-383.
Hanssen A.G., Langseth M. and Hopperstad O.S. (2000b). Static and dynamic crushing of circular aluminium extrusions with aluminium foam filler. *Int. J. Impact Eng.*, **24**, 475-507.
Harte A.M., Fleck N.A. and Ashby M.F. (2000). Sandwich panel design using aluminum alloy foam. *Advanced Engineering Materials*, **2:4**, 219-222.
Kanahashi H., et al. (2000). Dynamic compression of an ultra-low density aluminium foam. *Mater. Sci. Engng.*, **A280**, 349-353.
Lankford J. and Dannemann K.A. (1998). Strain rate effects in porous materials. *Porous and Cellular Materials for Structure Applications*, Materials Research Society, Warrendale, Symposium Proceedings **521**, 103-108.
Mukai T., et al. (1999a). Experimental study of energy absorption in a close-celled aluminum foam under dynamic loading. *Scripta Materialia*, **40:8**, 921-927.
Mukai T., et al. (1999b). Dynamic compressive behavior of an ultra-lightweight magnesium foam. *Scripta Materialia*, **41:4**, 365-371.
Paul A. and Ramamurty U. (2000). Strain rate sensitivity of a closed aluminum foam, *Materials Science and Engineering*, **A281**, 1-7.
Prakash O, Sang H and Embury J.D. (1995). Structure and properties of AL-SiC foam. *Mater. Sci. Eng.* **A199**, 195-203.
Santosa S., Banhart J. and Wierzbicki T. (2000). Bending crush resistance of partially foam-filled sections, *Advanced Engineering Materials*, **2:4**, 223-227.
Santosa S. and Wierzbicki T. (1999). Effect of an ultralight metal filler on the bending collapse behavior of thin-walled prismatic columns. *Int. J. Mech. Sci.*, **41:8**, 995-1019.
Simone A.E. and Gibson L.J. (1998a). Effects of solid distribution on the stiffness and strength of metallic foams. *Acta Mater*, **46:6**, 2139-2150.
Simone A.E. and Gibson L.J. (1998b). The effects of cell face curvature and corrugations on the stiffness and strength of metallic foams. *Acta Mater*, **46:11**, 3929-3935.
Sugimura Y, Meyer J, He M.Y., Bart-Smith H, Grenstedt J.L. and Evans A.G. (1997). On the mechanical performance of closed cell Al alloy foams. *Acta Mater.*, **45:12**, 5245-5259.

EXPERIMENTAL CHARACTERIZATION OF MIXED MODE FRACTURE TOUGHNESS OF POLYMER MATRIX COMPOSITE MATERIALS UNDER IMPACT LOADING

T. Kusaka[1], M. Masuda[2] and N. Horikawa[3]

[1] Department of Mechanical Engineering, Ritsumeikan University
1-1-1, Noji-Higashi, Kusatsu 525-8577, Japan
[2] Suzuki Motor Corporation
300, Takatsuka-cho, Hamamatsu 432-8611, Japan
[2] New Energy and Industrial Technology Development Organization
3-1-1, Higashi-Ikebukuro, Toshima-ku, Tokyo 170-6028, Japan

ABSTRACT

A novel experimental method has been developed to simply and precisely evaluate the mixed mode (I+II) interlaminar fracture toughness of polymer matrix composite materials at impact rates of strain. The MMF (Mixed Mode Flexure) specimen was employed for determining the critical energy release rate at the onset of macroscopic crack growth. The SHPB (Split Hopkinson Pressure Bar) system was used for measuring the dynamic load and displacement. The effects of strain rate and mode mixture on fracture behavior of unidirectional carbon-fiber/epoxy composite laminates were investigated on the basis of experimental results using the present method. The mixed mode fracture toughness clearly showed strain rate dependence. The mixed mode fracture toughness did not agree with the linear fracture criterion at higher strain rates, whereas it agreed with the linear fracture criterion at lower strain rates. The validity of the present method was confirmed by finite element analysis.

KEYWORDS

Composite Materials, Delamination, Fracture Toughness, Rate Dependence, Impact Strength, Mixed Mode, Fracture Criterion

INTRODUCTION

Delamination is one of the most serious defects in composite structures, because it will severely reduce the compressive strength of the material. Hence, the interlaminar fracture properties have been regarded as important in the development of new composite materials. Recent works have shown the availability of the linear fracture mechanics to this kind of problem, Friedrich (1989).

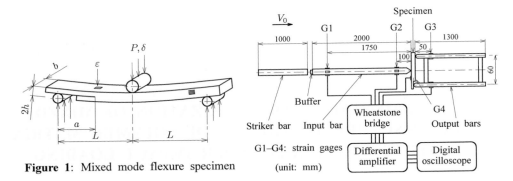

Figure 1: Mixed mode flexure specimen

Figure 2: Impact MMF testing apparatus

However, the effect of strain rate on fracture behavior of the materials have not fully characterized yet, especially at impact rates of strain, Cantwell & Morton (1991). The authors have, therefore, focused on the impact fracture behavior and clarified the pure mode I and II fracture behaviors of advanced polymer matrix composite materials, Kusaka et al. (1998), Kusaka et al. (1999), Kusaka et al. (2000).

In the present work, a novel experimental method has been developed to simply and precisely evaluate the mixed mode (I+II) interlaminar fracture toughness of polymer matrix composite materials at impact rates of strain. The effects of strain rate and mode mixture on fracture behavior of unidirectional carbon-fiber/epoxy composite laminates were studied on the basis of experimental results using the present method. The validity of the present method was also studied on the basis of computational results using finite element method.

EXPERIMENTAL PROCEDURE

Material

A 120 °C cure type carbon-fiber/epoxy composite material, HTA/112 (Toho Rayon), was investigated in the present work. Unidirectional panels of 20 plies were fabricated according to the prepreg manufacturer's recommended cure cycle with an autoclave. Coupon specimens of 8 mm width were machined from the panels with a diamond wheel. A 14 μm thick PTFE film was inserted at mid-thickness of the panels for introducing an artificial crack.

Experimental Procedure

The MMF (Mixed Mode Flexure) specimen, Russell & Street (1985), was employed for determining the mixed mode (I+II) interlaminar fracture toughness, as shown in Figure 1. The width, b, thickness, $2h$, crack length, a, and bending span, $2L$, of the specimen were about 8, 3, 15 and 60 mm, respectively. The displacement rate, $\dot{\delta}$, was varied from 0.05 mm/min to 5 m/s to study the effect of strain rate on fracture behavior. Static tests ($\dot{\delta} = 0.05$–500 mm/min) were conducted on a screw-driven testing frame. Impact tests ($\dot{\delta} \simeq 5$ m/s) were conducted on a SHPB (Split Hopkinson Pressure Bar) system, as shown in Figure 2. A small piece of solder metal was utilized as a buffer to suppress the flexural vibration of the specimen, Kusaka et al. (2000).

Static and impact mode I fracture toughness tests were carried out using the DCB (Double Cantilever Beam) and WIF (Wedge Insert Fracture) specimens, Kusaka et al. (1998). Static and impact mode II fracture toughness tests were carried out using the ENF (End Notched Flexure) specimen, Kusaka et al. (1999). The MMB (Mixed Mode Bending) specimen, Williams (1988), was used for determining the static mixed mode (I+II) fracture toughness for various mode ratio. The details, such as specimen geometries and data reduction scheme, of the DCB, WIF, ENF and MMB tests will be omitted on account of space consideration.

Data Reduction Scheme

The energy release rate, G, is given by the following equation.

$$G = \frac{\partial W}{\partial A} - \frac{\partial E_{els}}{\partial A} - \frac{\partial E_{kin}}{\partial A} \tag{1}$$

where A is the area of the crack. W is the work done by external force. E_{els} and E_{kin} are the strain and kinetic energies stored in the specimen, respectively. Each terms of Eqn. 1 must be generally determined with considering the kinetic effects for impact fracture toughness tests. However, the energy release rate, G, can be obtained in the same manner as static fracture toughness tests, when an adequate way of impact loading is used, as shown in the later section.

On the basis of the above discussion, the mixed mode energy release rate, G, for the MMF specimen was evaluated by the following equations, Kusaka et al. (2000).

$$G = \frac{21 P^2 a^2 C}{2b(2L^3 + 7a_C^3)} \equiv G^{load} \tag{2}$$

$$G = \frac{21 \varepsilon^2 a^2 C}{2b(2L^3 + 7a_C^3) D^2} \equiv G^{strain} \tag{3}$$

where P is the reaction force of the loading point. ε is the surface strain of the specimen near the crack tip. C $(= \delta/P)$ is the compliance of the specimen. D $(= \varepsilon/P)$ is the coefficient depending on the geometries and elastic moduli of the specimen. Equation 2 was mainly used for static fracture toughness tests. Equation 3 mainly used for impact fracture toughness tests. According to the finite element analysis, the mode ratio, G_{II}/G_{I}, which is theoretically 0.75, was 0.58 for the present MMF specimen.

EXPERIMENTAL RESULTS

Effect of Strain Rate on Fracture Toughness

Figure 3 shows the effect of strain rate on interlaminar fracture toughness. ● and ○ are the results of the MMF test ($G_{II}/G_{I} = 0.58$). ▲ and △ are the results of the DCB and WIF tests. ▼ and ▽ are the results of the ENF test. ■ is the results of the MMB test ($G_{II}/G_{I} = 0.78$, 3.0, 13). ●, ▲, ▼, ■ are the results for static tests using the screw-driven testing frame. ○, △, ▽ are the results for impact tests using the SHPB system.

As shown in Figure 3, the mixed mode fracture toughness, G_C, clearly showed strain rate dependence. The mixed mode fracture toughness, G_C, increased with increasing loading rate for $\dot{G} < 10^2$–10^3 J/m²/s. However, it decreased with increasing loading rate for $\dot{G} > 10^2$–10^3 J/m²/s. Consequently, a local maximum value was existed at $\dot{G} = 10^2$–10^3 J/m²/s. The mixed mode fracture toughness, G_C, at impact strain rates was about 20 % lower than the local maximum value at static strain rates.

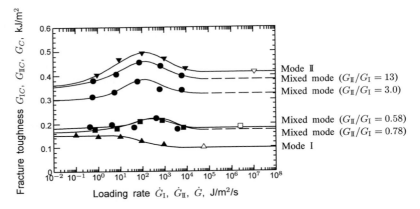

Figure 3: Effect of strain rate on interlaminar fracture toughness (HTA/112)

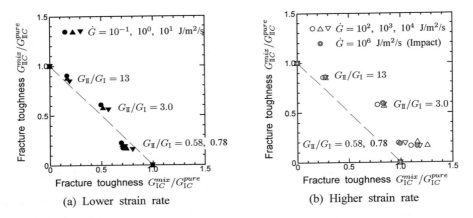

Figure 4: Effect of mode mixture on interlaminar fracture toughness (HTA/112)

Effect of Mode Mixture on Fracture Toughness

Figures 4 (a) and (b) show the effect of mode mixture on interlaminar fracture toughness for lower and higher strain rates, respectively. ●, ▲ and ▼ are the results for $\dot{G} = 10^{-1}, 10^0, 10^1$ J/m²/s, respectively ○, △ and ▽ are the results for $\dot{G} = 10^2, 10^3, 10^4$ J/m²/s, respectively. ◎ is the results for $\dot{G} = 10^6$ J/m²/s. The thin broken line represents the linear fracture criterion given by the following equation, Reeder (1993).

$$f(G_C) = \frac{G_{IC}^{mix}}{G_{IC}^{pure}} + \frac{G_{IIC}^{mix}}{G_{IIC}^{pure}} = 1 \qquad (4)$$

where G_{IC}^{mix} and G_{IIC}^{mix} are the mode I and II components of mixed mode fracture toughness. G_{IC}^{pure} and G_{IIC}^{pure} are the pure mode I and II fracture toughness.

As shown in Figure 4, the results for larger mode ratio, $G_{II}/G_I = 13$, approximately agreed with the linear fracture criterion regardless of strain rate. On the other hand, the results for smaller mode ratio, $G_{II}/G_I = 3.0, 0.78, 0.58$, did not agree with the linear fracture criterion at higher strain rates, $\dot{G} \geq 10^2$ J/m²/s, whereas they almost agreed at lower strain rates, $\dot{G} \leq 10^1$ J/m²/s.

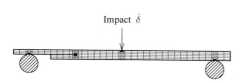

Figure 5: Finite element model

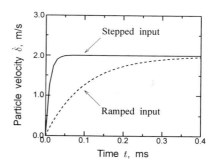

Figure 6: Impact loading for FEM analysis

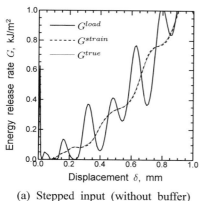

(a) Stepped input (without buffer)

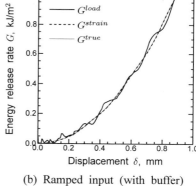

(b) Ramped input (with buffer)

Figure 7: Dynamic energy release rate calculated by FEM

FINITE ELEMENT ANALYSIS

Finite Element Model

Dynamic response of the MMF specimen was analyzed by the finite element code, MARC K7. The specimen was modeled with 4-nodes plane strain elements and assumed to be an orthotropic linear elastic body, as shown in Figure 5. Nonlinear boundary condition in conjunction with a contact algorithm was applied to the supporting points of the specimen. Two different types of impact loading were modeled and applied to the central loading point of the specimen as forced displacement, δ. Figure 6 shows the time derivative of displacement, $\dot{\delta}$, with time, t. The solid line (stepped wave) corresponds to the case without using the buffer in Figure 2. The broken line (ramped wave) corresponds to the case with using the buffer in Figure 2.

Finite Element Results

Figures 7 (a) and (b) show the finite element results for stepped and ramped inputs, respectively. The thin solid line represents the true dynamic energy release rate obtained from the local stress field near the crack tip, G^{true}. The thick solid line represents the apparent value given by Eqn. 2, G^{load}. The thick broken line represents the apparent value given by Eqn. 3, G^{strain}.

In the case of stepped input, the dynamic stress field was oscillated, and consequently the dynamic energy release rate, G^{true}, was also oscillated as shown by the thin solid line in Figure 7 (a). In the case of ramped input, however, the dynamic energy release rate, G^{true}, did not oscillate as shown by the thin solid line in Figure 7 (b). The apparent value, G^{load}, by Eqn. 2 was far from the true value, G^{true}, in the case of stepped input, whereas it agreed with the true value, G^{true}, to some extent in the case of ramped input. To the contrary, the apparent value, G^{strain}, by Eqn. 3 agreed well with the true value, G^{true}, even in the case of stepped input, as well as in the case of ramped input. The above results lead to the conclusion that the energy release rate, G, can be obtained in the same manner as static fracture toughness tests when an adequate way of impact loading is used.

CONCLUSION

A novel experimental method has been developed to simply and precisely evaluate the mixed mode (I+II) interlaminar fracture toughness of polymer matrix composite materials at impact rates of strain. The validity of the present method was confirmed by finite element analysis.

The mixed mode fracture toughness, G_C, clearly showed strain rate dependence. A local maximum value was existed at $\dot{G} = 10^2$–10^3 J/m^2/s. The mixed mode fracture toughness, G_C, at impact strain rates was about 20 % lower than the local maximum value at static strain rates.

For larger mode ratio, $G_{II}/G_I = 13$, the mixed mode fracture toughness, G_C, agreed with the linear fracture criterion regardless of strain rate. For smaller mode ratio, $G_{II}/G_I = 3.0$, 0.78, 0.58, it did not agree with the linear fracture criterion at higher strain rates, $\dot{G} \geq 10^2$ J/m^2/s, whereas it almost agreed at lower strain rates, $\dot{G} \leq 10^1$ J/m^2/s.

References

Cantwell W.J. and Morton J. (1991). The Impact Resistance of Composite Materials, Review, *Composites* **22**, 347–362.

Friedrich K. (1989). *Application of Fracture Mechanics to Composite Materials*, Elsevier.

Kusaka T., Hojo M., Mai Y.W., Kurokawa T., Nojima T. and Ochiai S. (1998). Rate Dependence of Mode I Fracture Behabiour in Carbon-Fibre/Epoxy Composite Laminates, *Composites Science and Technology* **58**, 591–602.

Kusaka T., Hojo M., Ochiai S. and Kurokawa T. (1999). Rate-Dependent Mode II Iterlaminar Fracture Behavior of Carbon-Fiber/Epoxy Composite Laminates, *Materials Science Research International* **5**, 98–103.

Kusaka T., Horikawa N. and Masuda M. (2000). Low-Velocity Impact Fracture Behaviour of Impact-Resistant Polymer Matrix Composite Laminates under Mixed Mode Loading, *Journal de Physique IV* **10**, 317–322.

Reeder J.R. (1993). A Bilinear Failure Criterion for Mixed-Mode Delamination, *ASTM STP 1206*, 303–322.

Russell A.J. and Street K.N. (1985). Moisture and Tempereture Effects on the Mixed Mode Delamination Fracture of Unidirectional Graphite/Epoxy, *ASTM STP 876*, 349–370.

Williams J.G. (1988). On the Calculation of Energy Release Rates for Cracked Laminates, *International Journal of Fracture* **36**, 101–119.

Impact Engineering and Application
Akira Chiba, Shinji Tanimura and Kazuyuki Hokamoto (Eds)
©2001 Elsevier Science Ltd. All rights reserved.

INCREASING THE MAXIMUM STRAIN MEASURED WITH ELASTIC AND VISCOELASTIC HOPKINSON BARS

Ramzi OTHMAN[1], Marie-Noëlle BUSSAC[2], Pierre COLLET[2], Gérard GARY[1]

[1] Laboratoire de Mécanique des Solides, UMR 7649, Ecole Polytechnique, 91128, Palaiseau.
[2] Centre de Physique Théorique, UMR 7644, Ecole Polytechnique, 91128, Palaiseau.

ABSTRACT

A new method has been developed for separating dispersive waves in elastic and viscoelastic rods from multi-point strain and velocity measurements. Knowing the basic waves, the stress, the strain, the displacement and the particle velocity can be calculated at any point of the bar. The method is based on the assumption of one dimensional and single mode dispersive wave propagation and it takes account of the wave dispersion. It is shown that the method is stable with respect to noise, so that the measuring time is increased considerably. Subsequently, the maximum strain which can be measured in a material tested with a classical SHPB (Split Hopkinson Pressure Bar) set-up is also increased and is no longer limited by the length of the bars. The method is illustrated here by applying it successfully to the analysis of a real test of aluminum honeycomb. It can also be applied to other kinds of one-dimensional and single-mode dispersive waves such as flexural waves in beams and acoustic waves in wave-guides.

KEYWORDS

Hopkinson bars, Wave separation, Viscoelastic material, Dynamic testing, Aluminium honeycomb.

INTRODUCTION

In the classical configuration, the loading time in the SHB (Split Hopkinson Bar) system is limited by the length of the bars together with the maximum measured strain in the specimen, because of the need to separate opposite waves propagating in the bar. Hence, for many materials, it is of no interest to carry out tests with the SHB apparatus at medium strain-rates. As mechanical testing machines are limited at much lower strain rates because of sensor oscillations, alternative solutions have been already investigated, in particular the wave separation technique. They are based on a two strain measurement and they take account of wave dispersion (Zhao & Gary, 1994, 1997 ; Bacon, 1999) or not (Lundberg & Henchoz, 1977 ; Park and Zhou, 1999). Bussac and al. (2001) showed that the noise is amplified on the rebuilt signals when using only two measurements. In this paper, a new separation method using N strain and P velocity measurements is presented. It is based on the Maximum of

Likelihood principle (Bussac and al., 2001 ; Othman and al., 2001). This method is illustrated with experiments performed on aluminium honeycomb.

THEORY

Wave separation in bars

Let us consider an elastic or viscoelastic bar of length L. In the case of single mode propagating longitudinal waves, and using the Fourier transform, stress, strain, displacement and velocity are expressed as follows (Lundberg & Blanc, 1988)

$$\varepsilon(x,t) = \frac{1}{2\pi}\int_{-\infty}^{+\infty}\tilde{\varepsilon}(x,\omega)e^{i\omega t}d\omega = \frac{1}{2\pi}\int_{-\infty}^{+\infty}\left(A(\omega)e^{-i\xi(\omega)x} + B(\omega)e^{i\xi(\omega)x}\right)e^{i\omega t}d\omega \tag{1a}$$

$$\sigma(x,t) = \frac{1}{2\pi}\int_{-\infty}^{+\infty}\tilde{\sigma}(x,\omega)e^{i\omega t}d\omega = \frac{1}{2\pi}\int_{-\infty}^{+\infty}E^*(\omega)\left(A(\omega)e^{-i\xi(\omega)x} + B(\omega)e^{i\xi(\omega)x}\right)e^{i\omega t}d\omega \tag{1b}$$

$$v(x,t) = \frac{1}{2\pi}\int_{-\infty}^{+\infty}\tilde{v}(x,\omega)e^{i\omega t}d\omega = \frac{1}{2\pi}\int_{-\infty}^{+\infty}\frac{\omega\left(-A(\omega)e^{-i\xi(\omega)x} + B(\omega)e^{i\xi(\omega)x}\right)}{\xi(\omega)}e^{i\omega t}d\omega \tag{1c}$$

$$u(x,t) = \frac{1}{2\pi}\int_{-\infty}^{+\infty}\tilde{u}(x,\omega)e^{i\omega t}d\omega = \frac{1}{2\pi}\int_{-\infty}^{+\infty}\frac{i\left(A(\omega)e^{-i\xi(\omega)x} - B(\omega)e^{i\xi(\omega)x}\right)}{\xi(\omega)}e^{i\omega t}d\omega \tag{1d}$$

where $A(\omega)$ and $B(\omega)$ are the Fourier components of the ascendant and descendant waves at origin, respectively. $E^*(\omega)$ is the complex Young's modulus and $\xi(\omega) = k(\omega) + i\alpha(\omega)$ is the complex wave number. The two parameters $E^*(\omega)$ and $\xi(\omega)$ are only related to the bar properties (geometry and material). In the following, it is assumed that they are known.

From strain and/or speed measurements, we want to recover $A(\omega)$ and $B(\omega)$ so that strain, stress, displacement and velocity can be calculated at any point of the bar, in particular at both ends.

We perform N strain and P velocity measurements on the bar. The corresponding record is modelled as the superposition of the exact measurement and a Gaussian white noise:

$$\hat{\varepsilon}_J(t) = \varepsilon(x_J,t) + W_J^\varepsilon(t), \quad J = 1,..,N \tag{2a}$$
$$\hat{v}_K(t) = v(y_K,t) + W_K^v(t), \quad K = 1,..,P \tag{2b}$$

The $N+P$ white noises are supposed to be independent. The amplitudes of the noise concerning strain and velocity are denoted $1/a_J^\varepsilon$ and $1/a_K^v$, respectively.

In order to estimate the two functions $A(\omega)$ and $B(\omega)$, the Maximum of Likelihood Method is used (Bussac and al., 2001 ; Othman and al., 2001). This method consists in writing that what is measured corresponds to the most probable event (a particular application is the least-square method). In our case, it leads to the minimisation of the following functional:

$$F = \int_{-\infty}^{+\infty}\left\{\sum_{J=1}^{N}(a_J^\varepsilon)^2(\hat{\varepsilon}_J(t) - \varepsilon(x_J,t))^2 + \sum_{J=1}^{N}(a_K^v)^2(\hat{v}_K(t) - v(y_K,t))^2\right\}dt \tag{3}$$

According to Parseval's theorem:

$$F = \int_{-\infty}^{+\infty}\sum_{J=1}^{N}(a_J^\varepsilon)^2\left|\tilde{\varepsilon}_J(\omega) - A(\omega)e^{-i\xi(\omega)x_J} - B(\omega)e^{i\xi(\omega)x_J}\right|^2 d\omega$$
$$+ \int_{-\infty}^{+\infty}\sum_{J=1}^{N}(a_K^v)^2\left|\tilde{v}_K(\omega) + \frac{\omega}{\xi(\omega)}\left(A(\omega)e^{-i\xi(\omega)y_K} - B(\omega)e^{i\xi(\omega)y_K}\right)\right|^2 d\omega \tag{4}$$

The function F is minimised when:

$$A(\omega) = \frac{h_2(\omega)E_1(\omega) - g(\omega)E_2(\omega)}{h_1(\omega)h_2(\omega) - g(\omega)g(\omega)} \quad (5a)$$

$$B(\omega) = \frac{h_1(\omega)E_2(\omega) - \overline{g(\omega)}E_1(\omega)}{h_1(\omega)h_2(\omega) - g(\omega)g(\omega)} \quad (5b)$$

where:

$$h_1(\omega) = \sum_{J=1}^{N} (a_J^\varepsilon)^2 \, e^{-i(\xi(\omega) - \overline{\xi(\omega)})x_J} + \left|\frac{\omega}{\xi(\omega)}\right|^2 \sum_{K=1}^{P} (a_K^\nu)^2 \, e^{-i(\xi(\omega) - \overline{\xi(\omega)})y_K} \quad (6a)$$

$$h_2(\omega) = \sum_{J=1}^{N} (a_J^\varepsilon)^2 \, e^{i(\xi(\omega) - \overline{\xi(\omega)})x_J} + \left|\frac{\omega}{\xi(\omega)}\right|^2 \sum_{K=1}^{P} (a_K^\nu)^2 \, e^{i(\xi(\omega) - \overline{\xi(\omega)})y_K} \quad (6b)$$

$$g(\omega) = \sum_{J=1}^{N} (a_J^\varepsilon)^2 \, e^{i(\xi(\omega) + \overline{\xi(\omega)})x_J} - \left|\frac{\omega}{\xi(\omega)}\right|^2 \sum_{K=1}^{P} (a_K^\nu)^2 \, e^{i(\xi(\omega) + \overline{\xi(\omega)})y_K} \quad (6c)$$

$$E_2(\omega) = \sum_{J=1}^{N} (a_K^\varepsilon)^2 \, e^{i\overline{\xi(\omega)}x_J} \tilde{\varepsilon}_J(\omega) - \frac{\omega}{\xi(\omega)} \sum_{K=1}^{P} (a_K^\nu)^2 \, e^{i\overline{\xi(\omega)}y_K} \tilde{v}_K(\omega) \quad (6d)$$

$$E_1(\omega) = \sum_{J=1}^{N} (a_K^\varepsilon)^2 \, e^{-i\overline{\xi(\omega)}x_J} \tilde{\varepsilon}_J(\omega) + \frac{\omega}{\xi(\omega)} \sum_{K=1}^{P} (a_K^\nu)^2 \, e^{-i\overline{\xi(\omega)}y_K} \tilde{v}_K(\omega) \quad (6e)$$

$A(\omega)$ and $B(\omega)$ are then calculated.

Assuming that mechanical values are constant in a bar cross-section, formulas 1 provide forces and displacements at any point, and in particular at bar ends.

The validity of the method is checked using the "left" part of the complete set-up presented in Fig.2, i.e. a single bar the right side of which is free from stress. The second optical extensometer is used to measure the displacement of the free bar end, to be compared with the reconstructed displacement using the presented method. A 40-mm long striker is launched on the left side at *3.03 m/s*. Using Eqns. 1, Eqns. 5 and Eqns. 6, stress and displacement are reconstructed at bar ends. Results are shown in Fig.1.

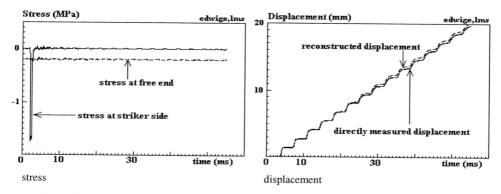

Figure 1 : reconstructed stresses at the ends of the bar and displacement at the free end
(The stress at the free end has been shifted down by 0.2 MPa to make the figure more readable)

The stress at the left side was almost zero as expected. At the right side, the stress became almost zero after the first incident wave. Compared to the amplitude of the impact stress, error on the rebuilt stress was less than *3.5%*. The reconstructed displacement was similar to the directly measured one. Relative error was less than *2.5%*.

APPLICATION TO SPECIMEN TESTING EXPERIMENT

Using two bars and a specimen between them, the same technique allows for the measurement of force an displacement at both specimen sides. Assuming then (and checking) the equilibrium of the specimen, stress, strain and strain-rate in the specimen can be calculated (like for the SHPB apparatus).

Experimental set-up

The experimental set-up is made of a 3.022-m input bar and a 1.865-m output bar. Both are made of Nylon and have a diameter of 40mm. Two strikers are used. The first (say striker 1) is 1 m long. It is made of the same material and has the same cross-section as the bar. The second one (say striker 2) is 1.8 m long. Its diameter is 50mm. It is made of steel. Three strain gauges are cemented on each bar. Gauge positions are $x_1^{in} = -2.619m$, $x_2^{in} = -1.514m$, $x_3^{in} = -0.611m$, $x_1^{out} = 0.428m$, $x_2^{out} = 0.817m$ and $x_3^{out} = 1.481m$. Additionnally, two optical extensometers are used for displacement measurements. These displacements are then numerically derived to obtain velocities at $y_1^{in} = -0.619m$ and $y_1^{out} = 0.220m$.

Two test are performed. In the first one the new method is successfully compared with the classical one in its range of validity. In the second, a significant increase of the maximum measured strain is demonstrated.

The first tes is done with the striker 1. Consequently, the first part of this test is a classical SHPB test. It can be then processed by the classical method (Zhao and al., 1997 ; Zhao and Gary, 1998). It then provides a comparison, for the first part of the test, between a commonly accepted method and the new one presented here.

Corresponding results are presented in Fig. 3. As shown in this figure, input and output forces are almost equal, and an equivalent stress-strain relation can be derived. This is done for the purpose of comparing the two methods in a condensed manner. It is indeed observed, as it is usual for Honeycomb specimens, that local buckling occurs at a single side of the specimen, and that the definition of a "strain" could be meaningless.

Figs. 2e and 2f show that the measurement (output force/average strain and stress/strain) given by the new method is very close to the classical one, for the common part.

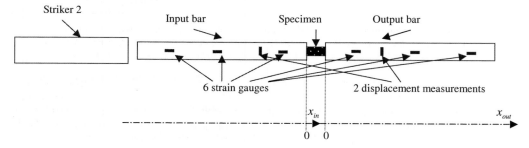

Figure 2 : experimental set-up

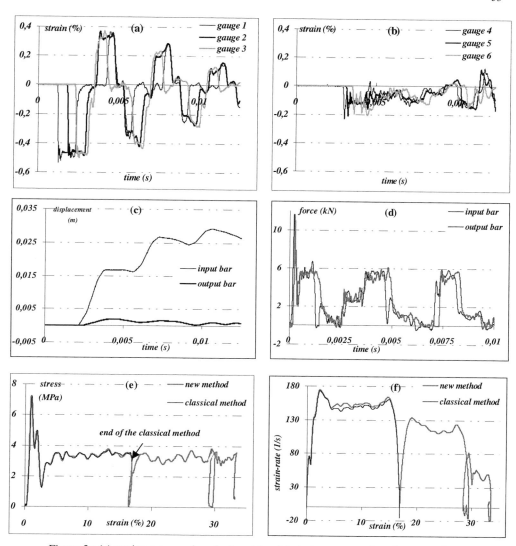

Figure 3 : (a) strains measured on the input bar (b) strains measured on the output bar
(c) displacement measured on the input bar (d) reconstructed forces at the specimen interfaces
(e) stress versus strain of the specimen (f) strain-rate versus strain of the specimen

The second test is performed with a high impedance striker (striker 2) by comparison with bar impedance. The corresponding measurements are presented in Fig 4. In such a case, the contact between the striker and the input bar would not stop if the input bar was long enough. Because of the fixed end of the output bar, it is seen, in fig. 4a, that the loading time lasts about 30 milliseconds. With such a loading, waves are always superposed so that the classical method cannot be used.

This test can then be performed at a lower average strain rate than the previous one. It shows that the output stress is not significantly lower than for the first test (see Fig. 5), as expected for aluminium Honeycomb.

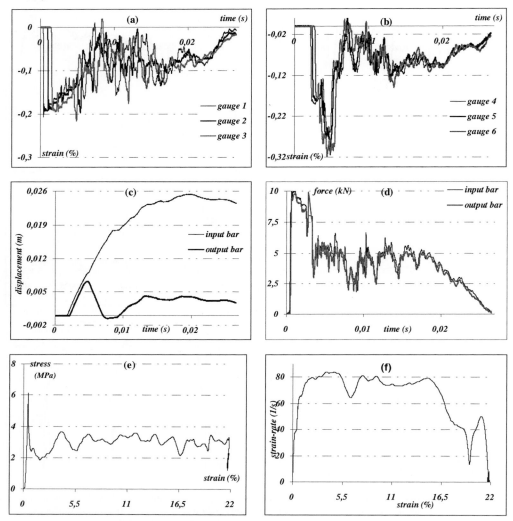

Figure 4 : (a) strains measured on the input bar (b) strains measured on the output bar
(c) displacement measured on the input bar (d) reconstructed forces at the specimen interfaces
(e) stress versus strain of the specimen (f) strain-rate versus strain of the specimen

This last test shows that using a striker, even a high impedance one, does not allow for a significant increase in available input energy (as the kinetic energy varies with the square of the speed). Consequently, a hydraulic actuator is under setting up in the authors laboratory. Together with the presented method for the separation of waves, it will make possible, in a very near future, the testing of various materials at medium strain rates.

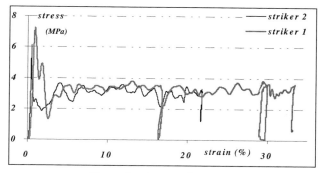

Figure 5 : stress-strain curve

CONCLUSION

A multi-point method (multi-strain and/or multi-velocity measurements) is presented for reconstructing one dimensional waves in bars. This method, explained in details in (Bussac and al., 2001), is exact when used with the single-mode dispersive propagation model commonly applied to Hopkinson bars. It yields consistent results (the inaccuracy due to imprecise measurements does not increase with time). It is illustrated here by applying it successfully to the analysis of a real test on a Nylon bar. It provides a significant increase in the observation time available when using measuring techniques based on the use of bars such as SHPB set-ups. The methods makes it possible to obtain precise measurements at medium strain rates in a test range in between that of mechanical testing machines and that of Hopkinson bars. It is illustrated here by testing aluminium Honeycomb at an average strain rate under 100/s up to 22% average strain. This limited step will be increased by the using a hydraulic actuator for the loading of the bars.

REFERENCES

Bacon, C. (1999) Separating waves propagating in an elastic Hopkinson pressure bar with three-dimensional effects. Int. J. Impact Engng. **22**, 55-69.

Bussac, M. N., Collet, P., Gary, G. and Othman, R. (2001) An optimisation method for separating and rebuilding one-dimensional dispersive waves from multi-point measurements. Application to elastic or viscoelastic bars. J. Mech. Phys. Solids, accepted for publication.

Lundberg, B. and Henchoz, A. (1977) Analysis of elastic waves from two-point strain measurement. Exper. Mech. 17, 213-218.

Lundberg, B. and Blanc, R. H. (1988) Determination of mechanical material properties from the two-point response of an impacted linearly viscoelastic rod specimen. J. Sound Vibration **137**, 483-493.

Park, S. W. and Zhou, M. (1999) Separation of elastic waves in split Hopkinson bars using one-point strain measurements. Exper. Mech. 39, 287-294.

Othman, R., Bussac, M. N., Collet, P. and Gary, G. (2001) Séparation et reconstruction des ondes dans les barres élastiques et viscoélastiques à partir de mesures redondantes. C. R. Acad. Sci. Série IIb **329(5)**, 369-376.

Zhao, H. and Gary, G. (1994) Une nouvelle méthode de séparation des ondes pour l'analyse des essais dynamiques. C. R. Acad. Sci. Série II **319**, 987-992.

Zhao, H. and Gary, G. (1997) A new method for the separation of waves. Application to the SHPB technique for an unlimited measuring duration. J. Mech. Phys. Solids **45**, 1185-1202.

Zhao, H., Gary, G., and Klépaczko, J. R. (1997) On the use of a viscoelastic split Hopkinson pressure bar. Int. J. Impact Engng. **19**, 319-330.

Zhao, H. and Gary, G. (1998) Crushing behaviour of Aluminium Honeycombs under impact loading, Int. J. Impact Engng. **21**, 827-836.

NEWLY DEVELOPED DYNAMIC TESTING METHODS AND DYNAMIC STRENGTH OF SOME STRUCTURAL MATERIALS

Shinji Tanimura and Koji Mimura

Division of Mechanical Systems Engineering, Graduate School of Engineering,
Osaka Prefecture University, 1-1, Gakuen-cho, Sakai, Osaka 599-8531, JAPAN

ABSTRACT

To improve safety during car crushing and to develop a high-speed plastic working and practical constitutive equation for computer codes, it is important to accurately obtain the entire tensile stress-strain curves for a variety of materials that cover a wide range of strain rates and strain. Based on numerical simulations and experiments, it is confirmed that the non-coaxial Hopkinson bar method, developed by Tanimura and Kuriu (1994), is effective at obtaining the stress-strain curve at high strain rates ranging from $10^2 \sim 10^4$ s^{-1} with accuracy and independent of flexural waves. The principles of the newly developed dynamic testing method based on a sensing block, developed by Tanimura, Mimura and Takada (1998), and these features are introduced and summarized. Some of the experimental results obtained by using this method are presented. It is demonstrated that the new system of the testing method is substantially useful for obtaining the entire stress-strain curves of many kinds of materials, and covering a wide range of strain rates from quasi-static strain rates to high strain rates on the order of 10^3s^{-1}, with good accuracy, such as that for a conventional quasi-static testing machine. Through numerical simulation using computer codes, the upper limits of the gauge length of specimens are presented. These limits are adaptable to dynamic testing for a wide range of strain rates.

KEYWORDS

Testing method, Dynamic strength, Strain rates, Numerical simulation, Experiment, Tensile strength, Material, Specimen, Gauge length

INTRODUCTION

Increasing the dynamic properties of materials, covering a wide ranges of strain rates and of strain, is desirable not only due to practical concerns regarding constitutive models for computer codes and for manufacturing but also for developing the data bases of a variety of materials to be used in the design of machines, cars, structures, etc.

In order to achieve material testing to obtain material behaviors that withstand strain rates on the order of $10^2 \sim 10^3$ s^{-1}, the Hopkinson bar method is usually used. To perform tensile test for sheet metals, the non-coaxial Hopkinson bar method is used, which includes setting two pressure bars as non-coaxials and applying dynamic tensile loads to specimens of sheet metal. This method was developed by Tanimura and Kuriu (1998), and involves a very simple clamping mechanism of using simple specimen. Moreover this method is able to obtain the entire stress-strain curves with high accuracy.

A new type of testing method, as well as the machine by which the tensile and compressive stress-strain curves are obtained for a wide range of strain rates from quasi-static to high rates on the order of 10^3 s^{-1}, were developed by Tanimura and others. The method and machine are both based on theoretical and experimental studies (1983, 1984, 1996, 1998). This newly developed testing method is effective and simple, i.e., much like a conventional testing machine.

In this paper, the results of studies and discussion of measuring principles and the measuring accuracy of the non-coaxial Hopkinson bar method, as well as that of the testing method based on the sensing block, are presented. The effectiveness of these various methods is also illustrated, and the results obtained by using these methods are presented.

To study the size of the test specimen, which can be adapted for testing for a wide range of strain rates, becomes an important subject to obtain the stress-strain curves of materials with sufficiently good accuracy and to be used obtained data in a manner similar to that of conventional testing machines.

We present the results of studies on the adaptable gauge length of test specimens for a wide range of strain rates, which are obtained by the numerical simulations and experiments. Authors would like to stress that these results should be valuable to improve the interchangeability of the dynamic testing results and standardization of the dynamic testing method for a wide range of strain rates.

ACCURACY OF THE NON-COAXIAL HOPKINSON BAR METHOD MEASUREMENT AND ITS APPLICATION

Accuracy of the measurement

A schematic view of the non-coaxial Hopkinson bar method, developed by Tanimura and others (1994, 1996), is shown in Figure 1. A detailed view of setting up a specimen for testing by this method is shown in Figure 2. The tensile stress-strain curve of sheet material set in non-coaxial bars can be obtained by the analysis following conventional Hopkinson bar methodology. Typical time-records of input, reflected and transmitted longitudinal waves, in the case of a mild steel sheet of 1 mm thickness, are illustrated in Figure 3. In Figure 3, the transmitted wave is shown by magnifying the vertical values ten times. Flexural waves are shown by dotted curve and are detected by two strain gauges cemented symmetrically with respect to the axis at the same position where the strain gages are also cemented; this is done in order to detect transmitted longitudinal waves by magnifying the vertical values (Figure 3).

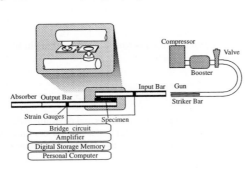

Figure 1: Schematic view of the non-coaxial Hopkinson bar method [Tanimura et al. (1994)].

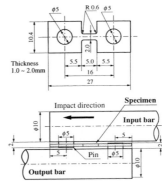

Figure 2: Detailed view of setting up a specimen according to a non-coaxial Hopkinson bar method.

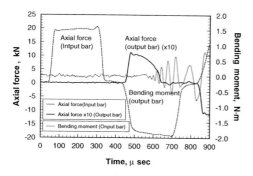

Figure 3: Typical time-records of input, reflected and transmitted longitudinal waves according to the non-coaxial Hopkinson bar method.

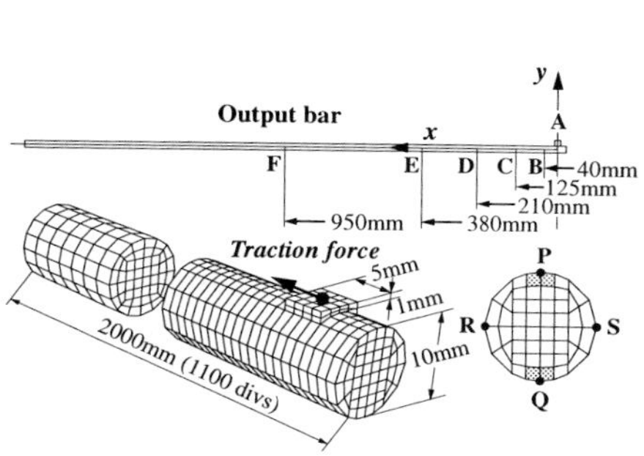

Figure 4: Evaluation points and finite element discretization of the output bar model.

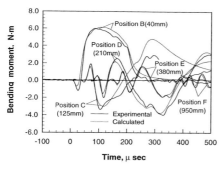
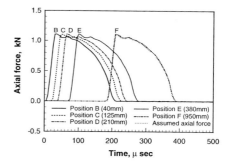

Figure 5: Comparison between calculated and experimentally obtained bending moments at points B to E.

Figure 6: Comparison of transmitted axial forces at points B to F.

To confirm the effect of these flexural waves on the measurement of the stress-strain curve, we analyzed the dynamic behaviors of the output bar by using MSC/DYTRAN code (MSC.Software, 1999, Figure 4). In the analysis, the recorded axial force, denoted as "output bar" in Figure 3, was applied on the top of small pin in the axial direction of the bar, as shown by the arrow in Figure 4. The analyzed bending moments at each point B ~ F on the output bar (Figure 4) are evaluated from the analyzed values of longitudinal strains at each point B ~ F, which are shown by dotted curves, denoted as "Calculated" in Figure 5. The recorded bending moments of the experiments at each point B ~ E are shown by solid curves, denoted as "Experimental". In this study we observed that the calculated values of the bending moments coincided fairly well with the experimental records. The maximum value, 0.14mm, of the lateral deflection at position A of the output bar was evaluated by the analysis during the testing time. The value of deflection was small, compared to the distance of 16 mm between the holes in specimen (Figure 2). This result shows that the deflection at position A of the bar caused by the bending moment may not have substantially affected to the value of the axial force to be applied to the specimen. The value of axial forces calculated from the axial strains by the analysis at each position B ~ F of the output bar (Figure 4), are shown in Figure 6, and also coincided with each other and with the input axial force at position A (shown by a solid curve as denoted "output bar" in Figure 3). These results show that the measurement of the axial force applied to the specimen by the non-coaxial Hopkinson bar method is of sufficiently good accuracy, independent of the mixture of bending moments. In an actual measurement of the axial force, by this method, the effect of the flexural waves on the measurement is further negligible, as the strain gauges are cemented onto the bar in order to cancel the flexural waves.

Application of this method

The stress-strain curves of A6000 Al alloy and OFHC Copper were obtained by this method. A6000 Al alloy has attracted interest recently due to its adoption as sheet metal for cars. The chemical composition (mass %) of the Al alloy is Si 1.02, Fe 0.18, Cu 0.01, Mn 0.07, Mg 0.58, Cr 0.01, Zn 0.02, Ti 0.03, Al the remainder. The specimen used was of 0.95 mm thickness and was treated by heating it to 170 °C for 60 minutes.

The quasi-static stress-strain curves of A6000 Al alloy for specimens of 0.95mm thickness and of various sizes of the gauge length (not included, R) are shown in Figure 7. The dynamic stress-strain curves of the alloy at 1000 s^{-1} strain rates for specimens, shown in Figure 2, of 0.95mm thickness and of 6.8 mm gauge length, were obtained by means of the non-coaxial Hopkinson bar method (Figure 8). We observed that the rate dependency of the strength of the material was almost negligible in comparison of the results (Figures. 7 and 8). Moreover, the apparent elongation, after the sample reached to the tensile strength, corresponding to the maximum stress of the stress-strain curve, became larger when the gauge length of the specimen was shorter than circa 7.6 mm.

Figure 9 shows the quasi-static tensile stress-strain curves of OFHC Copper for the study specimens, as shown in Figure 2. These latter were1.5mm thick and of various gauge lengths. As shown in Figure 9, the apparent elongation becomes remarkably larger when the gauge length is shorter. These phenomenon appear similar to those reported by Domont, C. and Levaillant, C. (1989), in which the elongation at the locally deformed portion, (at the bottom notch of R = 0.5mm)of a cylindrical specimen becomes 2.4 times that of a smooth specimen. Constraint at the end of the gauge length of the specimen, especially in the case of a short gauge, may effect the large apparent elongation (Figure 9). By making correction for the values of the

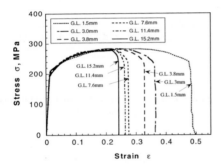

Figure 7: Effect of the gauge length of specimen on stress-strain relations in A6000 series aluminum alloy.

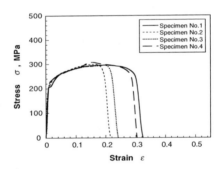

Figure 8: Dynamic stress-strain curves for A6000 series aluminum alloy obtained by means of the non-coaxial Hopkinson bar method.

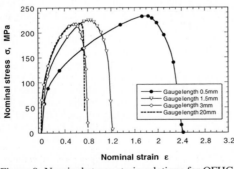

Figure 9: Nominal stress-strain relations for OFHC copper under quasi-static loading.

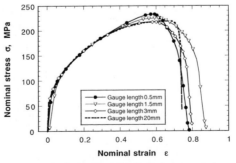

Figure 10: Compensated nominal stress-strain relations for OFHC copper under quasi-static loading.

apparent elongation given in Figure 9 the resultant stress-strain curves become as shown in Figure 10. Therefore, this is as if the value of strain at the maximum stress of each stress-strain curve coincided with the value of strain for the specimen with a gauge length 20 mm.

The obtained tensile stress-strain curves of OFHC Copper at strain rates from quasi-static to very high strain rates on the order of 10^4 s^{-1} obtained by means of the non-coaxial Hopkinson bar method, are illustrated in Figure 11. Here the values of the strains of the dynamic curves were corrected for according to the manner mentioned above.

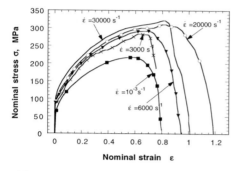

Figure 11: Compensated nominal stress-strain relations for OFHC copper under quasi-static and dynamic loading.

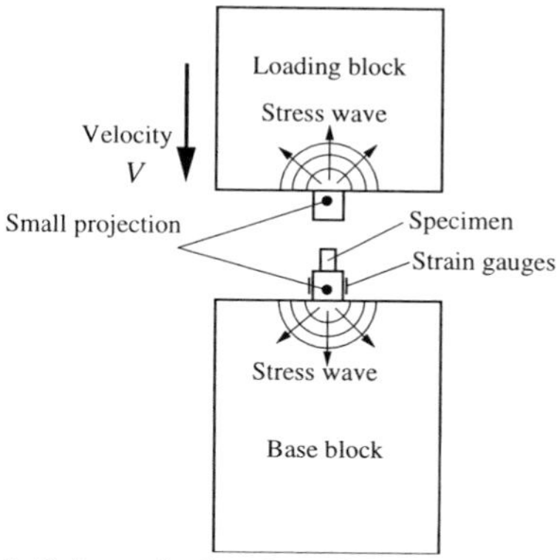

Figure 12: Schematic drawing of the sensing block method.

FEATURES OF DYNAMIC TESTING METHOD BASED ON SENSING BLOCK AND ITS APPLICATIONS

Principle and features of the method

The principle and technical features of the newly developed method, as distinguished from the conventional dynamic testing method (e.g. the Hopkinson bar method) are as follows; the example used in this case to apply this method involves a dynamic compression test.

(1) When a specimen is compressed dynamically by small projections made into bodies of a base block and a loading block by the relative velocity V (Figure 12), the generated dynamic load propagates into both the base block and the loading block through the small projections as spherical stress waves, attenuating the amplitude of the wave front in inverse proportion to the propagated distance from the center of the sphere. The stress waves are reflected on the boundaries of the blocks and interfere with each other. When the block size is sufficiently larger than the size of the test piece, the reflected stress waves are substantially not the cause of movement at the root of the projection, as compared with the velocity V. These phenomena lead to a measurement capable of determining the dynamic load, which is caused on the interface between the specimen and the top of the small projection. Substantially good accuracy is achieved for a sufficiently long measuring time, independent of the effects of the waves reflected from the boundaries of the blocks.

(2) The dynamic load generated on the interface between the specimen and the top of the projection propagates and reflects between the top and root of the small projection as elastic stress waves. When the projection size is small enough, fluctuation of the dynamic load caused by the propagation and reflection of the stress waves is negligible, compared with the variation of the generated dynamic load. By cementing strain gauges on the outer surface of the small projection, the dynamic load generated on the specimen can be measured with sufficiently good accuracy.

(3) The principle and features introduced above in (1) and (2) are applicable to both sides of the base block and the loading block. When both blocks are supported to isolate stress wave propagation substantially from bodies existing outside the blocks, a steady dynamic load can be applied to the specimen and an be measured with good accuracy, independent of the reflected waves from the boundaries of the blocks and

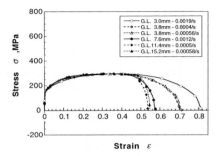

Figure 13: Dependence of quasi-staic stress-strain relations on gauge length for IF steel.

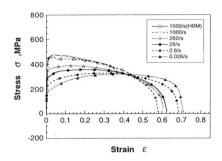

Figure 14: Nominal stress-strain relations for IF steel at various strain rates.

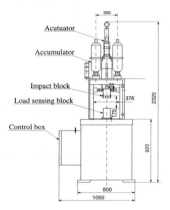

Figure 15: High-speed material testing system (Saginomiya Co. Ltd., TS-2000)

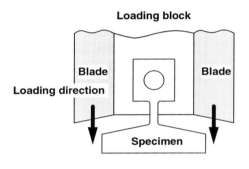

Figure 16: Loading method in a improved sensing block device.

interference from the bodies outside of the blocks.

A high Speed Material Testing System, based on the sensing block method (HSMTS, Figure 15), has been developed and is now on the market. The testing system is compact and is able to obtain the entire tensile and compressive stress-strain curves for a variety of materials, covering a wide range of strain rates from quasi-static to those on the order of 10^3 s^{-1}.

Results obtained by using the method

As examples, Figure 13 shows the quasi-static tensile stress-strain curves of IF2 sheet steel by cold rolling for a specimen of 1 mm thickness (Figure 2). The chemical composition (wt %) is C 0.002, Si 0.01, Mn 0.15, P 0.013, S 0.006, Al 0.031, Ti 0.054, and the remainder is Fe. The tensile stress-strain curves of the sheet steel for the specimen of 1 mm thickness and 3.8 mm gauge length obtained by using the HSMTS are shown in Figure 14, together with the result obtained by using the non-coaxial Hopkinson bar method introduced in chapter 2.

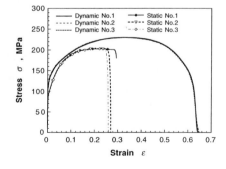

Figure 17: Nominal stress-strain relations for A5052 aluminum alloy at quasi-static (0.0014s^{-1}) and dynamic (1000s^{-1}) strain rate

In order to achieve tensile testing of sheet material by using the improved system of the HSMTS, the static and dynamic loads are applied directly to one side of the specimen, in a shoulder-like arrangement, as shown in Figure 16. This new method is quite effective at obtaining dynamic tensile stress-strain curves with very small oscillations on the obtained curves, especially for high-strain rate tests, with sufficient accuracy.

Some examples of the tensile stress-strain curves of A 5052 aluminum alloy at quasi-static and dynamic strain rates obtained by using the improved system of the HSMTS are shown in Figure17.

SPECIMEN SIZE TO BE USED FOR DYNAMIC TEST

The dynamic behavior of a cylindrical specimen with diameter D and length L were analyzed by using LS-DYNA code, when the displacement, in the axial direction, of the end surface of the specimen was fixed and the another end was pulled with velocity υ in the axial direction. The ratio L/D = 5 was chosen and the Johnson-Cook model for mild steel was assumed for the analysis.

The mean stress on the surface of the specimen end was evaluated for both ends, and the mean displacement between both end surfaces was obtained by the analysis for a specimen of length L and pulling velocity υ, during the deformation period. The solid curve in Figure 18 shows, as an example, the calculated stress-strain relation at a mean strain rate, obtained by the mean stresses and the mean displacement in the case of analysis for length L and the velocity υ. The broken curve in Figure 18 shows the assumed stress-strain curve by the Johnson-Cook model in the case of the analysis. Figure18 shows an example of conditions under which the maximum difference between the stress values of the calculated curve and the modeled curve is approximately 2 %. For the conditions of the analysis, the average strain rate during the deformation of the specimen can be estimated, and the results correspond to one point in Figure 19. By performing a series of analysis in this manner, varying the specimen length L and pulling velocity υ, relation between the gauge length of the specimen, that is specimen length L in this model, and the testing strain rate was obtained (Figure 19) for a wide range of strain rates. These relations correspond to the case of the testing error of 2%.

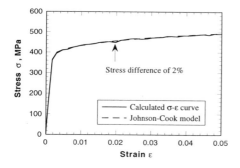
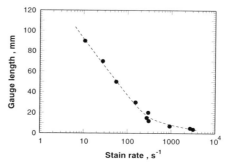

Figure 18: Stress difference between the simulated stress and the assumed stress based on the Johnson-Cook model.

Figure 19: Relation between the gauge length of specimen and the testing strain rate, corresponding to the case of a testing error of 2%.

CONCLUSIONS

The principles and features of a newly developed dynamic testing method based on a sensing block were introduced. A new method to apply dynamic load directly to a simple specimen was also developed. Based on the example of tensile stress-strain curves of IF 2 sheet steel, it was illustrated and demonstrated that the newly developed dynamic testing method is effective at obtaining the entire stress-strain curve of various materials, covering a wide range of strain rates from quasi-static to high strain rates on the order of $10^3 s^{-1}$, with sufficient accuracy. This can be done by simple operations much like those used in conventional quasi-static testing machines.

The accuracy of the measurement of the non-coaxial Hopkinson bar method was discussed. It was demonstrated that the tensile stress-strain curves of sheet metals at a high strain rate range of $10^2 \sim 10^4$ s^{-1} can be obtained with good accuracy by this method. To show the effectiveness of this method for high strain rates testing, tensile stress-strain curves of OFHC copper for a wide range of strain rates from 10^{-3} s^{-1} to 2×10^4 s^{-1} obtained by this method were used as examples.

By performing numerical simulations, the relation between the gauge length of a specimen and the testing strain rate, corresponding to a testing error of about 2 %, was obtained for a wide range of strain rates from 10 s^{-1} to 3×10^3 s^{-1}.

References

Tanimura S. and Kuriu N. (1994). *Proceeding of the 2nd Materials and Processing Conference* (M & P '94, JSME), **940:36**, 144-145.

Mimura K., Takagi S., Furukimi O., Obara T. and Tanimura S. (1996). Dynamic Deformation Behavior of Steel Sheet for Automobiles, *SAE Technical Paper, Society of Automotive Engineering,* **960019**, 1-5.

Tanimura S., Mimura K. and Takada S. (1998). Development of a Dynamic Testing Apparatus with a Sensing Block and Its Applications, *Proceeding of the 1998 Annual Meeting of JSME/MMD*, **5:98-5**, 303-304.

Mimura K., Hirata S., Chuman Y. and Tanimura S. (1996). Development of Dynamic Loading Device with Stress Sensing Block and Its Experimental Examination, *J. Soc. Mat. Sci. ,Japan*, **45:8**, 939-944.

Mimura K., Hirata S., Chuman Y. and Tanimura S. (1996). Development of a compact dynamic loading device with a stress sensing block and its application to the evaluation of stress-strain relations at high rates of strain, *JSME*, **62:603**, 2609-2614.

Tanimura S. (1984). Stress Wave Propagation in Elastic/Viscoplastic Cylindrical Bodies Containing a Spherical Cavity, *Acta Mechanica*, **51:1-2**, 1-13.

Tanimura S. and Aiba K. (1983). A New Device for Measuring Impact Load at a Contact Part, *JSME*, **49:448**, 1565-1571.

Tanimura S. (1984). A New Method for Measuring Impulsive Force at Contact Parts, *Experimental Mechanics*, **24:4**, 271-276.

Dumont C. and Lavaillant, C. (1989). Ductile Fracture of MEtals Investigated by Dynamic Tensile Tests on Smooth and Notched Bar, *Inst. Ohys. Conf. Ser.* **102** (*The Institute of Physics*), 65-72

MSC. Software Corporation (1999). MSC. Dytran user's manual, version 47, Los Angeles, USA.

EFFECTS OF STRIKER SHAPE AND ATTACHED POSITION OF STRAIN GAGE ON MEASURED LOAD IN INSTRUMENTED CHARPY IMPACT TEST

T. Kobayashi and S. Morita

Department of Production Systems Engineering, Toyohashi University of Technology,
Toyohashi, 441-8580, Japan

ABSTRACT

Instrumented Charpy impact test is widely used for the evaluation of toughness of many kinds of materials and small scale specimens. In the test, therefore, it is important to record an accurate impact load. Generally, one can obtain measured load in the instrumented Charpy impact test by multiplying the output signal from strain gage attached to the instrumented striker by load-calibration factor assuming a liner relationship between the strain gage signal and applied load. Although JIS or ISO describes about the instrumented striker, amplifier, data processing parameter and etc., detailed methods on the load measurement is scarcely described in any standard.
In the present study, two types of striker were used. The strain gages were attached to four positions in each striker. Instrumented Charpy impact test was carried out using these strikers in order to investigate the effect of gage position on actual impact load. By the finite element analysis, the effect of the strain gage position on the measured load was also investigated. As a result, it became clear that the accurate impact load was not measured around the end of slit which was introduced to release the constraining effect of elastic deformation of the gage position from surrounding hammer; the effect of the vibration of the hammer appeared strongly around this position. It is recommended to prevent the effect of such vibration by attaching the gage away from such position.

KEYWORDS

instrumented Charpy impact test, CAI system, strain gage position, instrumented striker, absorbed energy, FEM.

INTRODUCTION

One of the authors has already developed the CAI (Computer Aided Instrumented Charpy Impact Testing) system, where dynamic fracture toughness parameters are obtained simply from the analysis of the load-deflection curve of a single precracked specimen (Kobayashi T. *et al.* (1993)). In the test, therefore, it is important to record an accurate impact load. Generally one can obtain the measured load by multiplying the strain signal from attached strain gage on the instrumented striker by a load calibration factor assuming a linear relationship between the strain gage signal and applied load.

Recently, the instrumented Charpy impact test is used for the evaluation of toughness of many kinds of materials and miniaturized specimens (Kobayashi T. *et al.*, 2001). In those cases, a significant variation in the calibration factor has been reported because the Charpy specimen was changed from the standard steel specimen to another material or geometry (Kalthoff J. F. *et al.*, 1996; Kobayashi T. *et al.*, 2000; Marur P. R. *et al.*, 1995). Though a lot of methods of load calibration are proposed (Wilde G. *et al.*, 1996; Winkler S. and Boβ B., 1996), there is no report taking into consideration the change of material or geometry of the specimen. The elucidation of the mechanism that the load calibration factor changes by material or geometry of the specimen is important to measure accurate impact load and to enact the standard of load calibration method.

Although JIS or ISO describes about the instrumented striker, amplifier, data processing parameter and etc., detailed method on load measurement is hardly described in any standard. In the current standard of ASTM and ISO, there is no regulation on the accurate strain gage position for instrumentation (sometimes 11-15 mm from the tip is recommended).

In the present study, instrumented Charpy impact test was carried out using two types of striker in which the strain gages were attached to four positions in order to investigate the effect of striker geometry and strain gage position on actual impact load. The changes in strain with respect to the time were simulated by finite element analysis to explain the effect of the vibration of hammer on the measured load for the specified strain gage positions.

EXPERIMENTAL AND ANALYTICAL PROCEDURES

Instrumented Charpy Impact Test

An instrumented Charpy impact testing machine with 100 J capacity was used for the instrumented Charpy impact test. Instrumented Charpy impact test was carried out using two types of striker shown in Fig. 1. The hollowed striker is the conventional type. On the other hand, Non-hollowed striker was designed to prevent the strain localization in the portion of striker where the strain gage was attached (Kobayashi *et al.*, 2000). The semi-conductor strain gages were attached on both sides and the positions from the tip of instrumented striker were 15, 30, 45 mm and upper portion, respectively. The material used in this study was 6061-T6 aluminum alloy. Specimens for Charpy impact tests were machined according to the JIS Z2202 standard. The loading velocity was 4.5 m/s.

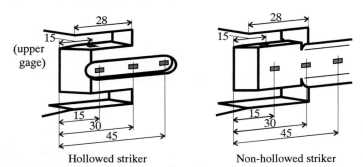

Figure 1: Schematic illustration of instrumented striker for instrumented Charpy impact test (mm).

Finite Element Analysis

To calculate strain fields in the instrumented striker, the ANSYS non-linear finite element code was used for the analysis. A whole finite element model of the instrumented Charpy hammer and the specimen are shown in Fig. 2. The hammer arm was disregarded in the model. The model was three-dimensional 1/4 size using the symmetric condition. Eight-noded brick elements were used in modeling the hammer and the specimen. The total numbers of the elements were 5,668 and 4,372 in Fig. 2 (a) and (b), respectively. Especially, the elements were refined in the regions of a tip of the striker and the strain gage point. The contact elements were formed at contact points between a surface of the instrumented striker and the specimen to calculate contact forces. The contact element that is a tetrahedral consists of the target surface being a triangular element on the surface and the contact n node on the other surface. Full-Newton-Raphson method was used for the convergence calculation. The instrumented striker and the specimen were modeled as a linear elastic solid. The elastic modulus of the instrumented striker was taken to be 210 GPa. The elastic modulus of the specimen used was 70 GPa.

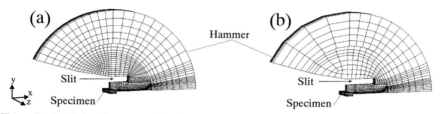

Figure 2: (a) Finite element model of Charpy hammer with specimen for low blow instrumented Charpy test. (b) Finite element model by introducing deep cutting slit.

RESULTS AND DISCUSSION

Effect of the Strain Gage Position on the Measured Impact Load

Figure 3 shows the typical load-deflection curves recorded from two types of instrumented striker for the V-notched 6061-T6 aluminum alloy Charpy specimen. The load-deflection curves recorded from gage position 15 mm in both types of striker are smooth. On the other hand, it is obvious that the vibration was superimposed on the load-deflection curves recorded from the other strain gage positions after the maximum load. The vibrations on the load-deflection curves are slightly reduced in non-hollowed striker. There are no differences between two types of striker geometries.

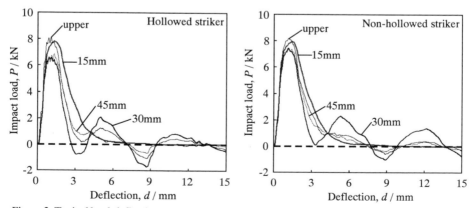

Figure 3: Typical load-deflection curves recorded from specified strain gage positions in hollowed and non-hollowed strikers for the V-notched 6061-T6 Al alloy Charpy specimen.

Figure 4 shows the comparison of dial energy with absorbed energies calculated from load-deflection curves of all strain gage positions. The absorbed energy calculated from load-deflection curve of gage position 15 mm decreases 2.66% compared with the dial energy. The strain gage attached to 15 mm recorded the accurate impact load history.

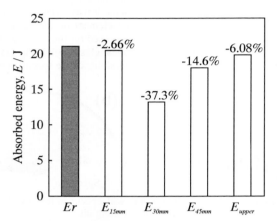

Figure 4: Comparison of dial energy, Er with absorbed energies, E_{15mm}, E_{30mm}, E_{45mm} and E_{upper} calculated from load-deflection curves recorded from specified strain gage positions in non-hollowed striker for V-notched 6061-T6 Al alloy specimen.

Figure 5 shows the prediction of the change in compressive strain along (x) and orthogonal to (y, z) direction of blow calculated by finite element analysis for specified strain gage positions. It can be seen that the strain of the x direction is symmetry for the time shown in Fig.5 (a). In Fig.5 (b), the vibrations were observed in the strain of the y direction recorded from gage positions 30 and 45 mm. These were affected by vibration of the hammer as shown in Fig.6. With respect to time of the vibration of displacement, the same manner is observed with the gage positions 30 and 45 mm shown in Fig.5 (b). The previous work (Yamamoto I. and Kobayashi T., 1993) reported that the hammer deforms elastically

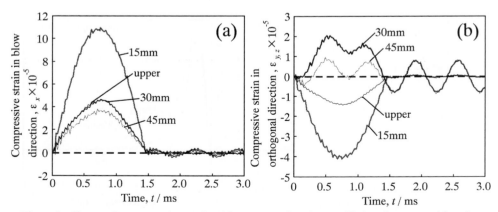

Figure 5: Changes in compressive strain with respect to time for specified strain gage positions in FEM analysis. (a)Compressive strains along the direction of blow (x-direction). (b)Compressive strains, orthogonal to the direction of blow for 15, 30, 45mm and upper positions, respectively, in y and z-directions.

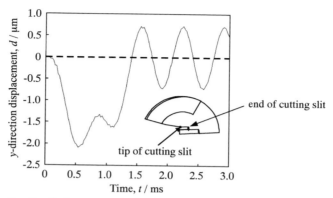

Figure 6: Variation of y-direction displacement at the tip of cutting slit with respect to time upon impact.

and then vibrates periodically. It is concluded that the both ends of the hammer are deformed conversely and the hammer edge portion (near the end of the slit), where the strain gage is attached, is bent by the natural vibration of the hammer. Consequently, it is recommended that the strain gage should be attached near the tip of striker.

Effect of the Slit Depth of Striker on the Measured Impact Load

Figure 7 shows the prediction of the change in compressive strain calculated by model of Fig.2 (b). The depth of slit is deeper than Fig.2 (a). The constraining of deformation of the gage position 30 mm from surrounding hammer was released by the deep slit. In Fig.7 (a), the strain of the x direction is symmetry for the time as shown in Fig.5 (a). In the strain of the y direction, the vibration was still observed in the strain-time curve obtained from gage position 45 mm as shown in Fig.7 (b). The accurate impact load was not measured around the end of slit which was introduced to release the constraining effect of deformation of the gage position from surrounding hammer; the effect of the vibration of the hammer appeared strongly around this position. However, it was possible to prevent the effect of such vibration

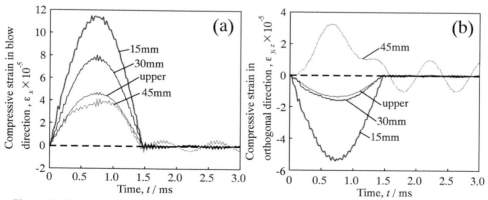

Figure 7: Changes in compressive strain with respect to time for specified strain gage positions in FEM analysis. (a) Compressive strains along the direction of blow (x-direction). (b) Compressive strains, orthogonal to the direction of blow for 15, 30, 45mm and upper positions, respectively, in y and z-directions.

by attaching the gage away from such position.

CONCLUSIONS

Instrumented Charpy impact test was carried out using two types of striker where the strain gages were attached to four positions in order to investigate the effect of striker geometry and strain gage position on actual impact load. The main results and conclusions are summarized as follows.

(1) According to the results, the load-deflection curves were slightly affected by the geometry of impact striker. The load-deflection curves recorded from gage position 15 mm in both types of striker were smooth. On the other hand, it is obvious that the vibration was superimposed on the load-deflection curves recorded from the other strain gage positions after the maximum load.

(2) The absorbed energies were estimated from load-deflection curves of all strain gage positions. The absorbed energy estimated from load-deflection curve of gage position 15 mm was approximately same as the dial energy. The strain gage position 15 mm recorded the accurate impact load history.

(3) The accurate impact load was not measured around the end of slit which was introduced to release the constraining effect of deformation of the gage position from surrounding hammer; the effect of the vibration of the hammer appeared strongly around this position.

REFERENCES

Kalthoff J. F., Walle E. van and Wilde G. (1996). Variation of the Sensitivity of Instrumented ISO/DIN and ASTM Tups and their Influence on the Determination of Impact Energies in Tests with Ductile Steels. In: *Evaluating Material Properties by Dynamic Testing, ESIS20*, Walle E. van (Ed). Mechanical Engineering Publications, London., pp.25-35.
Kobayashi T., Morita S. and Toda H. (2001). Fracture Toughness Evaluation and Specimen Size Effect. *Mat. Trans.* **42:1**,52-57.
Kobayashi T., Morita S., Inoue N. and Toda H. (2000). On the Accuracy of Measurement and Calibration of Load signal in the Instrumented Charpy Impact Test. In: *Pendulum Impact Testing: A Century of Progress, STP 1380*, Siewert T. A. and Manahan M. P. Sr. (Eds). ASTM, West Conshohocken, PA., pp.198-209.
Kobayashi T., Yamamoto I. and Niinomi M. (1993). Introduction of a New Dynamic Fracture Toughness Evaluation System. *J. Test. Eval.* **21**, 145-153.
Marur P. R., Shimha K. R. Y. and Nari P. S. (1995). A Compact Testing System for Dynamic Fracture Studies *J. Test. Eval.* **23**, 267-274.
Wilde G., Covic M. and Gregor M. (1996). Instrumented Impact Testing in Radioactive Hot Cells, Equation, and Software. In: *Evaluating Material Properties by Dynamic Testing, ESIS20*, Walle E. van (Ed). Mechanical Engineering Publications, London., pp.89-96.
Winkler S. and Boβ B. (1996). Static Force Calibration of Charpy Impactors. In: *Evaluating Material Properties by Dynamic Testing, ESIS20*, Walle E. van (Ed). Mechanical Engineering Publications, London., pp.37-44.
Yamamoto I. and Kobayashi T. (1993). Evaluation Method of Dynamic Fracture Toughness by the Computer-Aided Instrumented Charpy Impact Testing System. *Int. J. Pressure Vessels Piping*, **55**, 295-312.

A SPECTRAL METHOD FOR WAVE DISPERSION ANALYSIS APPLICATION TO AN ALUMINIUM ROD

Ramzi OTHMAN[1], Robert BLANC[2], Marie-Noëlle BUSSAC[3], Pierre COLLET[3], Gérard GARY[1]

[1] Laboratoire de Mécanique des Solides, UMR 7649, Ecole Polytechnique, F-91128, Palaiseau.
[2] TRANS.WAVES 150, cité Le Corbusier F-13008 MARSEILLE
(formerly Laboratoire de Mécanique et d'Acoustique du CNRS MARSEILLE France)
[3] Centre de Physique Théorique, UMR 7644, Ecole Polytechnique, F-91128, Palaiseau.

ABSTRACT

The spectral analysis of a single wave measurement (for instance strain measurement) on a rod exhibits multiple resonant frequencies. Wave celerity is inferred from the resonance position and attenuation is deduced from the resonance bandwidth. Dispersion is therefore computed on a countable set of frequencies. This method does not need any limitation of the loading pulse. When applied to an aluminium rod, the method allows for the measurement of the damping and the dispersion relations for a frequency range up to 50 kHz. It is found out in this case that damping is non zero. Results are checked using the wave propagation description in a real rod. They show to be of a high accuracy. This method can be applied to any one-dimensional and single-mode propagating waves (for example flexural waves in beams).

KEYWORDS

Hopkinson bars, Wave dispersion, Viscoelastic bars, Dynamics, Aluminium, Complex Young's Modulus.

INTRODUCTION

Wave dispersion in rods is due to geometric and (or) viscoelastic reasons. In elastic rods, because of the 3D-geometry, wave velocity decreases with the frequency. This phenomenon can be neglected when the ratio of the bar radius to the wavelength does not exceed 0.1 (Pochhammer, 1876 ; Chree, 1889 ; Davies, 1948). When bars are viscoelastic, the material characteristics depend on the frequency and the wave velocity varies with the frequency. Even though the geometry effects were absent, waves should be distorted and damped when propagating (Blanc, 1971). In the Split Hopkinson bar apparatus (SHB), experimental results are improved when wave dispersion is taken into account (Follansbee & Franz, 1983 ; Gorham, 1983 ; Gong & al., 1990). Wave dispersion in rods is also used to investigate the viscoelastic properties of materials (Blanc, 1971).

Dispersion may be experimentally determined by comparing the wave Fourier components measured on two points on the rod (Blanc, 1971 ; Lundberg & Blanc, 1988 ; Gorham & Wu, 1996 ; Bacon, 1998). Recently, Hillström & al. (2000) developed a multi-point method using least squares. In the present paper, a one-point method is developed using a spectral analysis of the resonant frequencies of the rod.

THEORY

Wave propagation

We consider a finite rod of length L (see Figure 1). ε denotes the longitudinal strain. A perturbation is generated at the left side of the rod. The right side of the bar is kept free from stress.
Henceforth, we assume that :
- Stress, strain and displacement are homogenous in a bar section.
- Only the first longitudinal mode of propagation is excited (Davies, 1948 ; Follansbee and Franz, 1983 ; Gong et al., 1990).

In this case, the Fourier transform of the strain can be expressed by (Lundberg and Blanc, 1988; Bacon, 1998):

$$\tilde{\varepsilon}(a,\omega) = A(\omega)e^{-i\xi(\omega)a} + B(\omega)e^{i\xi(\omega)a}, \qquad (1)$$

where

$$\xi(\omega) = k(\omega) + i\alpha(\omega) = \omega/c(\omega) + i\alpha(\omega), \qquad (2)$$

$\xi(\omega)$ is the complex wave number of the first mode of propagation and it takes account of geometric effects and (or) viscoelastic effects, $c(\omega)$ is the phase velocity and $\alpha(\omega)$ is attenuation.
The Fourier components of the strain at the left side (striker side, abscissa $-L$) of the bar are denoted $\theta(\omega)$. Considering the boundary conditions, i.e. $\tilde{\varepsilon}(0,\omega) = 0$ and $\tilde{\varepsilon}(-L,\omega) = \theta(\omega)$, it follows that :

$$A(\omega) = -B(\omega) = \frac{\theta(\omega)}{e^{i\xi(\omega)L} - e^{-i\xi(\omega)L}}. \qquad (3)$$

Replacing Eqns. 3 into Eqn. 1., the strain is expressed as a function of the complex wave number and of the boundary conditions at the left side of the rod

$$\tilde{\varepsilon}(a,\omega) = \theta(\omega)\frac{e^{-i\xi(\omega)a} - e^{i\xi(\omega)a}}{e^{i\xi(\omega)L} - e^{-i\xi(\omega)L}} = -\theta(\omega)\frac{\sin(\xi(\omega)a)}{\sin(\xi(\omega)L)}. \qquad (4)$$

Wave velocity

In this section, wave attenuation is neglected with respect to the wave number $k(\omega)$ ($\alpha(\omega) \ll k(\omega)$). With low damping materials, this hypothesis is valid for most of experimental applications. A correction must be done for high damping materials.

$$\xi(\omega) = k(\omega) + o(k(\omega)). \qquad (5)$$

Denoting $S(a,\omega)$ the strain spectrum, Eqns. 4 and 5 yield :

$$S(a,\omega) = |\tilde{\varepsilon}(a,\omega)| = |\tilde{\theta}(\omega)| \frac{|\sin(k(\omega)a)|}{|\sin(k(\omega)L)|} \quad (6)$$

The spectrum denominator becomes zero for the frequencies ω_n such as :

$$k(\omega_n) = n\pi/L, \quad (7)$$

where n is a relative integer. For each n, unless the angular frequency ω_n nullifies the numerator as well as the denominator, a resonance occurs. The corresponding resonant frequencies are associated to the local maxima of the spectrum, so they are easily assessed. Therefore, when

$$\theta(\omega_n) \neq 0 \text{ and } \forall p \in \mathbb{N}, a/L \neq p/n \quad (8)$$

the wave velocity is given by :

$$c(\omega_n) = \omega_n L/n\pi \quad (9)$$

Wave attenuation

In this section we reconsider Eqn. 4. The numerator and the denominator are developed as follows :

$$\sin(\xi(\omega)x) = \sin(k(\omega)x)\cos(i\alpha(\omega)x) + \cos(k(\omega)x)\sin(i\alpha(\omega)x) \quad (10)$$

We develop the sine and cosine to the second order of $\alpha(\omega)x$:

$$\cos(i\alpha(\omega)x) = 1 + \frac{\alpha^2(\omega)x^2}{2} + o(\alpha^2(\omega)x^2)$$
$$\sin(i\alpha(\omega)x) = i\alpha(\omega)x + o(\alpha^2(\omega)x^2) \quad (11)$$

Eqns. 4., 10. and 11. yield :

$$S^2(a,\omega) = |\theta(\omega)|^2 \frac{\alpha^2(\omega)a^2 + \sin^2(k(\omega)a)}{\alpha^2(\omega)L^2 + \sin^2(k(\omega)L)} \quad (12)$$

We assume that ω is close to a resonance value ω_n. The attenuation is considered to be constant in this vicinity: $\alpha(\omega) = \alpha_n = \alpha(\omega_n)$. The sine is developed in the vicinity of ω_n :

$$\sin(k(\omega)x) = \sin(k(\omega_n)x) + x\mu_n \cos(k(\omega_n)x)(\omega - \omega_n) - \tfrac{1}{2}x^2\lambda_n \sin(k(\omega_n)x)(\omega - \omega_n)^2 + o((\omega - \omega_n)^2) \quad (13)$$

where $\mu_n = \left.\frac{\partial k}{\partial \omega}\right|_{\omega=\omega_n}$ and $\lambda_n = \left.\frac{\partial^2 k}{\partial \omega^2}\right|_{\omega=\omega_n}$.

Let $\delta\omega_n$ be half the bandwidth at the half-height. A development to the second order, with respect to $\delta\omega_n$, yields a simple expression of wave damping coefficient:

$$\alpha_n^2 = \alpha^2(\omega_n) = \mu_n^2 \delta\omega_n^2 \quad (14)$$

EXPERIMENTAL HANDLING

Experimental set-up

A 3.019-m-long and 40-mm-diameter aluminium bar was considered. One strain gauge was cemented at $x = -1.505\,m$. Fives tests were carried out using a 15-cm-long aluminium striker having the same section as the bar (see Table 1). The measured strain in test 1 and its spectrum are shown in Figure 2. Signals were sampled with a frequency $f_s = 100kHz$. Fast Fourier Transforms (FFT) were performed on *Matlab 6.0* using 131070-point long signals.

Figure 1 : Simplified sketch of the experimental set-up

TABLE 1
VELOCITY OF THE STRIKER IN THE DIFFERENT TESTS

Tests	velocity of the striker (m/s)	Tests	velocity of the striker (m/s)
Test 1	15,62	Test 3	16,67
Test 2	16,17	Test 4	15,84
		Test 5	15,38

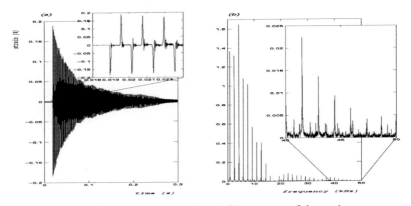

Figure 2 : (a) recorded strain of test 1 (b) spectrum of the strain

Longitudinal wave dispersion

For each test, the spectrum of the strain was computed using FFT. Firstly, we determined the local maxima, which correspond to the resonant frequencies. Considering Eqn. 9, the wave celerity is therefore assessed. Results of the five tests are superimposed in fig.3. The observed sensitivity to the noise of the measurements is very little. The wave number was interpolated over the interval 0-50kHz as a polynomial function using *Matlab 6.0*. Secondly, we determined the bandwidth of the resonance at the half-height of the peak. Knowing the derivative of the wave number with respect to the angular frequency, which was easily determined from the wave celerity, attenuation was calculated using Eqn. 14. The values were also interpolated over the same interval as a polynomial function. Results for damping obtained from the five tests show a stronger sensitivity to noise than those obtained for wave celerity : they are more scattered.

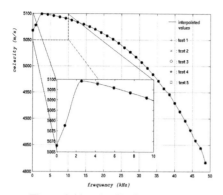
Figure 3 : longitudinal wave celerity

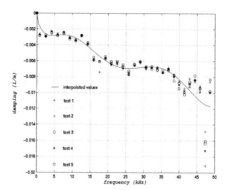
Figure 4 : longitudinal wave damping

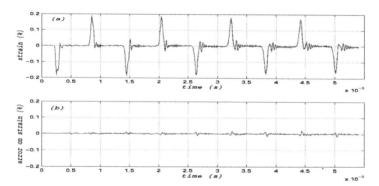

Figure 5 : *(a)* measured strain in test 1 (solid line) reconstructed strain using the dispersion relation (dashed line) *(b)* difference between the measured and the reconstructed strains

Wave shifting

In this section, the quality of the dispersion relation determined in the above section was evaluated. We considered the measured strain corresponding to test 1. The first incident wave (the first 0.06 ms, see Figure 5a) was extracted from the complete signal. The strain, after N round-trip in the bar, was then rebuilt, by transportation of the first incident wave, using the following equation :

$$\tilde{\varepsilon}^{(N)}(x,\omega) = \underbrace{\widetilde{\tilde{\varepsilon}_I^{(1)}(x,\omega)} \sum_{k=0}^{N-1} e^{-2i\xi(\omega)kL}}_{\text{Incident wave after N round-trip}} - \underbrace{\widetilde{\tilde{\varepsilon}_I^{(1)}(x,\omega)} \sum_{k=0}^{N-1} e^{-2i\xi(\omega)((k+1)L+x)}}_{\text{Reflected wave after N round-trip}} . \quad (15)$$

This equation is easily derived by developing the fraction in Eqn. 4. The measured and the rebuilt strains are then synchronised. The ascendant and descendant fronts start at the same time. The tail oscillations also appear as synchronised (Figures 5a and 5b). The phase difference between the two signals is very small. The wave celerity is therefore of a high accuracy. Damping is less accurate. The amplitude of the reconstructed strain is slightly greater than the amplitude of the measured one.

However, results are still satisfactory. The relative error is less than 4% after the first wave round-trip in the bar (i.e. after 6 m shifting) and less than 10% after three round-trips. This approach is then sufficient for classical SHPB applications where waves are shifted less than a length of the bar.

CONCLUSION

A new experimental method was developed to compute dispersion relation in finite rods. It relies on the spectral analysis of the measured strain at one point of the bar. We checked the validity of the method on an aluminium bar. The wave number is determined for a frequency range up to 50 kHz. The damping coefficient is slightly sensitive to noise, much more than the wave number. Despite this observation, the method shows a high accuracy when waves are shifted by less than 10m.

REFERENCES

Bacon, C. (1998) An experimental Method for Considering Dispersion and Attenuation in a Viscoelastic Hopkinson Bar. Exper. Mech. **38:4**, 242-249.

Blanc, R. H. (1971) Détermination de l'équation de comportement des corps viscoélastqiues linéaires par une méthode d'impulsion. Symposium franco-polonais, Problèmes de Rhéologie, Varsovie, IPPT Pan, W. NOWACKI Ed., 65-85.

Chree, C. (1889) The equations of an isotropic elastic solid in polar and cylindrical co-ordinates, their solutions and applications. Cambridge Phil. Soc. Trans., 14, 250-369.

Davies, R.M. (1948) A critical study of the Hopkinson pressure bar. Philosophical Transactions, A **240**, 375-457.

Follansbee, P. S. & Frantz, C. (1983) Wave Propagation in the Split Hopkinson Pressure Bar. J. Engng. Mater. Tech., **105**,61-66.

Gong, J.C. Malvern, L.E. & Jenkins, D. A. (1990) Dispersion investigation in the split Hopkinson pressure bar. J. Engng. Mater. Tech., **112**, 309-314.

Gorham, D. A. (1983) A numerical method for the correction of dispersion in pressure bar signals. J. Phys. E : Sci. Instrum., **16**, 477-179.

Gorham, D. A. & Wu, X. J. (1996) An empirical method for correcting dispersion in pressure bar measurements of impact stress. Meas. Sci. Technol., **7:9**, 1227-1233.

Hillström, L. Mossberg, M. & Lundberg, B. (2000) Identification of complex modulus from measured strains on an axially impacted bar using least squares. J. Sound Vibration, **230:3**, 689-707.

Lundberg, B. & Blanc, R. H. (1988) Determination of mechanical material properties from the two-point response of an impacted linearly viscoelastic rod specimen. J. Sound Vibration, **126:1**, 97-108.

Pochhammer, L. (1876) Uber die Fortpflanzungsgeschwindigkeiten kleiner Schwingung in einem ubegrenzten isotropen Kreiscylinder. J. für die Reine und Angewande Mathematik, **81**, 324-336.

Zhao, H. & Gary, G. (1995) A three dimensional analytical solution of the longitudinal wave propagation in an infinite linear viscoelastic cylindrical bar, application to experimental techniques. J. Mech. Phys. Solids, **43:8**, 1335-1348.

EVALUATION OF ACCURACY IN MEASUREMENT OF DYNAMIC LOAD BY USING LOAD SENSING BLOCK METHOD

S. Tanimura[1], T. Umeda[1], W. Zhu[2], and K. Mimura[1]

[1] Division of Mechanical Systems Engineering, Graduate School of Engineering, Osaka Prefecture University, 1-1, Gakuen-cho, Sakai, Osaka 599-8531, Japan
[2] Infineon Technologies Asia Pacific, Pte. Ltd., 168 Kallang Way, 349 253, Singapore

ABSTRACT

In recent years, increasing interest has been paid to the accurate measurement of the stress-strain relation at high strain rates in order to formulate precisely the constitutive equation of rate-sensitive materials. The Hopkinson pressure bar device is widely used for this purpose. However, the device should be long enough to prevent the immediate reflection of stress waves from the ends of the input and output bars. Another dynamic loading device is the newly developed Sensing Block (SB) system that is consisted of a small sensing projection and a relatively large mass block. In this study, the accuracy of the measured dynamic load by the SB system was examined numerically. The relation between the height of the sensing projection and the rising time of the applied load was clarified in order to satisfy certain criteria regarding the accuracy of each model of varying height and constant height/diameter ratio. The results ensure the adequate accuracy of the measured dynamic load as determined by the SB system.

KEYWORDS

Sensing block method, Accuracy of measurement, Strain rate dependency, Material testing, Numerical analysis, Structural analysis

INTRODUCTION

The behavior of inelastic materials is greatly affected by the loading state, and the accurate formulation of strain rate dependency into the constitutive equation for rate-sensitive materials is indispensable to simulations of the dynamic behavior of materials or structures. In order to acquire fundamental data needed for the formulation of the precise constitutive equation regarding rate-sensitive materials, many kinds of experimental devices have been developed. In the strain rate range above 10^2 s^{-1}, the Hopkinson pressure bar method is widely used for this purpose. However, the device should be long enough to prevent the immediate reflection of stress waves from the ends of the input and output bars.

Another dynamic loading device is the newly developed Sensing Block (SB) system that consists of a small sensing projection and a relatively large mass block. The important advantages of the SB system are its compact structure, negligible reflection disturbance, and long measuring time [Mimura et al. (1996a, 1996b); Tanimura et al. (1998)]. In the following section, the effects of the sensing projection size and the rising time of the applied load on the accuracy of the measured dynamic load when using the SB system were examined numerically.

SENSING BLOCK (SB) SYSTEM

The apparatus consists of a small sensing projection and a base block of a relatively large mass in comparison with the sensing projection. The schematic illustrations of SB and the magnification of its sensing projection are shown in Figures 1(a) and (b). The stress wave which travels from a specimen to the sensing projection creates an almost uniform stress state around the location of the strain gauge after some reflections between the upper surface of the sensing projection and the boundary surface with the base block. On the other hand, the stress wave which travels from the sensing projection to the base block is almost completely dissipated by expansion with a spherical wave front and by the high repetition of an oblique reflection on the base block's free surface. The stress wave reflected at the bottom surface of base block returns into the sensing projection with vanishing magnitude, and hardly affects the sensing projection's stress state. Therefore, it is possible to measure a dynamic load without the effect of the interference of the reflected stress wave for a relatively long time.

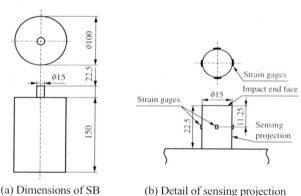

(a) Dimensions of SB   (b) Detail of sensing projection
Figure1: Example of configuration of SB (unit: mm)

MODELS OF ANALYSES

In order to simulate the transient dynamic response of SB, an explicit, nonlinear, finite-element code was used. In this code, the explicit central difference method is used to integrate the equation of motion in time, and the Updated-Lagrangian method is used to make a finite-element model.

Effect of Sensing Projection Size (Case A)

The finite-element models of sensing projections of various sizes were made to evaluate the effect of sensing projection size on the accuracy of measured dynamic load. Each sensing projection is considered an isotropic elastic material, and its elastic properties are steel-like, as shown in Table 1. In the following analyses, the sensing projection was modeled as a cylinder, and the ratio of its height h to its diameter d was fixed at 1.5, which was close to the value already successfully adopted for one of the prototypical apparatuses. This is probably an appropriate first step in determining the SB system's accuracy in estimating the dynamic load. A model of the sensing projection is shown in Figure 2, and the boundary conditions are shown in Figure 3. The maximum value of input pressure A_p was fixed at 100 MPa.

TABLE 1
MATERIAL PROPERTIES OF SB

Density ρ	Young's modulus E	Poisson's ratio v
7.89×10^3 kg/m^3	208 GPa	0.28

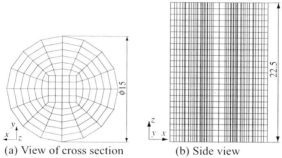

(a) View of cross section (b) Side view
Figure 2: Example model of sensing projection (Case A, unit: mm)

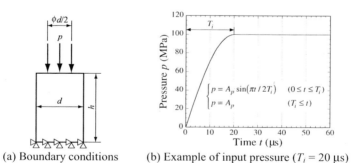

(a) Boundary conditions (b) Example of input pressure ($T_i = 20$ μs)
Figure 3: Boundary conditions and input pressure (Case A)

Effect of Damping Mainly Due to Viscoplastic Material (Case B)

Complete finite-element models, which included a SB, a hammer, and a specimen of viscoplastic material, were also made to take into account the damping effect due mainly to the viscosity of the specimen material. The material properties in Table 1 were used for the base block and hammer, as well as for the sensing projection. Regarding the model of the specimen material, the Johnson-Cook model without thermal coupling was adopted to give the dynamic yield stress σ_y as follows:

$$\sigma_y = (\sigma_0 + B\varepsilon^n)\left(1 + C\ln\frac{\dot{\varepsilon}}{\dot{\varepsilon}_0}\right), \tag{1}$$

where σ_0 is the initial quasi-static yield stress at the room temperature, and ε, $\dot{\varepsilon}$, and $\dot{\varepsilon}_0$ (= 1 s^{-1}) are strain, strain rate, and its standard value, respectively. B and n are work-hardening parameters, and C is the parameter concerning strain-rate dependency. The specimen material was assumed to be 1006 steel as shown in Table 2 [Meyers (1994)]. (Concerning ρ, E, and ν, the values in Table 1 were used.) All the parts were modeled as cylinders whose ratios of height to diameter were set to be 1.5. A model of case B is shown in Figure 4(a), and its boundary conditions are shown in Figure 4(b). The sizes of the base block and hammer were fixed, and the ratio of the diameter of the specimen to that of the sensing projection was also fixed at 1.5. The base block and the sensing projection was connected rigidly at the contact surface between them, and the contact surface between the sensing projection and the specimen and that between the specimen and the hammer were evaluated using friction. At the bottom surface of the base block, the node positions were fixed. In order to simulate the impact load, the velocity υ was applied to the hammer, and its rising time T_υ was assumed to be 10 μs as defined in Figure 4(b). A_υ is the magnitude of the maximum velocity. In each case, the total number of elements is identical (2688 or 18432) in all the models to simplify their manufacturing process.

TABLE 2
MATERIAL PROPERTIES OF SPECIMEN (1006 STEEL)

σ_0	B	n	C
350 MPa	275 MPa	0.36	0.022

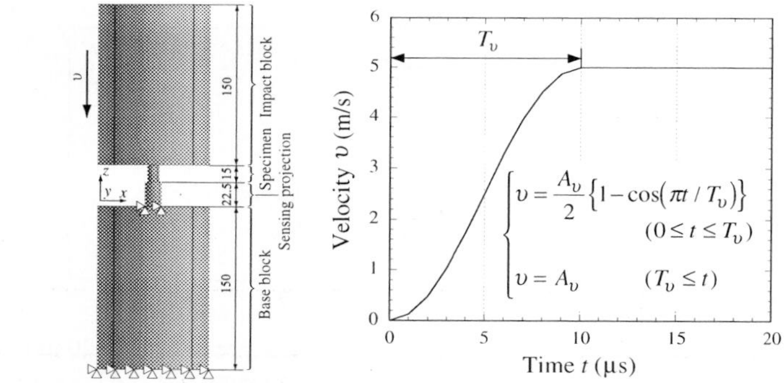

(a) Side view and boundary conditions (h/d = 1.5, unit: mm)   (b) Input velocity (A_υ = 5 m/s, T_υ = 10 μs)

Figure 4: Example model of SB (Case B)

RESULTS OF ANALYSES

In order to evaluate the accuracy of the measured load, the difference δ_σ and its maximum value δ_{max} between the input stress and the calibrated output stress are defined as follows:

$$\delta_\sigma(t) = (\sigma_g(t) - \sigma_i(t))/\sigma_i(t), \tag{2}$$

$$\delta_{max} = \max_{t_0 \le t \le t_1} |\delta_\sigma(t)|, \qquad (3)$$

where σ_i is the mean value of the input stress applied to the sensing projection, and σ_g is the calibrated value of the mean value of the output stress in the range $-0.25h \le z \le 0.25h$. t_0 and t_1 are the starting and ending times for the evaluation of δ_{max}.

Results of Case A

An example of the stress response curve is shown in Figure 5(a), and the resultant relation between h and T_i is shown in Figure 5(b). In case A, δ_{max} is evaluated from the time the input pressure reaches its maximum value A_p (= 100 MPa) to when the strain reaches 0.1. σ_g shows a transient vibration with no damping in Figure 5(a), and the period of its vibration corresponds to the 4th natural period T_n (see Table 3). For example, in Figure 5(b), the curves δ_{max} = 3% and 10% are drawn. The former shows that T_i should be longer than 170 μs to keep δ_{max} under 3 % in the case of the practical model h = 22.5 mm.

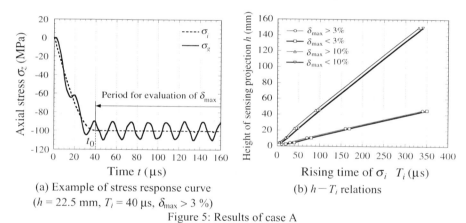

(a) Example of stress response curve
(h = 22.5 mm, T_i = 40 μs, δ_{max} > 3 %)

(b) $h - T_i$ relations

Figure 5: Results of case A

TABLE 3
RESULTS OF EIGENVALUE ANALYSIS TO THE 4TH NATURAL FREQUENCY (h = 22.5 mm)

Frequency number	Frequency (kHz)	Period (μs)	Mode
1	17.5	57.2	Bending around y-axis
2	17.5	57.2	Bending around x-axis
3	34.9	28.6	Torsion around z-axis
4	57.7	17.3(= T_n)	Stretching in z-direction

Results of Case B

An example of the stress response curve is also shown in Figure 6(a), and the resultant relation between h and T_i is shown in Figure 6(b). In case B, δ_{max} is evaluated after the first cycle of the oscillation in the range where the stress reaches the yield stress of the specimen material and the strain is smaller than 0.1. In Figure 6(c), the curve δ_{max} = 3 % is drawn as an example and shows that T_i should be longer than 25 μs in order to keep δ_{max} under 3 % in the case of the practical model h = 22.5 mm.

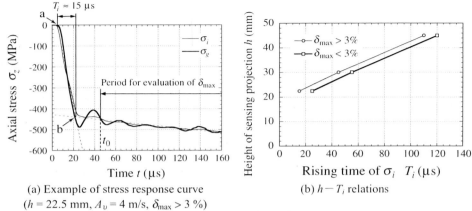

(a) Example of stress response curve
($h = 22.5$ mm, $A_v = 4$ m/s, $\delta_{max} > 3$ %)

(b) $h - T_i$ relations

Figure 6: Results of case B

CONCLUSIONS

The results of the analyses using the sensing projection model loaded with a constant pressure show that the rising time of the applied pressure should be longer than 170 μs to keep δ_{max} under 3 % in the case of a practical sensing projection whose height is 22.5 mm. However, the results of the analyses by using the complete model, which included a SB, a hammer, and a specimen of viscoplastic material, show that the rising time of the applied stress should be longer than 25 μs to keep δ_{max} under 3 % (except for the first vibration cycle after the stress reaches the yield stress of the specimen material) in the case of the practical model.

Acknowledgements

This work was supported by the Special Coordination Funds for Promoting Science and Technology (SCF), the Science and Technology Agency (STA), Japan, as one of the studies of the new research project "Enhancement of Earthquake Performance of Infrastructures Based on Investigation into Fracturing Process". This work was also supported by the Ministry of Education, Culture, Sports, Science and Technology, Japan through Grand-in-Aid for Scientific Research (B)(2)(10450050). These financial supports are gratefully acknowledged.

References

Meyers M.A. (1994). *Dynamic Behavior of Materials*, John Wiley & Sons, Inc., New York, USA.
Mimura K., Hirata S., Chuman Y., and Tanimura S. (1996a). Development of Dynamic Loading Device with Stress Sensing Block and Its Experimental Examination. *J. Soc. Mat. Sci., Japan.* **45**:8, pp.939-944 (in Japanese).
Mimura K., Hirata S., Chuman Y., and Tanimura S. (1996b). Development of a Compact Dynamic Loading Device with a Stress Sensing Block and Its Application to the Evaluation of Stress-Strain Relations at High Rates of Strain. *Transactions of Japan Soc. Mech. Eng.* **62**:603A, pp.2609-2614 (in Japanese).
Tanimura S., Mimura K. and Takada S. (1998), Development of a Dynamic Testing Apparatus with a Sensing Block and Its Applications, Proceedings of the 1998 Annual Meeting of JSME/MMD, No.98-5, **5**, pp.303-304 (in Japanese).

DYNAMIC SIMPLE SHEAR OF SHEETS AT HIGH STRAIN RATES

W. K. Nowacki

Institute of Fundamental Technological Research, Polish Academy of Sciences
Świętokrzyska 21, 00-049 Warsaw, Poland

ABSTRACT

The mechanical characteristics of thin metal sheets using in thin-walled constructions, such as bodies and energy absorbers of cars, buses, elements of air-planes, etc., subjected to dynamic impact depend on the metallurgical composition as well as on the production methods. It is indispensable to collect experimental data concerning this specific form of construction material. Tests in simple shear are a very important part for the experimental investigation of the constitutive relations. Such test is supplementary to other tests performed in tension, torsion as well as in compression. Simple shear test is particularly attractive, since the application of this loading path can assure large strains without occurrence of plastic instability. In this paper the results of dynamic plane simple shear are discussed. Use is made of a new double shear device in which loading and displacement are controlled in compression. The role of the special device is to transform the compression into a simple shear. The temperature field due to plastic shear is measured using a thermovision set. The relevant initial-boundary-value problem of the simple shear is formulated taking into account the case of finite deformations. The rate-independent constitutive relations for the coupled thermo-plasticity are used. The analytic solution is compared with the experimental data.

KEYWORDS

dynamic simple shear, sheet metal, high strain rate, thermo-plastic behaviour, numerical simulation

INTRODUCTION

In the static investigations of the thin sheets we can distinguish two approaches, based on different principles of the shear device. The first type of shear device of the one shear zone only, proposed by G'Sell et al (1983), was been performed for the investigation of the thin polymer sheets in the conditions of statical shear. The principle of this construction has been adopted for the testing of the steel sheets, cyclic shear of metallic sheets and for the shape memory alloy. In the second type of device for the static tests, Yoskida & Myauchi (1978), it is apply the „double shear" specimen. It is the specimen with two zones of shear. The fundamental attribute of this device is their symmetry with respect to the direction of loading. Both the above experimental techniques have their advantage and the inconvenience. In the first method, proposed by G'Sell, one can obtain the uniform zone of deformation field. First of all it is possible in this method to select one direction in which we can examined the behaviour of material. The symmetry of loading my needs the special very sophisticated carriages. The second method, proposed by Myauchi, is the simplest in use. In this method the rigorous conditions of symmetry may be executed. The exact fixation in grips must be assured. The inconvenience of this method is the fact that the homogeneity of the deformation field is not very good, comparing with the first method. The effects of free bounds of the specimen have the influence for the field of deformations and stresses.

In the case of dynamic loading, using the Kolsky bar, also called a Split Hopkinson Pressure Bar (SHPB) the scheme of the shearing device (see Figure 1), proposed by Myauchi, is more convenient for this kind of test. Regarding the miniature of the shear device, the scheme with the horizontal and vertical carriages, proposed by G'Sell, is practically unfeasible. A new shear device was used by Gary and Nowacki (1994) to perform tests under high strain rates on specimens having the form of slabs such as metal sheets. The loading and the displacements of this device are controlled by a Split Hopkinson Pressure Bar acting in compression. The role of the special device is to transform the compression into a plane shear (as in the Myauchi system). The dynamic simple shear tests so obtained is the only known method allowing to obtain a very good homogeneity of the remnant strain field over the total length of the specimen, without the localization of deformations as in the case of torsion of thin-walled tubes.

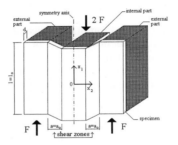

Figure 1: Scheme of shear testing – two shear zones

The relevant initial-boundary-value problem of the simple shear is briefly recalled, taking into account the case of finite deformations. The rate-independent constitutive relations are implemented in the case of adiabatic processes and with combined kinematic-isotropic hardening at moderate pressures. The analytical solution is compared with the experimental data.

EXPERIMENTAL TECHNIQUES

The shear device for investigations at high strain rates consists of two coaxial cylindrical parts, namely the external tubular part and the internal massive part. Both cylinders are divided into two symmetrical parts, and between them the sheet in testing is fixed. Two bands of the specimen between the internal and external parts of the device are in plane shear when these cylinders move axially one toward the other. Each band before test is rectangular and becomes very near parallelogram having the constant length and the constant height. The greatest double shear specimen height allowed by the shear device is $l_o = 30$ mm. The width value a_o is chosen with a view to satisfying two requirements: first the ratio a_o/d of the width to the thickness has to be small enough to avoid the buckling of the sheared zone ($a_o/d < 10$) and secondly the ratio a_o/l_o of the width to the height has to be sufficiently small to minimize the error due to the non homogeneity of the shear stress and strain at the two ends of the sample ($a_o/l_o<1/10$). In our case, the material sheet has a nominal thickness of 0.8 mm. The specimens can have different thickness. In our experimental investigations there is specimens made of the steel ES (French Standards), with chemical composition (wt%): C – 0.02%, Mn – 0.2%, S – 0.01%, P – 0.006%, Al – 0.04%. We take in our tests the ratio $a_o/l_o = 1/10$, $a_o/d = 3.75$, and we suppose that the results of test are good for both static and dynamic cases. It is important to mentioned that we use our shear device in the large range of strain rates: in the case of quasistatic tests from 10^{-3} s^{-1} to 10^4 s^{-1} in the case of dynamic tests at high strain rates. Adapting the same shear device in all experiments we can obtain certitude of the reliability of our experimental results.

The device with double shear specimen is placed between two bars of the SHPB. The mechanical impedance of the shear device and the SHPB must be the same to avoid the noise in the interface signal. The impulse is created by the third projectile bar: the usual compression technique. The impact velocity is measured by the set-up consisting of three sources of light and three independent photodiodes. The time interval during the crossing of a projectile by two rays of light is recorded by the time counter. We have to register the input, the transmitted and the reflected impulse: ε_i, ε_t and ε_r. Impulses are registered at the strain resistance gauges.

The specimen is deformed not only between the grips, in the gauge section. The part under the grips is partially deformed too. The transversal strike lines, marked on the specimen before the test, becomes after deformation the curve with the strike sections. Relative displacement of external and internal parts of shear device Δl is a sum of two terms: $\Delta l = \Delta l_g + 2\, \Delta l_s$, where Δl_g corresponds to the deformation in the simple shear defined as $\gamma = \Delta l_g/a_o$, and Δl_s correspond to the value of sliding under the grips. The mean shear stress is defined by $\sigma_{12} = F/2\, S_E$ ($S_E = l_o\, d = $ const). The estimations of the sliding value under the grips Δl_s can be obtained in the quasi-static test of loading-unloading - cf. Klepaczko et al (1999). We obtain directly the hardening function $\sigma_{12}(\gamma)$ of the examined specimen in simple plane shear. Using this method we can determined the surface $f(\sigma_{12}, \gamma, \dot\gamma) = 0$ if we assume that the time of transition of waves by the specimen is very small comparing with the total time of loading by the first pulse. The result of multiple reflections of loading waves in the specimen is the uniform distribution of strain and stresses in the whole gauge section.

The highest strain rate in the specimen can be obtained using only one bar of the SHPB system. We use the transmitted bar only and the shearing device is placed in the front of this bar. This technique, based on the principle given by Dharan and Hauser (1970), was described previously by Klepaczko (1994). In this new experimental technique the flat projectile strikes directly on the double shear specimen placed in the device on the front of transmitting bar, as shown in Figure 2. In this technique, the transmitting bar has the form of a tube. We have to register only the transmitted impulse ε_t and the velocity of the projectile. The set-up with three light axes makes it possible to determine the acceleration/deceleration of the projectile just before impact and, thus, an exact value of v_o can be determined.

The axial displacement Δl imposed on the double shear specimen by a projectile impact can be found as follows: $\Delta l(t) = \delta_A(t) - \delta_B(t)$, where $\delta_A(t)$ is the axial displacement of the internal part of the shearing device with respect to the external part, $\delta_B(t)$ is the displacement of the Hopkinson tube, given by

$$\delta_A(t) = v_o t \quad \text{and} \quad \delta_B(t) = c_o \int_0^t \varepsilon_t(\tau)d\tau \qquad (1)$$

where v_o is impact velocity of the projectile, c_o is the elastic wave speed in the Hopkinson tube and $\varepsilon_t(t)$ is the transmitted signal in the Hopkinson tube, measured by the strain gauge T_1. In experiment described in the paper Klepaczko et al (1999) the axial displacement $\delta_A(t)$ of the central part of the double shear specimen is measured as a function of time by an optical extensometer, acting as a contact-less displacement gauge.

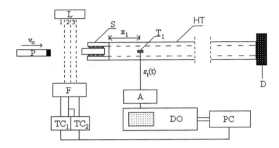

Figure 2: Configuration of experimental set-up for direct impact. S – shear device with a double shear specimen, HT - Hopkinson tube (transmitter bar), P - projectile, D - damper, L - laser, F - photodiodes, TC - time counter, T_1 - strain gauge, A - amplifier, DO - digital oscilloscope, PC - computer.

The mean shear strain and the mean strain rate can be found as

$$\bar{\gamma}(t) = \frac{1}{a_o}\left[v_o t - c_o\int_0^t \varepsilon_t(\tau)d\tau\right], \qquad \dot{\bar{\gamma}}(t) = \frac{1}{a_o}[v_o - c_o\varepsilon_t(t)]. \tag{2}$$

The shear stress can be expressed as

$$\bar{\sigma}_{12}(t) = \frac{S_T}{2S_E}E_o\varepsilon_t(t), \tag{3}$$

where E_o is the Young moduli of the tube, $S_T = \pi(d_2^2 - d_1^2)/4$, d_1 and d_2 internal and external diameter of the tube respectively. Thus, the shear stress in the double shear specimen is proportional to the current signal of the transmitted longitudinal wave ε_t.

The experiments made on the ES steel at low and moderate strain rate (quasistatic test in compression and dynamic compression test using SHPB system) are performed by Gary and Nowacki (1994). These results are completed by investigations in dynamic simple shear in the SHPB system and in the case of direct impact technique. The diagrams in the Figure 3 present the variation of the maximal stress versus logarithmic rate of deformation, for different kinds of experiments mentioned above.

MEASUREMENT OF TEMPERATURE DURING DYNAMIC SIMPLE SHEAR

Deformation processes always modify the temperature field of materials. The temperature of the surface of tested specimen is measured on the basis of infrared radiation detection. The measuring system (thermovision camera) has no contact with the specimen and this is an important advantage, particularly in dynamic tests. However, the dynamic shear deformation at high strain rates requires satisfying the conditions, which make the temperature measurements complicated. In the case of dynamic shearing the time of deformation process is very short, of the order of 100-500 µs. The specimen is deformed for a period, which is too short to enable the accompanying increase of temperature to be observed as a thermovision image. We can measure only the mean temperature of two shear zones of the specimen in the minimum time period of 10ms. The mean maximal increase of the temperature of the specimen of ES steel, subjected to simple shear at strain rate 1.05×10^3 s^{-1} was 60 K. The specimen's dimensions were: length $l_o = 20$ mm, width $a_o = 3$ mm and thickness $d = 0.8$ mm. The corresponding maximum of the equivalent deformation of the specimen was of the order of 75%.

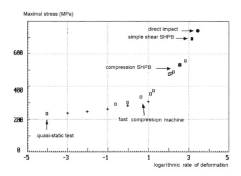

Figure 3: Comparison of the simple shear test with the simple compression test for ES steel sheets:
    □ - quasi-static and dynamic test in compression, ■ - simple shear: Gary and Nowacki (1994)
    + - shear test in hydraulic machine
    ● - direct impact: Klepaczko, Nguyen and Nowacki (1999)

The method of temperature measurements on the basis of infrared radiation detection was modified in order to satisfy the conditions of the dynamic simple shear. The temperature variation can be determined by analysing the amplified detector signal produced by the infrared radiation emitted from the specimen surface points which lie on the line parallel to the axis x^1, i.e. by analysing the signal of the termovision image line corresponding to the line in the middle of one of the shear zones in the specimen. Stopping one of the rotating prisms in the thermovision camera we can decrease the recording time of the image line. The registering time of this line was 625 µs. The thermal sensitivity of the camera is less than 0.1K. When one of the prisms is stopped then the temperature stabilization system in the thermovision camera does not work. For that reason an indicator of constant temperature was placed in the observation space of the camera. The temperature of this indicator needs to be known. This temperature is a reference level. It is possible to take records of two or more time sequences of radiation power at the shear zone by means of the camera sweeping from the left to the right side. Successive sequences start every 565 µs. In Oliferuk et al (2000) the experiments were carried out in order to measure the temperature of the specimen of carbon steel in the case of simple shear at the strain rate of the order of 10^3 s^{-1}. We can observe that the maximum heating of the shear zone occurs in the fourth time sequences.

COMPARISON OF SIMPLE SHEAR TEST WITH TENSION TEST

More recently, a new tension device was used by us to perform tests at high strain rates on specimens made of metal sheets. The loading and the displacements of this device are controlled by SHPB acting in compression. The role of the special device is to transform the compression into tension (inversion of motion). In the Figure 4 the scheme of the device for high strain rate tension test is shown.

The comparison of the result obtained in the simple shear test and tension test for ES steel at strain rate 100 s^{-1} is shown in the Figure 5a. We assume the Huber-Mises-Hencky yield criterion: $\gamma = \sqrt{3}\varepsilon_{11}$ and

$\sigma_{12} = \sqrt{3}\sigma_{11}$. In fact, the influence of stress components σ_{11} and σ_{22} must be taking account. The stress intensity is defined as: $\sigma_i = [\sigma_{11}^2 + \sigma_{22}^2 + 3\sigma_{21}^2 - \sigma_{11}\sigma_{22}]^{1/2}$.

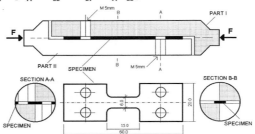

Figure 4: Device for high strain rate tension tests

We can observe that the hardening of the material is similar for different values of strain rate (the same effect is observed in the case of Tresca yield condition). In the case of Huber-Mises-Hencky (HMH) yield condition an important difference in stress increment is observed. The existence of a difference between simple shear tests and tensile tests after application of the HMH criterion was also observed by various authors. They explain this difference by the different systems of slip activated according to the type of loading and interaction of displacements in long and short distance. Also, for the metal like ES steel, the HMH yield criterion is not the best. In addition, in Figure 5b is shown a comparison of our results (IPPT-tests) with results obtained independently by Rusinek and Klepaczko (2001) – (LPMM-tests), where a different specimen geometry was used, with the shear zone twice shorter in the LPMM-case. Direct results are different. But, the results of LPMM, after correction by finite element method (LPMM-analysis) are in perfect agreement with our direct results, taking account the three-dimensional state of stress and the effect of sliding under grips of the shear device.

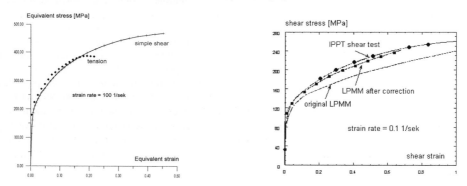

Figure 5: (a) Comparison of the shear and tension test, (b) Comparison of IPPT and LPMM tests

THEORETICAL SIMPLE SHEAR ANALISYS

The simple shear in the direction e_1 of the coordinate system (e_1, e_2) is defined by the relations:

$$u_1 = \gamma(t) x_2, \quad u_2 = u_3 = 0, \quad v_1 = \dot{\gamma} x_2, \quad v_2 = v_3 = 0. \tag{4}$$

The Cauchy stress tensor $\boldsymbol{\sigma}$ and the back stress tensor Π have the non-zero components: σ_{11}, σ_{12}, σ_{22} and Π_{11}, Π_{12}, Π_{22}. From the velocity field $\mathbf{v}$, the rate of deformation $\mathbf{D}$ and the material spin ω, can be determined as:

$$\mathbf{D} = \frac{\dot{\gamma}}{2}\begin{pmatrix} 0 & 1 \\ 1 & 0 \end{pmatrix}, \qquad \omega = \frac{\dot{\gamma}}{2}\begin{pmatrix} 0 & 1 \\ -1 & 0 \end{pmatrix}. \qquad (5)$$

Using the constitutive relations for adiabatic process for rate-independent materials with combined isotropic-kinematic hardening at moderate pressures, when the thermal expansion, the heat of elastic deformation and the heat of internal rearrangement are neglected, we obtain the following constitutive equations

$$\check{\mathbf{T}} = \beta \mathbf{L} \mathbf{D} - \frac{3j\mu\beta \, \mathbf{D}\cdot(\overline{\mathbf{T}}-\Pi)}{\sigma_Y^2 \, \mathcal{H}}\left[(\overline{\mathbf{T}}-\Pi)+\mathbf{P}\right], \qquad \rho_o c_v \dot{\vartheta} = (1-\pi)(\overline{\mathbf{T}}-\Pi)\cdot\mathbf{D}^p, \qquad (6)$$

where $j = \begin{cases} 1 & \text{if} \quad f=0 \quad \text{and} \quad \mathbf{D}\cdot(\overline{\mathbf{T}}-\Pi) \geq 0, \\ 0 & \text{if} \quad f=0 \quad \text{and} \quad \mathbf{D}\cdot(\overline{\mathbf{T}}-\Pi) < 0 \quad \text{or} \quad f<0, \end{cases}$

$\mathbf{T} = \beta\boldsymbol{\sigma}$, $\beta = \rho_o/\rho$ is the ratio of densities in the reference and actual configurations, $\overline{\mathbf{T}}$ is the deviatoric part of $\mathbf{T}$, $\check{\mathbf{T}}$ is the Zaremba-Jaumann rate, and tensor $\mathbf{P}$ is obtained by expressing the term $(\omega^p \mathbf{T} + \mathbf{T}\,\omega^p)$ as a function of $\mathbf{D}^p$, where ω^p and $\mathbf{D}^p$ are plastic part of ω and $\mathbf{D}$, respectively. $\mathcal{H}$ is the hardening function, $\mathbf{L}$ is the fourth order tensor of elastic moduli, σ_Y is the yield stress in simple tension, μ is the Lamé constant and f is the HMH yield criterion – cf. Nguyen and Nowacki (1997).
The change in the temperature ϑ is described by Equation (6)$_2$, where c_v is specific heat at constant volume. The term on the right-hand side represents the rate of energy dissipation and, therefore, $\pi < 1$. For numerous metals π takes the value from 0.02 to 0.1. Under the initial conditions that for $\gamma = 0$, stresses $\sigma_{11} = \sigma_{12} = \Pi_{11} = \Pi_{12} = 0$, we have the analytical solution - Nguyen and Nowacki (1997).

NUMERICAL SIMULATIONS OF THE EXPERIMENT

The program of finite element method was used to the numerical simulations of the formulated problem of quasistatic and dynamic simple shear of thin sheets. We assume the initial and boundary conditions similar to those used in the experiment. In the finite element method the deformation of the mesh in time is determined. At the same time, the components of the stress tensor σ_{12} and $\sigma_{22} = -\sigma_{11}$, the stress intensity $\sigma_i = (3/2 \, s_{ij} \, s_{ij})^{1/2}$ and the equivalent strain $e_i = (2/3 \, \varepsilon_{ij}^p \, \varepsilon_{ij}^p)^{1/2}$ are determined. Results of numerical simulation in the specimen subjected to simple shear are shown on the Figure 6.

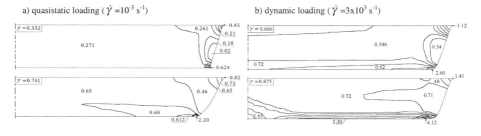

Figure 6: Numerical simulation. Equivalent deformation: quasistatic and dynamic loading.

The equivalent deformation field is shown for one half of the shear zone of the specimen, in view of the symmetry of deformation process with respect to the axis x_2. The results for quasistatic test are presented for different values of shear strain: $\gamma = 33.2\%$ and $\gamma = 74.1\%$. We observe the heterogeneity of the strain (and stress) fields at the free ends of the specimen, at the distance less than 6.6% of the total length when the strain is 74%, exactly as in the experiments. In the case of dynamic loading at high strain rates of the order of 3×10^3 s^{-1} we observe zones of strain localization. The beginning of the localization in the dynamic case occurs at lower nominal shear strains. A distinct shear band appears in the middle part of the zone with inclination 51° to the x_1 axis. In addition, an intense shear strain is observed in the specimen and the support areas of the shear device.

CONCLUSIONS

Considerable homogeneity of the permanent strain field at finite deformations over the total length of the specimens is observed in experiments of simple shear and in the results of simulation. The proposed method is the only known test providing, in the case of a thin sheet, homogeneous stress and strain fields in both the dynamic and static tests. Simple shear test is particularly attractive, since the application of this type of loading path can result in large strains without the occurrence of plastic instability.

Acknowledgments. This paper is supported by the Polish Committee for Scientific Research, KBN Project No. 7T07A011 16 on "Localisation of deformations in shear and tension of steel sheets".

REFERENCES

Dharan H.C.K., Hauser F.E. (1970) Determination of Stress-Strain Characteristics at Very High Strain Rates. *Exp. Mech.* **10**, 370-376.

Gary G. and Nowacki W.K. (1994). Essai de Cisaillement Plan Dynamique Appliquée a des Tôles Minces. *Journal de Physique* IV, **C8**, 65-70.

G'Sell C., Boni S. and Shivastava S. (1983). Application of the Plane Shear Test for Determination of the Plastic Behaviour of Solids Polymers at Large Strain. *J. of Mat. Sciences* **18**, 903-918

Klepaczko J.R. (1994). An Experimental Technique for Shear Testing at High and Very High Strain Rates. The Case of Mild Steel. *International Journal of Impact Engineering* **15**, 25 - 39.

Klepaczko J.R., Nguyen H.V. and Nowacki W.K. (1999). Quasi-static and Dynamic Shearing of Sheet Metals. *European Journal of Mechanics, A/Solids* **18**, 271-289.

Nguyen H.V., Nowacki W.K. (1997). Simple Shear of Metal Sheets at High Rates of Strain. *Archives of Mechanics* **49**, 369 - 384.

Oliferuk W., Kruszka L. and Nowacki W.K. (2000). Measurements of Temperature During Dynamic Shear Deformation of Carbon Steel. *Journal de Physique* 10, **Pr 9**, 243 - 248.

Yoshida K., Myauchi K. (1978). Experimental Studies of Mechanical Behaviour as Related to Sheet Metal Forming. In *Mechanics Sheet Metal Forming*, Plenum Press, New York, 19-49.

Rusinek A. and Klepaczko J.R. (2001). Shear Testing a Sheet Steel at Wide Range of Strain Rates and a Constitutive Relation with Strain Rate Dependence *Int. Journal of Plasticity* 17, 87-115.

STRENGTH AND FAILURE CHARACTERISATION AND BEHAVIOUR UNDER IMPACT LOADING

L.W. Meyer[1], L. Krüger[2] and T. Halle[3]

[1] Chair Materials and Impact Engineering
Chemnitz University of Technology, D-09107 Chemnitz, Germany
[2] Department of Mechanical and Aerospace Engineering,
University of California at San Diego, La Jolla, 92093-0411, CA, USA
[3] Institute of Materials and Impact Engineering
Chemnitz University of Technology, D-09107 Chemnitz, Germany

http://www.tu-chemnitz.de
lothar.meyer@wsk.tu-chemnitz.de

ABSTRACT

An overview is given of the present state of the art of high rate mechanical testing including mono, biaxial and triaxial loading states at Chemnitz. This includes strength, deformability and failure aspects of a tool steel for high speed cutting (HSC)- simulation at high strain rates and enhanced temperatures, a compression strength-rate dependence up to $\dot{\varepsilon} = 10^5$ s^{-1} and a presentation of a new constitutive equation describing adiabatic shear failure strain as a function of loading state, strain rate and temperature. Furthermore a new testing procedure is named to include the biaxial hydrostatic pressure influence on adiabatic shear failure initiation.

KEYWORDS

strain rate, temperature, dynamic yield stress, strength, tensile, compression, biaxial, compression-shear, hydrostatic pressure-shear, adiabatic shear failure, constitutive equation

INTRODUCTION

The properties of materials are related to the microstructure and the loading conditions. Apart from a temperature influence, loading conditions with respect to impact are determined mainly by the involved stress state and the deformation rate or strain rate.
Most technical production methods like cutting, forming, forging, stamping or impact processes like car crash or ballistics, take place under multiaxial and high strain rate conditions. From creep to high speed machining the strain rate varies from 10^{-10} s^{-1} to 10^2 for forging, up to 10^5 for cutting or stamping. Due to microstructural reasons combined with dislocation mechanics, the strength, deformability and

toughness changes with strain rate. Therefore, we may not and cannot conclude from the quasistatic material properties to the behaviour at high strain rates.

Therefore, it is a necessity, to explore the high rate material behaviour, not only monoaxial under compression loading, which is done world-wide with Hopkinson bars, but to include the wide range of biaxial or multiaxial loading states into the centre of scientific interest. As an example, how much the loading state controls the strength and failure behaviour, a pure high rate compression state is compared with the same level of equivalent stress, but splitted up into 90 % compression and 10 % shear loading, fig. 1.

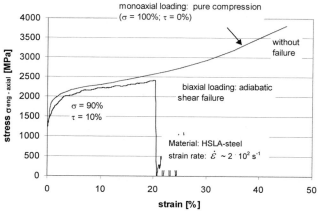

Figure 1: Comparison of different behaviour due to different loading conditions (monoaxial to biaxial)

Under pure monoaxial loading, the deformation reaches $\varphi > 1$, but with a slight transition to biaxial loading, an adiabatic shear failure changes the deformability and strength drastically, Meyer (1991), Meyer, Staskewitsch & Burblies (1994). To select proper materials for impact loading as well as to provide the simulation community with suitable constitutive equations, a centre of high rate material characterisation is build up at Technical University Chemnitz. A multitude of different loading conditions can be applied by a variety of special testing machines to explore the different material behaviour under extreme conditions of loading states and strain rates, fig. 2.

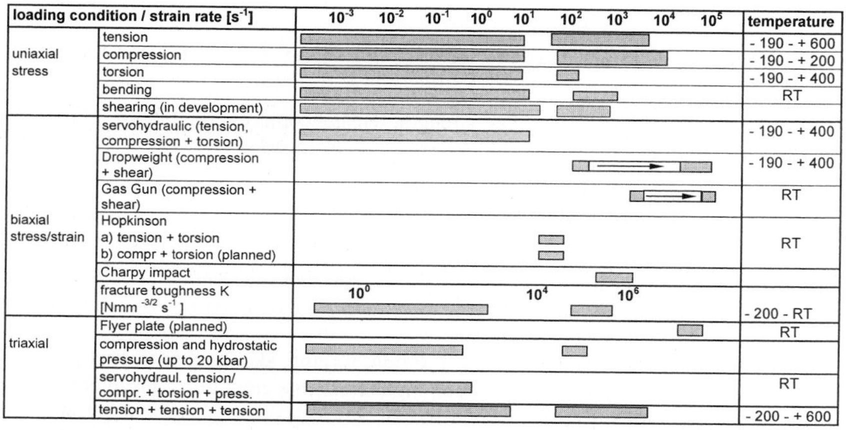

	loading condition / strain rate [s^{-1}]	strain rate range	temperature
uniaxial stress	tension		-190 - +600
	compression		-190 - +200
	torsion		-190 - +400
	bending		RT
	shearing (in development)		
biaxial stress/strain	servohydraulic (tension, compression + torsion)		-190 - +400
	Dropweight (compression + shear)		-190 - +400
	Gas Gun (compression + shear)		RT
	Hopkinson a) tension + torsion b) compr + torsion (planned)		RT
	Charpy impact fracture toughness K [Nmm$^{-3/2}$ s^{-1}]		-200 - RT
	Flyer plate (planned)		RT
triaxial	compression and hydrostatic pressure (up to 20 kbar)		
	servohydraul. tension/compr. + torsion + press.		RT
	tension + tension + tension		-200 - +600

Figure 2: Testing Facilities and Strain Rates at LWM, Chemnitz University of Technology (2001)

With small samples down to 1 mm diameter and 1 mm length, stress-strain-behaviour (without curve smoothing) which can be determined at very high strain rates from a miniatur-Hopkinson-set up, is reasonably undisturbed, fig. 3.

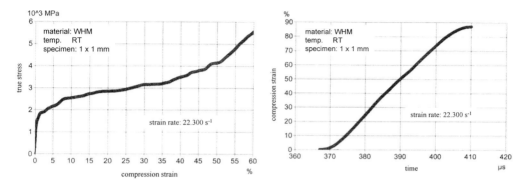

Figure 3: Stress-strain and strain time behaviour as example for testing at very high strain rates.

EXAMPLES OF TESTING AND RESULTS

Strength behaviour

With a servohydraulic machine and a rotating wheel set up, which is build up as a direct hopkinson bar, testing at $\dot{\varepsilon} = 3 \cdot 10^3$ s^{-1} and temperatures between RT and 600 °C is possible, Meyer, Halle & Abdel-Malek (2000), Meyer (2000), fig. 4.

Figure 4: High speed tension testing principle at elevated temperatures

The geometric area of elevated temperatures is short, therefore no corrections for different wave speeds are necessary, furthermore the strain is measured by a touchless electro-optical Zimmer-Extensiometer at the gage length. With the missing reflection interface between specimen and output bar, even spike-similar upper and lower yield stresses are determinable with very low disturbances.
The transfer of the results into Johnson-Cook types of constitutive equations yielded to differences of more than 150 MPa. Better agreements are reached with Armstrong-Zerilli types of equations, which include athermal, thermal activated, Ludwik and Hall-Petch terms. Because of no change in grain sizes by twins, the Hall-Petch term c_6 in equation (1) was set to be a constant:

$$\sigma = \Delta\sigma_G + c_1 \exp(-c_3 T + c_4 T \ln\dot{\varepsilon}) + c_5 \cdot \varepsilon^n + c_6 \qquad (1)$$

The constants are given in Tab. 1 (valid: $0 \le \varepsilon_p \le 9$ %; $10^{-4} \le \dot{\varepsilon} < 3 \cdot 10^3$; RT $\le$ T $\le$ 600 °C):

Table 1: Constants for Zerilli-Armstrong type of constitutive equation for tensile loading.

$\Delta\sigma'_G$	28,9101	c_4	0,00022
c_1	822,629	c_5	1199,042
c_3	0,00495	c_6	50,9102
n	0,380197		

A three-dimensional plot $R_{p0,2} = f(\varphi, \dot{\varepsilon})$ at different temperatures demonstrates, that the strain rate dependence at T = 600 °C reduces to very low amounts, fig. 5.

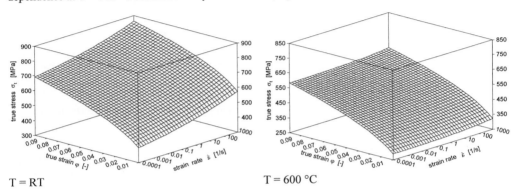

T = RT   T = 600 °C

Figure 5: Yield strength (0,2 %) (true stress) as a function of strain φ and strain rate $\dot{\varepsilon}$ at two temperatures of carbon steel C45E, normalised (ZA-model), Meyer & Halle (2000)

Compression testing at the same material proved, that a similar rate dependence with a monotonic increase of rate sensitivity exists, fig. 6.

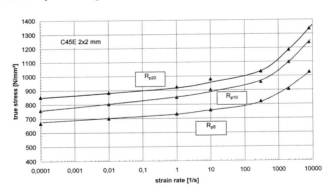

Figure 6: True compression flow stresses versus strain rate at RT of carbon steel C45E, normalised

Worth to notice is the <u>not</u> accelerated slope $d\sigma/d\dot{\varepsilon}$ above $\dot{\varepsilon} = 10^3$ s^{-1}, which indicates that no sign of a phonon drag is present at this steel and velocity.
The same is confirmed by results of an Titanium-alloy Ti-6Al-2Sn-2Zr-2Cr-2Mo-Si (Ti6-22-22S) of Krüger (2001) with compression hopkinson-results up to $\dot{\varepsilon} = 1,2 \cdot 10^4$ s^{-1} and additional flyer plate results Krüger, Meyer, Razorenov & Kanel (2000) at $\dot{\varepsilon} \approx 10^5$ s^{-1}, fig. 7.

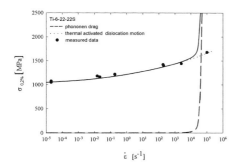 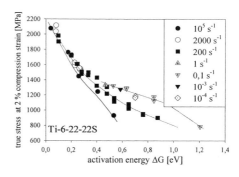

Figure 7: Measured yield stresses $\sigma_{0,2}$ under compression loading and description by thermal activated equations and/or dislocation drag, Krüger (2001)

Figure 8: True 2 % compression flow stress versus activation energy, Krüger (2001)

The used constitutive equation with reasonable constants of Burgahn, Vöhringer & Macherauch (1992) contains both, the thermal activated and the drag influence.
It is obvious, that the theory of phonon drag is not applicable here, all results indicate that only thermal activated flow is present, fig. 8. This can be concluded by the intersecting data points at different temperatures and velocities. This is a very important result, because it answers – for this alloy – the old question of the validity of the influence of phonon drag.

Adiabatic shear failure behaviour

Different loading states incorporate different failure behaviour. Beside the well known tensile stress initiated failure types of brittle cleavage or ductile void collapse, a third important type, the adiabatic shear failure determines many perforations. Only in some examples, f.e at precision stamping, exact cutting and exact punching, technical advantage is drawn by the reduced cutting energy. The effect of the self-concentrating shear band width, in high strength steels down to some µm, is known since the forties. But only in general , the influence of macro- or microbehaviour or of physical constants, fig. 9, can be described. Even today, details or quantitative dependencies of influencing parameters are rare.

One of the reasons for that is the lack of experimental data. To improve this situation, new testing techniques were established. Well known is Duffy´s torsion test (1991), which offers a relatively homogenous shear-stress state of a thin tube geometry. The hat shape test of Hartmann & Meyer (1981), which found recently more attention by Meyers et al (1994), Beatty et al (1992), Meyers et al (1992), Marquis (2001) and Clos, Schreppel & Veit (2000), forces the material to shear and the materials susceptibility to concentrate the shearing determines the instability or failure. This can be used nicely to study the microstructural changes in the developed shear zone. Another advantage of the hat shaped geometry, described in detail by Meyer & Krüger (2000a), is, that even materials, which are unwilling to build shear concentrations, are testable. The disadvantage is the non specified stress state in the shear zone, due to the stress concentrations at the edges and the non homogenous stress distribution across the shear zone.

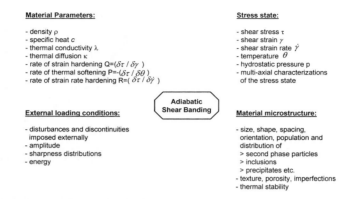

Figure 9: Parameters with influence on adiabatic shear banding

In order to overcome the limits of a relatively thin tube or the undefined stress state, Meyer (1991), Meyer (1994), Meyer, Staskewitsch & Burblies (1994).Meyer & Krüger (2000b) developed and proposed the shear/compression test with slightly inclined cylindrical compression specimen. Here a solid geometry is used and the stress state is much more clear, in principle like in a compression specimen, but with small additional shear components orthogonal to the main axial compression component.

Again, like in Duffy's torsion testing, a special stress state is offered to the material, and the properties of the materials determine the developing of the shear failure under this biaxial stress state. By choosing different inclinations of the compression specimen (up to 10°) the amount of biaxial shear component can be varied. For higher-strength materials the adiabatic shear failure is initiated earlier or later, fig. 10 or not at all, fig. 1.

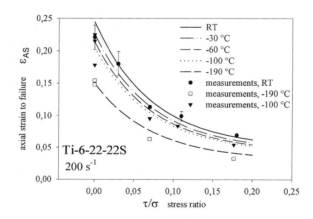

Figure 10: Comparison of measured failure strains with Meyer-Krüger failure equation (eq. 2)

This is depending on the susceptibility to adiabatic shear failure – under this certain stress state. Important and interesting is to notice, that even small changes of the stress state, just for a few percent, the failure behaviour can change drastically.

With the Meyer-Krüger-law, Krüger (2001), the axial strain to failure can be described by $\varepsilon_{AS} = f(\dot{\varepsilon}, T,$ stress state), eq. 2.

$$\varepsilon_{AS} = \left[A \left(\frac{\dot{\varepsilon}}{\dot{\varepsilon}_0} \right)^{n1} + B \exp\left(-C\left(\frac{\tau}{\sigma}\right)\right) \cdot \left(1 - D \ln\left(\frac{\dot{\varepsilon}}{\dot{\varepsilon}_0}\right)\right) \right] \cdot \left(1 + E\left(\frac{T}{T_s}\right)^{m1} \right) \quad (2)$$

Table 2: Constants of equation 2 for Ti-6-22-22S

Constants	A	B	C	D	E	n1	m1
	0,0139	0,08	13,61	0,104	8,0	0,0017	0,61

For Ti6-22-22S the equation (2) with the given constants describes well the experimental results, fig. 10. In some impact events f.e. in real penetrations the stress state is not only biaxial, but triaxial with a strong hydrostatic component.
To explore the material behaviour under this condition, the hat shaped geometry was converted into a symmetric form. The hydrostatic pressure is applied by a confining ring, which compresses the shear specimen orthogonal to the shear direction, up to 18 kbar. The alloy Ti6-22-22S exhibited improved shear deformability with high pressure components, whereas some other materials like tungsten alloy did not, fig. 11, Krüger (2001).

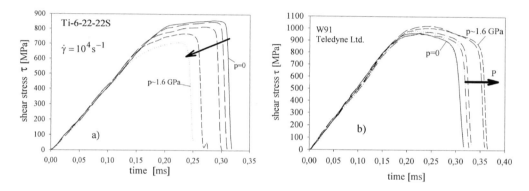

Figure 11: Influence of radial pressure on dynamic shear stress-time behaviour, ($\dot{\gamma} = 10^4 s^{-1}$)

SUMMARY

With new testing procedures and equipment very high strain rates and multiaxial loading states can be reached. Evidence is found, that the flow stress of a titanium alloy is influenced solely by thermal activated dislocation movements, even up to $\dot{\varepsilon} = 10^5 \, s^{-1}$.
The adiabatic shear failure behaviour is strongly dependent on the stress state, especially on the degree of biaxial loading.
With a new equation, which includes stress state, strain rate and temperature, the experimental measured failure behaviour can be modelled.
Additional hydrostatic pressures on shear zones change the deformability, depending on material microstructure.

REFERENCES

Beatty. J. et al (1992).Formation of Controlled Adiabatic Shear Bands in AISI 4340 High Strength Steel. Shock-Wave and High-Strain-Rate Phenomena in Materials, Dekker,. p. 645-656

Burgahn,F., Vöhringer O.and Macherauch,E (1992). Modelling of Flow Stress as a Function of Strain Rate and Temperature. Shock-Wave and High-Strain-Rate Phenomena in Materials, Dekker. p171-179

Clos,R. Schreppel,U and Veit,P (2000). Experimental Investigation of Adiabatic Shear Band Formation in Steels. Journal de Physique IV Volume 10, Pr9.Sept. 2000, p.257-262

Hartmann,K.H., Kunze H.D. and Meyer,L.W. (1981). Metallurgical Effects on Impact Loaded Materials. Shock-Wave and High-Strain-Rate Phenomena in Metals, Plenum Press New York. p. 325-337

Krüger (2001). Untersuchungen zum Festigkeits-, Verformungs- und Versagensverhalten der Legierung Ti-6-22-22S in Abhängigkeit von der Temperatur, der Dehngeschwindigkeit und dem Spannungszustand. PhD-Thesis TU Chemnitz.

Krüger,L. Meyer,L.W:, Razorenov,S and Kanel,G.I. (2000). Investigation of Dynamic Flow and Strength Properties of Ti-6-22-22S at Normal and Elevated Temperatures. Journal of Impact Engineering. (subm. 2000)

Marquis, F. (2001). Evolution of Microstructure and Strength during high-strain, high-strain-rate deformation of Tantalum and Tantalum Based Alloy. Shock-Wave and High-Strain-Rate Phenomena. Elsevier. p. 87-97.

Meyer, L.W. (1991). Adiabatic Shear Failure under Biaxial Dynamic Compression/Shear Loading. presentation at EUROMECH 282 Congress "Microscopic and Macroscopic Plastic Deformation Instabilities".Metz

Meyer, L.W. (1994). Failure under Biaxial Loading. EUROMAT 94, Conference Proceedings Vol. 2 eds. B. Vorsatz and E. Szöke. p. 377-384

Meyer,L.W. (2000). Charakterisierung des schlagdynamischen Werkstoffverhaltens. Tagungsband Werkstoffprüfung 2000 Bad Nauheim DVM. p. 63-70

Meyer, L.W.,Halle,T. (2000). Dynamische Prüfung von C45E und 40CrMnMo7 bei erhöhten Temperaturen und gleichzeitig erhöhten Dehngeschwindigkeiten unter Zugbeanspruchung. Tagungsband Werkstoffprüfung 2000 Bad Nauheim DVM. p. 349-358

Meyer,L.W., Halle T. Abdel-Malek,S. (2000). Tensile Testing. ASM Handbook No 8,Mechanical Testing and Evaluation, 10th edition, ASM International. p. 452-454

Meyer,L.W., Krüger,L. (2000a). Shear Testing with Hat specimen, ASM-Handbook no. 8, 10th Edition, Mechanical Testing and Evaluation, ASM International. p. 451-452

Meyer,L.W., Krüger,L. (2000b). Biaxial Compression Shear Testing and Failure Analysis. ASM-Handbook No. 8. Mechanical Testing and Evaluation ,ASM International p. 452-454

Meyer,L.W., Staskewitsch E., Burblies A. (1994). Adiabatic Shear Failure under Biaxial Dynamic Compression/Shear Loading. Mechanics of Materials 17. Elsevier. p. 203-214

Meyers M.A. et al (1992). High Strain, High-Strain-Rate Deformation of Copper. Shock-Wave and High-Strain-Rate Phenomena in Materials, Dekker, p. 529-542.

Meyers M.A. et al (1994). Evolution of Microstructure and Shear-Band Formation in α-hcp Titanium-Alloy. Mechanics of Materials 17. Elsevier. p. 175-193

TEMPERATURE AND STRAIN RATE EFFECTS ON THE TENSILE STRENGTH OF 6061 ALUMINUM ALLOY

K.Ogawa

Division of Aeronautics and Astronautics, Graduate school of Engineering,
Kyoto University, Yoshida Honmachi, Sakyo-Ku, Kyoto, Japan

ABSTRACT

Tensile strength of 6061-T6 aluminum alloy is presently investigated in the wide ranges of temperature from 77K to 473K and strain rate from 10^{-4}/s to 10^3/s. Temperature and strain rate effects on the stress strain relations are clarified and behaviors are discussed with the precipitation hardening and the dislocation intersection mechanism. At around the room temperature, strain rate effect on the flow stress is only minimal in the low strain rate range, while it increases at high strain rates beyond 10^3/s. At low temperatures, the flow stress depends on the strain rate in the whole range of strain rate investigated. The temperature and strain rate effect on the flow stress is strongly strain dependent, and can be understood in connection with the dislocation intersection mechanism. Viscous drag does not play an important role in the high strain rate deformation investigated, and the thermally activated process concept well explains the experimental results.

KEYWORDS

Split Hopkinson bar, High strain rate, Aluminum alloy, Temperature dependence, Thermal activation, Strain rate effect, Stress-strain relation, Tension test, Temperature-strain rate parameter

INTRODUCTION

Aluminum alloy is widely used for aircrafts because of its high strength-weight ratio and excellent corrosion resistance, and will be applied to ground vehicles for saving energy demands. Therefore, the dynamic deformation characterization is increasingly needed from the point of impact crashworthiness. At low strain rates it is usually confirmed that the thermally activated process controls plastic deformation, but it is still not evident how far it can be extended. It is not always reasonable to apply the viscous drag of high velocity dislocations in explanation of significant increase of strength at high strain rates. Clifton(1983) concluded that the viscous drag could not be expected at strain rates less than 10^4/s and Armstrong and Zerilli(1988) successfully correlated their Taylor test results on Armco iron with thermally activated flow. Follansbee(1988) pointed out that temperature and strain rate history effect often veiled deformation behaviors of soft materials controlled by thermal activation. Holt et al(1967) suggested that the strengthening effect of alloying is almost entirely due to an increase in the athermal stress component and the rate sensitivity would be relatively smaller for a strong alloy. Several authors have clarified the relatively insignificant strain rate effect on 6061-T6 aluminum alloy in the range of strain rate up to 10^3/s, while the upturn of the effect was reported at the strain rate beyond(Lindholm et al(1971), Kim & Clifton(1980)). These results were obtained only at the room temperature and it is, therefore, a fundamental approach to evaluate the strain rate effect on the strength in the wide range of temperature and strain rate.

In the present paper, temperature and strain rate effects on the tensile stress-strain relations on 6061-T6 aluminum alloy were experimentally clarified in the wide ranges of temperature from 77K to 473K and of strain rates from 10^{-4}/s to 10^3/s, and the deformation behaviors are discussed using the thermally activated process concept.

SPECIMEN AND EXPERIMENTAL PROCEDURES

The material used is A6061-T6 aluminum alloy and its chemical compositions are 0.49Si-0.24Fe -0.34Cu-0.01Mn-1.03Mg-0.06Cr-0.01Zn-0.02Ti-bal.Al. The material was extruded and solution treated at 793K, and was water quenched and tempered at 448K for 8 hours. The specimen was machined from the material and the tensile axis was in the extruded direction.

Quasi-static test in the strain rate of $10^{-4} \sim 10^{-2}$/s was performed using the universal testing machine, and the strain rate change test to evaluate the strain rate sensitivity of flow stress was carried out as well as the constant strain rate test. Split Hopkinson pressure bar apparatus shown in Fig.1 was used to perform the impact tensile test at the strain rate above 10^2/s. The striker was launched by compressed air and hit the yoke on the input bar. The stress waves in the input and the output bars were sensed by strain gauges cemented on the indicated positions in the figure, and were stored in the multi-channels transient recorder with sampling rate of 1 word/microsecond after amplified up to 60dB through the pre-amplifier. The stress, strain, and strain rate were analyzed in digital form by using the one-dimensional stress wave theory as usual.

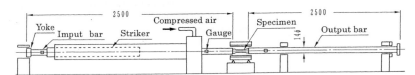

Figure 1: Split Hopkinson bar apparatus for impact tensile test.

EXPERIMENTAL RESULTS

Nominal stress-nominal strain relations obtained at static and dynamic strain rates are represented in Fig.2. An enlargement indicates abrupt stress response in the strain rate change during static strain rate deformation. When the strain rate is increased from $\dot{\varepsilon}_1$ to $\dot{\varepsilon}_2$, a slight increase was associated, but it is quickly diminished and a stress-strain relation returns to that at the strain rate $\dot{\varepsilon}_1$. This indicates that the flow stress at around the room temperature is almost independent of the strain rate as reported by many authors. Uniform deformation is expected up to the 7 to 8 % strain from these curves.

Fig.3 shows the relation between the flow stress at 5% strain to the strain rate for the respective temperatures. The flow stress increases with an increase of strain rate at low temperatures. At 289K, the stress shows almost constant in the low strain rate range, but it tends to increase in high strain rate range of 10^3/s. Solid lines in the figures are drawn according to the thermally activated process concept and will be explained afterward. The experimental results obtained by Yadev et al(1995) and Nicholas & Rajendran(1990), and the dotted lines will be also described.

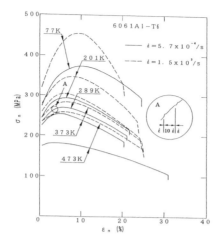

Figure 2: Static and dynamic nominal stress-strain curves at various temperatures

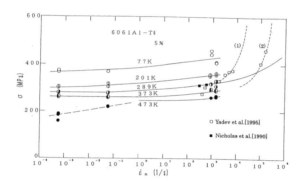

Figure 3: Strain rate dependence of stress for 5% strain at various temperatures

CONSIDERATIONS

Thermally Activated Process

The stress, σ, can be expressed in terms of the thermal, σ^*, and the athermal component, σ_μ, of the stress as follows.

$$\sigma = \sigma^* + \sigma_\mu \tag{1}$$

The strain rate, $\dot{\varepsilon}$, of a single thermally activated flow can be expressed as

$$\dot{\varepsilon} = b\rho_m L v_0 \exp\{-(H_0 - \int v^* d\sigma^*)/kT\} = \dot{\varepsilon}_0 \exp\{-H(\sigma^*)/kT\} \tag{2}$$

where b, Burger's vector, ρ_m, mobile dislocation density, L, average distance of short range obstacles, v_0, part of Debye-frequency, H_0, the activation energy, v^*, the activation volume, k,

Boltzman constant, and T is the absolute temperature, respectively. Since the activation energy is a function of σ^*, the stress , σ, is rewritten as in terms of the temperature-strain rate parameter ξ;

$$\sigma = \sigma_\mu + H^{-1}(\sigma^*) = \sigma(\xi,\varepsilon) \tag{3}$$

where

$$\xi = kT(ln\dot{\varepsilon}_0 - ln\dot{\varepsilon}) \tag{4}$$

The stress was modified by taking account of temperature dependence of the elastic constant.
In Fig.4 all true stresses measured at are plotted versus the temperature –strain rate parameter, and the experimental data fit smooth solid curves for the respective strains by adopting $ln\dot{\varepsilon}_0 = 14$. These curves predict the stress to the strain rate relations at temperatures and were already shown in Fig.3 for the strain of 5%. Dotted and chained curves will be explained afterward.

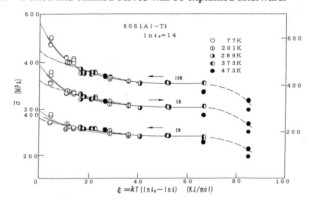

Figure 4: Relationships between the flow stress and the temperature-strain rate parameter

Precipitation Hardening

As was mentioned before, the precipitation hardening significantly increases the athermal component of stress, but still it may contribute some extent to an increase of thermal component of stress. So, first of all, it is relevant to consider dislocation cutting process through the precipitates. The activation energy for cutting the precipitates was evaluated by Seeger(1954) as follows;

$$H(\tau^*) = H_0(1-(\tau^*/\tau_0^*))^{3/2} \tag{5}$$

where $\tau_0^* = H_0/(4bl_0x_0)$, and l_0 is a distance between precipitates , x_0 is the equilibrium position, and the applied stress τ^* is assumed to be nearly the same magnitude of τ_0^*.
The flow stress is almost independent of temperature and strain rate for the large vales of ξ ($\xi \geq 40KJ/mol$), while it is significantly increased for the small values of ξ ($\xi \leq 40KJ/mol$), and will be discussed using Eqn.5. Substituting Eqn.5 into Eqn.2, the thermal component of stress, σ^*, can be expressed as follows.

$$\sigma^* = (H_0/v^*)\left\{1-(kT/H_0)^{2/3}(ln(\dot{\varepsilon}_0/\dot{\varepsilon}))^{2/3}\right\} = (H_0/v^*)\left\{1-(\xi/H_0)^{2/3}\right\} \tag{6}$$

The equation can be applied for temperatures for $T \leq T_c$. The critical temperature , T_c, represents the temperature where the thermal component of stress can be neglected , and is given as;

$$T_c = H_0 / k(\ln \dot{\varepsilon}_0 - \ln \dot{\varepsilon}) \tag{7}$$

Since the stress obtained at $\xi \geq 40 KJ/mol$ is apparently constant and can be interpreted as σ_μ. The stress and the temperature-strain rate parameter relations below T_c will be predicted by Eqn.6. Using the relation $H_0 = T_c k \ln(\dot{\varepsilon}_0/\dot{\varepsilon}) = \xi_c$ obtained from Eqn.7, the $\sigma - \xi$ can be drawn as chain lines in Fig.4 for the most appropriate values of H_0 and v^*. The value of the activation volume decreases with the strain, while the activation energy is constant and is much larger than the value for Al-Cu alloy($H_0 = 25 kJ/mol$)(Byrne et al(1961)). The chain lines do not fit the experimental results for the large strains, and also, in the precipitation hardening theory, the activation volume does not depend on the strain and work hardening is not significant. Therefore, the temperature and strain rate effects can not be understood by the mechanism of dislocation cutting through the precipitation.

Dislocation Intersection

Dislocation intersection is most probable deformation mechanism in FCC pure metals such as aluminum, and it may be also applicable for precipitation hardening alloy, since prismatic dislocation loops formed around large precipitates interact with glide dislocations(Kelly & Nicholson(1963)). In the case of pure aluminum, activation volume, v^*, was expressed as follows(Tanaka & Nojima(1978));

$$v^* \sigma^* = B \tag{8}$$

where, B is the experimental constant. The strain rate is now given as;

$$\dot{\varepsilon} = b\rho_m L v_0 \exp\left\{-(H_0 - \int_0^{\sigma^*} v^* d\sigma^*)/kT\right\} = \dot{\varepsilon}_0 \exp\left\{-B(\ln \sigma_0^* - \ln \sigma^*)/kT\right\} \tag{9}$$

It should be noted that the stress, σ_0^*, is the thermal component of stress at $T = 0$. Then, the thermal component of stress is written as;

$$\sigma^* = \sigma_0^* \exp(-\xi/B) = \sigma_0 \varepsilon^{1/2} \exp(-\xi/B) \tag{10}$$

Here, $\sigma_0^* = \sigma_0 \varepsilon^{1/2}$ (Zerilli & Armstrong(1987)), since dislocation density is increased with strain. Reasonably fitting curves for the respective strains can be drawn as shown by dotted lines in Fig.4 for σ_0 of 800MPa, and $B = 12 KJ/mol$. These curves agree fairly well with the experimental results, and evaluated activation volume at $T = 0$ ranges from 1.8×10^{-22} to 0.8×10^{-22} cm^3 for 2% to 10% strains, well corresponding to the value estimated from the experimental results on pure aluminum(Tanaka et al(1977)). The dislocation intersection is interpreted to control the deformation of the present alloy, and the thermal fluctuation can not overcome strong precipitates.

Viscous Drag Mechanism

Now, it is relevant to know how far we can extend the thermally activated process to higher strain rate range. Yadev et al(1995) performed the plate impact experiments on A6061-T6 alloy at room temperature. Their experimental results of flow stress at 6% strain are shown in Fig.3 together with the experimental results obtained by Nicholas & Rajendran(1990). Since their results at static strain rates were slightly higher than the present results, their dynamic data were shifted to compensate the difference. It is clearly seen that the flow stress upturns significantly beyond at the strain rate of 10^3/s. The phonon scattering and/or the phonon viscosity effect may cause viscous drag acting on high velocity dislocations at room temperature, and the flow stress is expressed as follows.

$$\tau = C\dot{\gamma}/(\rho_m b^2) + \tau_b \qquad (11)$$

where τ is the applied shear stress, τ_b is the back stress, $\dot{\gamma}$ is the shear strain rate, and C is the constant, respectively. Then, the flow stress can be given as:

$$\sigma = \alpha\dot{\varepsilon} + \sigma_G \qquad (12)$$

Using the results reported at the strain rates of 10^3/s~10^4/s, the dotted curve (1) is predicted from Eqn.12. The curve shows a significant upturn, but this does not fit on their data at higher strain rates. When their data at the strain rates of 10^5/s are used, the dotted curve (2) is derived. There is great discrepancy between their data at 10^3/s and the prediction. On the other hand, the solid curve predicted from the thermally activated process concept relatively agrees well with their data at around the strain rate of 10^4/s, at least. At much higher strain rate, probably of 10^5/s, the viscous drag mechanism may play a predominant role on the dynamic strength of present material.

CONCLUSIONS

Tensile stress strain relations of A6061-T6aluminum alloy are clarified in the temperatures from 77K to 473K and the strain rates from 10^{-4}/s to 1.5×10^3/s, and the following conclusion is obtained.
The stress strain relations depend on the temperature and the strain rate and are well understood by the thermally activated process concept of dislocation intersection mechanism where the temperature and strain rate parameter strongly depends on the strain.

References

Armstrong R.W. and Zerilli F.J. (1988). *Proc.DYMAT88,Jour.de Physique* 49,C3,529
Byrne J.G., Fine M.E. and Kelly A. (1961). *Phil.Mag.*,[8]6, 1119
Clifton R.J.(1983). *J.Appl.Mech.*,**50**,941
Follansbee P.S. (1988). *Proc.IMPACT'87*, Bremen,315
Holt D.L., Babcock S.G., Green S.J. and C.J.Maiden (1967).The Strain-Rate Dependence of the Flow Stress in Some Aluminum Alloys *Trans. ASM*,**60**,152-159.
Kelly A. and Nicholson R.B.(1963).*Precipitation hardening*, ed. Chalmers B., Pergamon Press. Oxford
Kim K.S. and Clifton R.J. (1980). Pressure-Shear Impact of 6061-T6 Aluminum *J.Appl.Mech., Trans.ASME*, **47**,11-16.
Lindholm U.S., Bessey R.L. and Smith G.V. (1971).Effect of Strain Rate on Yield Strength, Tensile Strength, and Elongation of Tree Aluminum Alloys *J. Mat., JMLSA*,**6**,119-133.
Nicholas T. and Rajendran A.M. (1990). *High Velocity Impact Dynamics*, ed. J.A.Zukas, Wiley,N.Y.,127-296.
Seeger A. (1954).*Phil.Mag.*,[7]**45**, 771
Tanaka K. and Nojima T. (1978). Flow Stresses of Al and Al Alloys *The 21st Jap.Cong.Mat.Res.*,1-5.
Tanaka K., Ogawa K., and Nojima T. (1977) *High Velocity Deformation of Solids*, ed. Kawata K. and Shioiri J., Springer-Verlag,Berlin. 98.
Yadav S., Chichili D.R. and Ramesh K.T. (1995). The Mechanical Response of A 6061-T6 Al/Al2O3 Metal Matrix Composite at High Rates of Deformation *Acta Metall. Mat.*,**43**,4453-4464.
Zerilli F.J. and Armstrong R.W. (1987). Dislocation-mechanics-based constitutive relations for material dynamics calculations *J.Appl.Phys.* **61**,1816-1825.

DYNAMIC ELASTIC-PLASTIC BUCKLING
OF AXIALLY LOADED CIRCULAR AND SQUARE TUBES

D. Karagiozova[1] and Norman Jones[2]

[1] Institute of Mechanics, Bulgarian Academy of Sciences,
Acad. G. Bonchev St., Block 4, Sofia 1113, Bulgaria
[2] Impact Research Centre, The University of Liverpool,
Brownlow Hill, Liverpool L69 3GH, UK

ABSTRACT

The dynamic buckling phenomena of elastic-plastic circular and square tubes subjected to an axial impact are studied from a stress wave propagation viewpoint. A numerical study reveals that the square tubes exhibit a greater variety of buckling initiation patterns in comparison to geometrically equivalent circular tubes. At impact velocities between 16 and 92 m/sec, the analysed circular tubes always buckle progressively (when the folds form sequentially), while the square tubes buckle either progressively or dynamically (when the entire length wrinkles before the development of large local displacements). This phenomenon can be attributed to both the structural differences between the two types of tubes and the stress wave propagation in these structural elements.

KEY WORDS

Dynamic plastic buckling, dynamic progressive buckling, square tubes, circular tubes, axial impact, stress wave propagation

INTRODUCTION

It has been reported recently in the literature that the transient deformations in circular cylindrical shells subjected to axial impact loads play an important role in the formation of the initial instability pattern of these shells for impact velocities above a certain value (Karagiozova & Jones (2000, 2001), Karagiozova et al. (2000)). It was revealed that particular combinations of the inertia properties of the shell and the hardening characteristics of the material cause either dynamic plastic or dynamic progressive buckling to develop under an axial impact. These theoretical findings are supported by several experiments on circular cylindrical tubes subjected to axial impact loadings at various velocities (Florence & Goodier (1968), Changeen et al. (1992), Murase & Jones (1993), Li Ming et. al (1994)). It was established that dynamic plastic buckling can develop only within a sustained plastic flow when the entire shell length is compressed axially beyond the material yield limit (Karagiozova &

Jones (2001)), so that the speed at which the plastic waves can propagate in a shell, becomes one of the major parameters controlling the type of buckling. The characteristic feature of dynamic plastic buckling, when small wrinkles develop along the entire shell length, is also present in square aluminium tubes (Yang & Jones (2001)) even for relatively low impact velocities in comparison to the similar phenomenon in circular tubes. The purpose of this study is to provide some insight into the role of the transient deformation process in square tubes and compare their response to equivalent circular tubes when subjected to various axial impact velocities.

NUMERICAL SIMULATION OF AXIAL IMPACT

A numerical simulation of the impact event was carried out using the FE code ABAQUS/Standard. Shell elements RS4 (1.85 mm x 1.185 mm) are used to model a square shell. Two planes of symmetry for the buckled shape are assumed so that one quarter of the shell is modelled. The load is applied as a point mass attached to the nodes of a rigid body, which has an initial velocity V_0. The contact between the shell and the striker and between the distal end of the shell and the rigid surface is defined using the 'surface interaction' concept together with a friction coefficient of 0.25 at both ends. The dimensions of the analysed square tubes are $L = 146$ mm, $b = 23.7$ mm and $h = 1.14$ mm and they are made of a strain rate insensitive material (Al 6063 T5) having Young's modulus $E = 71$ GPa, density $\rho = 2700$ kg/m^3 and a yield stress $\sigma_0 = 200$ MPa. The experimentally obtained stress-strain curve (Yang & Jones (2001)) is approximated as a piecewise linear function characterised by hardening moduli of 500 MPa and 209 MPa associated with the true strains $\varepsilon^p_e \in (0, 0.04)$, and $\varepsilon^p_e > 0.04$, respectively. In order to initiate non-symmetric buckling of the square tubes, initial imperfections with a magnitude of 0.1 mm have been introduced. It is assumed that the edges of the model along the entire tube are asymmetrically displaced. These initial imperfections don't affect the location of the wrinkles, which develop along the tube, but only trigger the expected mode of buckling for a square tube.

The axial impact behaviour of geometrically equivalent circular tubes is simulated with the same loading conditions. The equivalency between square and circular tubes was made by assuming that both tubes have an equal mass, equal length and an equal thickness. This led to a mean radius $R = 15.09$ mm for a circular tube. No initial imperfections are assumed for the equivalent circular tube since an axisymmetric buckling mode was considered.

STRESS WAVE SPEEDS

Karagiozova & Jones (2000, 2001) observed that the plastic wave speeds in circular tubes ($\sigma_x \neq 0$, $\sigma_\theta \neq 0$) depend on the stress state in the tube and that the elastic wave speed is an upper bound for the plastic waves in a strain rate insensitive material. The minimum speed at which the plastic waves propagate in these tubes shows a strong dependence on the hardening modulus, which is comparable to the corresponding proportional dependence in the uniaxial case. Using the assumption of a plane stress state ($\sigma_{xx} \neq 0$, $\sigma_{yy} \neq 0$ and $\sigma_{xy} \neq 0$) the longitudinal and shear plastic wave speeds in the axial direction of propagation ($n_x = 1$, $n_y = 0$) in a square tube were obtained by Karagizova (2001) as functions of the stress state and material hardening parameter, E_h, when assuming the von Mises yield criterion. It was revealed that these speeds decrease more rapidly with any growth of the shear stress, rather than with a decrease of the hardening modulus, and their maximum values are equal to the corresponding elastic wave speeds.

The dependence of the plastic wave speeds on the stress state in a circular cylindrical shell with a hardening characteristic $\lambda = E_h/E = 0.007$ are compared in Figure 1(a) with the longitudinal plastic wave speeds in a square tube ($\sigma_{xy} = 0.1\sigma_0$). The initial stress state of a square tube subjected to an axial impact is associated with $\sigma_y \approx \sigma_x/2$ (both negative), while the stress state in a circular cylindrical shell at $t = 0$ is associated with $\sigma_\theta \approx \sigma_x/2$ (both negative). These stress states correspond to the minimum value of the plastic wave speed (Karagiozova et al. (2000))

$$c^P_{min,Mises} = \pm (E/\rho)^{1/2}\{4\lambda[4\lambda(1-\nu^2) + 3(1-\lambda)]^{-1}\}^{1/2}. \qquad (1)$$

A variation of the stress wave speeds in both structures is presented in Figure 1(b) for this particular stress state. A significant difference in the plastic stress wave speeds is evident for the initiation of the deformation process when the plastic waves in circular cylindrical shells propagate at a lower speed than the corresponding waves in square tubes (point A in Fig.1 (a)). It is also evident that the stress waves propagate the plastic strains along square tubes at speeds considerably higher than those associated with uniaxial stress waves, $c^P_{uniaxial} = (\lambda E/\rho)^{1/2}$. It was shown by Karagiozova (2001) that the shear stress influences significantly the plastic waves speeds in square tubes regardless of the strain hardening characteristics. Plastic wave speeds of 1200 to 1400 m/sec, which propagate in the x-direction, were measured experimentally by Yang & Jones (2001) in some of the tubes examined in the present study.

BUCKLING OF SQUARE AND CIRCULAR TUBES

The experimentally observed final buckling shapes (Yang & Jones (2001)) for aluminium alloy square tubes are presented in Figure 2(b-e). One can see that only local deformations develop visibly in the first two tubes associated with the lower impact velocities, when the unbuckled part is not disturbed. This type of response is characteristic of progressive buckling (Jones (1989)) when the shell folds sequentially. In contrast, small amplitude wrinkles are observed along the entire length of the other two tubes, which have been subjected to higher impact velocities.

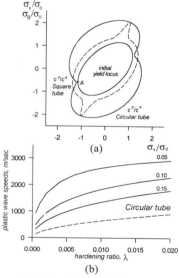

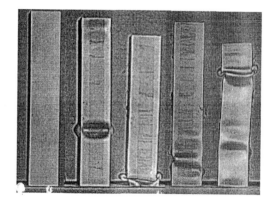

Figure 1: (a) Longitudinal plastic wave speeds ($n_x = 1$, $n_y = 0$) in square (----) and circular (—) tubes; (b) Longitudinal plastic wave speeds in a square tube for different shear stresses (σ_{xy}/σ_0) (—) and in a circular tube (----).

Figure 2: Final buckling shapes (Yang & Jones (2001)) (a) Initial, (b) N04 ($V_0 = 15.91$ m/sec, $G = 0.95$ kg), (c) N81 ($V_0 = 35.35$ m/sec, $G = 0.44$ kg), (d) N86 ($V_0 = 64.62$ m/sec, $G = 0.13$ kg), (e) N82 ($V_0 = 91.53$ m/sec, $G = 0.13$ kg).

The axial inelastic compressive strains dominate the tube response during the initial phase of deformation, so that the analysis presented in this study is restricted to these strains, which are shown in Figures 3(a-c) for tubes $N86$, $N82$ and $N04$, respectively. Shear stresses, σ_{xy} up to $0.2\sigma_0$ develop in these tubes reducing the speed of the longitudinal stress waves, which propagate the compressive strains in the x-direction. Initial axial plastic strains of about 0.08 propagating from the proximal end of tube $N86$ result only in small magnitude wrinkles, which develop within a sustained plastic flow near to this end (Figure 3(a)). The reflected plastic wave from the distal end of the shell carries smaller strains, which propagate with a larger speed, and until about $t = 0.16$ msec, the lateral inertia supports a shape with only small wrinkles developing along the shell, so that the entire shell is in a state of plastic compression. An analysis of Figure 3(a) shows that the longitudinal stress wave speed in the shell varies between 980 m/sec and 400 m/sec associated with $E_h/E = 0.003$. The lateral displacements near to the distal end start to grow rapidly for $t > 0.16$ msec causing elastic unloading followed by a reverse plastic loading.

A further increase of the impact velocity increases the axial and lateral inertia properties of the shell. Initial axial plastic strains of about 0.14 (Figure 3(b)) and large shear stresses are associated with an initial velocity of 91.53 m/sec in tube $N82$ causing the stress waves to propagate in the x-direction at speeds between 640 m/sec and 460m/sec. The lateral inertia cannot support the unbuckled shape for $t > 0.1$ msec and buckling then occurs within the primary plastic wave close to the proximal end where much larger strains develop in comparison to the strains near to the distal end. The strains, due to the reflected stress wave are large enough to cause buckling, which develops also within a sustained plastic flow. Small wrinkles start to develop from the distal end and the entire length of the shell is wrinkled but with strain localisation at $x \approx 120$ mm, as shown in Figure 3(b).

The propagation of the stress waves in tube $N81$ is similar to the process in tube $N86$. The development of the buckling shape for tube $N04$, although subjected to a relatively low impact velocity of 15.91 m/sec is also caused by transient deformations. The initial stress wave originating from the proximal end carries small strains propagating at 2300 m/sec approximately, and does not cause any bending of the shell wall. The reflected wave from the distal end propagates larger strains as shown in Figure 3(c) and at $t \approx 0.25$ msec reaches the proximal end. As no elastic unloading has occurred, this wave reflects from the proximal end and later causes a localisation of strains at $x = 0.04$ m where a local wrinkle develops.

An analysis of the influence of stress wave propagation on the initial buckling pattern of circular tubes reveals (Karagiozova & Jones (2001)) that the lower plastic wave speeds assist the development of strain localisation and progressive buckling. Thus, it could be expected that square tubes and circular tubes having the same inertia properties will respond by different buckling patterns because the plastic waves propagate at different speeds in both structures. In order to clarify the influence of the transient deformation process on the buckling of these structures, a comparison is made between the initiation of buckling and the final buckling shapes of the square tubes $N82$, $N86$, $N81$ and $N04$ and geometrically equivalent circular tubes $N82c$, $N86c$, $N81c$ and $N04c$. The numerically obtained final buckling shapes of these tubes are presented in Figure 4. One can see that all circular tubes buckle only progressively but with different wrinkle locations, while a larger variety of buckling shapes is observed for the square tubes. This response is due to a combination of the low stress wave speed determined by the particular material characteristics and the small radial inertia of the circular tube, which leads to a localisation of strains soon after the impact, a quick growth of the radial displacements and progressive buckling. The plastic wave speeds according to eqn (1) are $c^p_{min} = 493$ m/sec for $E_h = 500$ MPa and $c^p_{min} = 323$ m/sec when $E_h = 209$ MPa.

A comparison between the stress wave propagation behaviour in square and circular tubes for impact velocities of 15.91 m/sec, 35.35 m/sec and 64.62 m/sec are presented in Figure 3 (d-f). It is evident that in all cases the plastic waves propagate the strains faster in the x-direction of square tubes and

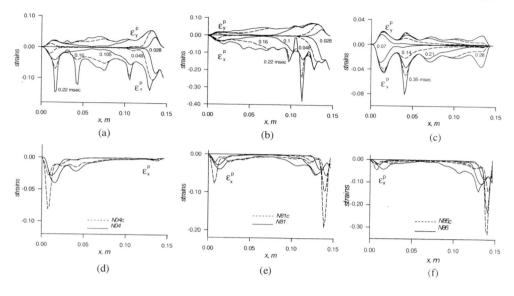

Figure 3: Strain distributions in square and circular tubes. (a) Tube $N86$, (b) Tube $N82$, (c) Tube $N04$; (d-f) Comparison between the stress wave propagation in square and geometrically equivalent circular tubes; (d) $t = 0.07$ msec and 0.14 msec, (e) $t = 0.028$ msec, $t = 0.06$ msec and 0.092 msec, (f) $t = 0.028$ msec, 0.049 and 0.105 msec

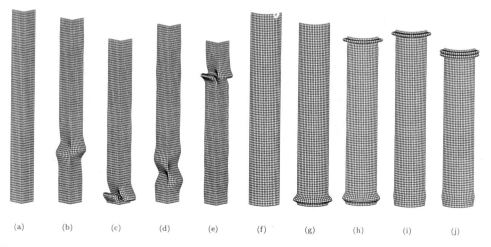

Figure 4: Final buckling shapes of square and geometrically equivalent circular tubes. (a,f) Initial; (b,g) $V_0 = 15.91$ m/sec, $G = 0.95$ kg ($N04$ and $N04c$); (c,h) $V_0 = 35.35$ m/sec, $G = 0.44$ kg ($N81$ and $N81c$); (d,i) $V_0 = 64.62$ m/sec, $G = 0.13$ kg ($N86$ and $N86c$); (e,j) $V_0 = 91.53$ m/sec, $G = 0.13$ kg ($N82$ and $N82c$)

during this short time the lateral inertia of a square tube is able to support the unbuckled shape. Only small wrinkles develop within a sustained plastic flow as axial compression dominates the deformation process. The slower plastic waves in a circular shell cause a localisation of the strains shortly after impact and a concomitant rapid growth of the radial displacements. In addition to the different speeds of the stress waves, which propagate in square and circular tubes, the circumferential stress plays an important role in the buckling process of circular cylindrical shells by introducing an additional irregularity along the tube. As soon as the radial displacements start to grow, the radial force, which results from the circumferential strains, decreases rapidly due to an increase of the local radius. However, this force remains unchanged in the shell section, which have not deformed radially. This influences additionally the localisation effects and as a result of the slower plastic waves and the variation of the radial force along the shell, a circular tube buckles more rapidly than a square one. The faster plastic waves, which propagate in a square tube allow for a larger part of the shells to be deformed within a sustained plastic flow in a shorter time, which leads to dynamic plastic buckling for the impact loadings studied.

CONCLUSIONS

A numerical study of the transient deformation process in elastic-plastic square tubes subjected to axial impact loadings reveals that the initiation of buckling is influenced significantly by stress wave propagation along the tube. It is shown that different types of buckling occur in square and geometrically equivalent circular tubes when subjected to identical axial dynamic loadings. This behaviour is due to the different speeds of the plastic waves, which propagate along these structures. The particular square tubes examined herein respond either by dynamic progressive or dynamic plastic buckling when subjected to impact velocities between 15.91 m/sec and 91.53 m/sec, while the circular tubes respond only by dynamic progressive buckling.

References

Chen Changeen, Su Xianyue, Han Mingbao, Wang Ren. (1992). An experimental investigation on the relation between the elastic-plastic dynamic buckling and the stress waves subjected to axial impact. Proc. Int. *Symposium on Intense Dynamic Loading and its Effects*, Chengdo, China, June, 9-12, 543-546.

Florence A L Goodier JN. (1968). Dynamic plastic buckling of cylindrical shells in sustained axial compressive flow. *J. Appl. Mech.*, **35:1**, 80-86.

Jones N. (1989). *Structural impact*, Cambridge University Press, Paperback edition, 1997

Karagiozova D, Jones N. (2000). Dynamic elastic-plastic buckling of cylindrical shells under axial impact. *Int. J. Solids and Struct*, **37:14**, 2005-2034.

Karagiozova D, Alves M, Jones N. (2000). Inertia effects in axisymmetrically deformed cylindrical shells under axial impact. *Int. J. Impact Engng.*, **24:10**, 1083-1115.

Karagiozova D, Jones N. (2001). Influence of stress waves on the dynamic plastic and dynamic progressive buckling of cylindrical shells under axial impact. *Int. J. Solids and Struct.* (In press)

Karagiozova D. (2001). Dynamic buckling of elastic-plastic square tubes under axial impact. Part I – Stress wave propagation phenomenon. IRC Report 199 /2001, The University of Liverpool

Li Ming, Wang Ren, Han Mingbao. (1994). An experimental investigation of the dynamic axial buckling of cylindrical shells using a Kolsky bar. *Acta Mechanica Sinica*, **10**, 260-266.

Murase K, Jones N. (1993). The variation of modes in the dynamic axial plastic buckling of circular tubes. In: NK Gupta, editor. *Plasticity and Impact Mechanics*. Wiley Eastern Limited, 222-237.

Yang C-C, Jones N. (2001). Dynamic axial inelastic buckling of aluminium square tubes, IRC Report No. 166/2001, The University of Liverpool

DYNAMIC BUCKLING BEHAVIOR OF FOAM-FILLED METAL TUBE

K. Mimura[1], T. Umeda[1], K. Yamashita[2], T. Wakamori[2] and S. Tanimura[1]

[1] Division of Mechanical Systems Engineering, Graduate School of Engineering,
Osaka Prefecture University, Sakai-city, Osaka 599-8531, JAPAN
[2] Sunstar Engineering Inc. R&D Center,
5-30-1 Kamihamuro, Takatsuki-city, Osaka 569-1044, JAPAN

ABSTRACT

Weight reduction in vehicles without decreasing their crashworthiness is an important subject in manufacturing automobiles to improve the fuel efficiency and eventually to save the finite petroleum resource. To use high-stiffness urethane foam as a filler for existing structural members is a possible way of improving the body stiffness and the energy absorption of the vehicle while decreasing the weights of its members by diminishing their lengths and/or thicknesses. In the present study, the dynamic buckling behavior and the energy absorption of thin metal tubes filled with urethane foam, as well as the mechanical property of urethane foam itself are experimentally investigated. In the experiments, a high-speed material testing system (Saginomiya Co. Ltd.) and a dynamic buckling test device employing the sensing block method proposed by Mimura & Tanimura (1996) in measuring dynamic loads are used. The compaction behavior of urethane foam is found to be successfully described by the constitutive equation proposed by Oyane & Shima (1976) for porous materials. It is also found that the strain rate sensitivity of the urethane foam in a strain rate range from $0.001s^{-1}$ to $450s^{-1}$ is described by a simple power law regarding a strain rate. In the dynamic buckling of foam-filled tubes, an axi-symmetric buckling mode is frequently observed even though the tube walls are very thin. It is also found that the energy absorption of the foam-filled tubes is drastically improved as compared with that for hollow tubes, especially in the small thickness range.

KEYWORDS

Dynamic buckling, Buckling mode, Energy absorption, Thin walled-tube, Urethane foam filler, Anticollision, Strain rate sensitivity, Porous material, Consolidation condition, Constitutive equation

INTRODUCTION

High-Stiffness urethane foam (Product name: Penguin foam, Sunstar Engineering Inc.) was developed to reinforce the structural members of vehicles, and accordingly to improve their energy absorption. Any inner space or gap in the frame-elements or side membrane-elements can easily be filled with the urethane foam since it is highly liquid before hardening, and the foaming and hardening processes progress at room temperature. In this study, in order to evaluate the effectiveness of this urethane foam when it is used as a filler, the dynamic buckling behavior (buckling modes) and energy absorption of thin-walled metal tubes filled up with urethane foam as well as the compaction behavior and strain rate sensitivity of urethane foam itself were investigated experimentally.

EXPERIMENTAL PROCEDURES

Three different kinds of urethane foams of which relative densities were 0.1, 0.2, and 0.33, respectively, were used in the experiments. Here, the relative density (ρ_r) means the ratio of apparent density of urethane foam to the true density (1150kg/m^3) of nonporous urethane. Compaction properties of the urethane foams were examined through both the uniaxial compression of cylindrical specimens of ϕ 20mm × 25mm and closed-die compaction of rectangular parallelepiped specimens of 15mm × 15mm in cross section and 50mm in height. These compression tests were carried out by using a hydraulic universal testing machine (Shimazu, REH-30) under quasistatic loading conditions. Furthermore, strain rate sensitivities of the urethane foams were evaluated through dynamic compression of small cylindrical specimens of ϕ 11mm × 7mm in the strain rate range from 0.001s^{-1} to 450s^{-1}. A high-speed material testing system (Saginomiya, TS-2000) employing the sensing block method proposed by Mimura & Tanimura (1996) for long-duration dynamic load measurement was used in this kind of dynamic compression test (see Figure 1).

In addition to the evaluation of the mechanical properties of the urethane foam, the dynamic buckling behavior and energy absorption of foam-filled metal tubes were experimentally investigated by using the accelerated drop-weight type dynamic buckling test apparatus [Mimura & Tanimura (1997)] shown in Figure 2. Only the free-fall of the impact block from a height of 3.0m was used to impact the specimens although this apparatus had the ability to accelerate the impact block to a maximum velocity of 20ms^{-1} with the help of high air pressure. The mass of the impact block (drop-weight) was about 50kg, and as is the case with the high-speed material testing system, this buckling test apparatus also employs the sensing block method for precise measurement of dynamic loads. In the experiments, both steel (STKM13A) and aluminum alloy (A1050TD) tubes with inner diameters of 34mm were employed, and their inner spaces were filled with urethane foams with relative densities of 0.1, 0.2, and 0.33. Tube thicknesses varied within a range of 0.2mm to 1.0mm in order to investigate their influence on energy absorption. Furthermore, extra tubes of 51mm and 64mm in inner diameter as well as standard tubes of 34mm in inner diameter were examined to clarify the size effect on the energy absorption. The combination of tube materials, diameters and thicknesses are summarized in Table 1.

In evaluating (or calculating) the energy absorption of the foam-filled tubes, a length of the tube should be carefully determined. Figure 3 illustrates typical deformation patterns of foam-filled tubes under dynamic loading. When the tube is subjected to a dynamic load, the progressive buckling mode appears as shown in Figure 3. The buckling zone is, however, limited to a part of the tube when the length of the tube is too long

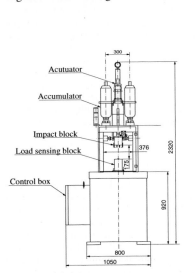

Figure 1: High speed material testing system (Saginomiya, TS-2000).

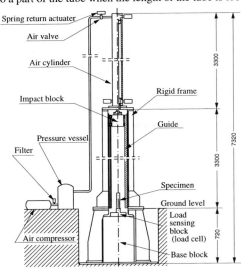

Figure 2: Dynamic buckling test apparatus employing the sensing block method.

TABLE 1
DIMENSIONS OF FOAM-FILLED METAL TUBES USED IN THE EXPERIMENTS.

Tube material	Inner diameter D_i, mm	Thickness t, mm
STKM13A	34.0	0.2, 0.3, 0.5, 0.8, 1.0
	51.0	0.3, 0.5
	68.0	0.4, 0.6
A1050TD	34.0	0.3, 0.5, 0.7, 1.0

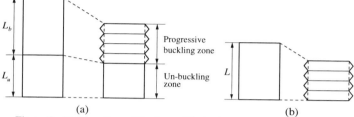

Figure 3: Deformation of the foam-filled tubes under dynamic loading.

because the kinematic energy of the impact block is not infinite (see Figure 3(a)). In this case, the total energy absorbed by the tube may be regarded as the sum of the energy absorbed by the buckling zone and that by the unbuckling zone where the deformation is considered to be more homogeneous. Since this study focused on the energy absorption for progressive buckling, which is frequently observed in car collisions, we arranged the length of the tube (L) so that the whole part of the tube suffers progressive buckling (see Figure 3(b)); the energy absorption of the tube was then evaluated by integrating a force-displacement curve from zero to the displacement where the tube bottomed out and the force showed a remarkable increase.

RESULTS AND DISCUSSIONS

Mechanical Properties of Urethane Foam

Since both the urethane foams and the foam-filled tubes discussed here are subjected to compressive force, compressive stress is considered to be positive throughout this paper. Figures 4, 5, and 6 respectively, show nominal stress-strain curves for the urethane foams of ρ_r=0.1, 0.2, and 0.33 at various strain rates. The initial yield stress in these urethane foams at the quasistatic strain rate of 0.001s^{-1} increases with the relative density, and shows the linear dependence on the relative density. The increase in flow stress with the strain rate is fairly large, and it becomes remarkable with increasing relative density; the dynamic-to-quasistatic stress ratio at a strain rate of 450s^{-1} is about 1.6 (ρ_r=0.1, 0.2) to 2.2 (ρ_r=0.33), and these values are almost

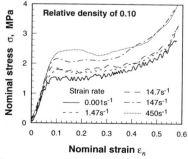

Figure 4: Nominal stress-strain relations for urethane foam of ρ_r=0.1 at various strain rates.

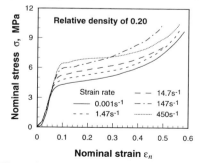

Figure 5: Nominal stress-strain relations for urethane foam of ρ_r=0.2 at various strain rates.

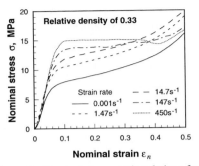

Figure 6: Nominal stress-strain relations for urethane foam of $\rho_r=0.33$ at various strain rates.

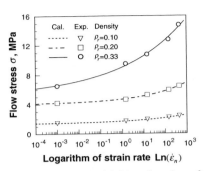

Figure 7: Strain rate sensitivities of urethane foam of three different relative densities.

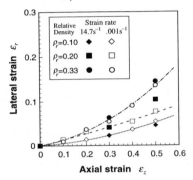

Figure 8: Relationship between axial strain ε_z and lateral strain ε_r in static and dynamic uniaxial compression of cylindrical specimens.

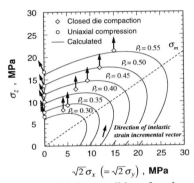

Figure 9: Consolidation condition of urethane foam depicted in the Rendulic plane.

equal to those for low carbon steels, which are known as 'high strain rate sensitivity materials'. Although strain hardening at a high strain rate of over 147s^{-1} is relatively small, this is considered to result from the collapse of the specimen by high-speed deformation. Figure 7 shows the strain rate sensitivities of the urethane foams. In this figure, the symbols express the stresses at the 2% offset strain after initial yielding; three different kinds of lines represent the approximation curves based on the equation, $\sigma = \sigma_0 + C\dot{\varepsilon}^h$. The supposed values in the approximation are $\sigma_0=1.38$MPa, $C=0.285$MPa, $h=0.204$ for $\rho_r=0.1$, $\sigma_0=4.05$MPa, $C=0.625$MPa, and $h=0.221$ for $\rho_r=0.2$, and $\sigma_0=5.17$MPa, $C=3.84$MPa, and $h=0.145$ for $\rho_r=0.33$. Obviously, the values of C for $\rho_r=0.33$ is significantly large as compared with those for the other relative densities reflecting a large increase in flow stress at high strain rates.

As mentioned previously, the urethane foams considered here were developed to reinforce existing structural elements; it will be poured into the gaps or inner spaces of structures and will be deformed under severe constraint. It is, therefore, of importance to clarify the consolidation condition of urethane foam under multi-axial loading conditions. In the present study, the constitutive equation proposed by Oyane & Shima (1976) for porous metals and arranged for ceramic powders by Shima & Mimura (1986) is assumed to describe the consolidation behavior of the urethane foam. The constitutive equation is expressed as

$$F = (1/2)\sigma'_{ij}\sigma'_{ij} + (1/27f^2)\sigma^2_{kk} - (1/3)(S\rho_r^n)^2 = 0,$$

where σ'_{ij} is deviatoric true stress, f is a function of relative density as $f = a(1-\rho_r)^m$, and a, m, n, S are material constants. In order to identify the material constants in the above equation, experimental data obtained by closed-die compaction (floating die method) as well as those by uniaxial compression were used. Furthermore the relationship between axial strain (ε_z) and lateral strain (ε_r) for cylindrical specimens shown in Figure 8 was measured to obtain the true stress in uniaxial compression. The identified values of the constants are $a=0.838$, $m=0.284$, $n=1.58$ and $S=43.41$MPa, respectively. Figure 9 shows the comparison between experimentally obtained true stresses at several compaction densities and calculated densification surfaces (or plastic potential surfaces) on the Rendulic plane (the plane containing the σ_z axis and the mean stress axis σ_m in the stress space). The arrows passing through the stress points in the figure denote the directions of the inelastic strain incremental vectors. As you can see, the calculated densification surfaces accurately express the experimental points; strain incremental vectors are found to be generally normal to the calculated surfaces.

Dynamic Buckling Behavior and Energy Absorption of Foam-filled Tubes

To estimate the effectiveness of urethane foam as a filler, dynamic compression of thin-walled tubes made of steel (STKM13A) and aluminum alloy (A1050TD) filled with urethane foam was performed. Typical load-normalized displacement curves for two thicknesses ($t = 0.2$mm and $t = 1.0$mm) of STKM13A tubes filled with urethane foam of various densities are shown in Figures 10(a) and (b). Here, the normalized displacement is obtained by dividing the displacement by the initial length of the tube. In these figures, the results for hollow tubes are also depicted as references. All of the buckles formed on the side surfaces of the foam-filled tubes were axisymmetric (or concertina), while those formed on the surfaces of the hollow tubes were non-axisymmetric (or diamond); obviously such a difference in the buckling mode results from the constraint on the inner surface of the metal tube caused by the urethane foam filler. The authors should point out that the plastic work to form an axisymmetric wrinkle (or buckle) is about 20% higher than that to form a non-axisymmetric wrinkle [Tanimura et al. (1999)]. In the case of 1.0mm-thickness tubes, the load fluctuation shown in the figure is much larger than that in the case of 0.2mm-thickness tubes because this fluctuation in loads is basically caused by the formation of wrinkles of the outer metal tubes; namely, the magnitude of the fluctuation is due to the metal tube's resistance to buckling. It is also found that the increase in loads due to

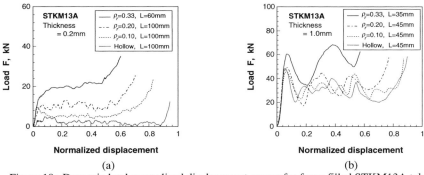

Figure 10: Dynamic load-normalized displacement curves for foam-filled STKM13A tubes;
(a) thickness of the tubes is 0.2mm and (b) thickness of the tubes is 1.0mm.

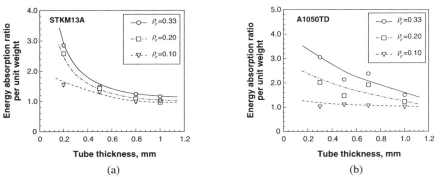

Figure 11: Variation of energy absorption ratio per unit weight with tube thickness;
(a) for foam-filled STKM13A tube and (b) for A1050TD tube.

urethane foam fillers grows larger with increasing relative density.

In the estimation of the energy absorption of a structural element, its efficiency associated with the element-weight seems of importance; high energy absorption with heavy weight is undesirable. This study therefore focuses on the energy absorption ratio per unit weight, which is defined by dividing an absorbed energy per unit weight of a foam-filled tube by that of the hollow tube. The variations of energy absorption ratio per unit weight with tube thickness are illustrated in Figure 11(a) for STKM 13A tubes and Figure 11(b) for A1050TD tubes. For both kinds of metal tubes, the efficiency of energy absorption is improved by the filler. However, a remarkable improvement is achieved by the high relative density filler of $\rho_r=0.33$ and the thin tube with a thickness of 0.2mm or 0.3mm; the energy absorption ratio on this condition

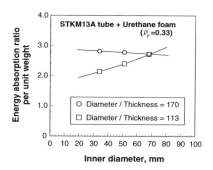

Figure 12: Size effect on energy absorption ratio per unit weight (STKM13A tube).

attains 2.8 for the STKM13A tube and 3.0 for the A1050TD tube. It is also found that the energy absorption ratios of A1050TD tubes with $\rho_r=0.2$ and 0.33 fillers is, in general, slightly higher than those of STKM13A tubes in the thickness range from 0.5mm to 1.0mm, while the energy absorption ratios of A1050TD tubes with $\rho_r=0.1$ filler is held very low even in tubes with thin walls. The relatively low density of aluminum alloy as compared with that of steel is probably the direct cause for this low energy absorption in the foam-filled A1050TD tube with $\rho_r=0.1$ filler. Finally, Figure 12 shows the size effect on the energy absorption for STKM13A tubes. Two different inner diameter-to-thickness ratios of about 113 and 170 were examined. The energy absorption ratio for a diameter-to-thickness ratio of 113 is found to be strongly affected by the inner diameter, and increases linearly with inner diameter although no significant size effect is observed in the case for a diameter-to-thickness ratio of 170. This should be investigated in further detail.

CONCLUSIONS

The consolidation condition and strain rate sensitivity of urethane foam developed as a filler to reinforce structural elements were experimentally investigated and then clarified. Constitutive relations of the urethane foam were also shown. Furthermore, foam-filled metal tubes were examined in regard to the dynamic buckling behavior and energy absorption as models of anti-collision elements in vehicles. It was found that the buckling modes of the foam-filled tubes was, in general, an axisymmetric type, and the energy absorption of the tubes was considerably improved by urethane foam fillers.

Acknowledgments — This work was supported in part by Grant-in-aid for Scientific Research (B)(2) 10450050, the Ministry of Education, Culture, Sports, Science and Technology, Japan, and also by the Special Coordination Funds for Promoting Science and Technology, the Science and Technology Agency, Japan, 'Enhancement of Earthquake Performance of Infrastructures Based on Investigation into Fracturing Process'. The authors are also grateful to Sunstar Engineering Inc., who supported this research and provided urethane foams.

References

Mimura, K., Hirata, S., Chumman, Y. and Tanimura, S. (1996). Development of Dynamic Loading Device with Stress Sensing Block and Its Experimental Examination. *J. of the Soc. of Mat. Sci., Japan* **45:8**, 939-944 (in Japanese).

Mimura, K., Doi, O., Chumman, Y. and Tanimura, S. (1997). Development of a Dynamic Buckling Test Device Employing a Load Cell Block with a Small Sensing Projection. *Trans. of Japan Soc. of Mech. Engineers* **63:605A**, 165-169 (in Japanese).

Shima, S. and Oyane, M. (1976). Plasticity Theory for Porous Metals. *Int. J. Mech. Sci.* **18** , 285

Shima, S. and Mimura, K. (1986). Densification Behaviour of Ceramic Powder, *Int. J. Mech. Sci.* **28:1**, 53-59.

Tanimura, S., Mimura, K., Ishikawa, Y. and Umeda, T. (1999). Dynamic Axial Crushing Test to Thin-Walled Circular Tube Specimen and Evaluation of Its Energy Absorption Ability. *Trans. of Japan Soc. of Mech. Engineers* **65:635A**, 1622-1628 (in Japanese)

STERILIZATION OF DRY POWDER FOODS BY SHOCKS

Kazuhito FUJIWARA[1], Tetsuyuki HIROE[1], Hideo MATSUO[1]
and
Makio ASAKAWA[2],
[1]Faculty of Engineering, Kumamoto University
2-39-1 Kumamoto, 860-8555, JAPAN
[2]Faculty of Education, Kumamoto University, JAPAN
2-39-2 Kumamoto, 860-8555, JAPAN

ABSTRACT

The sterilization for fungi and bacteria in some kinds of dry powder foods is hard to complete without special processing. Especially in spices the heat sterilization is restricted to minimum use, because the heating lowers the hot-taste essentially required in spices. The heating or the use of a sanitizer has to take effect only into fungi and bacteria, otherwise the volatile components containing flavor elements are degraded. The idea of the shock sterilization was proposed, and the development of the sterilization equipment and its performance was discussed. Experimental results showed the potential of the shock for the sterilization and the possibility making large contribution to food industry.

KEYWORDS

Sterilization, Shock Waves, Food, Dry Powder, Explosion

INTRODUCTION

Some kinds of dry foods have not been sterilized, since the dried condition cannot provide the reproduction environment for fungi and bacteria. Rather than that, the reason would be that the heating in flavor or the use of sterilizing agents is hard because the hot-taste components are degenerated in high temperature or the taste and flavor degrade. However, the danger of propagation of bacteria still is in rapping the processed food stocked under exhibitions for a couple of days. The mixing of the infected powder and gravy or juice from stuffs could make the reproduction environment. The electromagnetic microwaves and the steam heating are also unavailable for dry food because they base the moisture contained in foods on the sterilization. The fungi and bacteria generally and often exist on the surface of the spice particle, and there are many cases in which the sterilization is achieved, if only the surface temperature of the particle is raised. These sterilization techniques have also unwanted nature originated from the process time. The more heating operates on outer particles when it is applied from the outside. In this paper, the new sterilization is proposed and extended to the practical system to solve those difficulties. Shock sterilization of the liquid food has been studied by K. Fujiwara et al. (1996).

THE PRINCIPLE OF SHOCK STERILIZATION

The dry powder food is mechanically considered as the low-density material with low filling rate. Large compressibility is expressed on the low filling rate, and it is meant that the temperature rise by the compression is high for this fact. That is suitable for the sterilization if only the surface temperature of particles rises when they are cooled rapidly through the expansion after the shock compression. The short time compression make possible to do that, as this high temperature is fast generated in the void between powder particle, and the inside temperature of the particle itself rises more slowly than at the surface (Fig.1). On the other hand the quite short heating of food powders does not degrade any the taste and flavor because flavor components require some volatilization time. The impulsive compression and high pressure, one of what provides those requirements is the utilization of the shock wave. The shock wave generated by an explosion has the medium temperature rise only in the very short time (order of the microsecond in this study) in which the shock wave has passed through. It should be noticed that an advantage of using the explosion to generate a shock wave come from the fact that the large expansion follows the compressive shock wave. And, the second advantage using the shock wave is that the total amount of heating is nearly equal

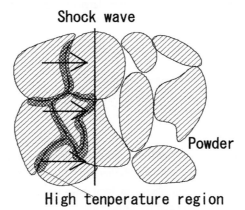

Figure 1: Feature of shock sterilization

throughout the compressed area, while the shock wave heats the food material locally. The third advantage is that the shock wave generates the high pressure and the high temperature simultaneously, and then the pressure while shock heating quenches the volatilization. In this study the explosion of the explosive powder was used as the energy source and the sealing and insulation of the food material

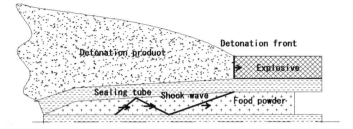

Figure 2: The basic idea of the shock operation

was required to prevent it from contaminated with dirty driver gases and to keep it germfree. An extra advantage is the use of explosion enables the sealing is enabled the sealing with the sterilization at the same instant. The successive energy release of the explosive has the deformation point of the seal move and sweep from one end to another end, and processes packing of the food. The basic idea is shown in Fig.2 schematically.

THE STERILIZATION SYSTEM

The strength of the sealing container (20mm diameter tube) itself is also required in order to enclose the food, while the large powdered food is sterilized. Here, using the copper that is rich in the ductility as a container material, it is compressed by the PETN explosive from the outside by impact, and the shock wave is generated in the food inside. The composition of this equipment is shown in Fig.3. Copper fine wires and electrodes are implemented to initiate the explosive by wire explosion. The aspect of this shock compression (Fig.1), which comes from the results of numerical simulation

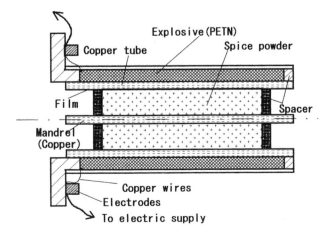

Figure 3: Shock sterilizer driven by detonation of explosive

described later, is similar to the powder compaction usually used in metals and ceramics except the order of power. While the explosive is detonated from one tube end, the sealing tube filled with the spice powder is compressed inside. The shock wave induced by the explosive explosion is reflected at the center of the tube, immediately after at the inside surface of the tube, and this repetition bring the high temperature and pressure. The heating period in this case is very short, since heated points move at the speed equal to detonation velocity. The tube Crushes and both ends of it are closed by its inertia, after shock wave passes the entire region of powders. A mandrel is applied at the center of the tube in order to prevent that the temperature excessively rises. The shape of the container after the recovery is shown in Fig. 4. The recovery was good even in only 2mm tube thickness for the 3mm explosive thickness. The 100 to 200mm length tubes with some geometrical ratios were tested, and it was found that the length and diameter not important to get stable recovery but the shape of tube ends to be closed.

Figure. 4: Recovered container

THE SHOCK STERILIZATION IN SPICES

Using a commercial white pepper and a commercial red pepper including some other spice components, their properties after the shock sterilization were observed and the effect of the sterilization was examined.

White pepper

The property did not differ from the before impact almost, although the powder was somewhat consolidated in the container as it was taken out. The change hardly could be viewed from the observation of the particle by the optical microscope. The degradation also hardly could be sensed either on perfume and hot-taste. As a result of the culture study (for 24 hours at 50°C) using 2 kinds of culture medium (nutrient agar medium, standard agar medium), even though the control sample exhibits about 20 millions colonies per samples of 1 g , the colonies are never found in samples after the impact. That is to say, it would be perfectly sterilized.

Red pepper mixture with other spice

The difference of it from the white pepper was that it contained some taste materials such as sesame and Japanese pepper, and the variation of particle size affected deformation characteristics. Especially, large particles having the brittle shell structure like sesame were fragile against the shock, and fats and oils contained in sesame degenerated the hot-taste of the pepper. On the sterilization effect, bacteria and fungi contained for about 0.12 millions per 1g perfectly died out after the impact. Fig.5 shows SEM images of the infected spice powder before the impact (a) and after the impact (b).

Germs were putted in the sample before the test only in the case of taking SEM so that bacteria cells could be found easily. In the control innumerable cocci (*staphylococcus*) were found on the surface of spice particles and adhered to them. After the impact cells were crushed and could not hold themselves on the surface.

(a) (b)
Figure 5: SEM images ((a) :Control, (b):After shock))

NUMERICAL SIMULATIONS

The compression process and the deformation process were simulated numerically by using the Lagrangian code of AutoDyne 2D. The calculation area was divided into 4 blocks and materials (Copper, light weight plastic instead of food powder and explosive PETN) were assigned on each block. The P-α model was adopted as the EOS for powder. Figure 6 shows the pressure contours (a) and the cross-section of the container after the impact. Light colors express the high pressure generally except near the detonation point because of the mismatch of color conversion into monochromatic image (Fig.6). The oblique shock wave converged into the center and high pressure was generated. The improvement in the shape of container ends made the recovery possible.

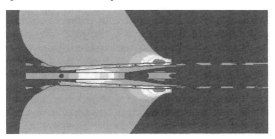

Figure 6: Simulation results (Pressure contour)

THE EXTENSION TO LOWER POWER SYSTEM AND CONTINUOUS SYSTEM

The shock sterilization was extended to the system driven by lower energy source for practical use. Electric energy was used for the energy source generating the shock wave in this system (Fig.7).

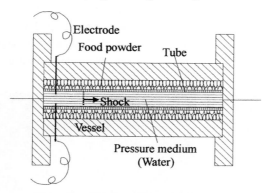

Figure 7: Shock sterilizer driven by electric discharge

The shock wave generated by the spark between two electrodes propagates in the pressure medium(water) and impulsively compresses the food powder through the vinyl tube. Almost hot-taste in the spice was not changed although bacteria decrease was limited to about 1/10 rather than the perfect sterilization due to the reduction of the operating energy. This unbroken vessel however makes the repeated operation and the continuous processing possible, and then the low efficiency in one shot was covered by them.

CONCLUSION

The new sterilization method utilizing the shock wave for dry foods, in which it is difficult to use usual heat sterilization, was proposed and its performance was examined. This sterilization is the method that instantaneously heats the food particle surface and sterilizes without degradation in small size powders. Several advantages of it were mentioned and the effectiveness was confirmed. At present, it is possible to do sterilization, air vent and the packing at one stroke when this method is applied to the batch processing system. Although in the current study, this system has not been optimized yet for many kinds of foods and more improvement is required to apply the system to the food production, it is sure of that the principle of this method contributes to the development of the food industry.

REFERENCE

K. Fujiwara, H.Matsuo, T.Hiroe and M.Asakawa(1996), The size effect on microorganisms damaged by shock, Journal of Japan Society of Biosci.Biotec.and Agrochem., **70**, 15-18.

MICROSTRUCTURAL EVOLUTION AND SELF-ORGANIZATION OF SHEAR BANDS

M.A. Meyers[1], Q. Xue[1], Y. Xu[2], and V.F. Nesterenko[1]

[1]University of California, San Diego, La Jolla, CA 92093, USA
[2] Institute of Metals Research, CAS, China

ABSTRACT

Hat-shaped specimens deformed in a Hopkinson bar and specimens recovered from the collapse of thick-walled cylinders were used to generate strain rates of approximately $10^4 s^{-1}$ and shear strains that could be varied between 1 and 100. Shear bands were generated in Al-Li, Ti, Ti-6%Al-4%V alloy, AISI 304L SS. Transmission electron microscopy reveals, for Al-Li and AISI 304L SS, a number of features that include the observation of grains with sizes in the nanocrystalline domain. An evolutionary model, leading from the initial grain size of 15 µm to the final submicronic (sub) grain size describes the microstructural changes, by rotational dynamic recrystallization.

The shear-band initiation, propagation, as well as spatial distribution were examined under different global strains. The shear bands nucleate at the internal boundary of the cylindrical specimens and construct a periodical distribution at an early stage. The evolution of shear-band pattern during the deformation process reveals a self-organization character. The experimental shear-band spacings are compared with theoretical predictions that use the perturbation analysis and momentum diffusion. A new two-dimensional model is proposed for the initiation and propagation that treats initiation as a probabilistic process with a Weibull dependence on strain; superimposed on this, a shielding factor is introduced to deal with the disactivation of embryos.

KEYWORDS

shear bands, self organization, dynamic recrystallization, recrystallization

1 EXPERIMENTAL TECHNIQUES

Shear bands were generated by two methods:
(a) The hat-shaped specimen method, Fig. 1(a), uses a split Hopkinson bar in the compression mode to generate large shear strains in a small region (~200µm thick). This method was developed by Meyer and Manwaring (1986) and has been successfully used to generate shear localization regions in a number of metals (e.g., Andrade et al., 1994).
(b) The explosive collapse of a thick-walled cylinder under controlled and prescribed conditions. The experimental procedure consisted of subjecting thick-walled cylinders to controlled collapse by means of explosives placed on the periphery, in a cylindrical geometry with initiation at one of

the extremities. Nesterenko and Bondar (1994) and Nesterenko et al. (1998) describe the procedure in detail. Selected dimensions of copper tubes sandwiched in the sample control the collapse of thick-walled cylinder specimen. Figure 1(b) shows the process of cylinder collapse, with the shear bands marked schematically. They have spiral trajectories and initiate at the internal surface of the cylinder. The strain rate imparted is on the order of $10^4 s^{-1}$. The collapse of the cylinder generates the highest shear strains along the internal surface. The unstable deformation, which is initially homogeneous, gives rise to shear bands along the internal surface. These bands grow inward to the thick-walled cylinder. The shear band spacing, L_i, length, l_i, and the edge displacements, δ_i, were measured at different global strains.

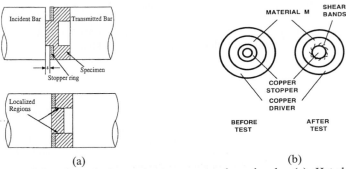

Figure 1. Experimental methods used to generate shear bands; (a) Hat-shaped specimen; (b) Thick-walled cylinder specimen

2 TRANSMISSION ELECTRON MICROSCOPY

The AISI 304 stainless steel exhibited, outside of the shear band, the structure characteristic of high-strain rate deformation, which had been systematically identified earlier by Staudhammer et al. (1981). It is characterized by twins and stacking faults, propitiated by the low stacking-fault energy of 304 SS. The microstructure inside of the shear band was radically different. Two principal domains could be identified:

(a) A region composed of nanoscale grains. Figure 2(a) shows these regions for 304 SS in both bright and dark field TEM. The grains are approximately 100-200 nm in diameter. These grains have clear boundaries and are equiaxed. This structure is similar to the ones observed in titanium (Meyers and Pak, 1986), copper (Andrade et al., 1994), Al-Li (Xu et al., 2001), and brass (Li et al., 2000). For an Al-Li alloy, the features shown in Figure 2(a) are analogous to the ones for 304SS. This has been attributed to a rotational recrystallization mechanism, which was proposed and quantitatively expressed by Meyers et al. (1997, 2001). Mataya, Carr, and Krauss (1982) were the first to analyze shear bands in stainless steel and they correctly identified the mechanism for the formation of these grains. Figure 3 shows the sequence of events leading to rotational dynamic recrystallization.

(b) A glassy region separated from the nanocrystalline region by an interface. Figure 4 shows this interface, with the glassy region at the right and the crystalline one at the left. High-resolution transmission electron microscopy confirms the amorphous nature of the material. This is a surprising finding and the first observation of a crystalline to amorphous transition in a shear band. Barbee et al. (1979) were able to produce the amorphous transition in 304 SS by sputter depositing it. However, this was only possible for a carbon concentration greater than 5 at.%.

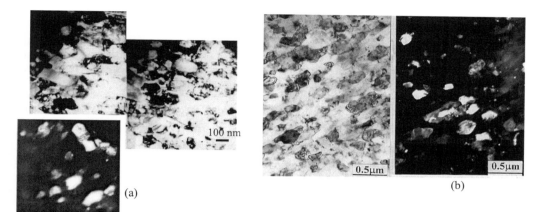

Figure 2: (a,b)Microcrystalline structure (d~100-200nm) and dark fields inside bands of (a)AISI 304 SS and (b)Al-Li alloy.

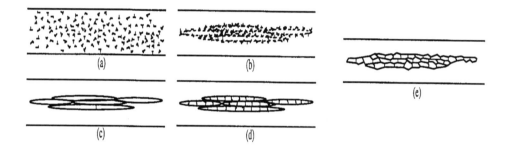

Figure 3: Schematic illustration of microstructural evolution during high-strain-rate deformation. (a) Randomly distributed dislocations; (b) Elongated dislocation cell formation; (c) Elongated subgrain formation; (d) Initial break-up of elongated subgrains; and (e) Recrystallized microstructure

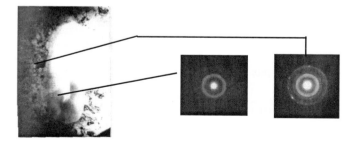

Figure 4: Interface between nanocrystalline and amorphous regions.

3 MEASURED AND PREDICTED SHEAR BAND SPACINGS

Most of the studies till this date have focused on a single band and assumed a one-dimensional configuration. Notable exceptions are the experimental contributions of Bowden (1980) in polymers and Shockey and Erlich (1981) in steels, and the analyses developed by Grady (1980), Wright and Ockendon (1996), Grady and Kipp (1987), and Molinari (1997).
Grady (1980) was the first to propose a perturbation solution to shear instability of brittle materials. Wright and Ockendon (1996) also developed a theoretical model, based on small perturbations. The predicted spacing is expressed as:

$$L_{WO} = 2\pi \left(\frac{m^3 kC}{\dot{\gamma}_0^3 a^2 \tau_0} \right)^{1/4} \tag{2}$$

Where a is the thermal softening coefficient; k is the thermal conductivity; C is the heat capacity; τ_0 is the shear flow stress; $\dot{\gamma}_0$ is the strain rate; and m is the strain rate sensitivity. Grady and Kipp (1986) extended Mott's (1947) analysis for dynamic fracture to deformation localization. Momentum diffusion was considered as the dominant mechanism of shear bands. The spacing is:

$$L_{GK} = 2 \left[\frac{9 kC}{\dot{\gamma}_0^3 a^2 \tau_0} \right]^{1/4} \tag{3}$$

Molinari (1997) modified the WO model by introducing strain-hardening effect:

$$L_{M'} = 2\pi \left[1 - \frac{3}{4} \frac{\rho c}{\beta \tau_0^2} \frac{n(1-aT)}{\beta a \gamma} \right]^{-1} \cdot \left[\frac{kCm^3(1-aT_0)^2}{(1+m)\dot{\gamma}_0^3 a^2 \tau_0} \right]^{1/4} \tag{4}$$

Where n is the strain-hardening index.
The measured spectra of shear bands in 304SS, CP titanium and Ti-6Al-4V alloy obtained by the thick-walled cylinder method are shown in Figure 5 at two global effective strains: 0.55 and 0.92 (for 304SS and CP Ti, Figs 5(a)-(f)); 0.13 and 0.26 (for Ti-6Al-4V, Fig. 5(g,h)). The numbers of shear bands at both the early and later stages are similar in SS 304 and Ti. There is no significant effect of grain size on shear band patterning for 304SS from Figs. 5(a)-(d).
Table 1 shows the comparison of experimental and predicted results. The grain size has only a minor effect on the spacing for 304SS, in the range investigated. The predicted results from Grady-Kipp model are roughly 20 times larger than measured values for Ti and 304SS, and twice for Ti-6Al-4V. On the other hand, the WO and M models provide reasonable estimates for Ti and 304SS but not for Ti-6Al-4V. The failure of these theoretical predictions for Ti-6Al-4V alloy suggests that some mechanisms of shear band development have not been included into the above theories. One very important factor that is not incorporated into these theories is the two-dimensional nature of interactions among growing bands, which increases with their size. This was recognized by Nemat Nasser et al. (1976) for parallel propagating cracks. A conceptually similar analysis is presented in the next section.

4 TWO-DIMENSIONAL MODEL FOR SHEAR BAND SPACING

The failure of the one-dimensional models to explain the evolution of shear–band spacing led to proposed model below. Initiation of shear bands is assumed to require a critical strain. Heterogeneous microstructural or surface effects (boundary geometry, defects, and orientation of grains, etc.) determine the range of strains in which the nucleation takes place. A heterogeneous

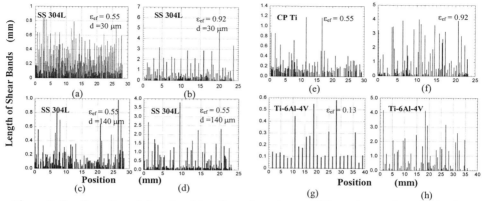

Figure 5: Configuration of shear-band lengths and spacings at different imposed strains for (a, b) AISI 304 SS, =30 μm ; (c, d) AISI 304 SS, d=140 μm; (e, f) CP titanium; (g, h) Ti-6Al-4V alloy

TABLE 1
PREDICTIONS AND EXPERIMENTAL SPACINGS FOR 304SS TI, AND TI-6%AL-4%V

Spacing(mm)	Exp. Data Initial level	L_{WO}(mm)	L_{GK}(mm)	L_{MO} (mm) (Without strain hardening)	L_{MO} (mm) (with strain hardening)
SS 304L	0.12	0.33	2.40	0.29	--
CP Titanium	0.18	0.29	2.13	0.24	0.64
Ti-6Al-4V	0.53	0.1	1.15	0.09	0.10

nucleation process, which is characterized as the selective activation of sites, will be assumed. The probability of nucleation is given by P (V_0,S_0), in a reference volume, V_0, or surface, S_0, depending on whether initiation occurs in the bulk or on the surface. It can be described by a modified Weibull distribution, using strain as the independent variable:

$$P(V_0, S_0) = 1 - \exp\left[-\left(\frac{\varepsilon - \varepsilon_i}{\varepsilon_0 - \varepsilon_i}\right)^q\right] \quad (5)$$

where ε_i am the critical strain below which no initiation takes place; ε_0 is the average nucleation strain (material constant); ε is the variable; and q is a Weibull modulus. For different materials the nucleation curve can have different shapes and positions, adjusted by setting q, ε_i, and ε_0. For Ti and Ti-6Al-4V, the mean nucleation strains are selected as 0.4 and 0.12, respectively, to best fit the experimental results; q was given values of 2, 3, 6, and 9, providing different distributions. Figures 6 (a) and (b) show the predicted distributions of initiation strains for Ti and Ti-6Al-4V, respectively.

Due to thermal softening inside bands, each growing band generates a shielded region around itself. Thus, the bands that actually grow can be a fraction of the total possible initiation sites. Three factors govern the evolution of self-organization: (a) the strain rate, $\dot{\varepsilon}$; (b) the velocity of growth of shear band, V; (c) the initial spacing, l. Different scenarios emerge, depending on the growth velocity V. The activation of embryos at three times, t_1, t_2, and t_3, is shown in Figure 7.

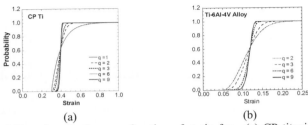

Figure 6. Probability of nucleation as a function of strain for (a) CP titanium; (b) Ti-6Al-4V alloy.

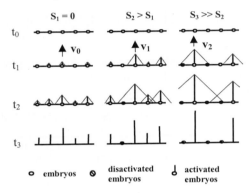

Figure 7: Two-dimensional representation of concurrent nucleation and shielding.

When V is low, the embryos are all activated before shielding can occur, and the natural spacing l_0 establishes itself. As V increases, shielding becomes more and more important, and the number of disactivated embryos increases. The shielded volume is dependent on the velocity of propagation of stress unloading and is given by k_1V, where $k_1<1$. The width of the unloaded region is $2k_1$ times the length of the shear band. The shielding effect can be expressed by S:

$$S = 1 - \frac{\dot{\varepsilon} L}{k_0 k_1 V} \quad (6)$$

k_0 defines the range of strains over which nucleation occurs. It can be set as $k_0 = 2(\varepsilon_0 - \varepsilon_i)$. It correctly predicts an increase in shielding S with increasing V, decreasing $\dot{\varepsilon}$, and decreasing L. For an extremely large velocity of propagation of shear band, the critical time is very small and the shielding factor is close to one, which means almost complete shielding. The probability of nucleation under shear band shielding, P (L), is obtained by multiplying Eqn. 5 by (1-S):

$$P(L) = \left(\frac{\dot{\varepsilon} L}{2(\varepsilon_0 - \varepsilon_i)V}\right)\left\{1 - \exp\left[-\left(\frac{\varepsilon - \varepsilon_i}{\varepsilon_0 - \varepsilon_i}\right)^q\right]\right\} \quad \text{if} \quad t_{cr} < \bar{t} \quad (7a)$$

$$P(L) = \left\{1 - \exp\left[-\left(\frac{\varepsilon - \varepsilon_i}{\varepsilon_0 - \varepsilon_i}\right)^q\right]\right\} \quad \text{if} \quad t_{cr} \geq \bar{t} \quad (7b)$$

When $S = 0$ no shielding effect exists and all nuclei grow. If, in other extreme case, $S = 1$, no nucleation can happen. Figure 8(a) shows predicted evolutions of nucleation probabilities as a function of increasing strain, for different values of the shielding factor, S: 0, 0.2, 0.4, 0.6. This simple model shows how the initial distribution of activated embryos can be affected by different parameters. Xue et al. (2001) present a more complete version of this analysis. It explains, qualitatively, the smaller spacing of shear bands in Ti, as compared to Ti-6Al-4V. The shear-band spacing, corrected for shielding, is represented by:

$$L_S = \frac{L_{WO}}{P(L)|_{l>\bar{l}}} = \frac{L_{WO}}{(1-S)}. \tag{8}$$

L_{WO} is the Wright-Ockendon spacing. This spacing is plotted as a function of plastic strain, at a fixed value of S, in Figure 8(b). It is clear that the shear-band spacing decreases with strain until a final, steady state value is reached (Fig.8 (b)). Using the calculated shielding factors S of 0.14 for Ti and 0.89 for Ti-6Al-4V, the corresponding values for L_S are 0.34 and 0.9 mm, respectively. These values are in agreement with the experimental results (0.18 and 0.53 mm, respectively)

During growth, an analogous selection process takes place, leading to a discontinuous increase in the spacing of the shear bands as their length increases, as described by Xue et al (2000, 2001). At a certain length l_i, the spacing is L_i; the growth becomes unstable at a critical length $l_{cr,i}$, and alternate SB's grow with a new spacing L_{i+1}; the other shear bands stop growing. The mathematical representation of the step function is shown as:

$$L = L_0 + \sum_{j=1}^{n} k'_j \cdot H(l - l_{cr,j}) \tag{9}$$

where $H(l-l_{cr,j})$ is a Heaviside function. The parameters k_j' can be expressed as:

$$k_j' = f(V, L_0, \varepsilon_i, \varepsilon_p) \tag{10}$$

Where V is the shear band propagation velocity, L_0 is the initial spacing, ε_i is the critical strain for initiation and ε_p is the critical strain for propagation.

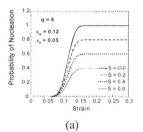
(a)

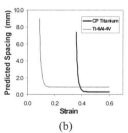
(b)

Figure 8: Probability of nucleation incorporating shielding factor; (a) schematic; (b) calculated shear-band spacing for titanium and Ti-6Al-4V alloy.

5 CONCLUSIONS

The microstructural evolution inside of adiabatic shear bands in AISI 304 SS and Al-Li was observed by transmission electron microscopy and it is suggested that rotational dynamic recrystallization is responsible for the generation of the recrystallized structure with grains in the 100-300 nm domain. An amorphous region was also found in AISI 304 SS. The evolution of multiple adiabatic shear bands was investigated in stainless steel, titanium, and Ti-6%Al-4%V alloy through the radial collapse cylinder technique under high-strain-rate deformation ($\sim 10^4$ s^{-1}). The shear-band spacing is compared with current theoretical predictions. These one-dimensional models can not provide accurate estimation of shear band spacing in Ti6%Al4%V alloy. The two-dimensional character of self-organization has to be considered. A novel analytical description, in which a distribution of embryos (potential initiation sites) is activated as a function of strain (greater than a threshold) according to a Weibull-type distribution, is proposed. The model incorporates embryo disactivation by stress shielding as well as selective growth of shear bands. The imposed strain rate, embryo distribution, and rates of initiation and propagation determine the evolution of shear band configurations.

ACNOWLEDGMENTS: Work supported by U.S. Army Research Office MURI Program No. DAAH004-96-1-0376 and by the Natural Science Foundation of China Grant No. 50071064.

6 REFERENCES

Andrade, U. R., Meyers, M. A., Vecchio, K. S., and Chokshi, A. H. (1994)., *Acta Met.,* **42**,3183-3195
Barbee, T. J. Jr., Jacobson, B. E., and Keith, D. L. (1979), *Thin Solid Films,* **63**,143-150.
Bowden, P. B. (1970), *Phil. Mag.,* **22**, 455-462.
Grady, D. E. (1980), *J. Geophys. Res.,* **85**, 913-924.
Grady, D.E., and Kipp, M.E. (1987), *J. Mech. Phys. Solids,* **35**, 95-118.
Li, Q., Xu,Y. B., Lai, Z. H., Shen, L. T., and Bai, Y. L. (2000), *Mat. Sci. Eng.,***A276**,127.
Mataya, M. C., Carr, M. J., and Krauss, G. (1982), *Met. Trans. 133A*,1263-1274
Meyer, L. W., and Manwaring, S. (1986), in *Metallurgical Applications of Shock-Wave and High-Strain-Rate Phenomena,* M. Dekker, ,pp. 657-674.
Meyers, M. A., and Pak, H.-r. (1986), *Acta Met.,* **34**, 2493.
Meyers, M. A., LaSalvia, J. C., Nesterenko, V.F., Chen, Y. J., Kad, B. K.(1997), *Proc. 3rd Intl. Conf. Recrystallization and Related Phenomena,* (REX'96). McNelley, Ed.,p.279.
Meyers, M. A., Xue, Q., Nesterennko, V. F., LaSalvia, J. C. (2001), *Mat. Sci. Eng.,* in press.
Molinari, A. (1997), *J. Mech. Phys. Sol.* **45**, 1551-75.
Mott, N. F. (1947), *Proc. Roy. Soc.,***189**, 300
Nemat-Nasser, S., and Keer L.M. and Parihar, K.S. (1978)., *Int. J. Solids Structures,* **14**, 409-430.
Nesterenko, V. F., and Bondar, M. P. (1994), *DYMAT J.*,**1**,243-251.
Nesterenko, V.F., Meyers, M.A., and Wright, T.W. (1998), *Acta Mat.,* **46**, 327-340
Nesterenko,V. F., Xue, Q., and Meyers, M. A. (2000), *J. Phys.*IV, **10**, 9-269-9-274.
Shockey D.A. and Erlich D.C. (1981), In *Shock Waves and High-Strain-Rate Phenomena in Metals* (eds. Meyers, M.A. and Murr, L. E.), Plenum Press, New York, pp. 249-261.
Staudhammer, K. P., Frantz, C. E., Hecker, S., S., and Murr, L. E.(1981*),* in *Shock Waves and High-Strain –Rate Phenomena in Metals,* M. Dekker, pp.91-112.
Stout, M.G., Follansbee, P.S. (1986), *Trans. ASME., J of Eng Mat. & Tech,* **108**, 344-53.
Wright, T.W., and Ockendon, H. (1996)., *Int. Journal of Plasticity,* **12**, 927-34.
Xu, Y. B., Zhong, W. L., Chen, Y. J., Shen, L. T., Liu, Q., Bai, Y. L. and Meyers, M. A. (2001), *Mat. Sci, Eng.,***A299**,287-295.
Xue Q., Nesterenko, V. F., and Meyers, M. A.(2000), in *Shock Compression of Condensed Matter-1999* (eds. M.D. Furnish et al.),AIP, NY,pp. 431-434.
Xue, Q., Meyers, M. A., and Nesterenko, V. F. (2001), *Acta Mat.,* submitted.
Xue, Q., Nesterenko, V. F, and Meyers, M.A. (2000), in *Shock Compression of Condensed Matter-1999* (eds. M.D. Furnish et al.), AIP, New York, pp. 431- 434.

# VISAR INTERFEROMETRY: SIMPLE IN DESIGN, VERSATILE IN APPLICATION

A BRIEF HISTORY OF VISAR DEVELOPMENT

Dr. William M. Isbell
ATA Associates
Santa Barbara, California, USA

ABSTRACT

As evidenced by the several hundred papers and reports written about VISARs and their applications over the past 30 years, this technique of measuring the velocity-time history of specimens set into sudden motion has become a standard in laboratories around the world. This paper presents a brief history of the development of VISARs from large optical bench instruments, administered by a cadre of trained scientists and technicians, to today's compact, pre-aligned system, used by students and researchers alike.

The paper describes the development of interferometer systems for shock wave and projectile measurements over a period of more than 35 years. Since manuscript page count limitations dictate only a sketchy treatment of several of the more interesting aspects of VISAR Interferometry, the reader is referred to the list of references for more detail.

The author apologizes in advance for omitting several promising VISAR configurations and many deserving references. Development of VISAR interferometry has been the product of many talented researchers, only a few of whom are mentioned in the paper.

KEYWORDS

VISAR, velocity interferometry, shock wave instrumentation, wave profiles, projectile motion

BACKGROUND

The Displacement Interferometer

VISAR Interferometry has, as its antecedents, two distinct types of laser interferometers; the first, a *displacement interferometer*, uses a low-power laser reflected from a specular (mirror-like) surface (see Figure 1a). Fringes are produced as the surface moved\s, one fringe for each $\lambda/2$ (~0.32 µm, for a HeNe laser of wavelength $\lambda = 632.8$ nm). Velocity is obtained by differentiating the displacement-time record.

Although the instrument is appealing in its simplicity, two problems prevented its wide-spread use.

1) Fringe frequency is proportional to velocity, producing a signal of ~300 MHz at 100 m/s. Since many tests require higher velocities, detector and recording bandwidth limitations became severe.

2) The mirror surface must remain specular under high pressure shock waves (or the intensity of the reflected light is decreased to unusability) and must not tilt substantially (or the reflected beam no longer enters the interferometer).

The Velocity Interferometer

The first problem, that of bandwidth, was solved by arranging the interferometer optics such that the fringes were proportional to *velocity* rather than to displacement (Barker and Hollenbach, 1965). This is accomplished by splitting the input beam and directing the beams along paths of unequal length. Upon recombination, interference fringes are formed.

Doppler-derived changes in frequency, produced by reflection of the beam from the moving surface, are received at the recombining beamsplitter (see Fig. 1b) at a slightly earlier time (order 1 ns) from the "Reference Leg" than were the signals from the "Delay Leg". While the surface is accelerating or decelerating, fringes are produced, proportional to the surface velocity.

$$V(t) = N(t) \cdot \tau/2 \quad (1)$$

$$\tau = \lambda c /(L_D - L_R) \quad (2)$$

where $V(t)$ = time varying velocity, $N(t)$ = time varying fringe count, τ = delay time between the two legs, λ = laser wavelength, c = speed of light, and L_D and L_R are Delay Leg and Reference Leg lengths, respectively. More rigorous treatments of basic VISAR equations and relationships can be found in the literature (Clifton 1970, Hemsing 1978).

By adjusting the length of the delay leg, fringe frequency can be controlled. The fringe constant, K, was introduced to represent the sensitivity of the instrument,

$$K = \lambda/2\tau \text{ (m/s/fringe)} \quad (3)$$

$$V(t) = K \cdot N(t) \quad (4)$$

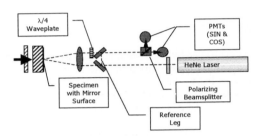

a) Displacement Interferometer

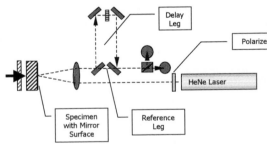

b) Velocity Interferometer

Fig. 1. The VISAR was preceded by the two interferometers shown here

A large K indicates that fewer fringes are produced for a given change in specimen velocity.

The second problem, that of keeping the mirror reflective and of not introducing tilt into the surface remained unsolved for several years, during which time hundreds of tests were conducted, primarily using gas guns to impact specimens held at the muzzle (Johnson and Barker 1969, Isbell and Christman 1970, Christman et al 1971, Asay 1975, Isbell 1993). Figure 2 (Isbell 1999) demonstrates the substantial amount of information which can be obtained in such a test.

- Elastic and Plastic Compressive Wave Velocities
- Elastic and Plastic Release Wave Velocities
- Point on the Hugoniot Equation of State
- Dynamic Material Strength (proportional to the "pullback" velocity)
- Thickness of the spall piece (derived from the ringing frequency in the spall signal)
- Dislocation Multiplication (derived from the damping in the spall signal)

Few, if any, other shock wave measurement technique furnishes such a wealth of information in a single test.

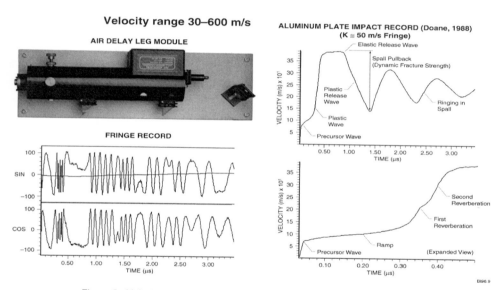

Figure 2. Velocity record from impact test (copper impacting copper at 38.9 m/s).

A BRIEF HISTORY OF VISAR DEVELOPMENT

The VISAR

The problems of mirrors and tilt were solved in the early 1970s with the introduction of the VISAR Interferometer (Velocity Interferometer System for Any Reflector) (Gillard et al 1968, Barker and Hollenbach 1972, Isbell 1976). By using a diffuse (Lambertian) surface, instead of a specular surface, reflected light was relatively unaffected by tilt (tilt to perhaps 5-10 degrees, depending on the material and the surface treatment). Moreover, high pressure shock waves no longer destroyed the (already destroyed) mirror surface and the maximum pressures at which the interferometer could be used was raised from a few hundred kilobars to several megabars.

Glass Etalon Delay Legs

By placing a glass "etalon" in the Delay Leg path (Fig. 3), the two images are at the same *optical* distance (allowing fringes from a diffuse source, where *spatial* coherence has been destroyed, but *temporal* coherence is maintained), but *geometrically* are at different distances (producing the necessary delay time).

Velocity interferometer systems have progressed from single photomultiplier (PMT) systems (SIN), to dual PMTs (SIN & COS), which are capable of distinguishing accelerations from decelerations, to a 3-PMT system, which added a beam intensity monitor (BIM) to adjust fringe signals modified in amplitude by shock waves affecting the reflecting surface of the target (Isbell 1981, Isbell and Fuller 1983).

In 1979, a modification of the 3 PMT system combined the SIN and COS signals with their previously unused conjugate beams, -SIN and –COS, to produce signals with amplitudes 2•SIN and 2•COS. This "push-pull" configuration (see Fig. 4) is in standard use today (Hemsing 1979).

Air Delay Legs

A limitation on instrument sensitivity was the length of glass which could be manufactured with sufficient wave transmission tolerance to maintain high fringe contrast. An 8.5" (21.6 cm) length of glass produced a K =200 m/s/fringe, sufficient for tests with a peak velocity of perhaps 100 m/s and with a velocity resolution of ~4 m/s. Tests at lower velocities or tests demanding greater velocity resolution required longer glass etalons, produced at sometimes prohibitive costs.

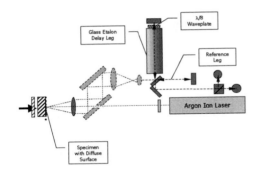

Fig. 3. Basic VISAR design, utilizing a glass etalon for the delay leg

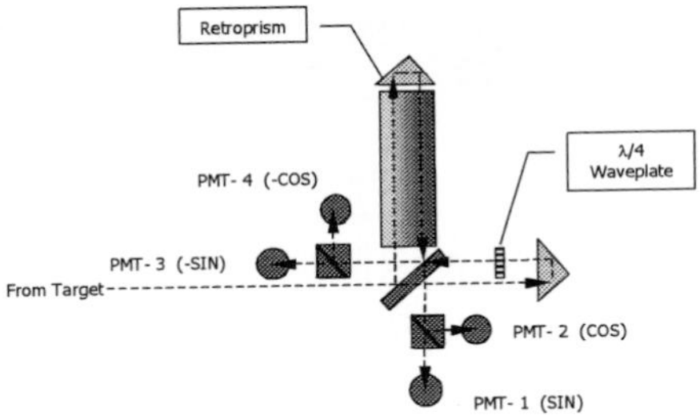

Fig. 4. Modified design makes maximum use of available light by combining the four conjugate beams (see text)

One solution to this limitation was the introduction of the "Air Delay Leg" (Amery 1976, Isbell 1980, Froeschner et al 1996), in which the beams were delayed by two, two-lens systems with slightly different focal lengths. This was later simplified and made more stable in the single two-lens configuration shown in Figure 5 (Isbell 1983, 1987, 1990).

By folding the air delay leg over a 2 m path, VISARs with fringe constants to K = 12 m/s/fringe were constructed. The velocity range was extended to peak velocities below 10 m/s with a velocity resolution of ~10-20 cm/s.

Another adaptation of the air delay leg design is shown in Fig. 6. This compact, "folded mirror" design has produced sensitivities to 40 m/s/fringe, with a resolution of 1 m/s. The design is used in the ATA Associates Model 605 VISAR and has the advantage of interchangeability with other, glass-type, etalons.

The problem of thin etalons was solved by the "Differential Etalon" system (Isbell 1996). The Reference Leg holds an etalon which is slightly shorter than the etalon in the Delay Leg. The difference in lengths between the two etalons produces the (small) time delay. Time resolution is 10-15 picoseconds for velocities >10,000 m/s. Foil velocities to >20,000 m/s have been measured with such a system.

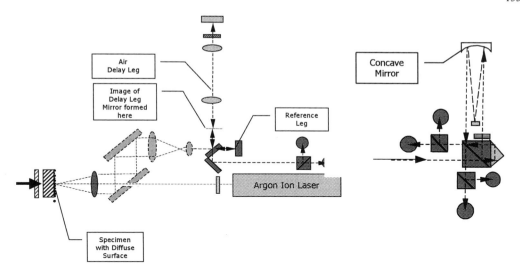

Fig. 5. The air delay leg lowered the velocity range to peak velocities less than 12 m/s

Fig. 6. Compact air delay leg folded mirror design

Increased Stability, Reduced Size

With the introduction in 1988 of the compact VISAR (Isbell 1988, Isbell 1991), VISAR systems were now reduced in size from a large optical bench, filled with individual components, to a single rack-mounted chassis, with all optical components cemented together in permanent alignment. The stability of this configuration was furthered enhanced by the development of the "Fixed Cavity VISAR" system by Sandia National Laboratory (Sweatt, Stanton, and Crump, 1990) and incorporated in the ATA Associates Model 605-FCV VISAR system.

VISAR systems were now pre-aligned, eliminating the need for periodic adjustments to maximize fringe contrast. This feature, and the reduction in size made possible by placing all photodetectors and electronics on the same chassis as the optical components, has provided interferometer systems suitable for operation by students and researchers who may not be interested in the VISAR itself, but are concentrated on their experiments.

The VISAR had come of age.

LASER ILLUMINATION

Because of the low reflectivity of diffuse surfaces, (typically < 10%), VISARs inherently require high power lasers. Additionally, the laser must have a relatively long coherence path length (>> than the length of the delay leg). The latter requirement dictates a *single frequency* laser, rather than a *single line* For argon ion lasers, this requirement is met by placing an etalon in the laser cavity.

Argon Ion Lasers

In the early days of VISAR development, only the argon ion laser met both requirements and, even today, most VISARs are illuminated with argon ion lasers. However, the size, the cost, the requirement for special AC power, and the necessity of cooling the laser resulted in examining alternative light sources.

Laser Diodes

The first laser diodes with sufficient power (~100 mW) for VISAR applications lased in the infrared and had difficulty producing sufficient coherence path length for standard VISAR systems. Nevertheless, the low cost and compactness of both the laser and the IR photodiode receivers motivated researchers to adopt this new form of illumination (Isbell 1988).

With the more recent introduction of visible laser diodes with sufficient power, coherence path length, and stability, VISAR researchers and manufacturers have incorporated them into their designs.

PHOTODETECTORS AND RECORDING SYSTEMS

Photomultipliers and Photodiodes

Requirements for nanosecond time resolution, linear output, and low noise have placed stringent requirements on photomultipliers (PMTs). Early instruments used bulky and costly PMTs, operating at -5000V. Modern VISARs use either lower cost PMTs with specialized voltage strings or use photodiodes, placed directly in the fiber optic cables connecting to the digitizing recorders (Froeschner 1996).

Streak Cameras

Higher time resolution is provided by using streak cameras to record VISAR data, at the expense, to some degree, in velocity resolution and in total recording time. The records are digitized and the individual lines averaged to obtain a single record.

This type of recording becomes attractive when picosecond time resolution is required, such as in tests of laser-driven foils or with specimens with small dimensions.

LIGHT COLLECTION SYSTEMS

Direct Beam Technique

Early VISARs used the "direct beam" configuration, in which the beam was transported through the air from laser to target and from target to VISAR. While this technique was light-efficient, the high intensity beams exposed to the air were a hazard to the eyes and skin. Additionally, it frequently is difficult to position the argon ion laser sufficiently close to the test apparatus.

Fiber Optic Cable Technique

Fiber optic cables, from laser to target and from target to VISAR, provide a convenient and safe method for transmitting the laser beams (Isbell 1988, 1996). With such a system, the laser, the test apparatus, and the recording equipment can be placed in convenient positions, tens of meters apart.

Two basic types of light collection systems are used with fiber optic cables. A small lens system is expended in each test, while a larger light collection system is placed behind a barrier and views the test through an expendable mirror. An advantage of the expendable lens technique is its ability to focus the laser spot to <100 μm, whereas with the longer focal length lens of the non-expendable system, spot sizes of few mm are typical.

Multiple Fiber Technique

One disadvantage of VISARs, that only a single point could be measured on the specimen surface, was solved by devising a multiple fiber probe system. As many as 5-10 fibers are arranged in a compact array and focused on the target, either though one lens or several. At the VISAR, the beams are sent on parallel, non-interfering paths through the optical system, then collected and sent to multiple recording systems.

VISAR CONFIGURATIONS

Over the past decade, the simple VISAR has been transformed into many specialized configurations. The following paragraphs outline two of the more useful and unusual designs.

"Line" VISAR

An imaging VISAR system with quadrature coded outputs consisting of a laser illuminated line on the experimental surface, optics to image the line through the interferometer to produce quadrature (push-pull) signals, and either a streak camera or other detectors for recording data. It provides the capability to continuously measure the velocity histories of many points during a single experiment.

Imaging "White Light" VISARs

An imaging white light interferometer, consisting of two imaging superimposing Michelson interferometers in series with the target, imprints the same delay, independent of ray angle, position, and wavelength (Erskine and Holmes 1996, Gidon and Behar 1988). This capability provides:

- Any illumination source may be used, including flash lamps and multi-wavelength sets of lasers.
- 2-dimensional maps of moving surfaces can be measured.
- Both radial and transverse velocities can be measured when the illuminating and viewing beams are non-parallel.

Figure 7 is a line drawing of a white light VISAR, as given in the referenced paper. If interferometer delays, τ_1 and τ_2, match within the coherence path length of the source, partial fringes are produced at the output which vary with $(\tau_1 - \tau_2)$.

SUMMARY

This brief history describes the chronology of the development of VISAR interferometry from a large, unwieldy instrument, difficult to align and maintain, to the present day VISAR—compact, permanently aligned, and used both by trained and neophyte scientists and students. As evidenced by the rapid pace of continued development, the end of this cycle is not yet in sight.

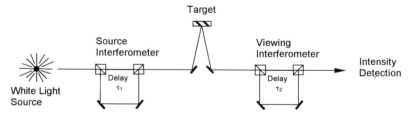

Figure 7. Line Drawing of a White Light VISAR

REFERENCES

Amery, B.T. (1976), "Wide Range Velocity Interferometer", 6th Symposium on Detonation, San Diego, CA,
Asay, J. R., (1975) "Shock and Release Behavior in Porous 1100 Aluminum", J. Appl. Phys. 46
Barker, L. M. and R.E. Hollenbach (1965), Rev. Sci, Instr. 36, 4208
Barker, L.M. and R.E. Hollenbach (1972), "Laser Interferometer for Measuring High Velocities of Any Reflecting Surface", J. Appl. Phys., Vol. 43, No. 11, November.
Christman, D. R., W. M. Isbell, and S. G. Babcock (1970), "Measurements of Dynamic Properties of Materials, Vol. V: OFHC Copper", General Motors Materials and Structures Laboratory, DASA-2501, July 1971 (AD728846)
Clifton, R. J. (1970), "Analysis of the Laser Velocity Interferometer", J. Appl. Phys., vol. 41, 3535
Erskine, D. J. and N. C. Holmes (1996), "Imaging White Light VISAR", 22nd International Congress on High peed Photonics and Photography, Santa Fe, New Mexico
Froeschner, K. E., et al (1996), " Subnanosecond Velocimetry with a New Kind of VISAR" 22nd International Congress on High peed Photonics and Photography, Santa Fe, New Mexico
Gidon, S. and G. Behar (1988), Multiple-Line Laser Doppler Velocimetry", Appl. Optics 27, 2315-2319
Gillard, C. W., G. S. Ishikawa, J. F. Peterson, J. L. Rapier, J. C. Stover, and N.L. Thomas (1968), Lockheed Report No. N-25-67-1 (unpublished)
Gooseman, D.R. (1975), J. Appl. Phys. vol. 45, p3516
Hemsing, W. F. (1979), "Velocity Sensing Interferometer (VISAR) Modification", Rev Sci Instr, 50(1), Jan.
Hemsing, W. F., (1991) A. R. Mathews, R. H. Warnes, M. J. George and G. R Whittemore, "VISAR: Line-Imaging Interferometer", 1991 American Physical Society Topical Conference, Williamsburg, VA, June 17-211
Isbell, W. M. and J.R Christman (1970), "Shock Propagation and Fracture in 6061-T6 Aluminum from Wave Profile Measurements", General Motors Materials and Structures Laboratory, DASA-2419, (AD705536)
Isbell, W. M. (1976), "The Versatile VISAR: An Interferometer for Shock Wave and Gas Gun Diagnostics", Proceedings, 26th Annual Meeting of the Aeroballistic Range Association,
Isbell, W. M. (1980), "Hypervelocity Impact Shock Propagation Measurements Using the VISAR Interferometer", Proceedings, 31st Meeting of the Aeroballistic Range Association
Isbell, W. M. (1981), "A Combined Displacement/Velocity Interferometer for Impact Measurements at 0.1 to 100 m/s", Proceedings, 32nd Meeting of the Aeroballistic Range Association
Isbell, W. M. (1983), "Laser Interferometry for Accurate Measurements of Projectile Motion", Proceedings, 34th Meeting of the Aeroballistic Range Association
Isbell, W. M. and P. W. W. Fuller (1983), "Wide Range, High Resolution Measurements of Projectile Motion Using Laser Interferometry", 27th Annual Meeting, SPIE and High Speed Photonics and Videography Conference, TR-16-83
Isbell, W. M. (1987), "Initial Tests of VISAR Interferometry to Measure E.M. Launcher Projectile Motion", Proceedings, 38th Meeting of the Aeroballistic Range Association
Isbell, W. M. (1988), "An Infrared VISAR for Remote Measurement of Projectile Motion", Proceedings, 39th Meeting of the Aeroballistic Range Association
Isbell, W. M. (1990), "Interferometric In-Bore Velocity Measurements of Electromagnetically-Launched Projectiles", Proceedings, 41st Meeting of the Aeroballistic Range Association
Isbell, W. M. (1991), "A Simplified, Compact VISAR: Concept and Construction", Proceedings, 42nd Meeting of the Aeroballistic Range Association
Isbell, W. M. (1996), "Extending the Range of the Third-Generation VISAR from 30 m/s to 30,000 m/s", Proceedings, 47th Meeting of the Aeroballistic Range Association
Isbell, W. M. (1999), "Modern Instrumentation for Measurements of Shock Waves in Solids", presented at the Japanese Shock Wave Symposium, Tokyo Japan
Isbell, W. M. (1993), in "Measurements of the Dynamic Response of Materials to Impact Loading", Doctoral thesis, Shock Wave Research Center, Tohoku University, Sendai Japan Revised version to be published 2002
Johnson, J. N. and L. M. Barker (1969), "Dislocation Dynamics and Steady Plastic Wave Profiles in 6061-T6 Aluminum", J. Appl. Physics, Vol. 40, p4321-4334
Sweatt, W.C., P.L. Stanton, and O.B. Crump, Jr. (1990), "Simplified VISAR System", SAND90-2419C, SPIE, Vol. 1346, July
Yoshida, K. (1996), private communication, National Institute for Material and Chemical Research, Tsukuba, Japan

EMISSION SPECTROSCOPY OF HYPERVELOCITY IMPACTS

D. H. Ramjaun, M. Shinohara, I. Kato and K. Takayama

Shock Wave Research Center, Institute of Fluid Science,
Tohoku University, 2-1-1 Katahira, Aoba, 980-8577 Sendai, JAPAN

ABSTRACT

The light flash emitted during the impacts of 5 km/s projectiles on thin bumpers has successfully been measured by a spectroscopic system in the ballistic range of the Shock Wave Research Center of Tohoku University. The system was used to acquire time-integrated spectra of the front jetting cloud during impacts of projectiles made of high-density polyethylene or magnesium on 2 mm thick aluminum bumpers. Preliminary time-resolved spectroscopy of the event was also done. The measurements were focused on the near UV region of the spectrum. Both impacts at normal and oblique angles were tested.

KEYWORDS

Hypervelocity impact, impact flash, spectroscopy, space debris, ballistic range

INTRODUCTION

For the last decades, researchers have been studying the impact phenomenon with the ultimate goal of designing better shielding for spacecrafts orbiting the earth against space debris. The exact mechanism by which the impacting and target material undergo fracture and ablation is however not clearly understood. Generally, it is agreed that upon impact, strong shock waves propagate both along the projectile and target. Simultaneously, rarefaction waves are created, satisfying the zero stress condition at the free surfaces of the material involved (see Figure 1). Fracture is believed to occur through a multiple-spalling process in which debris sizes are smaller with increasing velocity of the projectile. As the latter further penetrates the target, jets of fine particles are ejected from the sides at very high velocity, creating a briefly radiating cloud in the surrounding gas. On the opposite side of the bumper, a debris cloud is formed consisting of fragments from both the projectile and target. The energy partitioning mechanism by which the material is ejected and vaporized is not clear. Temperature measurements of the jetting vapor can therefore provide a better understanding of the physics of the impact vaporization.

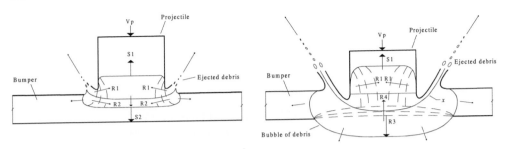

Figure 1: Early stages of hypervelocity impact, from Anderson and Mullin (1988)

Early studies by Jean and Rollins (1970) have shown that the radiation emitted from the fast jetting of debris from the front side consists mainly of atomic and molecular emission. This emission emission, if spectrally resolved, can give a lot of information about the nature of the impact: material involved, correlation of flash intensity with velocity and angle of impact. Careful analysis of these spectra can even lead to the temperature reigning in this region.

So, spectroscopic study of the impact flash can be very useful in understanding the impact phenomenon. There are many experimental techniques that are used to analyze the impact flash. One of the mostly widely used detectors is the photomultiplier tube. Though each method has its own advantages and disadvantages, the most efficient one, however, is to use high-resolution spectrometers. In the past, very few such studies have been done to analyze the jetting cloud. However, the work of Sugita and Shultz (1998) can be cited in which the emission lines of calcium were measured during impacts of quartz spherical projectiles on dolomite. Comparison of the experimental spectra with calculated ones yielded temperature estimations of the impact vapors formed in these experiments.

Similarly, in the Shock Wave Research Center, a study has been initiated in which spectroscopic measurements in the near UV region of hypervelocity impacts of high-density polyethylene projectiles on thin bumper shields have been carried out. The CN violet band has been chosen as candidate emitter because of its high emission intensity in the present experimental conditions and relatively straightforward numerical simulation. In this paper, the results obtained so far are presented. A streak system has also been used which time-resolved spectroscopy of the impact flash was acquired. However this technique is yet to be improved and only very low-resolution results were obtained so far with this device.

EXPERIMENTAL METHOD

Setup

The experiments have been carried out in the two-stage light gas gun of the Shock Wave Research Center of Tohoku University. Details of its operating principle can be found in Nonaka et al. (1998) Projectiles made of high-density polyethylene or magnesium and weighing 2 g are fired at around 5 km/s on 2.2 mm thick aluminium targets (Al 2017). Impact tests on different target materials such as FML (Fiber Metal Laminate) and composite (CFRP) have also been done. As shown in Figure 2 the emission radiated briefly upon impact is captured by a condenser lens placed at an angle of 15° and is guided by an optical fiber into a spectroscopic system and CCD camera. The radiation is collected in a solid angle making a 10 cm circle on the region of impact. It is time-integrated, i.e. the recording of the CCD camera is triggered when the projectile cuts the laser beam situated in the free-flight section and stops about 30 µs after impact. For acquiring time-resolved spectra, a Hamamatsu Photonics streak unit was added to the setup at the output of the spectrometer. Time-resolved spectra in the form of 1024×256 images are then recorded on the CCD camera. The duration of the recording in the streak configuration was fixed to 50 µs and the delay adjusted so as to record 30 to 40 µs after the projectile impacts the plate.

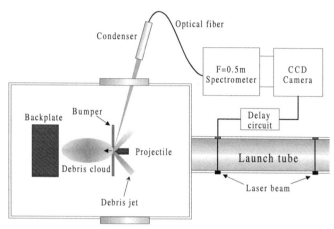

Figure 2: Optical setup for emission spectroscopy

General experimental conditions

Table 1 summarizes typical tests that have been carried out in the two-stage light gas gun facility. All the impacts were carried out for an average projectile velocity of 5 km/s in both normal and oblique angle impact configurations. The angles 51° and 64° were chosen after a statistical study by Voinovich et al. showing that for space vehicles orbiting the earth, impacts with debris occur mostly at these angles. Two different test gases were used in the normal impact tests: dry air and N_2, at an average pressure of 600 Pa. Magnesium projectiles have also been tested in either N_2, air or argon. Although different target materials have been tested, the quantitative results derived in this paper are only from impacts on aluminium bumpers.

TABLE 1
TYPICAL IMPACT TESTS

Test	Velocity, km/s	Gas
Normal	5.06	Air
Normal	5.23	N_2
Oblique 51°	5.13	Air
Oblique 64°	5.20	Air

DISCUSSION

Temperature determination method

In order to deduce the temperature of the impact flash, the $\Delta v = 0$ emission band belonging to $B^2\Sigma^+ \leftrightarrow X^2\Sigma^+$ electronic transition of CN have to be numerically simulated. For this purpose, a package, SPRADIAN by ISAS (refer to Fujita and Abe (1997) was used. It is expected that the temperature inferred from the CN spectrum will provide an estimate of the temperature around the impact point, in the surrounding jetting cloud. Several assumptions have to be made in the calculation of the spectrum and in its comparison to the experimental data. For instance, the impact is taken as an equilibrium process. Though this is certainly untrue, it simplifies greatly the calculation. Moreover, in the present case where temperature is determined from the time-integrated experimental radiation, it is expected that this value of temperature will be closer to the maximum attained in the gas near the impact region. Also, a single temperature is considered: the vibrational and rotational temperatures are assumed to be the same. Secondly, overlapping from other emission bands of N_2 and N_2^+ are ignored. This can be safely done since in the region analyzed, the contribution due to CN is very large compared to other species. Finally, line broadening due to pressure is taken into account in the calculation.

The relative intensities of the vibrational peaks of the $\Delta v = 0$ band of CN are very sensitive to temperature. The method of determination of the temperature from the measured spectra makes use of this characteristic by matching these peaks with values extracted from the calculated spectra. Only the first 4 vibrational levels are considered in this technique because the mutual overlapping of the higher levels makes the convergence of the calculation more difficult, and is superfluous.

Experimental results

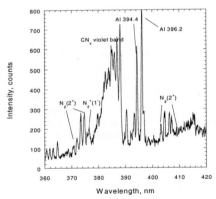
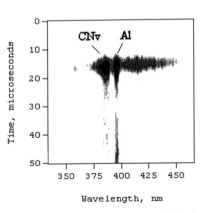

Figure 3: Normal impact of polyethylene projectile on aluminium bumper, v=5.06 km/s in dry Air

Figure 3 shows some of the experimental spectra of the impact flash in different conditions. In the experiments with polyethylene projectiles, the presence of CN dominated the spectra. Analysis of these spectra of CN for temperature determination is discussed below. Strong emission of the 394.4 and 396.2 nm Al lines, as well as some magnesium lines when using projectiles of this metal, could also be detected. Figure 3 also shows a streak image of the CN and alumnium emission measured in the normal impact of a polyethylene projectile on an Al bumper. Because of the technical limitations of the present streak setup, this measurement was done only in very low spectral resolution. However, it is found that within the present temporal resolution of the system, which is of the order of the microsecond, the onset of these bands seem to coincide. They last for a few microseconds after which, they quickly die down. However, there seem to be a slight rise in the intensity of the aluminium lines about 35 us later. This is probably due to some debris of aluminium causing stray emission by hitting the walls of the experimental chamber.

Using the method explained in the previous section, the average temperatures of the different impact tests when using polyethylene projectiles were determined and are summarized in Table 2.

TABLE 2
CALCULATED TEMPERATURES OF THE JETTING CLOUD

Impact type	Gas	Temperature, K
Normal	Air	8300±500
Normal	N_2	7700±400
Oblique	Air	8200±500

The average value of normal impacts of polyethylene projectiles on aluminium bumper in N_2 gas in the present experimental conditions is estimated to be around 7700 K (see Figure 4), lower than the same impact conditions in dry air, which is about 8300 K. This discrepancy is not clear since the mechanism of production of CN during these impacts is not fully understood. In general, it is reasonable to assume that during the impact event, CN is formed from exchange reactions of CH and N, themselves created from the dissociation of polyethylene $(CH_2)_n$ and N_2 respectively. In the case of dry air for instance, the presence of oxygen in the gaseous mixture may influence the reactions steps and in consequence, the impact temperature attained. Furthermore, the aluminium from the bumper will also readily react with the oxygen through a highly exothermic reaction, thus probably contributing to an overall raise in the temperature of the gas.

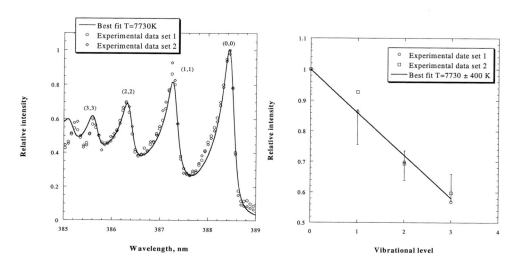

Figure 4: Best fit of vibrational peaks at T=7730 K, normal impact, v=5 km/s in N_2

In the case of oblique impacts, only shots in dry air were executed. Since no clear distinction could be made from the two different angles tested, data from these two were analyzed together. The temperature determined from these series of oblique impacts in dry air is 8200 K and within experimental error, is of the same order as the normal impact case. So, the present calculated results do not show any difference between the normal and oblique impact cases. As stated earlier, the method used here tends to give the highest temperature attained during the impact process. Certainly, the in-depth differences between the normal and oblique impact processes cannot be captured by the time-integrated measurements. A time-resolved analysis of the process by acquisition of time-resolved spectroscopy would have given valuable information on the formation of the jetting cloud in the oblique impacts but unfortunately, the present time-resolving capacity of the system in the streak configuration is severely limited. However, these preliminary time-resolved spectroscopy results are quite encouraging and improvement of this technique will certainly provide a new understanding of the impact phenomenon.

CONCLUSIONS

Time-integrated spectroscopy of the radiation briefly emitted from the jetting cloud during the hypervelocity impacts of polyethylene projectiles on aluminium bumpers have successfully been carried out. The emission spectra of the CN violet has been measured and analysis of these by comparison to calculation has yielded average temperatures of the jetting cloud in different experimental conditions. Tests in N_2 gas gave the lowest temperature as compared to tests in dry air. No difference of temperature between normal and oblique impacts could be detected by the present method. Preliminary time-resolved spectroscopy has also been carried out and results obtained so far are promising. Future work is in progress to measure microwave and radio-frequency emission during hypervelocity impacts.

REFERENCES

C.E Anderson and S.A Mullin (1988). Hypervelocity Impact Phenomenology: Some Aspects of Debris Cloud Dynamics. In *Impact: Effects of Fast Transient Loadings* ed. by Ammann et al., Balkema, Rotterdam, pp. 105-122.

B. Jean and T.L Rollins (1970). Radiation from Hypervelocity Impact Generated Plasma. *AIAA Journal* **8:10,** pp 1742-1748.

S. Sugita and P.H. Schultz (1998). Spectroscopic Measurements of Vapor Clouds due to Oblique Impacts. *Journal of Geophysical Research* **103:E8,** pp 19,427-19,441.

G. Jagadeesh, D.H. Ramjaun, I. Kato, K. Takayama, K.S. Raja and S. Shoji (2001). Experiments in Ballistic Range on Hypervelocity Crater Formation in Metallic/Non-Metallic Space Debris Bumper Shields. In: *39th AIAA Aerospace Sciences Meeting and Exhibit, Reno, Nevada, 8-11 January 2001*, AIAA Paper 2001-0750.

P. Voinovich and K. Takayama. On the Angle of Incidence by a Random Impact Event. *Shock Waves* Under process.

K. Fujita and T. Abe (1997). SPRADIAN, Structured Package for Radiation Analysis. *ISAS Report No. 669*, Institute of Space and Astronautical Science, Kanagawa, Japan.

FRAGMENT CREATION VIA LOW-VELOCITY IMPACT POSSIBLE IN SPACE

T Hanada[1] and T. Yasaka[1]

[1] Department of Aeronautics and Astronautics, Kyushu University,
10-1 Hakozaki 6-chome, Higashi-ku, Fukuoka, 812-8581, Japan

ABSTRACT

Authors have conducted laboratory impact experiments to understand dispersion properties of newly created fragments to be used in the eventual orbital dispersion and debris population models. Since there are no information concerned with low-velocity impact possible in space, especially in GEO, as far as authors know, these impact experiments have been conducted in the velocity range of 100–300 m/s with about 1 cm stainless steel or aluminum spheres and aluminum honeycomb sandwich panels with CFRP or aluminum face sheets. The visualization technique adopted herein is simply photographing fragment creation process from two directions, horizontally and vertically, with double flashing visible lights with time delays after the impact. By analyzing the images and the collected fragments, authors are investigating the three-dimensional dispersion properties of fragments. Authors will present some pictures and analyzed data to demonstrate dispersion properties of fragments created by low-velocity impact.

KEYWORDS

Orbital Debris, Spacecraft, Low-Velocity Impact, Fragment Creation

INTRODUCTION

Space structures may encounter impact phenomena in orbit. Apart from mission related collisions at rendezvous or deployment, the important phenomena to be considered are collision with exterior objects like meteoroids and artificial debris. These objects normally collide at very high speed, potentially causing grave effects to space structures and further creation of many more fragments, which can be threats to other spacecraft in other orbit. These hypervelocity impact phenomena have been extensively studied both theoretically and experimentally, McKnight & Brechin (1990), Bess (1975), Nebolsine, Lord & Legner (1983). Protection schemes of space structures, especially that of the space station is fairy well defined by now. The protection device is known as Whipple bumper, which is a thin wall placed in front of the main structure with a sufficient distance. At impact, much of the projectile kinetic energy is consumed during the material phase change process of both the projectile and the bumper wall, and the remaining energy is scattered into larger area by the time the projectile remnant reaches the structure wall. Due to this principle, the protection device is effective against a high velocity impact. At lower velocity, the projectile just penetrates the protection wall, and makes almost direct impact onto the main wall. This low-velocity impact is not well clarified in spite of a simpler mechanism of destruction. It is a purpose of the present paper to investigate this low-velocity impact through laboratory experiments, and then to establish a model to mathematically describe the phenomena. The model is constructed by comparison with those in hypervelocity impacts, which have been extensively investigated by now.

Collisions in low earth orbit (LEO) would be dominated by those at hypervelocity. However, lower velocity impact could happen in the space station, at lower frequency. Although the imparted specific energy is smaller, the damage caused could be fatal as well, depending upon the mass of the incoming projectile. We should remind that the space station will not be heavily armed by shield on the rear side where the low-velocity impact would occur, and the Whipple shield would not work any way. Higher orbit would be dominated by the low-velocity impacts, partly due to lower orbital velocity. Satellites in the geostationary earth orbit (GEO), in particular, will not be shielded. High specific cost to launch and highly functional exterior components, like solar arrays and antennas, make any kind of physical shielding impossible. They are naked to exterior incoming projectiles, and they are fragile to impacts at the lowest velocity.

LOW-VELOCITY IMPACT POSSIBLE IN SPACE

Impact velocity is defined by the difference of velocity vectors of two colliding objects. The scalar velocity is expressed in terms of the magnitudes of object velocities before impact, V_1 and V_2, and an angle between two vectors θ, as

$$V^2 = V_1^2 + V_2^2 - 2V_1V_2 \cos\theta$$

If the two objects are flying toward the same direction (θ = 0), the collision velocity becomes minimum. On the other hand, if θ = 180 or the head-on collision, V becomes maximum.

Impact rate is described by the product of object density, collision cross-sectional area and the impact velocity. Therefore, collisions at ram surface, where the head-on collision takes place, are most prominent and rearward collisions are scarce, when the external object distribution is random.

Practical useful case is that two objects are in near circular orbits of similar altitudes, when V_1 and V_2 are nearly identical. In LEO, flight paths of objects are very much uniformly distributed. Therefore, the average, or the expected velocity is $\sqrt{2}\, V_1$ that corresponds to an impact from side, and the value is about 10 km/s. Collisions onto the trailing side can have very small impact velocities, but the impact rate is low.

Artificial objects in GEO are distributed differently. Most of them are originally placed on the equatorial plane, and then, after termination of orbit control, they change orbital plane due to perturbation of lunar and solar gravitational attractions. The plane change in inclination occurs at a rate of about 0.8 deg/year at the beginning, but shifts the direction of the change after the inclination reaches the maximum of 15 deg. If an operating satellite in GEO is to be hit by another artificial object, the angle of collision must be less than 15 deg. Altitude and orbital eccentricity are also subject to various perturbations, but variation of these parameters is insignificant. All orbits are essentially circular at geostationary altitude. Objects in inclined orbits make a daily excursion in north-south direction relative to stationary operating satellites. Since the orbital velocity is about 3 km/s, the collision velocity is less than 800 m/s.

EXPERIMENTAL FACILITY

Figure 1 illustrates a facility at Kyushu University to conduct the low-velocity impact experiments. A single-stage air gun driven by a free piston was adopted to accelerate a projectile because of its simple and safe operation associated with low running cost. A high-pressure reservoir of 300 mm in diameter with a length of 700 mm was designed to store the pressurized air up to 75 kgf/cm^2 (7.5 MPa). The launch tube has a length of 4000 mm and an inner diameter of 25 mm. The vacuum chamber, with a degree of vacuum of 10^{-2} Torr, has a capacity of 600 mm in diameter with a length of 1000 mm enough to observe the fragment creation process at and after impact.

Projectiles to be launched were stainless steel spheres of 9 and 11 mm diameter, and aluminum alloy spheres of 1/8, 5/16 and 3/8 in. diameter. The air gun has a capacity to launch these projectiles at velocity of 100–300 m/s. Each projectile was centered in a sabot, which was removed at the sabot trap located in the forward direction of the launch tube. The speed the projectile before impact was measured by two pairs of visible laser and APD (Avalanche Photodiode) module units as illustrated in Fig.2.

The visualization technique adopted in this experiments is simply photographing fragment creation process from two directions, horizontally and vertically, using two CCD (Charge Coupled Device) cameras with double flash with time delays after the impact. Figure 2 also illustrates a schematic of the photographing system. The camera head features a high-resolution CCD array containing 1008 (H) × 1018 (V) light-sensitive elements (pixels), the size of which is 9 × 9 μm. Each exposure is started by a trigger signal generated by the digital oscillograph when the projectile intercepts the first laser light. After the CCD cameras take an exposure, the system makes two successive flash of 2 μs to capture a sequence of the fragment creation process. The images captured by this system enable us to measure velocity decrement of the projectile after impact and three-dimensional dispersion velocities of the fragments released from the target plate at impact.

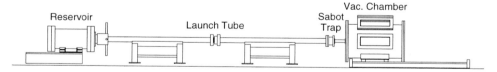

Figure 1: Low-velocity impact experiment facility.

LOW-VELOCITY IMPACT EXPERIMENTS

As typical structures of GEO satellites, a thin aluminum honeycomb sandwich shell with CFRP or aluminum face sheets, those are used for body structures or rigid antenna reflectors, were prepared for the low-velocity impact experiments. Target samples were mounted on a rigid frame, and then fixed in the vacuum chamber perpendicular or inclined to the trajectory of the projectile.

Figure 3 shows an example set of the images captured by the present photographing system. The target plate can be seen at the left of each image. The 11 mm diameter stainless steel sphere, moving from left to right at a speed of 185 m/s, shot through the target plate and created a thousand or more fragments. Identifying a pair of the same fragments according to fragment shape, then we can determine the fragment dispersion path to estimate the magnitudes and directions of dispersion velocities.

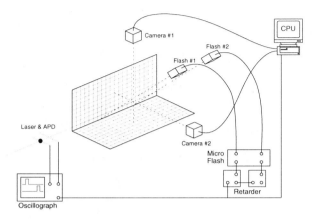

Figure 2: Photographing system using high-resolution CCD cameras with double flash.

(a) Image from camera #1.   (b) Image from camera #2.

Figure 3:   An example set of the images obtained from the experiment.

In the previous works, Yasaka (1995) and Hanada (1997), we had performed a series of low-velocity normal impact experiments and formulated the mathematical predictions of the fragment dispersion properties in a similar manner to what had been done in the case of hypervelocity impacts. The mass distribution was sufficiently curve-fitted by a power low with adjusted parameter values. The velocity distribution showed very good correlation with a model provided by Su (1990). We also indicated that the dispersion velocity profile could be expressed as the equation

$$V_f/V_p = 1.3\cos\phi \tag{1}$$

where V_f and V_p represent the fragment and the projectile velocities, respectively, and ϕ is the fragment dispersion path angle that is defined as an angle between the projectile incident and the fragment dispersion directions.   The proportional constant is a typical value for the normal impact.   The equation means that some fragments could be released at a faster velocity than the projectile impact velocity. It is noted that the previous works had been done by two-dimensional analysis.

It is a purpose of the present paper to investigate the three-dimensional dispersion properties of fragments. Figures 4 and 5 demonstrate the dispersion velocities after a normal impact and an inclined impact, respectively. The fragment velocities were normalized by the projectile impact velocity. It is difficult to identify the same fragment from a set of images from the cameras #1 and #2. Besides, it is also difficult to identify the same fragment on an image exposed by double flash. These difficulties may come from the inability to identify the small fragments. Therefore, we have not had data enough to provide the three-dimensional dispersion properties yet and have not established the mathematical predictions. The dispersion velocity profile of the fragments created by the inclined impact, however, can be explained like this. If the normal component of the projectile impact velocity contributes to fragment creation as well as a simple cratering equation, then we can obtain the circular velocity profile represented by the broken line in Fig.6. Deforming this circular velocity profile parallel to the target plate provides the elliptical velocity profile represented by the solid line. This elliptical velocity profile explains qualitatively the experimental data.

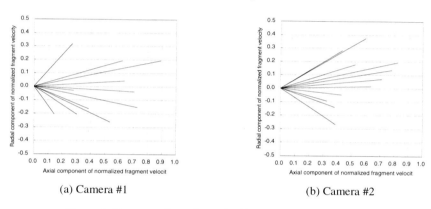

Figure 4: Fragment dispersion velocities after a normal impact.

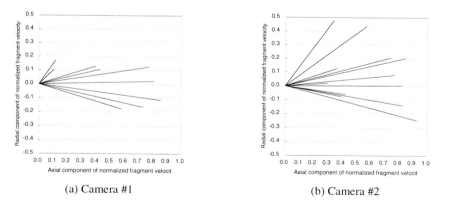

Figure 5: Fragment dispersion velocities after an inclined impact.

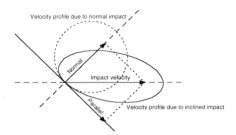

Figure 6: Dispersion velocity profile of fragments created by a inclined impact.

CONCLUSIONS

Research efforts of low-velocity impact phenomena in space are necessary to gather information to protect and assess space structures against possible interference of foreign objects, and then to establish a model to describe the phenomena mathematically. Although impacts in space are generally associated with very high-velocity impacts, low-velocity impacts are also possible, with less frequency.

In Kyushu University, a series of investigations has been carried out using an air gun to clarify some of these events. An impact velocity of artificial debris in GEO will be less than 800 m/s, and most possibly a few hundred meters per second. Normal and inclined impact tests on spacecraft structure panel have been conducted. The projectile penetrated the target plate and then created a thousand or more fragments. The images captured from horizontal and vertical directions using two CCD cameras with double flash have been examined carefully to investigate the tree-dimensional dispersion properties of fragments. The mathematical model has not been established because of lack of data. Authors will continue to conduct low-velocity impact tests.

References

McKnight D. S. and Brechin C. (1990). Debris Creation via Hypervelocity Impact, AIAA Paper 90-0084.

Bess T. D. (1975). Size Distribution of Fragment Debris Produced by Simulated Meteoroid Impact on Spacecraft Wall, NASA SP-379.

Nebolsine P. R., Lord G. W. and Legner H. H. (1983). Debris Characterization Final Report, PSI-TR-399.

Yasaka T. and Hanada T. (1995). Low Velocity Impact Test and Its Implications to Object Accumulation Model in GEO, *Advances in astronautical Sciences* **91**, 1029-1038.

Hanada T., Yasaka T. and Goto K. (1997). Fragments Creation via Impact at Low Speed, *Advances in Astronautical Sciences* **96**, 979-986.

Su S.-Y. (1990). The Velocity Distribution of Collisional Fragments and Its effect on Future Space Debris Environment, *Advances in Space Research*, **10:3-4**, 389-392.

RADIATION OBSERVATION OF HYPERVELOCITY SHOCK WAVES IN GASES

H. Honma [1], K. Maeno [2], T. Morioka [1] and H. Shibuya [1]

[1] Graduate School of Science and Technology, Chiba University,
[2] Faculty of Engineering, Chiba University
Yayoi-cho 1-33, Inage-ku, Chiba, 263-8522, JAPAN

ABSTRACT

The radiation phenomena are observed by using a couple of ICCD camera systems for 10 km/s shock waves in 66.7 Pa dry air and nitrogen gas in a shock tube experiment. Two-dimensional images of total radiation and time-resolved images of spectra are simultaneously obtained for shocked gas flows. The total and spectral radiation profiles along the tube axis are exhibited for air and nitrogen, and their remarkable features are compared with each other mainly in a view of the formation process of nitrogen atoms.

KEYWORDS
Radiation, Shock Wave, ICCD, High-Temperature Air and Nitrogen, Imaging Spectroscopy

INTRODUCTION

The radiation phenomena behind hypervelocity shock waves in low-density gases have been investigated during the past decades in association with reentry problems of space vehicles (Park (1989), Sharma & Gillespie (1991), Fujita et al. (1998)). The shock tube has been employed as one of the experimental tools, which can be reproduce the radiation phenomena ahead of reentering space vehicles. Various kinds of techniques have been also developed in many laboratories to observe the radiation phenomena behind strong shock waves in shock tube experiments. In our laboratory (Honma et al. (1998), Morioka et al. (1999a), Morioka et al. (2000)), a couple of image intensified CCD (ICCD) camera systems have been used since 1997 to observe simultaneously the two-dimensional image of total radiation and the time-resolved spectral image of high-temperature air, which is produced by a free-piston, double-diaphragm shock tube. The previous experiments were carried out, principally, to observe the nonequilibrium characteristics of radiation behind planar shock waves (Morioka et al. (2000)) and the two-dimensional profiles of radiation for Mach reflection on wedges (Morioka et al. (1999b)) in low-density air. The range of shock speed was 5 km/s - 13 km/s for the initial pressure of 133 Pa - 13.3 Pa. Recently, the low-pressure tube is replaced by new one with a high vacuum system to reduce the impurity level in test gases. In the present paper, some preliminary results are shown on radiation observation for 10 km/s planar shock waves in dry air and nitrogen gas of 66.7 Pa in a use of the replaced tube to look at the equilibrium flow zones rather than the nonequilibrium zones.

EXPERIMENTAL APPARATUS AND DIAGNOSTIC SYSTEM

Figure 1 shows a layout of the diagnostic system for observing the radiation from shocked gases in the test section of the shock tube, which is illustrated vertically in the center of the figure. The cross section of the test section is 40 mm x 40 mm square. The observation window of the test section is made from quartz to keep transparency in the UV range, and its viewable area is 40 mm x 150 mm rectangular for taking photographs of two-dimensional profiles of radiation. The test section can be evacuated to 10^{-3} Pa with a vacuum tank by a turbo-molecular pump system, and its rate of leakage is about 0.01 Pa/min. The test gas is air or nitrogen of 66.7 Pa in the initial pressure. The humidity of the test gas is measured approximately as 0.01 % by a gas chromatography. The purity of supplied nitrogen is 99.9995 %.

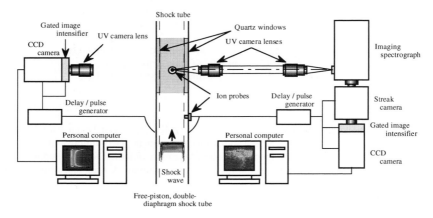

Figure 1: Layout of the shock tube and the diagnostic system

A couple of ICCD camera systems are illustrated in the left- and right-hand sides of the shock tube, similar to the actual layout. The left system consists of a UV camera lens, an ICCD camera, a delay/pulse generator, and a personal computer, and the right system consists of two UV camera lenses, an imaging spectrograph, a streak camera, an ICCD camera, a delay/pulse generator, and a personal computer. The effective wavelength range is from 200 nm to 800 nm for both systems. The left system is used to obtain a two-dimensional profile of total radiation in the test section like a usual picture, and the right system is used to obtain a time-resolved spectra at a point of the test section. Two-dimensional diode array of each CCD camera is 512 x 512 pixels with 12-bits brightness level or 1000 x 1018 pixels with 10-bits brightness level. The magnification factor of an image intensifier is about 10,000 for a photon. The spatial resolution in a former image of total radiation depends on the distance of the ICCD camera from the central axis of the shock tube, and it is fixed to be about 0.2 mm/pixel throughout the experiment. The gate time of the image intensifier is fixed to be 10 ns. The image may be considered as time-frozen, since the 10 km/s shock wave moves only in a distance of 0.1 mm for 10 ns. In the latter image of time-resolved spectra, the time-resolution depends on a combination of the entrance slit width and the sweep speed in the streak camera, and it is estimated to be 67 ns. This time-resolution corresponds to the space-resolution of 0.67 mm for 10 km/s shock waves. The spectral resolution depends on a combination of the entrance slit width and the grating in the spectrograph, and it is estimated to be 1 nm. An 80 nm wavelength range can be covered on an image in a shot of the shock tube experiment. The shock velocity is measured by using ion probes mounted on the side walls of the test section. Considering the shock velocity and the delay times of two pulse generators, one can obtain the spatial relationship between both images of total radiation and spectra.

RESULTS AND DISCUSSION

Figure 2 shows a couple of typical images of total radiation for incident shock waves of 10 km/s in air and nitrogen of 66.7 Pa (0.5 Torr). The shock waves are moving leftwards and can be observed as slightly curved but sharp fronts. The bright streak lines appear along the upper and lower walls, and the small bright zones on the upper wall are considered to exhibit sparks of the ion probe mounted on the upper wall. The contact zone between hot test gas and cold helium as a driver gas appears as smooth profiles with unstable characteristics.

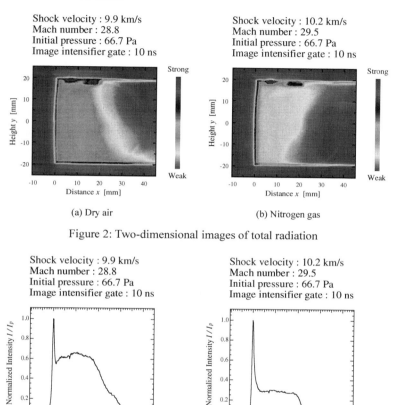

Figure 2: Two-dimensional images of total radiation

Figure 3: Axial profiles of total radiation

Figure 3 shows the distributions of total radiation intensity along the tube axis, which are obtained by integrating the intensity in the range of ±2 mm around the axis on the images of Fig.2. The integrated intensity is normalized by a peak value in each figure in order to compare the profiles with each other. The resultant profiles show some reproducible features for most of the data in spite of the first impression of the two-dimensional image on unstable nature of the contact zone. A zone of nearly uniform radiation can be seen clearly between the sharp peak and the gradual descent of radiation. The peaking, leveling and descending zones are well known as the zones of, respectively, nonequilibrium radiation, equilibrium radiation, and contact mixing. The origin of the distance is defined by the position of the radiation peak. The length of the hot flow region may be considered as nearly 20 mm along the tube axis. The equilibrium radiation zones are clearly established for all gases, though there are some discrepancies between air and nitrogen. The intensity ratio of the equilibrium radiation to the peak is larger for air than for nitrogen. A narrow valley is found at the mid-slope of the peak for a shock wave in air, but it is not found in nitrogen. The gas temperature of the equilibrium zone is estimated as an order to 10,000 K. A spectroscopic study follows in the next articles.

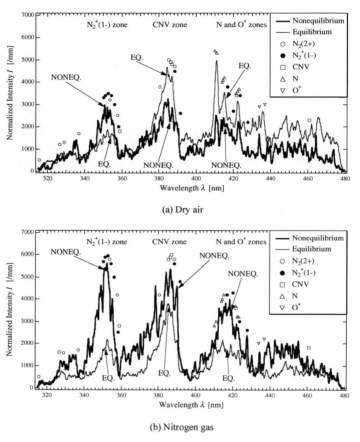

Figure 4: Nonequilibrium and equilibrium radiation spectra in a wavelength range of 315 nm to 480 nm

Figure 4 shows the time-integrated radiation spectra from 315 nm to 480 nm for air and nitrogen. Considering the conversion of the time-axis to the x-axis of Fig.3, the corresponding range of integration is selected from $x = -2$ mm to 0 mm for nonequilibrium radiation and from $x = 4$ mm to 14 mm for equilibrium radiation. The former range is selected to look at the spectra of the initial stage, and the latter range is selected in the middle part of the equilibrium zone. The spectral intensity is plotted after normalizing by the width of integration (per mm) to compare the intensities of the different zones with each other. The bold and fine lines correspond to radiation spectra from, respectively, nonequilibrium and equilibrium zones. The symbols in Fig.4 indicate the wavelength positions of the corresponding molecular or atomic line spectra. The positions of the wavelength on an image are calibrated by using the line spectra of Hg, Cd, Fe, and Sr lamps. As easily seen in the figure, three peaks attract our attention. The peaks correspond to the band spectral zones of $N_2^+(1-)$, CNV and N from the left side. For air, nonequilibrium radiation intensity of $N_2^+(1-)$ zone is lager than equilibrium one, while the tendency is reversed in CNV and N zones. For nitrogen, the tendency is the same in three zones. That is, the bold line exceeds the fine line over the whole range.

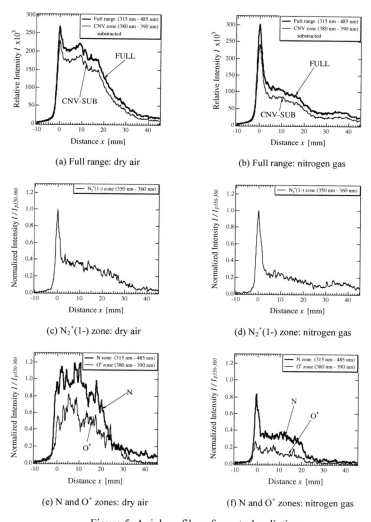

Figure 5: Axial profiles of spectral radiation

Figure 5 shows the distributions of the spectral radiation intensity along the tube axis, which are obtained by integrating the time-resolved spectral intensity in a wavelength width. In Fig.5(a) or 5(b), the bold line corresponds to the total intensity integrated for full range of Fig.4 and the fine line corresponds to the total intensity subtracted the effect of the CNV zone between 380 nm and 390 nm. The resultant profiles exhibit similar features to the total intensity profiles in Fig.2, though a narrow valley cannot be recognized for air. The effects of the CNV zone on the total radiation profile are not serious for both gases. In Figs.5(c) to 5(f), the wavelength width for integration is 10 nm in the spectral zones of $N_2^+(1-)$, N and O^+. The intensity is normalized by a peak value of $N_2^+(1-)$ for each gas. The radiation profile of $N_2^+(1-)$ zone for air is similar to that for nitrogen. That is, the sharp peak is followed by the gradual decrease of intensity. However, the radiation profiles of N and O^+ zones for air exhibit remarkable differences with those for nitrogen. In Figs.5(e) and 5(f), the bold and fine lines correspond to the radiation profiles of, respectively, N and O^+ zones. For air, the frontal peaks are not found in their profiles, which form hat-like mountains. On the other hand, for nitrogen, the profile of N zone is similar to that of $N_2^+(1-)$. The profile of O^+ zone for nitrogen does not mean the existence of O^+, but is shown only as a reference in comparison with air.

The above results suggest us that the discrepancy of total radiation profiles along the tube axis between air and nitrogen may be mainly attributed to the formation process of nitrogen atoms. That is to say, the delayed formation of nitrogen atoms may contribute to form a narrow valley in the mid-slope of the total radiation peak for air, while the quick formation of nitrogen atoms may contribute to form the larger ratio of the peak to the plateau for nitrogen. The detailed discussion will be reserved in future works, since the present study is only preliminary in views of the limited wavelength range of spectra, the lack of the radiation model analyses, etc..

CONCLUSION

The present paper aims at showing a technique for radiation observation of hypervelocity shock waves in low-density gases. A couple of ICCD camera systems are used to obtain simultaneously the two-dimensional images of total radiation and the time-resolved images of radiation spectra in a shock tube experiment. Some preliminary results are shown for 10 km/s planar shock waves in dry air and nitrogen gas of 66.7 Pa in a use of the new tube with a high-vacuum system. The total and spectral radiation profiles along the tube axis are exhibited for air and nitrogen, and are discussed mainly in a view of the formation processes of nitrogen atoms. The present experimental apparatus and diagnostic system are very useful for observing the radiation phenomena of hypervelocity shock waves in various gases and are expected to give us valuable data in future works.

References

Fujita K., Abe T., and Suzuki K (1998). UV Radiation Impact on Heating Rate of Super-Orbital Reentry Vehicle. *Proceedings of 21st International Symposium on Space Technology*, Paper No.98-d-39P.

Honma H., Morioka T., Sakurai N., and Maeno K. (1998). A Shock Tube Study on Nonequilibrium Radiation of Strong Shock Waves in Low-Density Air. *JSME International Journal, Series B* **41:2**, 390-396.

Morioka T., Sakurai N., Maeno K., and Honma H. (1999a). Nonequilibrium Radiation Spectra Behind Strong Shock Waves in Low-Density Air. *Rarefied Gas Dynamics*, Ceradues Editions, **2**, 345-352.

Morioka T., Suzuki Y., and Honma H. (1999b). Radiation Observation of Strong Shock Wave Reflection in Air. *Proceedings of 22nd International Symposium on Shock Waves* **2**, 1201-1206.

Morioka T., Sakurai N., Maeno K., and Honma H. (2000). Observation of Nonequilibrium Radiation Behind Strong Shock Waves in Low-Density Air. *Journal of Visualization* **3:1**, 51-61.

Park C. (1989). A Review of Reacting Rates in High Temperature Air, *AIAA Paper* 89-1740.

Sharma S.P. and Gillespie W.D. (1991). Nonequilibrium and Equilibrium Shock Front Radiation Measurements. *Journal of Thermophysics and Heat Transfer* **5:3**, 257-265.

DEVELOPMENT OF A NEW BALLISTIC RANGE FOR THE INVESTIGATION OF OBLIQUE IMPACTS ON SOLID AND LIQUID SURFACES

M. Shinohara, I. Kato, Y. Hamate and K. Takayama

Shock Wave Research Center, Institute of Fluid Science, Tohoku University, Sendai, Japan

ABSTRACT

A compact two-stage light gas gun used for ballistic range research is under construction at the Shock Wave Research Center (SWRC) of the Institute of Fluid Science, Tohoku University. This gun can be rotated from zero to ninety degree by fixing its center of rotation at the center of the test section. In this paper, the characteristics of this facility are outlined. A numerical simulation was conducted by Random Choice Method (RCM) in order to assess the performance of such a compact two-stage light gas gun. To validate the numerical scheme, its results were compared with experimental results which were collected from previous ballistic range results already conducted at the SWRC. For achieving precise controlling of the ignition of the new system, the results of a preliminary experiment was described, in which smokeless powder was ignited by the irradiation of a pulsed laser beam on micro-explosives in a prototype propellant chamber.

KEYWORDS

Ballistic range, Two-stage light gas gun, Oblique impact, Random Choice Method, Detonation

1. INTRODUCTION

1.1. Background

The design of space debris bumper shields has become one of the most important research topics in the field of hypervelocity impact study and shock wave dynamics as well. With the increasing number of launching space vehicles and communication satellites, space debris of large and minute sizes are monotonically being increased. Orbits of larger debris are well catalogued after monitoring them from the ground so that space crafts can avoid collision with them. The collection with debris of diameter smaller between 10mm and 100mm are avoided by maneuvers upon in-situ detection of them. However, small debris would inevitably hit space vehicles and space structures and there is no technology to avoid their impact. Their speed is estimated to range from low speed to 14 km/s and hence their impact energy becomes tremendously high enough to penetrated outer wall of space vehicles.

In order to prevent from fatal damages possibly caused by such impacts, bumper shields has been proposed to protect the space vehicles from their penetrations (Mizuno 2000). Oblique impacts are more probable cases depending upon the angle of orbital inclination between the space vehicles and the debris. In the Shock Wave Research Center (SWRC) of Tohoku University, we have previously tested with three impact angles, which revealed that the process of penetrations appeared to be significantly different upon the impact angle. This is a motivation of the installation of the compact oblique two-stage light gas sun.

This facility is also planned to simulate large scaled impact phenomena which happened to occur 65 million years ago when asteroid impacted obliquely on earth (Takayama et al. 2001). Authors believe that such a gigantic gas-liquid-solid impact should be experimentally and numerically simulated. Complex interactions of three-phase material should be examined from the point of view of impact study and also shock wave dynamics. The proposed facility accommodating such three-phase layers in horizontal position inside its test section achieve oblique impacts at muzzle speed of maximum 5 km/s.

1.2. Objectives

The characteristics of the compact two-stage light gas gun facility are numerically simulated by using Random Choice Method (Matsumura et al. 1990). Being a compact facility, the duration of time from the ignition of the propellant to the moment when a projectile is released from the gun muzzle is too short to oscillate pulse laser in due time for constructing holographic interferometry. For this purpose, electric signals should be delivered to oscillate the pulse laser in advance and with a proper delay time from this signal to the ignition of the propellant. We have tried various ignition systems aiming to make the ignition delay time minimized and repeatable (Nagayasu 2001) and this attempt was partially successful. In this paper, a 10 mg silver azide pellet was ignited by irradiation of pulsed YAG laser beam in stoichiometric oxyhydrogen gas mixtures and its detonation wave successively ignited propellant. This arrangement minimized the delay time of the operation of the compact two-stage light gas gun.

2. OBLIQUE TWO-STAGE LIGHT GAS GUN

2.1. Outline

The two-stage light gas gun is shown in Fig. 1. Figure 1 shows two cases in which guns are positioned vertically and horizontally. The gun barrel and other elements will be cleaned in horizontal position then tilted to a specified inclined angle by using a hydraulic system. The test chamber of 400 mm dia. view field and made of stainless steel was fixed in the frame and was supported by two bearings to permit its rotation. Every time after fixing the inclination angle, test models are placed inside the test chamber.

The observation is carried out by using double exposure holographic interferometry. Shock waves in air and water will be simultaneously visualized with keeping the water surface at the center plane of the test chamber. As fundamental features of two-stage light gas guns are already documented in Matumura et al (1990), detailed explanation of gun components will be omitted here.

This two-stage light gas gun has an optional extension tube of 270 mm in length which can be inserted between the diaphragm at the high pressure coupling and the projectile and inside which hydrogen will be filled. Its fill pressure will be described by turning with the results of RCM simulation. In the optional hydrogen tube a strong shock wave will be created which initiate relatively slow acceleration against the projectile. The state which will be established between its reflected shock wave and the rear surface of the projectile will provide high temperature so that ideally speaking higher sound speed and higher driver pressure are established. However, this feature will not be presented here.

2.2. Ignition System

As abovementioned, a small facility has a fatal demerit in its synchronization with external systems such as visualization or other time controlling functions in operating with another gun or a shock tube. In conventional ballistic range experiments at the SWRC, a signal of arrival of the projectile was collected by its passing through light beams of CW semiconductor lasers, which were arranged so as to cross the free flight chamber. Then when projectiles passed through the beams signals were time-accurately triggering other systems. In this two-stage light gas gun, such a system was found impossible to be applicable.

The duration time in which the projectile flies from the free flight chamber to the test chamber is too short to synchronize the visualization system. As shown in Fig. 2 the trigger signal has to be put in before the ignition. The process of conventional ignition system is shown in Fig. 3. An electric trigger signal initiates the igniter, which will then burn black powder and subsequently smokeless powder. Black powder is needed to supply ignition energy to initiate the combustion of the smokeless powder is not very consistent and its jitter time varies widely. Hence to minimize the jitter time of the initiation and obtain higher degree of reproducibility the ignition system needed to modify. The proposed ignition system consists of a 10 mg silver azide pellet supplied by Chugoku Kayaku Co. Ltd, which was inserted at the end wall of the propellant chamber. In the chamber, a few gram of smokeless powder was placed and then stoichiometric oxyhydrogen gas mixtures were filled.

3. ASSESSMENT OF PERFORMANCE

3.1. Random Choice Method (RCM)

In order to optimize operating conditions, the gun operation was numerically simulated by the Random Choice Method (RCM) (Matumura et al. 1990). This method solves the quasi-one-dimensional unsteady Euler equation flows with combination of the equation of motion of the projectile by properly assuming wall friction coefficient, heat transfer loss, and pressure losses at the diaphragm section. However, the real gas effects behind the shock wave which is created in front of the piston and the co-volume effect at the high pressure coupling were neglected.

The combustion model of the smokeless powder is similar to one documented in Matsumura et al (1990). The parameters used in this model are determined empirically (Matsumura et al. 1990). This method has been used intensively at the SWRC and proven to accurately predict the launcher performance for various launch conditions. A condition is selected as follows:
- Kind of propellant was smokeless powder of the mass from 20 to 100 [g].
- Helium gas at the pressure from 600 to 100 [kPa] is filled in pump tube of 30 [mm] inside dia. and 1800 [mm] long.
- Rupture pressure of diaphragm placed between pump tube and launch tube of 10 [mm] inside dia. and 1000 [mm] long was 200 [MPa].
- Material of piston and projectile was high-density polyethylene and their mass was 198 [g] and 0.46 [g], respectively.

3.2. Numerical Results

Figure 4 shows the result of RCM simulation. It is found that for a given condition the muzzle speed is affected by propellant mass and its optimized amount would be 60 g. The present RCM simulation is useful for the estimation of characteristics of the compact two-stage light gas gun.

4. IGNITION SYSTEM

A 20 mm i.d. and 200 long propellant chamber was designed and constructed as shown in Fig. 5. Stoichiometric oxyhydrogen mixture was filled and ignited the explosion of the silver azide pellet. The resulting detonation wave was expected to ignite smokeless powder contained in that chamber. Two pressure transducers (PCB Model 113A03) were installed on the side wall of the chamber at 120 mm and 150 mm from the end where the silver azide pellet was placed. The blast wave velocity created in the chamber was estimated from these pressure records. 10 mg silver azide pellets having cylindrical shape of 1.5 mm in dia. and 1.5mm height was pasted on the edge of 0.6 mm dia. optical fiber. Upon the irradiation of a pulsed Nd-YAG laser beam it was simultaneously exploded and induced detonation wave in the chamber.

The experiment was conducted at first at the initial mixture gas pressure of 101 kPa. The averaged detonation velocity was 2.142 km/s, whereas the C-J detonation wave speed at this initial pressure is 2.844 km/s. Hence the detonation wave was not fully generated. Similar trends were obtained when enhancing initial fill pressures, however, the degree of reproducibility was improved fairly good. In order to promote a turbulent combustion in oxyhydrogen mixture a wire mesh was placed inside the chamber so as to shorten the ignition delay time of detonation waves. Figure 6 shows a typical pressure history for this case. The initial pressure was 101 kPa and the grid size of the wire mesh was about 2 mm and averaged wave velocity of 3.03 km/s, which indicates that by the insertion of the wire mesh a detonation wave was successfully produced. The wave propagation was also well reproducible within an error of 3 %.

5. SUMMARY

The results so far obtained are summarized as following:
(1) RCM simulation indicates to achieve muzzle velocities of about 7 km/s.
(2) Detonation waves were generated in a tube at relatively short distance by introducing a wire mesh.
(3) The ignition of smokeless powder by the exposition of detonation wave was achieved by enhancing initial pressure and using a finer grain size of smokeless powder.

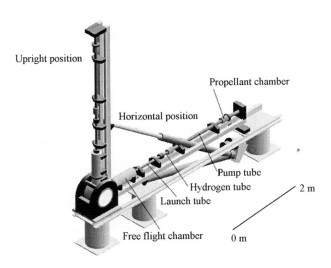

Figure 1: Schematic diagram of an oblique two-stage light gas gun

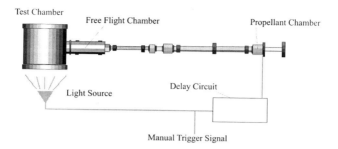

Figure 2: Trigger system in new ballistic range

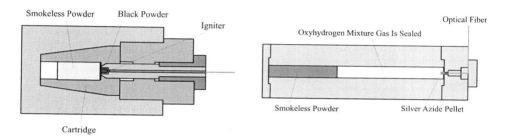

Figure 3: Ignition system of conventional and new ballistic range
Left: conventional system, right: new system

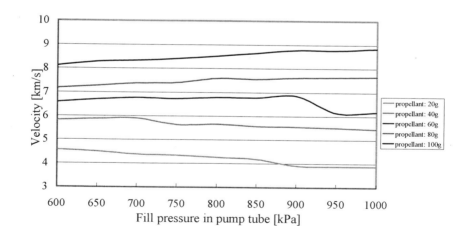

Figure 4: Muzzle velocity profiles obtained by RCM

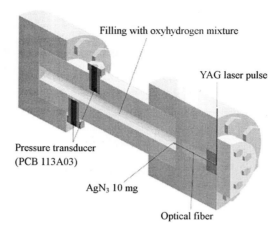

Figure 5: Prototype of the propellant chamber

References

Matsumura T., Inoue O., Gottlieb J.J. and Takayama K. (1990). A Numerical Study of The Performance of A Two-stage Light Gas Gun. *Rep. Inst. Fluid Science* **1**, 121-133
Mizuno H. and Takayama K. (2000). *Master Thesis of Graduated School of Tohoku University.*
Nagayazu N. (2001). *Dr. Thesis of Tohoku University*
Takayama K., Voinovich P., Timofeev E., Merlen A. and Kaiho K. (2001). Underwater shock-and ground seismic waves due to a catastrophic impact (to be presented ISSW23).

NON-IDEAL DETONATION PROPERTIES OF AMMONIUM NITRATE

A. Miyake

Department of Safety Engineering, Yokohama National University,
Yokohama 240-8501, JAPAN

ABSTRACT

In order to obtain a better understanding of the non-ideal detonation characteristics of ammonium nitrate, detonation velocity and pressure of low density prilled ammonium nitrate were measured in steel tubes with different diameters and wall thicknesses. It was found that the tube diameter has a much larger influence on the non-ideal behaviour of the detonation velocity and pressure than the confinement. The extrapolated detonation velocity and peak pressure to the infinite diameter were calculated as 3.83 km/s and 3.1 GPa respectively, which values coincide with the ideal detonation parameters for ammonium nitrate as predicted by the thermohydrodynamic TIGER code with the JCZ3 equation of state. Although the adiabatic exponent of the detonation gas products (γ) remained constant when the confinement became heavier, it showed a remarkable change with the increase of the charge diameter. The extrapolated γ value to the infinite diameter was obtained as 2.90 and it coincided with the γ calculated with the TIGER code.

KEYWORDS

Non-ideal detonation, Detonation velocity, Detonation pressure, Ammonium nitrate, TIGER code, The adiabatic exponent

INTRODUCTION

Ammonium nitrate (AN) is an explosive known for its non-ideal detonation behaviour. This is shown, for example, by the detonation velocity which does not easily reach theoretically predicted values. Explosives behave non-ideally between the critical diameter (d_c) below which a steady detonation wave cannot be sustained, and the minimum diameter (d_m) above which the detonation is ideal. AN is a typical "non-ideal" explosive because it has a large value for d_m and relatively small value for d_c. For most practical conditions they will never reach the ideal values as predicted by thermohydrodynamic theory. Generally the non-ideal detonation behaviour is explained by the relatively low decomposition rate of AN, which causes a wide reaction zone, in combination with lateral heat losses and rarefaction waves which distinguish the decomposition reactions (Cook, 1958).

It is the purpose of this investigation to obtain a better understanding of the non-ideal detonation behaviour of AN. A series of experiments were carried out to study the influence of the diameter and the confinement of the explosive charge on the detonation properties in the non-ideal region. In this paper the detonation velocity and pressure were measured simultaneously in steel tubes with different diameters and wall thicknesses using ionization probes and piezo-resistive manganin gauges. Observed detonation velocity and pressure were compared with the theoretically predicted values made by the thermohydrodynamic TIGER code and non-ideal detonation behaviour was discussed.

MATERIAL

A readily available highly porous type of AN (ANFO quality) with well-defined properties and the additional advantage that, due to the high sensitivity, the dimensions of the experiments could be kept within reasonable limits was used in this investigation. It consists of low density prills with a diameter between 0.35 to 1.00 mm. The purity is about 99% and the sample is treated with anti-caking agent. The nitrogen content is about 33.8 % and the loading density is 800 to 850 kg/m^3.

THEORETICAL CALCULATION

The ideal detonation parameters of AN was calculated by the thermohydrodynamic theory describing the state attained behind the detonation front. We used the TIGER code in combination with the BKW and JCZ3 equations of state to calculate the Chapman-Jouguet parameters (Cowperthwaite & Zwisler, 1973). The results for a density of 850 kg/m^3 are shown in Table 1 with the calculation results with KHT code with KHT EOS by Tanaka (Tanaka, 1983). As C-J parameters vary with the EOS of detonation gas products, we used the values calculated with the TIGER code with JCZ3 EOS in the discussion.

Table 1 Ideal detonation parameters of AN at a density of 850 kg/m^3
as calculated with several EOS

Name of code	EOS	C-J velocity [km/s]	C-J pressure [GPa]	C-J temperature [K]	Heat of detonation [cal/g]
TIGER	JCZ3	4.01	3.45	1660	352
TIGER	BKW	4.71	4.93	1080	354
KHT	KHT	4.23	3.94	1290	354

EXPERIMENTAL

Figure 1 shows the experimental set-up of the detonation velocity and pressure measurement. In all experiments the AN was confined in steel tubes. The inner diameters varied between 50 and 300 mm and the wall thicknesses between 5 and 30 mm. The length of the tube was 1 m. The AN was initiated by a booster over the whole cross section of the tube. The booster had a thickness of about 50 mm and was made of GX-1 dynamite. To obtain a more or less flat detonation wave for the larger diameters multi-point initiation technique using detonation cords were used to initiate the booster.

In order to measure the stable detonation velocity four ionization probes were mounted at each 50 mm from the bottom plate and the detonation velocity was calculated by the distance of the probes and the arrival time difference of the shock wave. The accuracy of the detonation velocity measurement is considered as within 3 %.

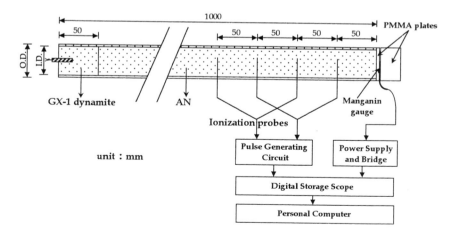

Figure 1 Experimental set-up of detonation velocity and pressure measurement of ANFO

Pressure profile was recorded with the commercially available piezo-resistive manganin gauge (Dynasen, MN4-50-EK), which was embedded within the PMMA plates to minimize the short-circuiting effect. The peak pressure was calculated by using the impedance mismatching technique (Miyake, van der Steen & Kodde, 1989). The accuracy of the pressure measurement is considered as within 5%.

RESULTS AND DISCUSSION

Two series of steel tube tests were carried out with the low density prilled AN. In the first series of experiments the effect of confinement on the detonation parameters was investigated. In the second series the diameter effect was studied.

Detonation velocity

Experimental results of detonation velocity are summarized in Figure 2. In the figure the dashed lines indicate the ideal detonation velocity (4.01 km/s) as calculated by the TIGER code with JCZ3 EOS at a density of 850 kg/m^3. At first to investigate the confinement effect steel tubes with a constant diameter of 100 mm and wall thicknesses between 50 and 30 mm were used. Although the detonation velocity increased with the increase of the wall thickness between 5 and 20 mm, it converge to a constant value of about 3.3 km/s. This value is estimated as 85 % of the theoretically predicted values at the same density.

Secondly the diameter of the steel tube was varied between 50 and 300 mm while the wall thickness was kept constant at 10 mm. The detonation velocity increased remarkably with the increase of the diameter and the highest velocity was measured as 3.70 km/s at the diameter of 300 mm. This value is estimated as 95 % of the calculated value.

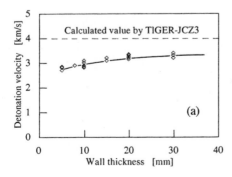
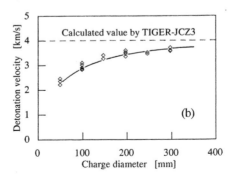

Figure 2 Detonation velocity as a function of [a] wall thickness (charge diameter is 100 mm) and [b] charge diameter (wall thickness is 10 mm).

Peak pressure

The results of the pressure measurements for different tube diameters and different confinements are summarized in Figure 3. The dashed lines indicate the ideal detonation pressure calculated by the TIGER code with the JCZ3 EOS at 850 kg/m^3, which is 3.45 GPa. In tubes with an inner diameter of 100 mm, the peak pressures increase from 1.9 to 2.8 GPa with the increase of the wall thickness from 5 to 30 mm. These values are estimated as 58 and 92 % of the calculated values respectively.

As the wall thickness is kept constant at 10 mm the peak pressures increase from 2.0 to 3.2 GPa with increasing diameter of the tube from 50 to 300 mm. These values are estimated as 45 and 96 % of the calculated values respectively at each experimental density.

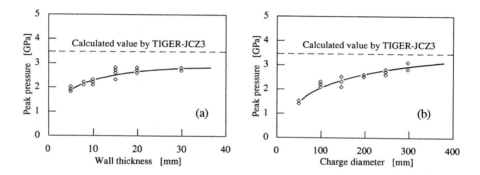

Figure 3 Peak pressures a function of [a] wall thickness (charge diameter is 100 mm) and [b] charge diameter (wall thickness is 10 mm).

Non-ideal detonation behaviour of AN

Experiments with low density AN prills show that the ideal detonation could be attained if the diameter of AN charge is large enough. For this type of AN a diameter of about 300 mm is needed to reach these conditions. By extrapolating the linear relation of the detonation parameters vs reciprocal diameter diagram, the ideal detonation velocity is determined as 3.83 km/s and 3.1 GPa for the peak pressure. These values show a good agreement with the ideal parameters as calculated by the TIGER code with the JCZ3 EOS.

From the classical detonation theory the C-J pressure is obtained as a linear function of square of the C-J velocity as a following formula;

$$P_{CJ} = [\rho_0 / (\gamma+1)] D_{CJ}^2 \tag{1}$$

where, P_{CJ} : C-J pressure, ρ_0 : initial density, γ : adiabatic exponent of gas products, D_{CJ} : C-J velocity

Table 2 shows the γ values of each experimental condition using Eqn (1). Although γ is almost constant with the increase of wall thickness, it increases with the increase of the charge diameter, and reaches 2.90 at an infinite diameter, which coincides with that of usual explosives.

Table 2 γ values of detonation gas products of AN

Wall thickness [mm]	Charge diameter [mm]	Detonation velocity [km/s]	Peak pressure [GPa]	γ [-]
5	100	2.72	1.34	2.53
10	100	3.06	2.40	2.24
15	100	3.18	2.63	2.19
20	100	3.23	2.74	2.16
25	100	3.27	2.80	2.17
30	100	3.29	2.85	2.15
infinite	100	3.40	3.07	2.13
10	50	2.29	1.35	2.22
10	100	3.06	2.24	2.47
10	150	3.32	2.53	2.62
10	200	3.45	2.68	2.69
10	250	3.52	2.77	2.71
10	300	3.58	2.83	2.76
10	infinite	3.83	3.12	2.90

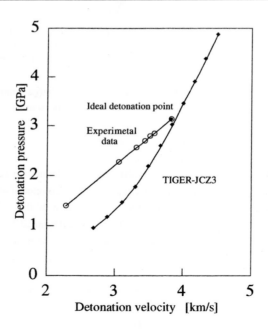

Figure 4 Relation between the non-ideal detonation velocity and pressure (for a constant density) and between the ideal detonation velocity and pressure (varying density)

In Figure 4 the non-ideal peak pressures are given as a function of the non-ideal detonation velocities. A linear relation exists between these two parameters. A linear relation is certainly not found as only ideal detonation conditions, for different densities, are compared. The double circle in the figure denotes the conditions for a density of 850 kg/m^3 and an extrapolation of the non-ideal experimental results nearly coincides with this point (Miyake et al., 1992).

CONCLUSIONS

From the detonation velocity and pressure measurement of AN, following conclusions can be drawn:

[1] The charge diameter has a much larger influence on the non-ideal behaviour of the detonation velocity and pressure than the confinement.
[2] The extrapolated detonation velocity and peak pressure to the infinite diameter were calculated as 3.83 km/s and 3.1 GPa respectively, which values coincide with the ideal detonation parameters for ammonium nitrate as predicted by the thermohydrodynamic TIGER code with the JCZ3 equation of state.
[3] Although the adiabatic exponent of the detonation gas products (γ) remained constant when the confinement became heavier, it showed a remarkable change with the increase of the charge diameter. The extrapolated γ value to the infinite diameter was obtained as 2.90 and it coincides with the γ calculated with the TIGER code.

REFERENCES

Cook, M.A. (1958). *The science of high explosives*, Rheinhold, USA

Cowperthwaite, M. and Zwisler, W.H. (1973). *TIGER computer program documentation*, SRI Publ. No.Z106, Stanford Research Institute, USA

Miyake, A., van der Steen, A.C. & Kodde, H.H. (1989). Detonation velocity and pressure of the non-ideal explosive ammonium nitrate, In *Proceedings of the 9th International Symposium on Detonation*, 560-565, Portland, USA

Miyake, A., Ogawa, T., Saito, S. & Yoshida, N. (1992). Non-ideal detonation properties of ammonium nitrate. *Journal of Industrial Explosives Society, Japan*, **53**, 67-74.

Tanaka, K. (1983). *Detonation properties of condensed explosives computed using the Kihara-Hikita-Tanaka equation of state*, National Chemical Laboratory for Industry, Japan

STUDY ON THE TREATMENT OF TOXIC WASTES BY CHEMICAL EXPLOSION

Takehiro Matsunaga[1], Ken-ichi Miyamoto[1], Mitsuaki Iida[1], Atsumi Miyake[2] and Terushige Ogawa[2]

[1]National Institute of Advanced Industrial Science and Technology,
  Higashi, Tsukuba, Ibaraki, Japan
[2]Yokohama National University, Hodogaya, Yokohama, Japan

ABSTRACT

The destruction technology of toxic waste, such as PCB, CFC or explosive materials, by using explosion reaction, was investigated. Three kinds of organic halides - tetrachloroethylene, 1,3,5-trichloro-benzene and 9,10-dichloroanthracene - were destructed with high destruction and removal efficiencies. Polychlorinated-dibenzo-p-dioxins (PCDDs) and other toxic compounds were not detected in the products within the detection limit of the instrument used.

KEYWORDS

Explosive, Toxic waste, Explosion products, PCB, Application of explosive, Waste treatment

1. INTRODUCTION

Produce and use of polychlorinated biphenyl (PCB) has been practically prohibited. Carbon paper, transformers and condensers containing PCBs are being stored in each facility that has ever been using them. However, it is almost impossible to keep storing them forever. Therefore, it is necessary to develop and introduce proper treatment methods to decompose them.

The purpose of this study is to develop a treatment method for toxic wastes using explosion reaction that generates high temperature of more than 3000 K and super high pressure of more than 10 GPa.

Advantages of using explosion reaction are following;
(1) It is possible to treat solid wastes as well as liquid.
(2) Almost all organic compounds cannot exist in the form of molecules under such high temperature and pressure condition. They should transform into inorganic compounds such as water, carbon dioxide and so on.
(3) By adding metal peroxides to explosives, it is possible to control chemical equilibrium and to transform halogen into harmless metal halide.

2. EXPERIMENT

An explosion chamber used in this study is shown in Fig.1. The explosion chamber can resist the pressure generated from about 30 g of an explosive. And after explosion, temperature and pressure in the chamber can be measured and all gaseous and condensed residue can be recovered.

Tetrachloroethylene, 1,3,5-trichlorobenzene and 9,10-dichloroanthracene were used as model compounds. Pentaerythritol tetranitrate (PETN) was used as an explosive. Metal peroxide such as sodium peroxide, barium peroxide, calcium peroxide, potassium peroxide, and manganese dioxide was added in order to trap the halogen of the model compound and convert it into metal halide. Such an additional compound was simultaneously expected to improve gaseous products more cleanly, because the compound evolved oxygen.

Three kinds of arrangement of the sample compound, explosive and metal peroxide were investigated, as shown in Fig.2. After explosion, the gaseous products and the residue were analyzed by using several methods.

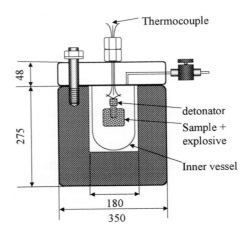

Fig.1　Explosion Chamber　(Unit: mm)

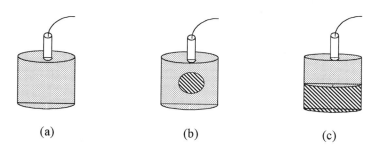

Fig. 2 Arrangement of organic halide, explosive and halogen trapping agent (metal peroxide). (a) All three components were premixed. (b) The mixture of organic halide and metal peroxide was wrapped by aluminum foil and the explosive surrounded them. (c) The explosive was attached to the mixture of organic halide and metal peroxide.

3. RESULTS

Within the limit of our experiments, more than 99.9% of the decomposition rate was obtained for each compound under suitable conditions. Poly-chlorinated-dibenzo-p-dioxins (PCDDs) and other toxic compounds were not detected in the products within the detection limit of the instrument used. Manganese dioxide and calcium peroxide were effective for halogen trap. Typical results are summarized in Table 1.

TABLE 1
TEST CONDITIONS AND RESULTS

Composition	Sample weight [g]	Sample arrangement*	DRE [%]****	Efficiency of halogen trapping [%]
PETN/C_2Cl_4/MnO_2	30/10/5	(a)	> 99.9999	—
PETN/C_2Cl_4/MnO_2	30/10/10	(a)	> 99.9999	—
PETN/C_2Cl_4/CaO_2	30/10/10	(a)	> 99.9999	—
PETN/C_2Cl_4/CaO_2	30/10/20	(a)	> 99.9999	—
PETN/TCB**/CaO_2	30/20/26	(a)	99.999619	79
PETN/TCB**/CaO_2	30/10/15	(a)	99.999921	81
PETN/TCB**/CaO_2	30/2/4	(a)	99.99858	65
PETN/TCB**/CaO_2	30/20/26	(b)	99.9642	85
PETN/TCB**/CaO_2	30/2/4	(b)	99.999641	40
PETN/DCA***/CaO_2	30/10/3	(c)	99.9992	89
PETN/DCA***/CaO_2	15/10/3	(c)	99.9977	74

*: refer to Fig. 2, **: 1,3,5-Trichlorobenzene, ***: 9,10-Dichloroanthacene,
****: Destruction and Removal Efficiency

ESTIMATION OF SHOCK SENSITIVITY BASED ON MOLECULAR PROPERTIES

M. Koshi, S.Ye, J.Widijaja, and K.Tonokura

Department of Chemical System Engineering, University of Tokyo,
7-3-1 Hongo, Bunkyo-ku, Tokyo 113-8656, JAPAN

ABSTRACT

Relationship between shock sensitivity and rate of vibrational excitation of energetic materials was investigated. When a molecular crystal receives a shock (or impact), lattice vibrations (phonons) are excited at first. Phonon energy then must be converted to molecular vibrations before bond breaking can occur. This energy transfer process is expected to be the rate-determining step for the explosion. The rate of the energy transfer from phonon to molecular vibration is evaluated on the basis of a simple theory coupled with experimental measurements of line width of Raman spectra at very low temperatures. Good correlation was found between the sensitivities derived from drop-hammer test and the rate of energy transfer. The rates of energy transfer were also compared with wedge test results. In this case, loading pressures were very high, and effects of pressure on the energy transfer have to be considered. Taking the increase of anharmonic coupling into account, the rates of energy transfer can be compared with the induction distances in wedge tests.

KEYWORD

Shock sensitivity, Energetic material, Raman Spectra, Phonon, Vibration, Energy transfer, Wedge test, Drop hammer test

INTRODUCTION

Understanding of shock initiation of explosives is essential for the safe handling of these materials. Such knowledge is also needed for the design of new high performance explosives. Shock initiation of explosives is quite complicated process. Whenever a mechanical excitation is created in explosive materials, the excess mechanical energy is eventually dissipated into a bath consisting of the low frequency mode of lattice vibrations (phonons). Before a detonation wave can begin, bonds of molecules in the crystal must break. Therefore, the initial energy in the phonons must be deposited to molecular vibrations that are responsible to the bond breaking.

Most of secondary explosives are molecular solids consisting of large organic molecules. Because secondary explosives are stable molecules with large energy barriers to chemical reaction, a sizable amount of energy must be transferred from the phonons to the molecules' internal vibrations. Phonons generally have frequencies less than 200 cm^{-1}, whereas molecular vibrations relevant to bond breaking have frequencies greater than 1000 cm^{-1}. It is clear that phonon energy must be converted to higher vibrations by multi-phonon up pumping.

Dlott and Fayer (1990) have studied multiphonon up conversion processes associated with shock induced chemical reaction. They derived a simple expression for the phonon-vibron energy transfer rate. In their treatment, the dominant mechanism for up-pumping is anharmonic coupling of excited phonon modes with low frequency molecular vibrations. These low frequency vibrational modes are termed as "doorway mode". The mechanical energy given by shock is quickly deposited in the external modes and redistributed among all the phonons. This "phonon rich region" is attained within ≈1 ps. Vibrons are not excited at this stage. In the up-pumping zone, the energy in phonon modes is transferred to the vibrations by multiphonon up-pumping processes, and the two baths (phonons and "doorway" vibrations) equilibrate within ≈100ps. As the vibrational modes in the doorway states are excited, higher vibrational states are then excited by V-V (vibration to vibration) up-pumping processes. As a result, reactivity is enhanced and bond breaking due to thermal decomposition occurs. This is the ignition zone and extends to ≈1 ns behind the shock front. A series of exothermic reactions will occur after the endothermic bond breaking in the reaction zone, and finally a C-J (Chapman-Jouget) state is realized.

Fried and Ruggiero (1994) have calculated the total energy transfer rate from phonons into a given vibron band in terms of the density of vibrational states and the vibron-phonon coupling. They estimated the phonon upconversion rates for several explosives such as TATB (1,3,5-triamino-2,4,6-trinitrobenzene), RDX (1,3,5-trinitro-1,3,5-triazacyrohexane), and HMX (1,3,5,7-tetranitro-1,3,5,7-tetraazacyclooctane). The estimated energy transfer rates in pure unreacted materials are found to be several times greater for the sensitive explosives than the insensitive explosives. They also showed that the energy transfer rates estimated at the vibrational frequency of $\omega=425$ cm^{-1} linearly correlated to the sensitivities derived from drop hammer tests. It is noted that the vibrational frequency of $\omega=425$ cm^{-1} corresponds to nitro group motion. Finding of such correlation between energy transfer rate and shock sensitivity is very important for the prediction and understanding of the shock sensitivity of energetic materials.

In the present study, we estimated the energy transfer rates by applying a simple formula given by Tokmakoff et al. (1993). Rate constants were calculated for RDX, HMX, PETN(Pentaerythritol tetranitrate), DNB(dinitrobenzene), TNT(2,4,6-trinitrotoluene), NQ(nitroguanidine), and PN (propylnitrate). Vibrational frequencies required for the calculation were obtained on the basis of *ab-initio* molecular orbital calculations. Experiments were performed to measure the Raman line width in order to determine the life times for phonons and vibrons of explosives at T=0 K. Results of these experiments were used to evaluate the energy transfer rates. We also investigated the correlation between energy transfer rates and the shock or impact sensitivities derived from wedge tests and drop hammer tests.

MULTIPHONON UP-PUMPING RATE

According to Tokmakoff et al., we define doorway modes as the molecular vibrations located just above the phonon cutoff frequency Ω_{max}. The large molecules have doorway modes whose frequency Ω is in the range of $\Omega_{max} \leq \Omega \leq 2\Omega_{max}$. In this case, the excitation of doorway mode vibration needs simultaneous absorption of two phonons. Absorption of further phonons transfers energy into higher modes until the IVR (Internal Vibrational Redistribution) threshold energy (E_{IVR}) is reached. Once the molecule is excited up to E_{IVR}, the energy is randomized rapidly by IVR. Large molecule such as HMX has very low frequency vibrational modes involving NO$_2$ rocking motions. Frequencies of these modes are below Ω_{max}, and they are amalgamated with the phonons. Amalgamated vibrations behave much like phonons in that they can be directly excited by the shock. The rate-limiting step in up pumping is the energy transfer from the phonon bath (including amalgamated vibrations) to the doorway modes, and it is expected that this energy transfer rate correlate to the shock sensitivity. Therefore, we estimated the energy transfer rate to the doorway states and compared with the experimental sensitivities.

Dlott and Fayer showed that the relaxation equations for the phonon and vibrational quasi-temperatures (θ_{ph} and θ_{vib}, respectively) were given by

$$\frac{\partial \theta_{ph}}{\partial t} = \frac{\kappa(V)}{C_{p,ph}}[\theta_{vib} - \theta_{ph}] \qquad \frac{\partial \theta_{vib}}{\partial t} = \frac{\kappa(V)}{C_{p,vib}(\theta_{vib})}[\theta_{ph} - \theta_{vib}] \qquad (1)$$

where, $\kappa(V)$ is the specific volume dependent energy transfer parameter that characterize the rate of energy flow between phonons and vibrations, $C_{p,ph}=6k_B$ is the constant pressure heat capacity of the phonon mode, k_B is the Boltzmann constant and $C_{p,vib}(\theta_{vib})$ is the constant pressure heat capacity of the vibrational modes, which depends on the vibrational quisi-temperature. The energy transfer parameter at ambient pressure, $\kappa(V_0)$, can be calculated by using following relation:

$$\kappa(V_0) = \frac{j\hbar\Omega}{\tau(0)\theta_e} \qquad (2)$$

where $\tau(0)$ is the lifetime of the doorway mode at T=0 K, j is the number of doorway modes, Ω is the vibrational frequency of the doorway state, θ_e is the equivalence temperature, the temperature at which the rate of up-pumping into the doorway mode is equal to the low temperature rate of relaxation of the doorway mode by the two phonon emission. θ_e is defined by the relation

$$n_{\Omega/2}(\theta_e) - n_\Omega(\theta_e) = 1 \qquad (3)$$

where $n_\Omega(\theta)$ is the occupation number of phonons at frequency Ω and temperature θ. For $\Omega=200$ cm^{-1}, $\theta_e=300$ K. The lifetime $\tau(0)$ is related to the specific volume dependent anharmonic coupling term $<V^{(3)}(V)>$ and the two-phonon density of states at the doorway mode frequency Ω, $\rho^{(2)}(V)$ through the following equation:

$$\frac{1}{\tau(0)} = \frac{36\pi^2}{\hbar}<V^{(3)}(V)>\sigma^{(2)}(V) \qquad (4)$$

Tokmakoff et al. used values of $\tau(0)=3$ps, $j=1$ and $\Omega=200$cm^{-1} as representative values for some explosives such as RDX and HMX. These values were actually obtained for naphthalene. In the present study, the value of $\tau(0)$ is experimentally checked by measuring low temperature line width of Raman spectra of RDX, HMX and NTO(5-nitro-1,2,4-triazol-3-one). Behind a shock front, pressure and density effects increase the rate of up-pumping into the doorway modes. Tokmakoff et al. formulated such a pressure and density effects. The anharmonic coupling term is increasing with compression due to the decrease in intermolecular separation. With the exp-6 intermolecular potential, this effect is evaluated as

$$\frac{<V^{(3)}(V)>}{<V^{(3)}(V_0)>} \approx (1-\Delta R_0)\exp(13\Delta R_0)(1-\frac{3}{2}\Delta R_0) \qquad (5)$$

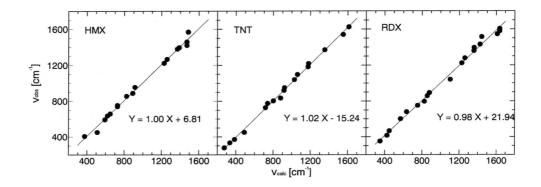

Figure 1: Comparison of measured and calculated vibrational frequencies

where ΔR_0 is the change of equilibrium inter molecular distance induced by compression:

$$\Delta R_0 = 1 - \left(\frac{V}{V_0}\right)^{1/3} \quad (6)$$

The specific volume-dependent energy transfer parameter behind a shock front is given by

$$\frac{\kappa(V)}{\kappa(V_0)} = \frac{|<V^{(3)}(V)>|^2}{|<V^{(3)}(V_0)>|^2} (1-\Delta V)^{-2} \exp(\Gamma_0 \Delta V) \quad (7)$$

where $\Delta V = 1 - V/V_0$ and Γ_0 is the bulk Grüneisen parameter at ambient pressure.

VIBRATIONAL FREQUENCIES AND VIBRATIONAL DENSITY OF STATES

The vibrational level structure at the doorway region is expected to be very sensitive to the energy transfer rates. As can be seen in Eqn. 2., the energy transfer rate is directory proportional to the number of vibrational modes in the doorway region. In the present study, we assumed $\Omega_{max}=200$ cm^{-1} and $E_{IVR}=700$ cm^{-1} for all the energetic materials, according to the model proposed by McNesby and Coffey (1997). In other word, the doorway region in the present model is in between 200 and 700 cm^{-1}. Normal mode vibrational frequencies of the energetic materials are calculated by *ab-initio* molecular orbital method for PETN, HMX, RDX, TNT, DNB, NQ and PN. Geometries of these molecules are optimized at first and then normal mode vibrational frequencies and Raman intensities of each vibrational mode were calculated by using GAUSSIAN 98 program at the B3LYP/6-31G(d) level of theory. Calculated frequencies are scaled by a factor of 0.9613 (Foresman & Fish, (2000)). Calculations were performed for the isolated molecules, and therefore, the calculated vibrational frequencies are not necessary to match the vibrational frequencies of solid exegetic molecules. However, intermolecular interaction in a molecular crystal is expected to be weak, so that the vibrational frequencies in molecular solid are not so much different from an isolated molecule. In order to confirm this, we measured the Raman spectra of energetic materials at room temperature, and measured vibrational frequencies are compared with the calculated vibrational frequencies of isolated molecules. Figure 1 shows some of the results of such comparison. As can be seen in the figure, measured and calculated frequencies are in good agreement

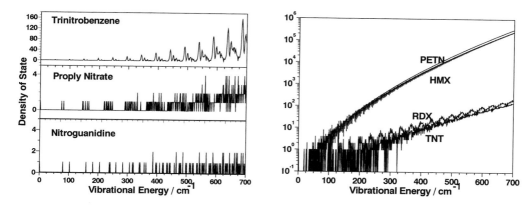

Figure 2: Vibrational density of states for selected explosive molecules

with each other. Therefore we used vibrational frequencies of isolated molecules to predict the energy transfer rates in molecular solid.

Density of states of energetic materials is calculated by using direct counting method. Results are shown in Figure 2. It is noted that the explosives with higher sensitivity have larger vibrational density of states. Density of states in the region of doorway modes (200-700 cm^{-1}) seems to be significantly related to the shock sensitivities of secondary explosives.

EVALUATION OF VIBRATIONAL LIFE TIME AT T=0 K, $\tau(0)$

For the evaluation of energy transfer parameter $\kappa(V_0)$ in Eqn. 2, one can use the experimental value of $\tau(0)$. Dlott et al. and Tokmakoff et al. assumed $\tau(0)$=3ps for RDX and HMX. This value of $\tau(0)$ is obtained for naphthalene, and there is no experimental value for secondary explosives. We determined the lifetime of RDX and HMX (highly sensitive explosives) and NTO (rather insensitive explosives) by measuring line width of Raman spectra at near 0 K.

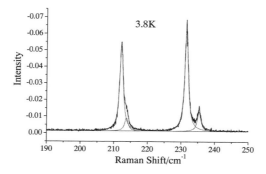

Figure 3: Raman spectra of RDX with Lorenzian fitting curves

Sample crystal was cool down by using cryogenic refrigerator for optical measurements (PS24SS, Nagase & CO., Ltd). Temperature of the sample can be controlled from 3.5 K to 300 K with the interval of 0.1 K. Sample crystal is then irradiated by 514.15 nm radiation from the output of Coherent

Inova 70-5 Ar⁺ laser. The laser power was in the range of 0.24-0.4 W. Raman spectra of the sample were recorded by using double monochromator (CT-1000D, 0.2 cm^{-1} resolution, Nippon Bunko Ltd.). The samples (RDX, HMX, and NTO) were obtained from the Chugoku Corporation of Chemical Pharmacy, and used with 3-time re-crystallization from a saturated ethanol or acetone solution (in case of RDX and HMX). In the case of NTO water was used as a solvent. Figure 3 shows example spectra of RDX obtained at 3.8 K. It is found that all the lines observed could be fitted to the Lorenzian line shape, as shown in Figure 3.

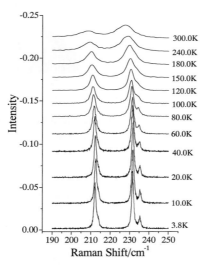

Figure 4: Temperature dependence of the line shift and width of RDX Raman spectra

The line position shifts to lower wave number as the temperature increases. The amount of shift largely depends on the type of mode. The line width is also changed with temperatures, as expected. Line boarding at low temperature is caused by two factors: relaxation by dephasing and population relaxation (Pinan et al., (1998)). Line width caused by dephasing may depend on temperature as T^2, whereas line width depends linearly on temperature in case of population relaxation caused by three phonon processes or the vibrational relaxation to two phonon states. It is noted that contributions from pure dephasing to the line width vanishes at T=0 because phonon population vanishes at T=0. The line width is determined only by the energy transferred from one vibron (or phonon) to two phonons at T=0. Therefore we can derive the rate of population relaxation from the line width at T=0. Figure 4 shows an example of the temperature dependence of Raman spectra of RDX. By extrapolating the line width to T=0, the value of $\tau(0)$ could be determined. Lifetimes for phonon modes in 25.7- 69.7 cm^{-1} regions are 5.5-9.0 ps. On the other hand, lifetimes of vibrational band are 2.7-5.7 ps in 115.3-675.0 cm^{-1} regions. Although different mode gives different values of $\tau(0)$, the averaged value in the doorway region is 4 ps. This value is rather close to a value for naphthalene (3ps). We extended these measurements to HMX and NTO. Tentative results showed that the lifetime at T=0 in the doorway region is 2.2-4.7 ps for HMX and 1.4-5.9 ps for NTO. These results combined with the value for naphthalene suggested that the rate of population relaxation at T=0 roughly the same for many molecules of this size. Although detailed studies on the mode dependence of the energy transfer rate are clearly needed, constant value of $\tau(0)$ can be used for the rough estimation of the energy transfer rate for many secondary explosives, as pointed out by Tokmakoff et al.

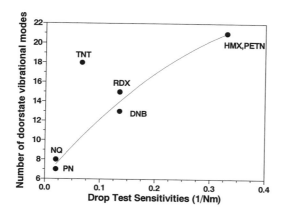

Figure 5: Correlation between impact sensitivity and number of vibrational mode in doorway regions

DROP HAMMER SENSITIVITIES AND ENERGY TRANSFER RATE

If we assume that the value of $\tau(0)$ is approximately the same, the energy transfer parameters in Eqn. 2 is directly proportional to the numeber of doorway modes, j. In the present study, j is defined as the number of vibrational modes in the region of 200 to 700 cm^{-1} and the values are evaluated from the results of *ab-initio* molecular orbital calculations. Resulting j values are 21, 21, 18, 15, 13, 8 and 7 for PETN, HMX, TNT, RDX, DNB, NQ, and PN. It is interesting to compare these numbers to impact sensitivity as measured by the drop hammer test. Formation of hot spot caused by adiabatic compression of small void can play important role in the drop hammer test initiation. Even though, energy transfer rate may have correlation to the sensitivity because thermal reaction also affected by the energy transfer rate. As pointed out by Fried and Ruggioro, strong correlation between shock and impact sensitivity is expected. In fact, Fried and Ruggioro found very good correlation between drop hammer sensitivity and the energy transfer rate at the vibrational frequency correspond to NO$_2$ vibrations. McNesby and Coffey also found the similar correlation between vibrational relaxation rates and the drop hammer sensitivity.

We define the drop hammer sensitivity as $(H_{50}W)^{-1}$, where H_{50} is the 50% explosion height and W is the weight of the drop hammer. There is a serious problem in the comparison with drop hammer sensitivity. Reported data in the literatures usually have very large scatter. In order to get more reliable data, we simply compiled data as many as possible, and averaged values are used for comparison. Data sources are: Dobratz (1981), Gibbs and Popolate (1980), Meyer (1987), Kato (1982) and Hasue (1983). Figure 5 shows the comparison of the drop hammer sensitivities with the number of vibrational modes in the doorway region. Good correlation is found except for TNT. It is possible that this kind of correlation could be caused by accidental coincidence. To confirm this correlation, accumulation of more reliable data on impact sensitivity is required.

COMPARISON OF SHOCK SENSITIVITY WITH ENERGY TRANSFER RATE

We compare the energy transfer rate with the results of wedge test as a measure of the shock sensitivity. In the wedge test, strong shock is applied to explosives and the pressure behind shock is extremely high. Therefore the energy transfer rate has to be calculated by using Eqn. 7. Induction distance data in the

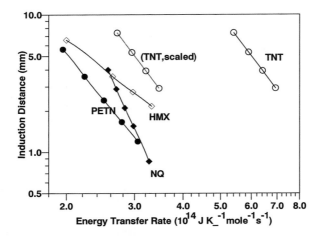

Figure 6: Induction distance in wedge test as a function of energy transfer rate

wedge test was taken from the book of Gibbs and Popolate. This book also gives us the shock Hougniot data. Pressure and a value of $\Delta V=1-V/V_0$ are calculated by using these Hougoniot data. In figure 6, the induction distance in wedge test and the energy transfer rate behind shock are compared. The valu of $\kappa(V_0)$ is assumed to be $j \times 2.5 \times 10^{12}$ JK^{-1}mole^{-1}s^{-1} for all explosives. This value is obtained with $\Omega=200$ cm^{-1} and $\theta_e=300$ K and $\tau(0)=$ 3ps. This is clearly very crude approximation. Nevertheless, it seems some correlation between induction distance and the energy transfer rate. It is noted that TNT is very large deviation in Figure 6, and this is similar in Figure 5. If we reduced j value of TNT to $j=9$ so that the correlation in Figure 5 is satisfied, agreement is improved in Figure 6.

References
Dlot D.D and Fayer M.D. (1990). Shocked Molecular Solids: Vibrational Up pumping, Defect Hot Spot Formation, and the Onset of Chemistry. *J. Chem.Phys.* **92**, 3798
Dobratz B.M. (1981). *LLNL Explosives Handbook*, Univ. of California Press, Berkeley
Fried L.E. and Ruggiero A.J. (1994). Energy Transfer Rates in Primary, Secondary, and Insensitive Explosives. *J. Phys. Chem.* **98**, 9786
Gibbs T.R. and Popolate A. (1980) *LASL Explosive property Data*, Univ. of California Press, Berkeley
Hasue K. Hirano S. Ogawa Y. Okazaki K. and Nakahara S. Pendulum Friction Tests on Some Explosives (III). *J. Jpn Explosives Soc.* **44**, 69
Kato K. Shimizu H. Fukuda T. Yoneda K. and Asaba T. Detonability of a Composite Propellant Containing Nitramine. *J. Jpn Explosives Soc.* **43**, 375
McNesby K.L. and Coffey C.S. (1997). Spectroscopic Determination of Impact Sensitivities of Explosives. *J. Phys. Chem. B* **101**, 3097
Meyer R. (1987). *Explosives* (3rd Eddition) Weinheim, New York
Pina J.P. Ouillon R. Ranson P. Becucci M. and Califano S. (1998). High Resolution Raman Study of Phonon and Vibron Bandwidths in Isotopically Pure and Natural Benzene Crystal. *J. Chem. Phys.* **109**, 5496
Tokmakoff A. Fayer M.D. and Dlott D.D. (1993). Chemical Reaction Initiation and Hot-Spot Formation in Shocked Energetic Materials. *J Phys. Chem.* **97**, 1901

DEEP PENETRATION OF TRUNCATED-OGIVE-NOSE PROJECTILE INTO CONCRETE TARGET

X.W.Chen and Q.M.Li[*]

School of Civil and Structural Engineering
Nanyang Technological University, Nanyang Avenue, Singapore 639798
[*] Corresponding author, E-mail: cqmli@ntu.edu.sg; Fax: (+65) 7910676

ABSTRACT

The penetration of a truncated-ogive-nose projectile into semi-infinite concrete targets is studied in this paper. Based on a general definition of the projectile nose factor and the dynamical cavity expansion model, the nose factor of the truncated-ogive-nose projectile is formulated in the present paper to give a theoretical prediction of penetration depth of truncated-ogive-nose projectile in concrete. Theoretical prediction is in excellent agreement with the test results.

KEYWORDS

truncated-ogive-nose projectile, concrete target, deep penetration, nose factor

INTRODUCTION

Impact dynamics of reinforced concrete (RC) and plain concrete targets by non-deformable projectile has been investigated experimentally in an extensive range of both civil and military applications. Various empirical formulae, e.g. the Army Corps of Engineers formula (ACE), the UKAEA formula (i.e., the Barr formula) and the National Defense Research Committee (NDRC) formula, etc., have been proposed to predict the local effects of a "hard" missile impact, i.e., penetration depth X, perforation limit e and scabbing limit h_s [Li and Chen(2001)]. However, majorities of the empirical formulae on penetration depth into concrete target were formulated from curve fitting test data and are dimensionally dependent, and therefore, do not provide any physically based descriptions and lead to difficulties for general comparison among various empirical formulae. In addition, these empirical formulae define the projectile nose factor in rough average scale and most of them are limited to small penetration depth, e.g., $X/d < 4$ for NDRC formula.

Relatively high impact velocities between 400 and 1000 m/s are associated with events, such as, attacks on buildings by kinetic energy weapons, fragment perforation and impacts of tornado-generated projectiles, etc., where X/d could be quite large. Systematic studies on penetration of concrete with ogive-nose projectile have been conducted by Forrestal, et al.(1994, 1996) and Frew, et al.(1998), which covered a broad range of concrete strengths for striking velocities up to 1km/s (X/d up to 70) until nose erosion becomes excessive. An empirical equation for penetration depth of ogive-nose projectiles into concrete targets was developed based on the dynamic spherical cavity expansion model, which is also capable of predicting the target resistance and the projectile motion during penetration.

Qian, et al.[2000] studied the truncated-ogive-nose projectile penetration into semi-infinite concrete targets by using the resistant force formula in dynamic cavity expansion model. They introduced a resistance constant $c' = 1 + K\dfrac{d_1^2}{d^2}$ (where d_1 and d are the diameters of the truncated area and projectile shank, respectively), which considered truncation effect of the ogive-nose projectile on penetration. However, it used an unverified assumption on initial impact force. Meanwhile, curve-fitting determination of coefficient K with limited test data and the complicate formulation restrict its applicability in a broad range of projectile-target analysis.

In the present paper, a general definition of projectile nose factor is introduced. Nose factor of truncated-ogive nose is formulated analytically to suggest a theoretical prediction of penetration depth. Comparison between experimental results and theoretical prediction shows excellent agreement.

NOSE FACTOR N^* OF NON-DEFORMABLE PROJECTILE

A non-deformable projectile with arbitrary nose shape, as shown in Fig. 1, impacts a target at normal incidence with velocity V_0 and proceeds to penetrate the target medium at rigid-body velocity V.

The dynamic cavity expansion analysis yields the following relation between the normal compressive stress σ_n on the projectile nose and the normal expansion velocity v [e.g., Forrestal, et al.(1992), Chen and Li(2001a)],

$$\sigma_n = AY + B\rho v^2 \tag{1}$$

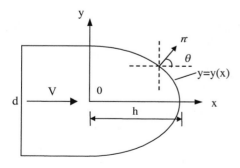

Fig. 1 Nose shape of an arbitrary non-deformable projectile

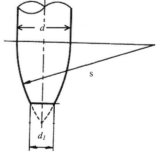

Fig. 2 Nose geometry of a truncated-ogive--nose projectile

where Y and ρ are yielding stress and density of target material, respectively. A and B are dimensionless target material constants.

The particle velocity at nose-target interface caused by the rigid-body velocity V of a projectile is

$$v = V\cos\theta . \tag{2}$$

A large amount of impact tests demonstrated that the interface friction of projectile-target can be ignored in penetrations into soil and concrete targets [Forrestal, et al.(1992,1994,1996)]. Thus, the tangential stress on the nose is not considered here. The resulting axial resistant force on the projectile nose can be integrated from the normal compressive stresses, i.e.,

$$F = \frac{\pi d^2}{4}\left(AY + BN^*\rho V^2\right). \tag{3}$$

If the nose shape can be represented by a nose shape function $y = y(x)$ for an arbitrary nose shape, as shown in Fig. 1, we have [e.g., Jones and Rule(2000)]

$$N^* = \frac{8}{d^2}\int_0^h \frac{yy'^3}{1+y'^2}dx . \tag{4}$$

where the integrals are performed in the nose surface A_n and h is the height of nose.

For ogive-nose shape [Forrestal, et al.(1992)],

$$N^* = \frac{1}{3\psi} - \frac{1}{24\psi^2}, \tag{5}$$

where ψ is the caliber-radius-head (CRH) of ogive projectile, and $0 < N^* \leq \frac{1}{2}$ as $\psi \geq \frac{1}{2}$.

For truncated-ogive-nose in Fig.2 [Chen and Li(2001a)]:

$$N^* = \psi^2\left[2\cos^4\phi_0 - \frac{8}{3}\left(1-\frac{1}{2\psi}\right)(2+\sin\phi_0)(1-\sin\phi_0)^2\right] + \zeta^2, \tag{6a}$$

$$\phi_0 = \sin^{-1}\left[1-\frac{1}{2\psi}(1-\zeta)\right], \text{ where } \zeta = \frac{d_1}{d} \text{ and } \psi \geq \frac{1}{2}. \tag{6b}$$

Ogive-nose is actually a special case of truncated-ogive-nose ($\zeta = 0$).

N^*, termed as "nose factor" in the present paper, comes from the geometry integral of nose shape and reflects the geometry characteristics of the projectile nose to influence the target resistance. Obviously, the larger the value of N^*, the more blunt the nose.

DYNAMIC CAVITY EXPANSION MODEL

After introducing the nose factor N^* of arbitrary projectile, there exists no intrinsic distinction between non-deformable projectile penetrating concrete target with a truncated-ogive-nose (or any arbitrary nose) and with an ogive-nose that has been studied by Forrestal, et al.(1994,1996) and Frew, et al.(1998). The following formulation follows Forrestal, et al.(1994)'s model.

Forrestal, et al.(1994) showed that parameter B in eqns. (1) and (3) depends mostly on the compressibility of the target material and B had a narrow range, e.g. $B = 1.0$ for concrete, $B = 1.1$ for aluminium targets and $B = 1.2$ for soil targets. By contrast, AY had a broad range, depending mostly on the shear strength of the target materials. Forrestal, et al.(1994) took $AY = Sf_c$, where S is a dimensionless empirical constant depending on the unconfined compressive strength of concrete [Forrestal, et al.(1994,1996), Frew, et al.(1998)].

Thus, the empirical equation for axial force takes the form

$$F = \frac{\pi d^2}{4}\left(Sf_c + N^* \rho V^2\right), \tag{7}$$

which is limited to penetration depth $X > kd$. For $X < kd$, the process is dominated by surface cratering. Post-test observations of soil and concrete targets show that the cavity after penetration is a conical region, namely crater, followed by a tunnel region. The value of k is between 1.5 and 2.5 based on the studies by Qian, et al.(2000b), and $k = 2.0$ in Forrestal, et al(1994, 1996)'s tests and Frew, et al.(1998)'s tests.

The final penetration depth given by Forrestal, et al.(1994) is

$$X = \frac{2M}{\pi d^2 N^* \rho}\ln\left(1 + \frac{N^* \rho V_1^2}{Sf_c}\right) + kd, \quad X > kd, \tag{8}$$

where,

$$V_1^2 = \frac{MV_0^2 - \frac{\pi k d^3}{4}Sf_c}{M + \frac{\pi k d^3}{4}N^* \rho}. \tag{9}$$

Based on experimental results, the dependence of dimensionless number S on f_c was suggested by Forrestal, et al.(1994,1996) and Frew, et al.(1998), as shown in Fig. 3,

$$S = 82.6 f_c^{-0.544} \quad (f_c \text{ in MPa}). \tag{10}$$

RESULTS AND DISCUSSION

All experimental data tested by Qian, et al.(2000) are predicted according to the above formulation and are presented in Fig.4.

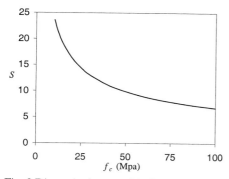

Fig. 3 Dimensionless empirical constant *vs* unconfined compressive strength.

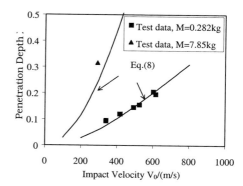

Fig. 4. Penetration data and model prediction.

Basically, there are two different mass truncated-ogive-nose projectiles reported by Qian, et al.(2000), i.e., M=0.282kg and M=7.85kg. Their nose factors are $N^* = 0.294$ and $N^* = 0.159$, respectively, according to eqs.(6a,b) and the geometry of the projectile. The mass density of concrete target is $\rho = 2300$ kg/m^3. Figure 4 shows that the analytical prediction agrees excellently with experimental measurement of the penetration depth. Thus, the dynamic cavity expansion model is applicable for the penetration of a non-deformable projectile with arbitrary nose into concrete target if a general nose factor N^* is introduced.

The present paper studied the deep penetration of truncated-ogive-nose projectile into concrete target by introducing a nose factor N^*. In addition to the projectile nose shape, other projectile geometrical dimensions, such as the shank diameter of a projectile, are equally important for penetration process. Recently, a systematic research on the non-deformable projectile penetrating target was conducted [Chen and Li(2001a,b) and Li and Chen(2001)], which suggested two dimensionless numbers, i.e., the impact function (I) and the geometry function of projectile (N) based on dynamic cavity expansion theory, dimensional analysis and careful study of experimental results. The proposed formula on predicting penetration depth shows excellent agreement with penetration tests on metals, concrete and soils in a broad range of impact velocities and projectile geometry.

CONCLUSIONS

General definition of the nose factor of arbitrary projectiles is introduced, which is applicable for predicting the penetration depth in concrete medium by a non-deformable projectile. The nose factor of a truncated-ogive-nose projectile is formulated in present paper. Using the dynamic cavity expansion model, the theoretical prediction of penetration depth of truncated-ogive-nose projectile into concrete has excellent agreement with the test results.

ACKNOWLEDGEMENTS

The first author also would like to acknowledge the PhD scholarship granted by the School of Civil and Structural Engineering, Nanyang Technological University.

REFERENCES

Li Q.M. and Chen X.W.(2001). Dimensionless formulae for penetration depth of concrete target impacted by a non-deformable projectile. Submitted to Int. J. Impact Engng..

Chen X.W. and Li Q.M.(2001a). Deep penetration of a non-deformable projectile with different geometrical characteristics. Submitted to Int. J. Impact Engng..

Chen X.W. and Li Q.M.(2001b). Transition from non-deformable projectile penetration to semi-hydrodynamic penetration. Prepare for publication.

Forrestal M.J. and Luk V.K.(1992). Penetration into soil targets. Int. J. Impact Engng. 12, 427-44.

Forrestal M.J., Altman B.S., Cargile J.D., and Hanchak S.J.(1994). An empirical equation for penetration depth of ogive-nose projectiles into concrete targets. Int. J. Impact Engng. 15:4, 395-405.

Forrestal M.J., Frew D.J., Hanchak S.J. and Brar N.S.(1996). Penetration of grout and concrete targets with ogive-nose steel projectiles. Int. J. Impact Engng. 18:5, 465-76.

Frew D.J., Hanchak S.J., Green M.L. and Forrestal M.J.(1998). Penetration of concrete targets with ogive-nose steel rods. Int. J. Impact Engng. 21, 489-97.

Jones S.E. and Rule W.K.(2000). On the optimal nose geometry for a rigid penetrator, including the effects of pressure-dependent friction. Int. J. Impact Engng. 24, 403-415.

Qian L.X., Yang Y.B. and Liu T.(2000). A semi-analytical model for truncated-ogive-nose projectiles penetration into semi-finite concrete targets. Int. J. Impact Engng. 24, 947-55.

PECULIARITIES OF PENETRATION INTO AN ELASTOPLASTIC PERTURBED TARGET

Yu. K. Bivin and I. V. Simonov

Institute for Problem in Mechanics RAS, 117526 Moscow, Russia

ABSTRACT

Laboratory-scale test data on the high-velocity penetration of steel spheres into a weak strength target perturbed by an added static or dynamic stress field and some qualitative analysis of the results are reviewed. The efforts are focused on observation and measurement of the cavity shapes, splashes, trajectories, and mainly depths. The comparison with the penetration into the unloaded half-space enable us to estimate significance of the pertubations.

KEYWORDS

High-velocity deep penetration, plastic target, static or dynamic perturbing stresses.

INTRODUCTION

Penetration into elastoplastic media was reviewed by Backman and Goldsmith (1978) and then a number of works has been done. Ricochet of solid sphere over water and sand was first examined by Soliman, Reid, and Johnson (1976). Further laws of refraction and its relation to ricochet *like Snella's ones in optics* has been discovered by an empirical fit of the experimental results on oblique crossing over the air-water and air-grounds interfaces by steel spheres (Bivin (1981); Bivin, Chekin, and Simonov (1994)). The group-of-bodies effects are also examined studied in particular, a phenomenon of the partial ricochet was described in Bivin (1993) and Simonov and Bivin (1999). In spite of great achievements in high-velocity impact-penetration, there are many "blank spaces" in this area. Thus the only investigated case the nonstandard target boundary conditions that is familiar to the authors, is the shots in parallel to the free target surface (Johnson and Daneshi, 1977). Hence the penetration into a perturbed target is the comparatively new problem.

This paper is offered a review of experimental results on the influence of added stress fields in a solid with a relatively low yield stress on the steel ball penetration process. These fields were produced by different way: statically, by compression of the target perpendicular to the penetration direction, in particular, performing some residual internal stresses or dynamically, by other moving bodies or by modification of the boundary conditions: changing the target front surface relief, covering this surface by a steel plate, or inserting the new free sur-or-interfaces.

The principle objective was to gain rather primary understanding of the perturbation effects in the penetration mechanics than a quantitative analysis of them.

Operating media were plasticine with a density 1.49 g/cm^3 and quartz sand, 1.64 g/cm^3, in maximum grain size of 0.63 mm. They occupied the volume of $30 \times 30 \times 80$ cm^3 large enough to consider the target as a half-space at central normal incidence. The experiments were accomplished by a single-stage light-gas gun system with 10-mm or 30-mm in caliber. Either steel sphere alone with $d = 10$ mm in diameter and density 7.8 g/cm^3 or the group of them arranged in layerwise pattern $n \times 7$, where n is the number of layers and $d = 9.5$ mm, were launched at velocities $V = 200 - 500$ m/s. The balls in a group were mounted in a very light wooden cup with setting of seven spheres in one layer. The balls/cup separation was performed by using a separator installed at the muzzle of the barrel. It also served to cut off spent gases from the target free surface. Usual techniques such as photodiodes couple to oscillogrph measured initial velocities with the $\approx 1\%$ relative error. The steel spheres behave in the tests as rigid projectiles, their elastoplastic deformations were very small.

It turns out that the sets of tests were made in different seasons under the different temperature conditions indoors by keeping for at least four hours before the test in refrigerator, at the constant temperature being close to the average indoor's one. This led to essential variation in the dynamic yield stress τ_d. Namely, proceeding from direct measurements in the cone-penetration tests, τ_d linearly ranges over 1.7 kg/cm$^2 \leq \tau_d \leq 6.2$ kg/cm^2 when the temperature T varies from $25.5°C$ to $18°C$ in the experiments. But it does not matter as far as each set of the tests was carried out under the constant temperature conditions and they do not compare with each other. The penetration depth of the steel ball H into plasticine and sand scaled by factor nd are shown in Figure 1-5 versus the impact velocity V under nonstandard conditions. In all Figures, the dots mark the depths H_0 for the nominal penetration into the unloaded half-space. Let us pass to details.

PLASTICINE TARGET PERTURBED STATICALLY OR DYNAMICALLY

The crosses in Figure 1 correspond to the one ball penetration when the static 0.07 MPa compression of the plasticine target perpendicular to the line of direction is realized. This value is a little less than the static yield stress of this material at the temperature $T = 22°C$, but the correspondent dynamic quantity equals 0.4 MPa. The white circles here mean that the target had been preliminary squeezed over 20% of its initial volume in the same direction in order to set up some internal residual stresses. The triangles refer to the normal impact onto the plasticine plate of thickness greater by $3.5d$ than the corresponding nominal penetration depth.

We see that the static compression slightly decreases the depth H. On the contrary, the residual stresses having tensile feature due to squeezing decrease a little the resistance force. The rear free target surface causes a reflected tensile wave, which prolongs the penetration process. It is evident then that the h-thickness plate perforation demands a lower velocity than that is required for the h-penetration into the half-space under the same conditions.

The added dynamic compression in the target are performed by other projectiles shot along with the investigated ball. First the pair, a cup-like 30 mm-in-diameter cylinder and the ball, put before the shot in the 5 mm depth lune on the cylinder leading face is used as a projectile. In the first test, the mass cylinder made from a magnesian alloy was $m = 20$ g and the specific mass loading on its cross section S was in 1.6 times less than that of the ball. Therefore, the ball separates from the cylinder right after the collision and goes ahead, but the medium around

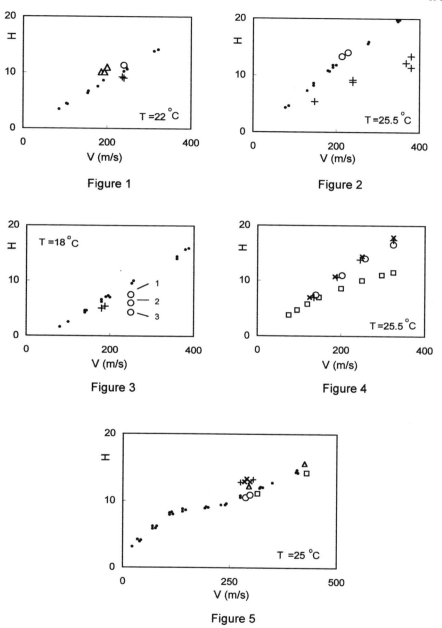

Figure 1-5: Normalized penetration depth vs. the initial velocity

($T = 25.5°C$) is perturbed by a dynamic stress field caused by the impact and deceleration motion of the cylinder. The crosses in Figure 2 are shown decrease of the ball penetration compared with the nominal case. In the second test, $m = 57.4$ g and the pair moves without separation (the white circles). The closeness to the nominal penetration is explained by approximate equality of the principal parameter $k = m/(SC_x)$ for the pair and for the ball, where C_x is the drag coefficient. Thus we prove experimentally that the penetration depth depends only on the parameter k at the constant velocity. Also, the condition that this parameter for the leading body in a group arranged in the single file is greater than that for the rest group moving non-separately, is necessary (but not sufficiently) for its separation from the group.

Second the group of identical bodies, say spheres, arranged in layerwise pattern is also used to set up a dynamic perturbation in the target. The number of ball's layers n is shown near the white circles in Figure 3 referred to the leading central ball penetration. The deflections from the nominal case induced by the fairly symmetric strong dynamic stresses are much greater than those for the mentioned above static preloading, even though the strength of medium here was well higher because here $T = 18°C$. Note that any peripheral ball moves along a curve trajectory and penetrates more deeply due to asymmetry and because the medium around it tests a smaller additional compression.

INFLUENCE OF THE BOUNDARY CONDITION VARIATIONS

To consider the influence of nonstandard boundary conditions to the penetration, the following experiments were performed. At first the target surface was covered with a $2d$-thickness steel plate containing a $1.4d$-in-diameter hole. The ball passes centrally through it and hits directly the plasticine at $T = 18°C$. The plate restricts the target material flow and hence it generates an additional compression-type stress field, which interferes with the moving body. This effect marked by crosses in Figure 3 does not turn out to be so much detectable as in the mentioned case of the mobile cylinder.

Then shots are made in parallel to the target free surface when $T = 25.5°C$. The comparison of the trajectory length to the nominal depth, is presented in Figure 4. The curves differ by initial distance from this surface $l = kd$. The cases $k = 0.5$, 1.0, 1.5, 2.0 are marked by squares, white circles, straight and oblique crosses, respectively. When the initial velocity increases or distance l decreases, the value H increasingly differs from H_0 having always less values. The explanation is evident. Although the free surface generates the tensile reflected waves, it also causes a curve of the trajectories and flying of the ball out into atmosphere. But when the ball stops in the target, the total run becomes a little larger than H_0 due to interaction with an additional tensile stress field. The data show that even if $l = 2d$, the cavity axis is not a straight line, i.e., the influence of the free lateral surface is still appreciable. On the other hand, the tests on a cylindrical target with $5d$ in diameter when the shots are made along its axis, show the nearly nominal depths. The only distinction then consists in the cavity shape. To eliminate this independently from the target lateral surface fixing, the diameter must be at least five times greater than that of the cavity shaped.

Set of experiments is carried out with the shots along the axis of a cylindrical hole into the target at $T = 25°C$ in order to study the influences of complex relief on the penetration process. In addition to nominal incidence, the two new parameters, diameter D and depth h of the hole, appear now. It turns out that the value H is equal to H_0 when $D > 2d$ and $V = 370$ m/s whatever h is. The cavity shape differs from that of produced in the half-space only along distance $\approx 2d$ from the hole bottom. As depth H as cavity shape are very close to those at nominal penetration and $D = 0.5d$. Only the cavity wall becomes rough because of traces of the air pinched initially in the hole and then forcing its way. If $D = d$, the air noticeably increases the resistance to the ball motion resulting in a lowering of the depth while the body does not go into contact with the wall. The main distinction in the cavity shape is its wideness near the bottom and absence of the splashes.

The two contradict effects exist at the enbedment into a layered separated plates: Any front surfaces of the plates are responsible for a shock wave at the impact, which gives a contribution to the resistance force and hence decreases the complete penetration depth. On the other hand, any distal side leads to a reflected tensile wave increasing the effect. Let us here discuss the result of the one test, wherein the $1.5d$, $1.5d$, $2d$, $2.7d$, and $4d$ thicknesses of the five plasticine plates with distance $1.5d$ between, $T = 15°C$, and $V = 300$ m/s were chosen. The nominal penetration depth is $9d$. For the layered target it equals $9.7d$, which corresponds well with penetration into the equivalent continuous plate of the $12d$ thickness. It means that the two mentioned effects quench each other almost completely.

PENETRATION INTO THE PERTUBED AND DISTURBED SAND

Let us discuss now the experiments with the shots throughout a hole in the quartz sand. In this case, a standard target preparation procedure is required for the qualitative results because sand relates to a medium with an internal structure and memory effect.

To keep the sand particles from drop to the bottom of an artificial hole in the sand, a polyethylene tubule with 30 mm outer diameter and wall's width 5 mm and with the bottom enclosed by a 0.1 mm-thickness polyethylene film is used. The sand was prepared hand in hand; the depth of the tubule edge downsinking h are varied. In the one case, the tubule is forced down into the well prepared sand; in the other, it is filled up to the given depth h and then the sand together with the tubule are squeezed in ordinary way.

The results for the penetration depth along the hole axis are plotted in Figure 5 under the different tubule-sand preparation conditions. The white circles, squares, oblique and straight crosses, and triangles refer to the cases when the tubule is filled up at $h = 5d$ and $10d$ or forced down up to $h = 3d$, $5d$, and $10d$, respectively. The forcing down leads to increase of H, especially at $h = 3d$, $5d$, compared to the nominal penetration. This effect diminishes for a large depth of forcing, because weight of the upper sand layer changes the strength property of the sand. This is also confirmed by the test data when the tubule is put in without any forcing. Then the penetration depth turns out a little less then H_0.

The splash-effect differs from the case of plasticine in the following way. It is hardly detectable for the plasticine at $2d \times 5d$ hole sizes up to the velocity 350 m/s, whereas large splashes in the sand are watched at 300 m/s. On the other hand, the essential ejection appears at $h = 10d$ starting with 430 m/s.

CONCLUSION REMARKS

The experimental results presented provide a reasonable measure of changes in the penetration process due to generation of the different perturbations in targets. They prove the theoretical view that an additional dynamic or static stress field with the compression or tensile feature applied to a elastoplastic target lead to decrease or increase of the resistance force against inertial motion of a body. For now, we can point out and roughly compare the three different physical mechanisms which are responsible for the variation in the resistance force and, as a consequence, in the penetration depth: the yield stress variation, change in the contact pressure mean value ΔP and of the interface square. The results enable us to state that the shear stresses are not the primary reason in our experiments with the balls. This follows from the comparison of the target/projectile responses to the static and dynamic additional loading when orders of the shear stress variations are approximately the same, but the detected deviations in the depths differ very much. Among the last two mechanisms, the change in the wetted ball surface square proved to be the lower meaning over the used range of velocities when one judges by pictures of cavities behind the ball. Thus the variation of the contact pressure gives the main contribution in the resistance force.

It seems clear that there is much yet to be learned about the perturbation of projectiles in grounds under the terrestrial and extraterrestrial conditions. It is hoped that the full theoretical-numerical analysis of the nonstandard penetration will follow later, and that the results advanced here can be used for setting up of the more comprehensive experiments in this domain as well as for sophistication of both constitute equations and simulation of processes in terradynamics.

Acknowledgement —This research was supported by the Russian Foundation for Basic Research under Grant N 99-01-01256. We are also grateful to K. Yu. Osipenko for technical help in preparation of this paper.

REFERENCES

Backman M.E. and Goldsmith W. (1978). The mechanics of penetration of projectiles into targets. *International Journal of Engineering Sciences* **16:1**, 1-99.

Bivin Yu.K. (1981). Motion direction change of a rigid ball at its crossing an interface. *Izvestia Academii Nauk, MTT [Mechanics of solids]* **4**, 105-109.

Bivin Yu.K. (1993). Oblique entry of a group of projectiles in an elastoplastic medium. *Izvestia Academii Nauk of Russia, MTT [Mechanics of solids]* **4**, 170-173.

Bivin Yu.K., Chekin B.S., and Simonov I.V. (1994). Refraction and penetration of solid by projectiles of the canonical and optimal shapes. In: *Proceeding of IUTAM Symposium on Impact Dynamics*, Beijing, China. Peking University Press, 1-6.

Johnson W. and Daneshi G.H. (1978). The trajectory of a projectile when fired parallel and near to the free surface of a plastic solid. *International Journal of Mechanical Sciences* **20:4**, 255-263.

Simonov I.V. and Bivin Yu.K. (1999). Ricochet laws and cavity formation by group of projectiles or particles' flow. In: *Proceeding of 3-rd International Symposium on Impact Engineering "Impact Response of Materials and Structures"*, Singapore. Oxford University Press, 517-522.

Soliman A.S., Reid S.R., and Johnson W. (1976). The effect of spherical projectile speed in ricochet of water and sand. *International Journal of Mechanical Sciences* **18:6**, 279-284.

EXPLOSIVE LOADING OF DUAL HARDNESS STEEL PLATES

J.Buchar[1], J.Voldrich[2] and S.Rolc[3]

[1]Department of Physics, Mendel University of Agriculture and Forestry, 613 00 Brno, Czech Republic

[2]Research Center, West Bohemia University, Plzen, Czech Republic

[3]Military Technical Institute, Brno, Czech Republic

ABSTRACT

The experimental investigation of the behaviour of the circular plates made from the dual hardness steel under shock loading has been performed. The shock loading follows from the detonation of an explosive charge, which has been in contact with the plate. The different thickness of single platelayers has been used. The given arrangement has been numerically simulated by the use of LS DYNA 3D finite element code. It has been found that the behaviour of plate can be described in terms of the Johnson Cook constitutive equation together with the criterion of material failure.

KEYWORDS.

Steel plate, explosive loading, dynamic plasticity, spall, numerical simulation, failure criterion

INTRODUCTION.

The study of the dynamic response of structures to some shock loading is mostly studied on axisymmetric structures, especially plates, because of simplicity of the experimental method and possibility to use some analytical method. The use of these methods is limited namely to elastic deformation of plates. These methods can be also used for the plastic deformation of plates. One of the first papers on this topic is probably that one written by Wierzbicki (1964). There are many other papers like Calder et al(1971), Bodner and Symonds (1979), Idczak (1985), Pennetier and Renard (1998) and Klosowski et al. (2000). The more complex approach to the solution of this response enables the use of numeric methods like finite elements method – see e.g. Dinis and Owen (1977). The most of papers has been focused on homogeneous plates. In the given paper we have studied the dynamic response of circular plates made from steel of dual hardness to the explosive loading. The dual hardness steel is more and more used in the design of the structures which should be resistant to a shock loading. The origin of the shock resistance of this steel is similar to that of classical layered structures like metal – ceramics etc.

EXPERIMENTAL DETAILS.

The circular plate (100 mm in diameter) freely supported have been loaded by the explosive charge (cylinder, 30 mm in diameter, 40 mm of height) of TNT (density 1630 kg/m^3, detonation velocity 6930 m/s, Chapman – Jouqet pressure – 21 GPa) - see Fig.1.

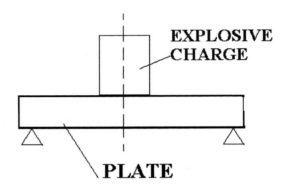

Fig.1 : Schematic of the experiment.

The plate has been made from the dual hardness steel. This structure consists of the two layers of different steels. In our experiments the dual hardness steel consists of the layer of a tool steel (TENAX) - upper layer just below the explosive charge and layer of the armour steel (2P) – second layer . The chemical compositions of the given steels are given in Table 1.

TABLE 1.

Chemical composition of the used steel (in wt. %) .

STEEL	C	Mn	Si	P	S	Cr	Ni	Mo	W	V	Ti
TENAX	0.68	0.43	0.65	0.019	0.014	1.28	-	-	2.04	0.16	0.01
2P	0.28	1.25	1.45	0.016	0.006	0.22	0.81	0.25	-	-	-

The total thickness of the plate has been kept 6 mm. Three different plates has been used :
- 2 mm of TENAX steel and 4 mm of the 2P steel
- 3 mm of TENAX steel and 3 mm of the 2P steel
- 4 mm of TENAX steel and 2 mm of the 2P steel

From the experiments a damage of the plates has been evaluated. All experiments have been carried at the room temperature.

MATERIAL BEHAVIOUR AND MODELLING.

To be able to describe the various phenomena taking place under high strain rates generated by the explosive loading it is necessary to characterise the behaviour of materials. In this paper numerical simulations have been performed and compared with experimental data using a coupled material model and ductile damage, developed for impact and penetration problems. These loading are characterised by the strain rates closed to those achieved during the detonation loading. The characterisation involves not only the stress – strain response at large strains, different strain rates and temperatures, but also the accumulation of the damage and the mode of failure. The stress – strain behaviour of materials is described in terms of the engineering model proposed by Johnson and Cook (1985). In this model the equivalent stress is expressed as :

$$\sigma_{eq} = (A + B\varepsilon^n)(1 + C\ln(\dot{\varepsilon}^*))(1 - T^{*m})$$

$$T^* = \frac{T - T_r}{T_m - T_r} \qquad \dot{\varepsilon}^* = \frac{\dot{\varepsilon}}{\dot{\varepsilon}_o} \qquad \dot{\varepsilon}_o = 1/s \qquad (1)$$

where A,B,C,n,m are material constants, ε is the accumulated plastic strain, dot denotes the derivation with respect to the time, T is the absolute temperature, index r refers to the room temperature (300 K) and m denotes the melting point. The values of the given constants have been obtained using static tensile test, Hopkinson Split Bar test and Taylor test. The values of material constants are given in Table 2.

TABLE 2
MATERIAL PARAMETERS IN EQ.(1)

STEEL	A (MPa)	B (MPa)	C (1)	n (1)	m (1)
TENAX	1440	492	0.011	0.24	1.03
STEEL 2P	1210	773	0.014	0.26	1.03

To describe ductile fracture, Johnson and Cook proposed a model including the effect of stress triaxiality, temperature, strain rate and strain path on the failure strain. The fracture strain is expressed as

$$\varepsilon_f = [D_1 + D_2 \exp(D_3 \sigma^*)][1 + D_4 \dot{\varepsilon}^*][1 + D_5 T^*]$$

$$\sigma^* = \frac{\sigma_m}{\sigma_{eq}} \qquad (2)$$

where D_{1-5} are the material constants and σ_m is the mean stress. These equations are implemented in LS DYNA 3D finite element code. The model enables to describe linear elasticity, initial yielding, strain hardening, strain rate hardening, damage evolution and fracture. The parameters in Eq. (2) have been determined from the spall experiments. Their values are given in Table 3.

TABLE 3.
MATERIAL PARAMETERS IN EQ.(1)

STEEL	D_1 (1)	D_2 (1)	D_3 (1)	D_4 (1)	D_5 (1)
TENAX	0	1.07	-1.22	0.000016	0.63
STEEL 2P	0.1	0.93	-1.08	0.000014	0.65

The behavior of the TNT detonation gas products, the Jones - Wilkins - Lee (JWL) equation of state has been used, together with the programmed burn model - the detonation velocity has been assumed to be 6930 m/s]. The JWL equation has the form :

$$p = A\left[1 - \frac{\omega}{R_1 V}\right]\exp(-R V_1) + B\left[1 - \frac{\omega}{R_2 V}\right]\exp(-R_2 V) + \frac{\omega E}{V}$$

Where p is the detonation pressure, V is the relative volume and E is the internal energy density. The parameters are

A=272.7 GPa, B= 3.231 GPa , R_1 = 4.15, R_2 = 0.95, ω = 0.3

Initial density of the explosive was 1630 kg/m^3.

EXPERIMENTAL RESULTS.

The plates with the rear face of the thickness 2 and 3 mm exhibit only a plastic strain. The profile of the permanent shape of the plate with rear face made from steel TENAX (2 mm) is displayed in Fig. 2.

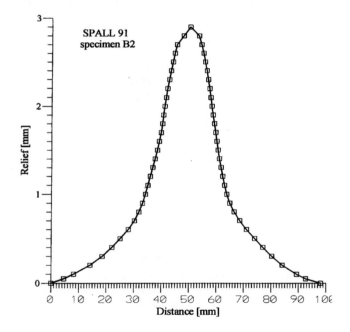

Figure 2 : The shape of the backing face (free surface) of the loaded plate.

The shape of the loaded surface is very similar. If we take the plate composed from rear plate of thickness 4 mm and backing plate of thickness 2 mm , the spall damage in the backing plate (steel 2P) has been observed. The spall occurs at about 0.2 mm from the free surface.

NUMERICAL SIMULATION.

Owing to the symmetry only one quarter of the problem displayed in Fig. 1 has been solved. The 29 400 brick elements have been used. The contact between the explosive charge and the plate has been described using of the model No. 15 which is described in LS DYNA 950 manual.

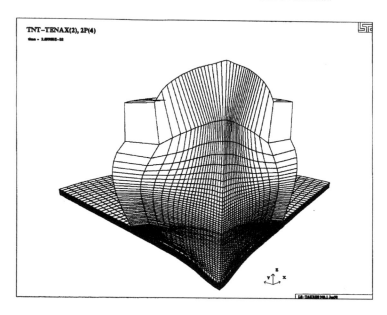

Figure 3 : Computed shape of the loaded plate.

In Fig. 3 the computed shape of the plate is displayed. The computed pressure at the interface between the explosive and plate as well as the maximum of the loaded surface deflection x_{max} exhibit satisfactory agreement with the experimental ones as shown in Table 4.

TABLE 4.

Compunted and expermimental found quantities.

PLATE	x_{max} – computed (mm)	x_{max} – experimental (mm)	p – computed (GPa)	p– experimental (GPa)
2 mm TENAX ↓ 4 mm STEEL 2P	3.12	2.91	14.05	14.60
3 mm TENAX ↓ 3 mm STEEL 2P	2.92	2.87	13.98	15.10
4 mm TENAX ↓ 2 mm STEEL 2P	2.47	2.53	14.10	13.90

The computation also enables to simulate the spall damage. The position of the spall is very closed to that experimentally found but the extent of the computed damage is higher. It means the model of material damage should be improved.

CONCLUDING REMARKS.

This paper presents a research project in progress where the main objective is to study the possibility of the design of protective structure against to some detonation action from steel of dual hardness. Experimental results indicate that these steel plates represent a reasonable protective structure even against to the detonation of the explosive charge in contact with the plate. It means that these plates should be resistant against blast waves following from the detonation of much more larger explosive charges than that used in the given paper. The response of the given plate can be predicted by the numerical simulation using of standard Johnson Cook equations. Even if the total number of parameters required in these models seems be considerably, they can be identified from common material tests. The agreement between numerical and experimental results is satisfactory at least.

ACKNOWLEDGEMENT

The authors would like to express their gratitude for the support of the Czech ministry of education under research project LN00B084.

REFERENCES

Bodner S.R. and Symonds P.S. (1979). Experiments on viscoplastic response of plates to impulsive loading. *J.Mech.Phys.Solids* **27**, 91 – 113.
Calder C.A., Kelly J.M. and Goldsmidt W. (1971). Projectile impact on an infinite, viscoplastic plate. *Int.J. Solids Struct.* **7**, 1143 – 1152.
Dinis L.M.SS. and Owen R.J. (1977). Elastic – viscoplastic analysis of plates by the finite element method. *Comp.Struct.* **8**, 207 – 215.
Idczak W. (1985). Approximate solutions in the dynamics of inelastic membranes. *Eng.Trans.* **33:4**,519 – 535 (in Polish).
Johnson G.R. and Cook W.H. (1985). Fracture characteristics of three metals subjected to various strains, strain rates, temperatures and pressures. *Engng Fracture Mech.* **21**, 31 – 48.
Klosowski P., Woznica K. and Weichert D. (2000). Comparison of numerical modelling and experiments for the dynamic response of circular elasto – viscoplastic plates. *Eur.J.Mech. A/Solids* **19**,343 –359.
Pennetier O. and Renard J. (1998). Structures minces face a une explosion application a la gestion de risques industriels. *Mécanique Industrielle et Matériaux* **51 :2**, 67 – 69.
Wierzbicki T. (1964). Bending of rigid visco – plastic circular plate. *Arch.Mech.* **16:6**, 1183 –1195.

A UNIFYING FRAMEWORK OF HOT SPOTS FOR ENERGETIC MATERIALS

K. Yano, Y. Horie, and D. Greening

Los Alamos National Laboratory,
Los Alamos, NM 87545, USA

ABSTRACT

A one dimensional model of hot-spots is proposed to consider various regimes of ignition scenarios using a common platform. The model contains the most advanced features of spherical void collapse models. They include: viscoplastic heating, phase change, gas phase heating, finite-rate chemical reactions, and heat transfer between the locally heated zone and the surrounding mass. Test calculations, based so far on cyclonite (RDX) data, show that the model behavior is comparable to that of spherical void collapse model under shock loading condition. Results also show that chemical initiation under various mechanical excitation as well as thermal heating can be understood in a unified fashion; the initiation thresholds are summarized in terms of total energy deposited to the hot-spot, size of the hot-spot, and the rate of the energy deposition.

KEYWORDS

Explosive materials, Hot-spot ignition, Unifying framework

INTRODUCTION

The concept of "hot-spot" or the region of high energy localization was introduced by Bowden and Yoffe (1952) to explain the ignition of explosive materials when the mean field conditions are inadequate. For solid explosives, various mechanisms have been proposed for hot-spot formation, such as pore collapse, shear banding, friction, fracture, and jetting (Davis 1981; Dienes 1995; Kang et al., 1992; Massoni et al., 1999).

Currently much is unknown about how the formation of hot-spot is related to the initial material state and loading conditions. As a result, modeling of hot-spot formation involves a priori choice of a mechanism, which typically involves a model geometry that may not be representative of actual hot-spot geometry. But at present, there is no incontrovertible evidence to prefer one mechanism over the others. In addition, single-mechanism approach can be problematic for complex systems like polymer-binder explosives (PBX).

Nevertheless, if one examines various models setting aside the mechanisms, a unifying framework can be identified. The framework consists of: (1) a region where mechanical energy is highly localized, (2) thermal mass surrounding the region described in (1), (3) creation or existence of the region occupied by gas, (4) heat flow between the regions, and (5) reaction chemistry in the gas, at the solid/gas interface, or both. In the present investigation, a hot-spot model was developed using the framework described above without being specific about the hot-spot formation mechanisms. Rather, attention is focused on whether the model yields results that are consistent with other mechanistic models by ranging over the parameter space, and how the model parameters affect the ignition conditions.

HOT-SPOT MODEL

Figure 1 shows an idealized model of hot-spot based on the unifying framework. The model is one dimensional and consists of three components: solid with a region of localized heating, gas cavity, and solid-gas interface. The solid is assumed to be incompressible. Thus, the solid moves with a common particle velocity, and the motion is determined by the balance of external loads, gas pressure, and configurational stress which represents the effect of contact force that may develop at the interface. The temperature is determined by the analytic solution of heat conduction equation for a fixed length domain. The heat is added uniformly to the zone of energy localization. It is noted that the width of the zone δ $(=b-a)$ and the rate of volumetric heating ϕ_s are free parameters, at present. In this way, the question of coupling between energy localization and the state of material is set aside for the moment.

The gas equations are obtained based on the assumption that average behavior is adequate for the present purpose. Reaction chemistry is assumed to be single step described by the Arrhenius kinetics model with pressure dependency. The equation of state is prescribed by a modified form of the ideal gas equation of state. Heat conduction between the gas phase and the interface is also considered.

At the interface, conservation conditions of mass, momentum, and energy are imposed. It is assumed that the interface is infinitely thin. Because of localized heating in the solid phase, a mass flux is produced at the interface as a function of the interface temperature. Part of the flux may go through instantaneous chemical reaction. Depending of the modeling needs, the localized heating can be generated at the interface with a known surface heating rate $\phi_{s,i}$.

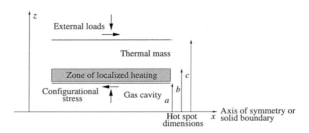

FIGURE 1. A schematic of the model structure. It consists of gas cavity, solid-gas interface, and solid with a zone of localized heating. Thick arrows represent stresses acting on the solid.

Mathematically, the model system consists of 7 ordinary differential equations and 10 algebraic equations. The differential equations were numerically integrated by Runge-Kutta scheme. The model contains all the features found in spherical pore collapse model by Kang et al. (1992). Major difference is that in their model, the energy localization is coupled with the model geometry whereas it is prescribed by free parameters in the present model. Details of the model are found in Yano et al. (2001).

MODEL SIMULATIONS

RDX was chosen for model simulations in order to compare the results with those of spherical void collapse calculation by Kang et al. (1992). Three sets of simulation results are presented in this section. The first two are results of ignition simulations with high and low heating rate. The last result presents a parametric study of critical ignition condition for RDX.

Figure 2 is a schematic of the conditions for ignition simulation where the rate of heating is high. The heating rate is indicative of the energy dissipation rate produced by spherical pore collapse under shock loading (Kang et al., 1992; Bonnet and Butler, 1996; Massoni et al. 1999). The magnitude of applied load and physical dimensions are from Kang et al. (1992). Figure 3 (a) and (b) show the results of the ignition simulation. At time $t = 0$, 1 GPa of external load is applied and the gas pore starts to collapse. At the same time, heat addition is started to the heating zone. The temperatures of the interface and the gas phase rise steadily. At 0.15 μs, the heat addition is terminated. The interface temperature starts to drop due to the termination,

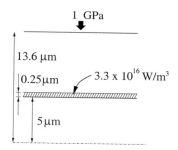

FIGURE 2. Conditions for ignition simulation with high rate of heating. The heating rate is indicative of energy dissipation rate of spherical pore collapse under shock loading condition.

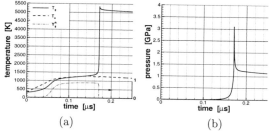

(a)   (b)

FIGURE 3. Results of ignition simulation for conditions shown in Fig. 2. (a) Time evolution of temperatures at the interface (thick dashed line) and in the gas phase (thick solid line). Thin dash-dotted line is the time evolution of reactant mass fraction in the gas phase. (b) Pressure evolution in the gas phase.

whereas the gas phase temperature keeps on rising because of exothermic chemical reaction in the gas phase. At about 0.17 μs, ignition takes place and both gas temperature and pressure rise in almost discontinuous manner. After the ignition, the gas pressure rapidly drops due to the expansion of the gas. Qualitatively and quantitatively, the ignition behaviors are very similar to those presented by Kang et al. (1992).

The conditions for the second simulation are shown in Fig. 4. In this simulation, the heat is added to the interface, and the heating rate is an estimate of frictional energy dissipation caused by low speed ($\sim$ 10 m/s) impact. Figure 5 (a) and (b) show the simulation results. The heat addition is maintained during $0 < t < 0.35$ ms for this calculation. Qualitatively, the results are similar to those of previous simulation, except the time scale for ignition is on the order of sub-millisecond. A plateau appeared in the pressure profile, because the gas pressure reached the external loading pressure, and the gas started to expand slowly. The plateau is not seen for higher external pressures.

Finally, the third results presents critical conditions for ignition. The idea of critical energy input is often pursued as a parameter to describe ignition conditions. One good example is the $p^2\tau$ criterion for explosives under impact loading where p is the shock pressure and τ is the time duration of the pressure (Walker,1985). Lee (1998), however, pointed out that the concept of critical energy alone is not sufficient to explain various ignition scenarios and that time factor over which the energy stimulus takes place must be considered. In this simulation, the critical energy deposited to the heating zone was calculated as a function of the size of heating zone and the rate of heating.

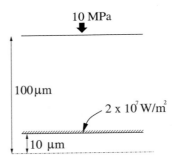

FIGURE 4. Conditions for ignition simulation with low rate of heating. The heating rate is comparable, in estimate, to frictional energy dissipation caused by low velocity impact.

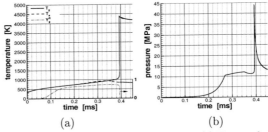

FIGURE 5. Results of ignition simulation for conditions shown in Fig. 4. (a) Time evolution of temperatures at the interface (thick dashed line) and in the gas phase (thick solid line). Thin dash-dotted line is the time evolution of reactant mass fraction in the gas phase. (b) Pressure evolution in the gas phase.

Figure 6 shows the results of parametric study. The symbols represent the points of critical conditions. For a given heating zone size and a heating rate, any energy above each symbol leads to ignition, and any energy below does not lead to ignition. The profiles are characterized by the existence of minima and convergence for large δ. In the left branch of the minimum point, less energy is required as δ gets larger. This is because heat loss at the interface due to conduction becomes smaller as the size of hot spot increases. On the other hand, in the right branch more energy is required for larger δ simply because of the volume increase. The curve approaches to linear profile for large values of δ as the effect of thermal conduction becomes negligible. The convergence behavior for large δ indicates that the ignition occurs once the interface temperature reaches the critical value (thermal ignition).

Three curves in Fig. 6 were obtained by curve fitting with the following fitting function.

$$E_{cr} = E_\infty + k_\infty[1 - \exp(-\alpha\phi_s\delta)]\delta + \frac{C_1}{(\phi_s\delta)^n}\exp(-C_2\phi_s\delta) \tag{1}$$

where E_{cr} is the critical energy and E_∞, k_∞, α, C_1, C_2, and n are fitting parameters. Values of the parameters used for the fitting are listed in Table 1.

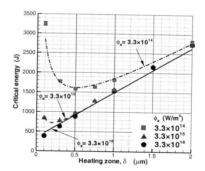

FIGURE 6. Critical condition for ignition in terms of total energy deposited to the heating zone as a function of the size of heating zone and heating rate. The symbols represent data obtained from parametric study, and the curves are the result of fitting using Eq.(1).

TABLE 1
VALUES OF PARAMETERS IN EQ.(1) USED FOR CURVE FITTING

variable (units)	value
E_∞ (J)	350
k_∞ (J/m)	1.142×10^9
α (1/W)	1.407×10^{-8}
n (—)	0.6
C_1 (J·W^n)	8.434×10^7
C_2 (1/W)	1.686×10^{-9}

CONCLUSIONS

An idealized model based on the unified framework described here produced ignition behaviors that are in general agreement with those of mechanism-specific models (Kang et al. 1992; Massoni et al. 1999). This agreement is seen over wide range of the parameter space, showing a consistent variation of time to ignition over orders of magnitude, depending on energy deposition rate and its duration. Further, a parametric study of critical ignition conditions shows results consistent to the observation by Field et al. (1992) on effective hot-spot ignition (hot-spot size: $0.1 - 10$ μm and duration of heating: $10^{-5} - 10^{-3}$ s). The results also show that critical ignition conditions are summarized in terms of total energy deposited to the hot-spot, the size of the hot-spot, and the rate of energy deposition.

REFERENCES

Bonnet D.L. and Butler P.B. (1996). Hot-spot ignition of condensed phase energetic materials. *Journal of Propulsion and Power* **12**, 680-690.

Bowden F.P. and Yoffe A.F. (1952). *Initiation and Growth of Explosion in Liquids and Solids*, Cambridge University Press, Cambridge

Davis W.C. (1981). High Explosives: the Interaction of Chemistry and Mechanics. *Los Alamos Science* **2**, 48-75.

Dienes J.K. (1995). A Unified Theory of Flow, Hot Spots, and Fragmentation, with an Application to Explosive Sensitivity. In: Davison L., Grady D.E., and Shahinpoor M. (eds.), *High-Pressure Shock Compression of Solids II*, Springer, New York

Field J.E., Bourne N.K., Palmer S.J.P., and Walley S.M. (1992). Hot-spot ignition mechanisms for explosives and propellants. *Philosophical transactions of the Royal Society of London* **A339**, 269-283.

Kang J., Butler P.B., and Baer M.R. (1992). A Thermomechanical Analysis of Hot Spot Formation in Condensed-phase, Energetic Materials. *Combustion and Flame* **89**, 117-139.

Lee P.R. (1998). Theories and Techniques of Initiation. In: Zukas J.A. and Walters W.P. (eds.), *Explosive Effects and Applications*, Springer, New York

Massoni J., Saurel R., Baudin B., and Demol G. (1999). A Mechanistic Model for Shock Initiation of Solid Explosives. *Physics of Fluids* **11**, 710-736

Walker E.H. (1985). Derivation of the P^2T detonation criterion. *Proceedings of the 8th Symp. (Int'l) on Detonation*, 1119-1125.

Yano K., Horie Y., and Greening D. (2001). *A Unifying Framework for Hot Spots and the Ignition of Energetic Materials*, Los Alamos National Laboratory LA-13794-MS

FRAGMENT ACCELERATION BY THE DETONATION OF HIGH EXPLOSIVE CHARGES WITHOUT AND WITH ALUMINIUM CONTENT

M. Held

EADS-TDW, Schrobenhausen, Germany

ABSTRACT

Aluminized and not aluminized high explosive charges achieve in the 12 kg warhead weight class or 4,5 kg high explosive mass equal fragment velocities and therefor fragment performances. The sensitivity and vulnerability values are marginal different and are not at all excluding one of the investigated types of high explosive charges. Nothing is dangerous, either by friction, drop test etc. or with regard to the survivability tests with bullet or norm projectile impacts or sensitivity to shock loads. The marginal higher sensitivity in the gap test for the not aluminized high explosive charge can be seen also as an advantage for an easier initiability. Only one point is a little different, the factor two larger diameter and the longer duration of the reaction products of the detonated charge. But this benefit depends on the attacked targets.

KEYWORDS

Aluminized explosives, fragment acceleration, detonation properties, warhead tests, sensitivity, survivability, comparison table with weighing factors

INTRODUCTION

The questions arises again and again, can increase a content of aluminum powder to the high explosive charge the fragment velocities of a cylindrical charge either of a smooth casing for natural fragments or of preformed fragments. These questions were investigated in a detailed research program whereby simultaneously the different detonation properties without any and with an aluminum content are been tested.

The Gurney constants are presenting the energy factors per weight of the high explosive charges which are used for the acceleration of attached materials, like fragmenting casings. The comparison of these Gurney-constants (Kennedy - 1969) for the different high explosive compositions without and with aluminum content give partially contradicting results (table 1). The composition "Torpex" with 18 % aluminum content instead of RDX shows with the Gurney constant of 2710 m/s a little higher value compared to Composition B with 2680 m/s. But exactly the same value has Cyclotol with a bit larger amount of RDX. The aluminum containing compositions H6, HBX 1 and HBX 3 have smaller values compared to Composition B after the table 1.

TABLE 1 Comparison of the Gurney constants for not aluminized and aluminized high explosive charges (Kennedy, 1969)

Type of Charge	TNT	RDX	Al	Wax	ρ_0 [g/cm^3]	$\sqrt{2E}$ [m/s]
Comp. B	40	59	-	1	1.68	2680
Cyclotol	30	70	-	-	1.71	2710
H6	31	47	22	5	1.71	2620
HBX1	38	40	17	-	1.69	2470
HBX3	29	31	35	5	1.81	2230
RDX/W	-	95	-	5	-	2680
RDX/W	-	97	-	3	-	2800
Torpex	40	42	18	-	1.81	2710
TNT	100	-	-	-	1.59	2130
Tritonal	80	-	20	-	1.72	2320
B/RDX/W	Binder 2	94	-	4	1.70	?
B/RDX/Al/W	2	78	16	4	1.75	?

TESTED HIGH EXPLOSIVE CHARGES

The composition of the not containing aluminum high explosive charge for this investigation contains 2 % binder, 94 % RDX and 4 % Wax, shortly B/RDX/W 2/94/4 and for the aluminum containing high explosive charge 2 % binder, 78 % RDX, 16 % aluminum and 4 % wax, or shortly B/RDX/Al/W 2/78/16/4. 16 % of RDX content are replaced by an aluminum powder. All high explosive charges are produced from granulates by pressing with 3000 bar with achieved densities of 1.70 g/cm^3 for B/RDX/W, respectively 1.75 g/cm^3 for B/RDX/Al/W (table 2).

TABLE 2: Pressed investigated charge types

Type		B/RDX/W	B/RDX/Al/W
Composition	[%]	2/94/4	2/78/16/4
Pressure	[kbar]	3	3
Density	[g/cm^3]	1.70	1.75

DETONATION PROPERTIES

The RDX crystals have already a negative oxygen balance on its own. Therefore the aluminum will not directly react as a reaction partner in the detonation zone (Held 1997). Therefore the aluminum powder is reducing the amount of immediately reacting molecules. The aluminum particles on its own will be compressed by the shock or detonation wave, but contributing on the reaction after the detonation front. This is the reason that the detonation velocity of 7900 m/s is 6 % reduced, compared to the high explosive charge without aluminum of 8400 m/s.

Therefor the initial pressure is reduced by the content of 16 % aluminum powder. But the aluminum particles are reacting behind the detonation front with radicals and some reaction products and therefore are increasing the temperature which means higher product velocities, also with higher densities of the aluminum-oxide containing reaction products. These products of explosives are expected to work longer against attached materials like casings and fragments. To measure this qualitatively a special arrangement was used for this purpose. The charges of 96 mm diameter and 100 mm length were arranged in 30 mm distance to a mild steel plate of 30 mm thickness. The test setup and the achieved results with the punched through holes are shown in Fig. 1. The normal cast TNT/RDX charge 35/65 has bulged the plate a little and created some scabbing. The pressed charge B/RDX/W 2/94/4 under investigation without aluminum shows a stronger bulging and a scabbing with few stress cracks in the plate. The aluminized pressed high explosive charge B/RDX/Al/W 2/78/16/4 creates the strongest bulging and maybe less scabbing by the less strong shock wave, but longer duration. The simple comparison test demonstrates that charges without aluminum have stronger shock waves in the near field, but reduced pressure duration, whereas aluminized charges have reduced pressure, but longer duration.

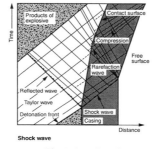
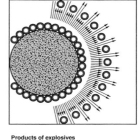

Fig.1: Deformation of a 30 mm mild steel plate by three different charge types

Fig.2 Acceleration of fragments in two steps, shock wave and aerodynamically

The acceleration of fragments can be split in two steps. The first acceleration step is given by the shock wave on the outside surface, whereas the fragment jumps from zero velocity up to the free surface velocity. The shock wave runs then back as a release wave which is now arriving on the inner surface and if the pressure is internally still high it comes up a second shock wave, and pushing again the free surface. The velocity is jumping up by the multiple reflection of the shock wave (Fig.2 left.

The pushing force is the second step on the acceleration of the fragments by the expansion of the detonation products, like an aerodynamic drag load (Fig. 2 right). The density and the velocity of the reaction products are the responsible quantities for this pushing phase. In repetition: In aluminized high explosive charges the shock pressure is reduced, but the pushing phase is stronger and longer.

The optimum size of the aluminum particles in the high explosive charge is on its own not very well defined. If they are very fine, then the aluminum content is reduced by the oxygen layers around the individual particles. But they are then reacting relatively fast behind the detonation front. If the particles are larger than the aluminum content is higher corresponding to the smaller oxidized surfaces. But the reaction times are much longer. They are therefore sometimes not more really fully reacting in the phase where the high explosive charge is under high pressure by the given dimensions and not expanded too much and cooled down by the rarefaction fans, coming from all the free surfaces. This means in very small charges a content of aluminum is not really helpful. In very big charges with strong confinements, like sea-ground-mines, with charge weights of 500 kg to 1000 kg high explosive charges with a large aluminum content are very useful in the underwater detonations (Held - 1997).

The question is open, what part of the acceleration of a fragment casing or of preformed fragments of a charge of few kilograms is caused directly by the "shock wave", introduced in the material, and by the stream of reaction products, which are expanding and are "aerodynamically" pushing additionally the fragments.

WARHEAD TESTS

The answer should be given by 12 kg fragmenting warhead tests of 125 mm diameter with 4.5 kg high explosive charge weight without and with aluminum. Once was measured the fragment "velocities" over a distance of 3 m by the impact, respectively perforation flashes of the fragments on aluminum target plates (Held - 1999). The tests are done under ambient temperature and under minus and elevated temperatures (-46 °C and +100 °C). The measured velocities are presented in Fig. 3. The fragments along the warhead length have naturally little different velocities. Both charges achieve equal velocities at ambient and +100 °C temperature. The aluminized high explosive charge has maybe given under minus temperature a little higher mean value compared to the charge without aluminum. The wide band means not a few firings but the scatter of the velocity of the fragments from different tiers along the warhead. This diagram with the measured fragment velocities with the warhead temperature as parameter shows no preference for one or the other charge type.

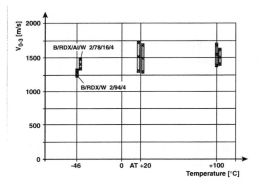
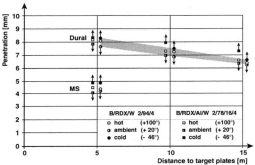

Fig. 3 Measured velocities by impact flashes with a HYCAM camera for the two investigated high explosive charges at 3 warhead temperatures (3 m distance)

Fig. 4: Penetration in semi infinite targets of aluminum alloy and mild steel plates as a function of distance of the three investigated temperatures

Further it was measured the "penetration" of the selected preformed fragments in semi-infinite thick aluminum plates at 5 m, 10 m and 15 m and also in mild steel plates at 5 m distance. The diagram (Fig. 4) shows really equal results in 5 m distance. Maybe a small preference is given for the aluminized high explosive charge, especially at cold temperature at the larger distances.

Astonishing is a result that the fragment velocity measurement has given little lower velocities for the cold charges. But the penetration diagram shows that just all the cold firings are on top compared to the warheads at ambient and hot temperatures. The arrows give the dispersion of the data and all the values are in the scattered area.

HANDLING AND SURVIVABILITY TESTS

Really no decision can be made from the performance of the two different high explosive charge types. Therefore in addition some handling safety and survivability tests on so-called 1 kg test samples in 1 mm steel casings and 1 mm cover plates were undertaken.

The different charge characterization tests are summarized in table 3. Hereby is added for a better comparison the normal cast composition B charges with 39 % TNT, 60 % RDX and 1 % Wax.

TABLE 3: Summary of the different test results together with some additional data of mechanical and handling sensitivity tests

Test charge type	TNT/RDX/W 39/60/1	B/RDX/W 2/94/4	B/RDX/Al/W 2/78/16/4
Density	1.70 g/cm^3	1.70 g/cm^3	1.75 g/cm^3
Detonation Velocity	7900 m/s	8400 m/s	7900 m/s
Impact test	5 N•m	4 N•m	10 N•m
Friction test	36.0 kp	36.0 kp	24.0 kp
Combustion Temperature	229° C	227° C	227° C
Thermal stability	71 - 75° C	120° C	120° C
Thermal conductivity (cal/cm • s • K)	2.6 • 10^{-3}	0.65 • 10^{-3}	0.95 • 10^{-3}
Compressive strength	110 kp/cm^2	600 kp/cm^2	340 kp/cm^2
Gap - test	23 kbar	16 kbar	17 kbar
Cu - projectile impact test	810/840 m/s	520/580 m/s	625/660 m/s
Standardised fragment charge (7mm)	974 m/s	1006 m/s	1014 m/s

SUMMARY OF THE TESTS IN A WEIGHING TABLE

To get a quality index for the pressed charges, not aluminized and aluminized, the different achieved test values were multiplied with weighing factors. The individual figures were then summarized to the decision figures in table 4.

The sum of the non aluminized high explosive is 84.8 points and of the aluminized high explosive charge 87.8. But the dominant difference is the diameter of the reaction products with 5 points more for the aluminized charge. And this seems rectified for an anti-air warhead by the fact that the longer duration of the reaction products gives in some cases also easier burning probability of fuels which comes out of the fuel tanks by hits with the fragments. So these increased incendiary effect is definitely one advantage of the aluminized high explosive charge in this scenario.

This table with the by the author used weighing factors is a possible analyzing procedure of the test results for these two types of high explosive charges. If somebody takes some values more into account then the weighing factors can be easily changed, to fit these data to his requirements or priorities.

TABLE 4: Assessment of the investigated high explosive charges with weighing factors especially for anti air warheads

Tests	B/RDX/W 2/94/4 Assessment	B/RDX/Al/W 2/78/16/4 Assessment	Weighting factor Grade	B/RDX/W 2/94/4 Grade	B/RDX/Al/W 2/78/16/4 Grade
Fragment Perforrmance	1	1.03	10	10	10.3
Standardised fragment velocity	1	1	3	3	3
Detonation velocity	1.1	1	3	3.3	3
Diameter of Products of HE	1	2	5	5	10
Incendiary	1	1	5	5	5
Density	1.05	1	4	4.2	4
Compression	2	1	2	4	2
Thermal conductivity	1	1.5	2	3	2
Sensitivity (Friction 36/24; Impact 0.4/1.0)	1	1.1	5	5	5.5
Survivability (case1.2/1; WH 1/1.1; Norm 1/1.2)	3.2	3.3	10	32	33
Gap - test	1.06	1	5	5.3	5
Manufacturing costs	1	1	5	5	5
				Σ 84.8	87.8

REFERENCES

Held M. (1997), "The Physical Phenomena of Underwater Detonations", Journal of Explosives and Propellants, R.O.C., 13, 1-12

Held M. (1999), "Testing of Fragmentation Warheads", Journal of Explosives and Propellants, R.O.C., 15, 1-22

Kennedy D.R. (1969), "The Elusive $\sqrt{2E}$", Paper on 21. Annual Bomb and Warhead Section Meeting, Picatinny Arsenal

CONSIDERATION OF THE BUMPER SHIELD FOR HYPERVELOCITY IMPACT USING SMOOTHED PARTICLE HYDRODYNAMICS METHOD

D. Watanabe[1] and Y. Akahoshi[2]

[1] Course of Mechanical Engineering, Graduate School, Kyushu Institute of Technology, 1-1, Sensui, Tobata, Kitakyushu, 804-8550, Japan
[2] Department of Mechanical Engineering, Faculty of Engineering, Kyushu Institute of Technology, 1-1, Sensui, Tobata, Kitakyushu, 804-8550, Japan

ABSTRACT

As space development progresses space debris, which collides with space structures, such as fragments of rockets and so on becomes a very important problem. The average relative impact velocity is about 8.7km/s. A usual bumper shield is whipple shield which consists of double metal wall structures. In this study, we observed the second debris cloud generated by the hypervelocity impact with a bumper. Then we carried out numerical simulation and experiments to develop a bumper shield. A standard hydrocode is classified into the two methods . The one is Eulerian method in which the fixed coordinate system is used at any time. The other is Lagrangian method in which the varying material coordinate is employed. Both methods are suffered from some mesh limitations. Smoothed Particle Hydrodynamics method (SPH) is one of meshless Lagrangian numerical techniques which offer more advantages than mesh-based Lagrange and Euler methods. The analysis domain is described as an assembly of smoothed particles in SPH method. SPH approximations are quite flexible and can be applied to a large deformation problem and so on. Mesh-based Lagrangian technique is principally suffered from a problem of mesh tangling, which will not appear in the meshless SPH method. In this paper, we develop three-dimensional hypervelocity impact analysis code using SPH method and simulate the second debris cloud generation. The mass distribution and the scatter angle distribution are compared between the numerical result and the experimental results.

KEYWORDS

Space Debris, Bumper Shield, Smoothed Particle Hydrodynamics, Meshless

INTRODUCTION

Orbital debris is a very serious problem for the space development. Because of its high speed, the impact of space debris to manned space vehicles and international space station is very dangerous. Then many agencies study protections against impacts of orbital debris. There are two ways to protect the structures in space. The first way is to avoid the orbital debris, more than 10cm in diameter, which are monitored by SSN (Space Surveillance Network), where the number of the monitored orbital debris is more than 8000 (see Toda (1998)). On the other hand, the bumper shield has been developed for the un-monitored orbital debris less than 1cm in diameter, by NASA (see Christiansen et al. (1999) and Cour-Palais et al. (1993)), NASDA (see Shiraki et al. (1998) and Shiraki (2000)) and ESA (see Destefanis et al. (1999)), where duralumin plate, MLI, Nextel, Kevlar and aluminum mesh are used as bumper shield structures. Formation of debris cloud has a close relation with bumper shield performance and it is thought that debris cloud should be spreaded more widely to improve the bumper shield performance. Therefore production of debris cloud is studied from the viewpoint of both experiment and simulation.

HYPERVELOCITY IMPACT EXPERIMENTS

The experiments have been made using the Two-Stage Light Gas Gun (see Tsutsumi (1999)) installed in Satellite Venture Business Laboratory, Kyushu Institute of Technology, Japan. Figure 1 shows configuration of the target and the capture material made of polystyrene plates. The projectile is made of polyethylen (diameter=10mm, length=15mm, mass=1g) and the target is made of Al2024-

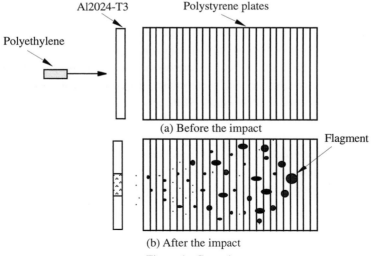

Figure 1 : Capturing system

T3(thickness=3mm). After the impact, many fragments are generated by fracture of both the projectile and the target. These fragments are then captured by a capture material located behind the target. Positions and masses of these fragments are measured.

Table1 shows experimental results, where impact velocity, maximum fragment mass and number of fragments are listed. It is found that mass of the maximum fragments becomes smaller and the number of the fragments increase as the impact velocity is larger.

TABLE 1
EXPERIMENTAL RESULT

Shot No.	Velocity [km/sec]	Maximum fragment mass[mg]	Number of fragments
TS00-018	1.61	409.7	26
TS00-028	4.54	89.6	551

HYPERVELOCITY IMPACT SIMULATIONS

We develop three-dimensional hypervelocity impact analysis code using Smoothed Particle Hydrodynamics method (see Monagan & Gingold (1983)). SPH method is an numerical technique for modelling many large deformation and dynamic problems, including hypervelocity impact problems. The feature of the SPH method is expression of an object as a cluster of particles. A particle interacts with other particles in smoothing length h. The interaction is weighted by the smoothing function W(x-x',h). The smoothing function uses the cubic B spline kernel. The value of a function f(x) can be expressed by the surrounding particles using the following kernel estimates:

$$<f(x)> \approx \int f(x')W(x-x',h)dx' \tag{1}$$

where f is a function of the three-dimensional position vector x. The equations of continuum mechanics are the conservation laws formulated in a lagrangian system. These equations can be expressed by the SPH formulation, as proposed by Libersky et al.(1993).

$$\frac{d\rho_i}{dt} = \rho_i \sum_j \frac{m_j}{\rho_j}\left(U_i^\alpha - U_j^\alpha\right)W_{ij,\beta} \tag{2}$$

$$\frac{dU_i^\alpha}{dt} = -\sum_j m_j \left(\frac{\sigma_i^{\alpha\beta}}{\rho_i^2} - \frac{\sigma_j^{\alpha\beta}}{\rho_j^2} + \Pi_{ij}^\alpha\right)W_{ij,\beta} \tag{3}$$

$$\frac{dE_i}{dt} = \frac{\sigma_i^{\alpha\beta}}{\rho_i^2}\sum_j m_j\left(U_i^\alpha - U_j^\alpha\right)\left(\frac{\sigma_i^{\alpha\beta}}{\rho_i^2} + \frac{1}{2}\Pi_{ij}^\alpha\right)W_{ij,\beta} \tag{4}$$

where $W_{ij,\beta}$ stands for the spatial derivative of W with respect to coordinate β. ρ, E, U^α, $\sigma^{\alpha\beta}$ $\alpha\nu\delta$ Π^α are the scalar density, specific internal energy, the velocity components, the stress tensor, and artificial viscosity, respectively.

COMPARISON OF EXPERIMENTAL RESULTS AND NUMERICAL RESULTS

Table 2 shows analysis condition, where impact velocity, configration of projectile and target, number of SPH particle, and areal density are listed. Figire 2 shows spatial distribution of 25μs after impact, the analysis condition of which is the same as the experimental condition of Shot No. TS00-028. The mass distribution and the scatter angle distribution are compared between experimental results and numerical results in the following.

TABLE 2
ANALYSIS CONDITION

	Diameter [mm]	Thickness [mm]	Number of particle	Velocity [km/sec]	Areal density [g/cm^2]
Projectile	10	15	8559	1.61(TS00-018)	-
				4.54(TS00-028)	
Target	50	3	54915	0	10.84

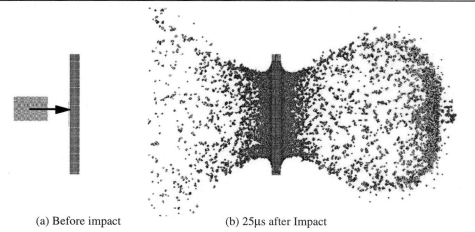

(a) Before impact   (b) 25μs after Impact

Figure 2 : Spatial distribution of numerical simulation under condition of TS00-028

Mass Distribution

Figure 3 shows the mass distribution of fragments for the experimental results and numerical results, where the y-axis is cumulative number of fragments and the x-axis is normalized mass (fragment mass divided by total mass of fragments). In the mass distribution, fragments from the projectile are not included. It is noted from both the experimental results and numerical results that larger fragments are found in the lower velocity impact (TS00-018) than in the higher velocity impact (TS00-028). The total number of fragments in the higher velocity impact is larger than that of the lower velocity impact.

Scatter Angle Distribution

Figure 4 shows the scatter angle distribution of fragments for the experimental results and numerical results, where the r-axis is mass of fragments and the θ-axis is scatter angle of fragments. It can be said from the experimental results and the numerical results that the fragments are spreaded more widely in the higher velocity impact (TS00-028).

CONCLUSIONS

The following conclusions are obtained.
1) The target are more broken as impact velocity increases.

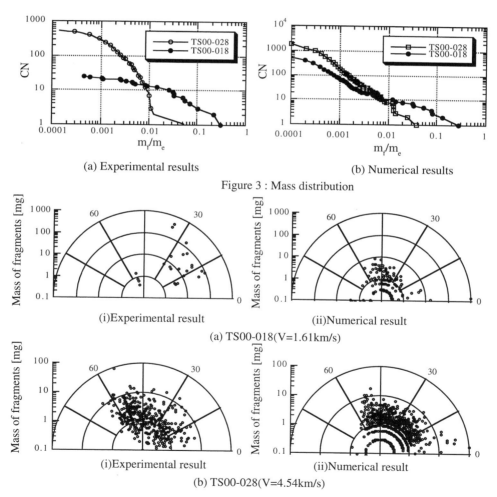

(a) Experimental results (b) Numerical results

Figure 3 : Mass distribution

(i) Experimental result (ii) Numerical result

(a) TS00-018(V=1.61km/s)

(i) Experimental result (ii) Numerical result

(b) TS00-028(V=4.54km/s)

Figure 4 : Scatter angle distribution

2) Larger fragments are found in the lower velocity impact than the high velocity impact.

3) The number of fragments in the higher impact velocity is larger than the lower velocity impact.

4) The fragments are more spreaded in the higher velocity impact.

5) The almost same tendencies are seen in the experimental results and the numerical results.

REFERENCES

Benson D.J.(1991). Computional methods in Lagrangian and Eulerian hydrocodes, *Computer Method in Applied Mechanics and Engineering* **99**, 235-394.

Christiansen E.L., Keer J.H., Delafuente H.M. and Schneider W.C.(1999). Flexible and Deployable Meteoroid/Debris Shielding for Space Craft. *International Journal of Impact Engineering* **23**, 125-136.

Cour-Palais B.G., Piekutowski A.J., Dahland K.V., Poormon K.L.(1993). Analysis of the Test on Nextel Multi-Shock Shields. *International Journal of Impact Engineering* **14**, 193-204.

Destefanis R., Faraund M. and Trucchi M.(1999). Columbus Debris Shielding Experiments and Bllistic Limit Curves. *International Journal of Impact Engineering* **23**, 181-192.

Hayhurst C.J. and Clegc R.A.(1996). Cylindrically Symmetric Simulations of Hypervelocity Impacts on Thin Plates. *Proceedings of Hypervelocity Impact Symposium*, 337-348.

Libersky L.D., Petschek A.G., Carney T.C., Hipp J.R. and Allahdadi F.A.(1993). High Strain Lagrangian Hydrodynamics, Journal of Computational Physics **109**, 67-75.

Monagan J.J., Gingold R.A.(1983). *Journal of Computational Physics* **52**, 429-453.

Shiraki K., Sakashita T. and Noda N.(1998). JEM Protection Shield Performance Evaluation for Hypervelocity Debris Impacts. *Proceedings of the JSASS/JSME Structure Conference*, 177-180.

Shiraki K.(2000). Doctral Thesis (Kyushu University).

Toda S. (1998). Trend in Space Debris Related Researches and Activities. *Proceedings of the JSASS/JSME Structures Conference*, S-1 - S-4.

Tsutsumi T., Okumura K., Matsushita N., Tanaka M. and Akahoshi Y.(1999). Improvement of Two-Stage Light Gas Gun, *Proceedings of Annual Meeting of Japanese Aerospace Society Western Branch*, 1-4.

MEASUREMENT OF THE SECOND DEBRIS CLOUDS IN HYPERVELOCITY IMPACT EXPERIMENT

Yasuhiro Akahoshi[1] and Kenichi Kouda[2]

[1] Department of Mechanical Engineering, Faculty of Engineering, Kyushu Institute of Technology,1-1, Sensui, Tobata, Kitakyushu, 804-8550, Japan
[2] Course of Mechanical Engineering, Graduate School, Kyushu Institute of Technology,1-1, Sensui, Tobata, Kitakyushu, 804-8550, Japan

ABSTRACT

Space debris in orbit is a very serious problem for space development. Because of its high speed, the impact of space debris to manned space vehicles and international space station is very dangerous. A lot of agencies, companies and universities have studied space debris and developed the shields against space debris impacts. The orbital debris, more than 10 cm in diameter, is observed from the ground, and that of less than 1cm in diameter is protected by the developed bumper shield. But this bumper shield can not protect the current International Space Station from space debris between 1cm and 10cm in diameter. The production mechanism of second debris clouds generated after the first bumper shield impact has not been clarified in details so far. To develop the more light and effective shield against impacts of space debris, control of the second debris cloud distribution is required. A purpose of this study is to construct mass distributions, three-dimensional distributions and velocity distributions of fragments of second debris clouds with wide impact velocity range. We carried out experiments in hypervelocity impact using two-stage light gas gun installed in Kyushu Institute of Technology. As a result, we can obtain the correlation data between impact velocity, and mass and three-dimensional spatial distributions of fragments.

KEYWORDS

Space Debris, Hypervelocity Impact, Second Debris Clouds, Mass Distributions, Three-Dimensional Spatial Distributions, Two-Stage Light Gas Gun

INTRODUCTION

Space debris in orbit is a very serious problem for space development. Because of its high speed, the impact of space debris to manned space vehicles and international space station is very dangerous. A lot of agencies, companies and universities have studied space debris and developed the shields against space debris impacts (see Christiansen et al.(1999) and Toda (1998)). The orbital debris, more than 10 cm in diameter, is observed from the ground, and that of less than 1cm in diameter is protected by the developed bumper shield. But this bumper shield can not protect the current International Space Station from space debris between 1cm and 10cm in diameter. The production mechanism of second debris clouds generated after the first bumper shield impact has not been clarified in details so far (see Shiraki et al.(1998) and Shiraki (2000)). To develop the more light and effective shield against impacts of space debris, control of the second debris cloud distribution is required. A purpose of this study is to construct mass distributions, three-dimensional distributions and velocity distributions of fragments of second debris clouds with wide impact velocity range. We carried out experiments in hypervelocity impact using

Figure 1 : Two-stage light gas gun installed at KIT

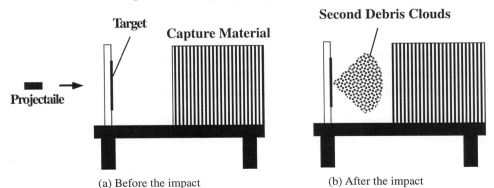

(a) Before the impact (b) After the impact

Figure 2 : Capturing system in vacuum chamber

TABLE 1
RESULTS OF EXPERIMENT

Shot number	Velocity (km/s)	Thickness of target (mm)
TS00-050	1.96	2
TS00-053	2.57	2
TS00-054	3.70	2
TS00-045	4.92	2
TS00-018	1.61	3
TS00-013	2.47	3
TS00-004	3.83	3
TS00-028	4.54	3
TS00-049	1.57	4
TS00-051	2.65	4
TS00-034	3.65	4
TS00-041	4.78	4

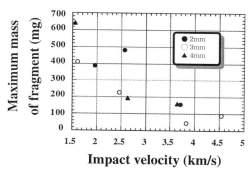

Figure 3 : Maximum mass of fragment vs. impact velocity

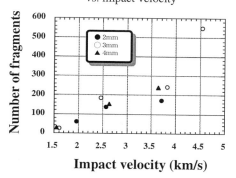

Figure 4 : Number of fragment vs. impact velocity

two-stage light gas gun installed in Kyushu Institute of Technology. As a result, we can obtain the correlation data between impact velocity, and mass and three-dimensional spatial distributions of fragments.

METHOD OF EXPERIMENTS

In this study, all the experiments were done using two-stage light gas gun, shown in Figure 1, installed in Kyushu Institute of Technology, where the diameter and the length of the launch tube is 10 mm and 1 m (see Tsutsumi (1999)). After the impact, many fragments are generated by fracture of both projectile and target. These fragments are then captured by a capture material located behind the target, and their positions and masses are measured individually. The distributions described in this paper were constructed by these manners (see Nakamura (2001)). The projectiles used in these experiments are made of cylindrical polyethylene of 1g, the targets are made of Al2024-T3(thickness:2,3,4 mm) and a capture material is polystyrene plate. Figure 2 shows the configuration of the target and the capture material for the experiments.

RESULTS OF EXPERIMENTS AND DISCUSSIONS

Table 1 shows the thickness of used targets and the impact velocities in the experiments. In the following, let us discuss the experimental results in details.

Maximum Mass and Number of Fragments

At first we focus on maximum mass among the captured fragments and the number of them. Figure 3 shows maximum mass among the captured fragments at each thickness of the target and each impact velocity. It is found that larger fragments are generated at lower impact velocity. And Figure 4 shows the number of fragments at each thickness of the target and each impact velocity. It is noted that more fragments are generated at higher impact velocity. These figures indicate that fracture state of the first bumper plate depends largely on the impact velocity.

Mass Distribution

Mass distribution was fitted by Parabola Law using an expression of Eqn.1 (see Mcknight (1991)).

$$\log(CN) = a + b\{\log(m/m_e) + c\}^2 \qquad (1)$$

Figures 5 and 6 show mass distribution in low and high velocity. The y-axis is cumulative number of fragments and the x-axis is logarithmical normalized mass (fragment mass divided by total mass of fragments). The comparison of the mass distribution expression is shown in Table 2. The mass distribution curve is expressed by the following form where CN and m_e are cumulative number and eject mass. It is found that both the results are successfully fitted by the Parabola formulation.

TABLE 2
COMPARISON OF CONSTANT IN PARABOLA LAW

Shot Number	Impact Velocity (km/s)	Thickness of Target (mm)	a	b	c
TS00-018	1.61	3	1.35	-0.198	3.02
TS00-028	4.54	3	2.75	-0.563	3.79

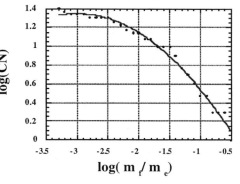

Figure 5 : Mass distribution with parabola law at shot number TS00-018 (Impact velocity 1.61km/s)

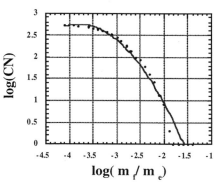

Figure 6 : Mass distribution with parabola law at shot number TS00-028 (Impact velocity 4.54km/s)

Three-Dimensional Distribution

Three-dimensional distributions are shown in Figures 7 and 8, where the thickness of the target is 3 mm, and the impact velocity is 1.61 km/s and 4.54 km/s. However these three-dimensional distributions do not show real positions of the fragments in flight but show captured positions of the fragments. In these figures, the left side figure is the front view and the right side figure is the side view. In the side view, the fragments advance from the left side to the right side. The fragments are shown by the spheres and their sizes are standardized. The color of each spere corresponds to the mass of fragments, and large fragments are black and small fragments are white.

Large fragments are distributed in the right side of the side view. On the other hand, many small fragments are distributed in the left side of the side view. It is found in Figure 8 that the front view is distributed in a quadrilateral state. This shape corresponds to skyline of the capture material. Consequently, it is neccesary to use larger capture material when the impact velocity is high.

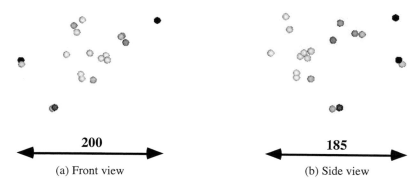

Figure 7 : Three-dimensional distribution of fragments in capture material at shot number of TS00-018 (Impact Velocity 1.61km/s)

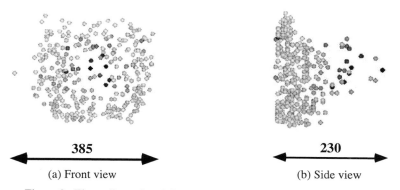

Figure 8 : Three-dimensional distribution of fragments in capture material at shot number of TS00-028 (Impact Velocity 4.54km/s)

CONCLUSIONS

The following conclusions are obtained.
1) Maximum mass and the number of the fragments generated after the impact were measured in the experiments with the different impact velocity. Maximum mass of the fragments decreases with higher impact velocity. On the other hand, the number of fragments increases with higher impact velocity. C onsequently, when the impact velocity increases, the fragments are smaller.
2) The data of mass of the fragments was fitted by Parabola law and mass distributions were constructed at low and high impact velocity.
3) The fragments were captured by capture material, and then each mass and captured positions were measured . From these data, three-dimensional spatial distribution was constructed. This distribution is useful to understand flight direction of each fragment.

FUTURE PROGRAM

The following future plan is under consideration.
1) In this experiment, the thickness of used targets were 2mm, 3mm and 4mm. In future, the other thickness of the targets will be tested.
2) Construction of velocity distribution is needed to make clear the formation of second debris clouds. We have a plan to install flash X-ray equipments and to detect fragments at the different time.

REFERENCES

Christiansen E.L., Keer J.H., Delafuente H.M. and Schneider W.C.(1999). Flexible and Deployable Meteoroid/Debris Shielding for Space Craft. *International Journal of Impact Engineering* **23**, 125-136.

Mcknight D.S. (1991). Determination of Breakup Initial Conditions. *Journal of Spacecraft* **28:4,** 470-477.

Nakamura T. (2001). Master Thesis (Kyushu Institute of Technology).

Shiraki K., Sakashita T. and Noda N.(1998). JEM Protection Shield Performance Evaluation for Hypervelocity Debris Impacts. *Proceedings of the JSASS/JSME Structure Conference*, 177-180.

Shiraki K.(2000). Doctral Thesis (Kyushu University).

Toda S. (1998). Trend in Space Debris Related Researches and Activities. *Proceedings of the JSASS/JSME Structures Conference*, S-1 - S-4.

Tsutsumi T., Okumura K., Matsushita N., Tanaka M. and Akahoshi Y.(1999). Improvement of Two-Stage Light Gas Gun, *Proceedings of Annual Meeting of Japanese Aerospace Society Western Branch*, 1-4.

DEVELOPMENT OF SELF-SHIELDING BUMPER AGAINST SPACE DEBRIS

Hidehiro Hata[1], Yasuhiro Akahoshi[2], Tomonori Tsunetomi[1], and Sinya Kawakita[1]
[1] Course of Mechanical Engineering, Graduate School, Kyushu Institute of Technology, 1-1, Sensui, Tobata, Kitakyushu 804-8550 Japan
[2] Department of Mechanical Engineering, Faculty of Engineering, Kyushu Institute of Technology, 1-1, Sensui, Tobata, Kitakyushu 804-8550 Japan

ABSTRACT

The bumper shield installed in Japanese Experiment Module (JEM) of International Space Station (ISS) can protect the pressure wall from penetration of space debris less than 10mm in diameter under severe restriction of allocated area density. If the space debris between 10mm and 100mm collides and penetrates the pressure wall, pressure in the pressurized module will start to decrease rapidly. It is said that the crews should escape from JEM in two minutes before the pressure in JEM drops to 0.7 bar. However it seems to be not enough time to escape from it, considering the influence of the impact shock on the crews' body. So we propose a self-shielding bumper as one of the alternative bumpers against space debris between 10mm and 100mm. This self-shielding bumper prevents rapid pressure drop and maintains a certain pressure until crews escape from the depressurized module. In this concept self-shielding materials are carried to the penetration hole by the air flow after the penetration event and partly close it. To evaluate a feasibility of this bumper shield with the self-shielding function, the experiments have been conducted using several shielding materials under atmospheric and vacuum environment. And the hypervelocity impact experiments have been conducted using arrangement of some styrofoam balls. The result of the experiments shows good performance of self-shielding. It is found that leak time with self-shielding material is tens larger than that of non-shieldings.

KEYWORDS

Space Debris, Self-Shielding Bumper, Two-Stage Light Gas Gun, Hypervelocity Impact

INTRODUCTION

The bumper shield installed in JEM of ISS can protect the pressure wall from penetration of space debris less than 10mm in diameter. We have studied enlargement of protection ability of bumper shield (see Akahoshi & Kouda (2001), Akahoshi & Tanaka (2001), and Watanabe & Akahoshi (2001)), but it is not easy to enlarge the protection capability. If the space debris between 10mm and 100mm, orbital elements of which are unknown, collides and penetrates the pressure wall, pressure in the pressurized module will start to decrease rapidly. It is said that the crews should escape from JEM in two minutes before the pressure in JEM drops to 0.7 bar. However it seems to be not enough time to escape from it, considering the influence of the impact shock on the crews' body. So we propose a self-shielding bumper as one of the alternative bumpers against space debris between 10mm and 100mm. This self-shielding bumper prevents rapid pressure drop and maintains a certain pressure until crews escape from the depressurized module, with allowance of some penetration in the pressure wall.

Generally speaking self-shielding is flat-tire prevention material, which can shield only a comparatively small hole. Application of this idea to space was already proposed by Christiansen et al. (1999) and Lyons (1998). In our proposed concept self-shielding materials with some finite size are carried to the penetration hole by the air flow after the penetration event and partly close it. To evaluate a feasibility of this bumper shield with the self-shielding function, the experiments have been conducted using several shielding materials under atmospheric environment and vacuum environment. In addition the hypervelocity impact experiments have been conducted using arrangement of some styrofoam balls.

TRIAL EXPERIMENT OF SELF-SHIELDING

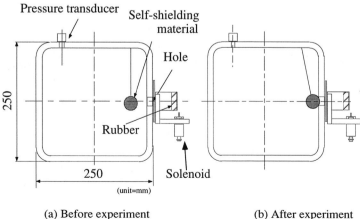

(a) Before experiment   (b) After experiment

Figure 1 : Configuration of test chamber

Figure1 shows experimental test chamber with transparent side windows through which movement of a self-shielding material can be obeserved optically. The self-shielding material is hung by a thread as shown in Figure 1 because it is suffered from gravity on the earth and its gravitation is compensated. The distance between the penetration hole in a duralmin plate and the thread is about 50 mm. The introduced penetration hole is 20 mm in diameter, where the hole of 20mm in diameter would be made by impact of space debris of 10 mm in diameter. The test chamber is filled with nitrogen gas of 0.2 MPa at the experiment under atomospheric environment. Namely, pressure difference between inside and outside of the test chamber is 0.1 MPa. In order to hold this pressure difference, a pipe with a diaphragm made of rubber are attached in the front of the penetration hole as shown in Figure 1. The rubber will be broken by a pin, which is activated by a solenoid coil. The volume of the test chamber is 1.404×10^{-2} m^3. On the other hand, the volume of JEM is 8.535×10^2 m^3. Therefore the size of the test chamber is about 1/60000 of JEM, and the penetration hole of 20 mm in diameter corresponds to the penetration hole of about 800 mm in JEM. Here let us consider that the hole of 20 mm is modeled as the hole penetrated by space debirs of 10 mm in JEM, and its neighborhood including the hole is closed up. But it will be made in this time the case where 20 mm of diameter holes was made in JEM. Therefore the hole neighborhood is being imitated. The times till the internal and external difference in pressure of test chamber reaches the following three pressures are measured.

Time A is time till the difference pressure reaches 0.07 MPa. Within this time a crew should escape from penetrated module to other modules.

Time B is time till the difference pressure reaches 0.05 MPa. A human will be under low oxygen condition at this pressure.

Time C is time till the difference pressure vanishes.

Experiment Under Atmospheric Environment

Table 1 shows kinds of self-shielding material used in this experiment. In CASE 1 no self-shielding

TABLE 1
KIND OF SELF-SHIELDING MATERIALS

CASE No	Self-shielding material	Diameter(mm)	Density(g/cm^3)
1	Non-shilding	-	-
2	Super-ball	30.06	6.516×10^{-1}
3	Styrol foam	28.00	4.350×10^{-2}
4	Soft tennis-ball	65.00	1.909×10^{-1}
5	Styrol foam coverd with airbag sheet	28.50	1.044×10^{-1}

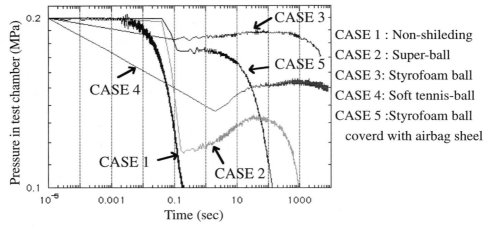

Figure 2 : Time history of pressure in the test chamber

material is used. These experiment of except CASE 1 do criterion the experimental result of CASE 1. The styrofoam ball of CASE 3 moves more quickly than the super-ball of CASE 2. Because the styrofoam ball of CASE 3 is lighter than the super-ball of CASE 2. Soft tennis ball of CASE 4 is easily deformed and is expected to be fitted to the hole. The styrofoam ball of CASE 5 is covered with airbag sheet, where it is expected that heat damage to the styrofoam ball is reduced.

Figure 2 shows typical experimental results. It can be said that Time A of CASE 3 is about 100 times larger than that of CASE 1, and the styrofoam ball has excellent performance from the viewpoint of Time A.

Experiment in Vacuum Environment

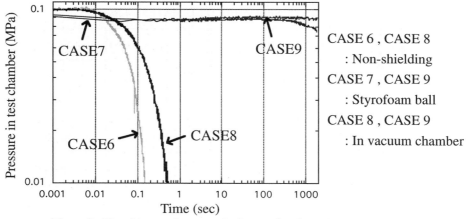

Figure 3 : Time history of pressure in the test chamber

In the previous sub-section, the experiments were done in the atmosphere. In this sub-section, the test chamber is put in vacuum environment and filled with normal air of 0.1 MPa. Here only the styrofoam ball is employed as a self-shielding material, which shows excellent performace in atomospheric environment. Figure 3 shows typical experimental results.

In CASE 6 and CASE 8 no self-shielding material was used. In CASE 7 and CASE 9 styrofoam ball was employed as a self-shielding material. And in CASE 8 and CASE 9 experiments were done in the vacuum environment. Therefore gauge pressure is used in CASE 6 and CASE 7, on the other hand, absolute pressure is used in CASE 8 and CASE 9. It is found that there is no great difference in results between vacuum and atmospheric environment, and the styrofoam ball shows excellent performance in vacuum environment as well as atmospheric environment.

HYPERVELOCITY IMPACT EXPERIMENT OF SELF-SHIELDING

In this chapter whether the styrofoam ball shows similar performance under hypervelocity impact or not is tested. Here a projectile is launched using two-stage light gas gun installed at Hypervelocity Impact Laboratory, Satellite Venture Business Laboratory, Kyushu Institute of Technology. As wellknown the second debirs cloud and shock wave are induced in a hypervelocity impact test. Then some styrofoam balls are arranged as shown in Figure 4 (a). The front plate of the test chamber was made of Al 6061 of 2 mm in thickness .

Figure 4 (b) shows arrangement of styrofoam balls after the impact test. It can be seen that the balls move and one of them covered the penetration hole. Figure 5 shows time history of pressure in the test chamber. It is found that Time A with self-shielding materials is about ten times larger than that of non-shielding.

(a) before experiment

(b) after experiment

Figure 4 : Arrangement of self-shielding materials

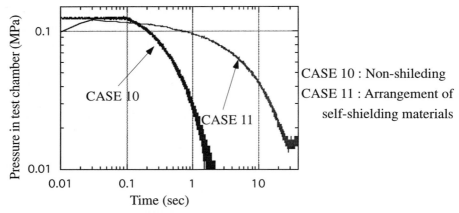

Figure 5 : Time history of pressure in the test chamber

CONCLUSION

In this study several kinds of self-shielding materials are tested from the viewpoint of self-shielding performace. The following conclusions are obtained.
(1) Styrofoam ball showed a excellent self-shielding performance in the experiment in both the atmosphere environment and the vacuum environment.
(2) Under hypervelocity impact environment the second debris cloud and shock wave have large influence on performace of the self-shielding materials. So then arrangement of styrofoam balls was proposed to increase self-shielding performace totally. The penetration hole was successfully covered with one of the styrofoam balls which moved in the leaking air flow.

REFERENCES

Akahoshi Y. and Kouda K.(2001). Measurement of the Second Debris Clouds in Hypervelocity Impact Experiment. *presented at 4th International Symposium on Impact Engineering.*

Akahoshi Y. and Tanaka M.(2001). Development of Bumper Shield Using Low Density Materials. *presented at 4th International Symposium on Impact Engineering.*

Christiansen E.L., Cour-Palais B.G. and Friesen L.J.(1999). Extravehicular Activity Suit Penetration Resistance. *International Journal of Impact Engineering* **23**, 113-124.

Lyons F.(1998). Hypervelocity Impact Testing of EMU Alternative Materials, JSC 28445.

Watanabe D. and Akahoshi Y.(2001). Consideration of the Bumper Shield for Hypervelocity Impact Using Smoothed Particle Hydrodynamics Method. *presented at 4th International Symposium on Impact Engineering.*

DEVELOPMENT OF BUMPER SHIELD USING LOW DENSITY MATERIALS

Yasuhiro Akahoshi[1] and Makoto Tanaka[2]

[1] Department of Mechanical Engineering, Faculty of Engineering, Kyushu Institute of Technology,
1-1, Sensui, Tobata, Kitakyushu, 804-8550, Japan
[2] Information Science Laboratory, Tokai University, 1117, Kitakaname,
Hiratsuka, Kanagawa 259-1292, Japan

ABSTRACT

In order to avoid critical accidents due to space debris, this paper proposes an extra bumper shield called an inflatable bumper. It consists of duralumin (Al2024-T3), high strength fiber Vectran and inflatable polyurethane. Beacause of this inflatability, it is compactly stored during its transportation. And after the installation outside of the space structures, this bumper is then inflated by the effect of expandable polyurethane. This method saves the amount of the transportation and the involved cost. In this paper the experiments, using several polyerethanes with different areal density, have been conducted, compared and evaluated from the viewpoints of the penetrated areal density, the total thickness penetrated. As the results of the experiments, it is found that the protection ability of the proposed structure depends on the areal density of polyurethane layers.

KEYWORDS

Space Debris, Multi-Layer Structure, Vectran and Polyerethane

INTRODUCTION

In 1998 the construction of the International Space Station(ISS) has started in the Low Earth Orbit(LEO), whose altitude is 400km. It is a start of the new era of space activity and inspires more space activity. Also, the space debris, which occurred due to space activity and orbit in the LEO, seems to increase and this causes the increasing probability of the collision between space debris and ISS.

The Japanese Experimental Module (Kibo), a part of the module in the ISS will be constructed from Oct. 2003. In order to prevent Kibo from the collision with space debris, the bumper shields are supposed to be installed outside of Kibo. This bumper is based on Whipple bumper and improved effectively. The current bumper shield has the ability to stop the space debris of 10mm in diameter, and of 7km/s at an impact velocity (see Shiraki et al.(1998) and Shiraki (2000)). On the other hand, ISS can maneuver beforehand in order to avoid the collision to the space debris, more than 100mm, observed from the earth. However, the space debris, from 10mm to 100mm, are not able to be observed from the earth. And there is not useful strategy against such space debris. In order to solve this problem, there is a way to install extra-bumper shield outside the space structure, but considering current cost and quantity of the transportation, it seems to be difficult to construct more heavy structures.

In this paper therefore we propose an inflatable bumper, which can be installed outside of the module, and satisfied with the lightness and compactness. This bumper is a multiple layer bumper and consists of duralumin, high strength fiber material Vectran (see Sugishima (1998), and Hatta et al.(1999)), and inflatable polyurethane. The advantage of this bumper is that it is transported compactly except for duralumin and achieves lower cost during the transportation. Once this bumper is installed outside the module, polyurethane inflates and makes thicker structure. From the previous research using same types of materials, it is confirmed that if distance of the intervals between each layer is larger, the multiple layer bumper is more protectable (see Christiansen (1993) and Cour-Palais & Crews (1990)).

In order to evaluate the proposed bumper, several experiments were conducted with different areal densities of the polyurethane. And this paper states the differences of the protectable ability of the polyurethanes themselves against second debris occurred by the collision between projectile and duralumin.

HYPERVELOCITY IMPACT EXPERIMENTS

The experiments have been conducted using Two-Stage Light Gas Gun, shown in Figure 1, installed in the Satellite Venture Business Laboratory, Kyushu Institute of Technology, Japan (see Tsutsumi (1999)). The projectiles were made of polyethlene (diameter=10mm, length=20mm, mass=1g), and these impacted the proposed multiple layer bumpers. Figure 2 shows the proposed a multiple layer bumper. This consists of duralumin, Al2024-T3 on the front, and high strength fiber Vectran (areal density=$1.3kg/m^2$, thickness=2.5mm) and inflatable polyurethane in turns. Each interval of the Vectrans are 25mm. And in order to evaluate the difference of the areal densities of the polyurethane, the total areal

Figure 1 : Two-stage light gas gun installed at KIT

density of the proposed bumper is united to be 21kg/m², therefore the total thicknesses are changed respectively. The impact velocities of the projectile are approximately 2 km~4km/s. Table1 shows the detail of the experimental conditions and results, where penetrated A.D. means the areal density of penetrated Vectrans and Polyurethanes.

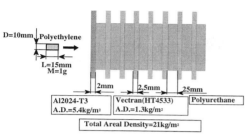

Figure 2 : Experimental Configuration

TABLE1
EXPERIMENTAL CONDITIONS AND PENETRATED AREAL DENSITIES

	Polyurethane	A.D. (kg/m²)	Experiment 1		Experiment 2	
			Impact Vel.(km/s)	Pen. A.D. (kg/m²)	Impact Vel.(km/s)	Pen. A.D. (kg/m²)
A	EGR-3	1.5	2.8	16.6	3.59	18.1
B	ECQ	0.35	2.6	15.3	3.89	16.95
C	EL-68H	0.75	2.7	14.35	3.40	15.65
D	EFF	0.525	2.7	13.23	3.47	15.05
E	18-02	0.5	2.7	14.9	3.71	16.2
F	EMH	0.75	2.81	20.5	3.4	16.4
G	ECS	0.55	1.7	12.8	3.66	17.05
H	EMT	1.5	2.2	20.9	3.35	20.9
I	EL-23-SS	1.375	2.3	14.8	3.51	17.48

EXPERIMENTS WITH VECTRAN AND POLYURETHANES

Figure 3 shows comparison of penetrated areal densities among experiments A - F, in which the impact velocity of 2.7 km/s and 3.5 km/s is almost the same one another. It is found that the areal densities of 3.5 km/s is larger than that of 2.7 km/s. This means that velocity of 3.5 km/s does not reach the bottom of ballistic limit curve around 3 km/s.

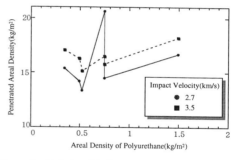

Figure 3 : Relationship between A.D of Polyurethane and Penetrated A.D.

EXPERIMENTS WITH ONLY VECTRAN

In the previous section, expriments were made where polyuretanes were inserted between Vectran sheets. In this section we focus on only Vectran sheets as shown in Table 2. The Vectran sheets with different areal density are prepared. The one is HT4533 of which areal density is 1.3 kg/m². The other is HT0544 of which areal density is 0.176 kg/m². This density is much smaller than that of HT4533, then 7 layers of HT0544 are used as a sheet. As a result, equivalent areal density is 1.232 kg/m².

The penetrated areal density is the same between J and L, and the penetrated areal density of K is a little smaller than that of J and L. It is thought that this means that the bottom of the ballistc limit curve for these experiments is between 3.20 km/s and 3.43 km/s. Comparing the results of the previous section, the protection ability of polyurethane could be small.

TABLE 2
EXPERIMENT WITHOUT POLYURETHANE

	Fiber	Areal Density (/layer)	Impact Vel. (km/s)	Pen. A.D. (kg/m^2)
J	Vectran (HT4533)	1.3	3.43	14.5
K	Vectran (HT4533)	1.3	3.60	13.2
L	Vectran (HT4533)	1.3	3.20	14.5
M	Vectran (HT0544)	0.176×7	3.58	12.26

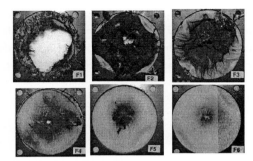

Figure 4 : Vectrans(HT0544) after Experiment M

Figure 4 shows Vectran sheets after the experiment M, where F on the pictures denotes a front face The first layers are penetrated widely and the size of the penetration hole is 65 mm. The second layers swell in the opposite direction of the impact. A lot of fragments are captured at the third layers. The 4th layers and the latter are not deformed so much. The maximum fragment is stopped at the 6th layers. From the viewpoint of the penetrated areal density, it can be said that the stacked thin Vectran sheets are better than thick one.

CONCLUSIONS

In this study, two types of experiments are made. The one is experiment with Vectran sheets and polyurethanes. The other is experiment with only Vectran sheets. In the former experiments, it can be said that the penetrated areal density increases as the impact velocity is higher in the range between 2.7 km/s and 3.5 km/s. In the latter experiments, it can be said that polyurethanes do not have an ability of protection, and that finer Vectran sheets have more protection ability at the same areal density.

REFERENCES

Christiansen E.L., Keer J.H., Delafuente H.M. and Schneider W.C.(1999). Flexible and Deployable Meteoroid/Debris Shielding for Space Craft. *International Journal of Impact Engineering* **23**, 125-136.

Christiansen E.L.(1993). Design and Performance Equations for Advanced Meteoroid and Debris Shields. *International Journal of Impact Engineering* **14**, 145-156.

Cour-Palais B.G. and Crews J.L.(1990). A Multi-Shock Concept for Spacecraft Shield. *International Journal of Impact Engineering* **10**, 135-146.

Hatta H., Udagawa A., Higuchi K., Yokota R. and Sugibayashi T.(1999). Foaming Behavior of Polyurethane to be Applied to Inflatable Space Structures Under Low Pressure Environment. *J. Soc. Mat. Sci.Japan.* **48**, 49-55

Shiraki K., Sakashita T. and Noda N.(1998). JEM Protection Shield Performance Evaluation for Hypervelocity Debris Impacts. *Proceedings of the JSASS/JSME Structure Conference*, 177-180.

Shiraki K.(2000). Doctral Thesis (Kyushu University).

Sugishima H.(1998). Properties and uses of Fully Aromatic Polyester Fiber Vectran, *The Japan Society of Fiber Science and Technology*, 12-14.

Toda S. (1998). Trend in Space Debris Related Researches and Activities. *Proceedings of the JSASS/ JSME Structures Conference*, S-1 - S-4.

Tsutsumi T., Okumura K., Matsushita N., Tanaka M. and Akahoshi Y.(1999). Improvement of Two-Stage Light Gas Gun, *Proceedings of Annual Meeting of Japanese Aerospace Society Western Branch*, 1-4.

IMPACT ENERGY ABSORPTION AFFECTED BY THE CELL SIZE IN A CLOSED-CELL ALUMINUM FOAM

T. Mukai[1], S. Nakano[2], T. Miyoshi[3] and K. Higashi[2]

[1] Osaka Municipal Technical Research Institute, Morinomiya Joto-ku,
Osaka 536-8553, Japan
[2] Department of Metallurgy and Materials Science, College of Engineering,
Osaka Prefecture University, Sakai, Osaka 599-8531, Japan
[3] Shinko Wire Co., Ltd., Tsuruhara, Izumisano, Osaka 598-0071, Japan

ABSTRACT

Metallic foams also have a potential for absorbing impact energy and enhancement of the absorption energy can be achieved by the extent of plateau strain and an increase in the plateau stress. In this study, modification of the structure in a closed-cell aluminum foam was performed for the crashworthiness. The edge length of the modified foam is obviously reduced from that of the conventional aluminum foam, while the aspect ratio of the wall thickness against the edge length increases by the present modification. A novel Split Hopkinson Pressure Bar was used to estimate the dynamic compressive properties. As a result of the compression tests at a dynamic strain rate ($\sim 1 \times 10^3$ s^{-1}), the plateau stress of the modified foam was found to exhibit a remarkable increase as compared to that of a conventional foam without increasing the relative density. It was found that the absorption energy per unit volume (W) of the modified foam was ~ 30 % higher than that of the conventional foam.

KEYWORDS

Closed-cell, Aluminum foam, Cell size, Compression test, Plateau stress, Energy absorption

INTRODUCTION

Recently, there is a high interest in using light-weight metallic foams (e.g., Al and Mg) for automotive, railway and aerospace applications where weight reduction and improvement in comfort are required (Gibson & Ashby 1998). Metallic foams also have a potential for absorbing impact energy during the crashing of a vehicle either against another vehicle or a pedestrian. Although the metallic foam has a high possibility for the application of structural use in automotives, limited data has been reported for the dynamic mechanical response in metallic foams. Very recently, several data are available for Aluminum foams. For example, Lankford & Dannemann (1998) examined an open-cell 6101 Aluminum, DUOCEL, and found the plateau stress was independent of strain rate. Deshpande & Fleck (2000) have also demonstrated the plateau stress is independent of strain rate both for the open-cell aluminum of DUOCEL and a closed-cell foam of Alulight. Hall et al. (2000) also reported that the plateau stress is independent of strain rate for a closed-cell 6061 aluminum foam. On the other hand, Mukai et al. (1999) investigated the closed-cell aluminum foam, ALPORAS, and reported the plateau stress exhibited the strain rate sensitivity. Dannemann & Lankford (2000) also demonstrated that the plateau stress in ALPORAS exhibited a strain rate sensitivity and pointed out the importance of the existence of the gas inside the each cell. Thus, there is no universal agreement with the strain rate sensitivity of the plateau stress. In this study, the strain rate sensitivity of a closed-cell aluminum is investigated. To effectively absorb the impact energy, a material is required to exhibit an extended stress plateau. Thus, enhancement of the absorbed energy can be achieved by the extent of plateau strain and an increase in the plateau stress. Grenestedt (1998) has investigated the influence of imperfections such as wavy distortions of cell walls on the stiffness of closed-cell aluminum alloys. Sanders and Gibson (1998) have also pointed out that the reduction in Young's modulus of aluminum foams is due to cell wall curvature and corrugation. Therefore, it is noted that controlling the structure of cell walls is important for the enhancement of energy absorption. In this study, enhancement of absorption energy in a closed-cell structure has been performed by an increase in the aspect ratio of cell-wall thickness against the cell-edge length with the reduction of cell size. The absorbed energy is estimated and compared for the closed-cell aluminum with different cell sizes.

STRUCTURE OF CLOSED-CELL FOAM

The material used in the present study is a modified aluminum foam, ALPORAS (denoted as #M) which was produced by Shinko Wire Co. Ltd., Japan. The aluminum foam was manufactured by a batch casting process. The chemical composition of the modified foam is Al-1.42Ca-1.42Ti-0.28Fe (by mass%), which is the same as the conventional ALPORAS. Typical structure of the foam is shown in Fig. 1 (a). The diameter of the cells was measured to be 2.5 ~ 3.0 mm according to the method prescribed by ASTM for the measurement of grain diameter in polycrystalline materials, and the relative density were about 0.106(#M) and 0.155(#I). The referenced material is a conventional foam(denoted as #C) the structure of which is shown in Fig. 1 (b). The relative density(0.105) of the foam is very close to the present foam, however the average diameter(~ 4.5 mm) is larger than that of the present foam. In order to characterize the structures of two foams, optical microscope has been used and measured the apparent edge length(denoted as L) and the thickness of cell walls for any 200 edges. The schematic illustration for the measurement is shown in Fig. 2. The average value of the measured edge length and wall thickness has been reported elsewhere (Miyoshi et al. 1999). It has been found that the thickness and length were reduced by the present modification. The aspect ratio of the wall thickness against the edge length has been also estimated for the closed-cell foam, #C(=0.052) and #M(=0.058). It is noted that the aspect ratio in the modified foam is essentially higher than that in a conventional foam.

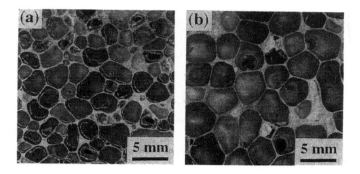

Fig. 1 Typical cell structures of (a) a modified foam(#M), and (b) the close-celled aluminum foam(#C) of a conventional ALPORAS (Miyoshi *et al.* 1999).

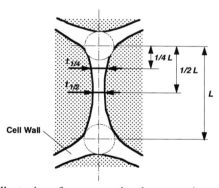

Fig. 2 Schematic illustration of a cross-sectional structure in a close-celled aluminum.

COMPRESSIVE BEHAVIOR

Compression tests were performed to evaluate the absorption energy. Specimens with a dimension of 16x16x11 mm were cut from each aluminum foam. The compression tests were carried out at a strain rate of ~ 1.3×10^3 s^{-1} by the split Hopkinson pressure bar composed by a maraging steel. The compressive behavior was evaluated by a one-wave calculation (Gray 2000). To examine the stress equilibrium, an as-extruded magnesium alloy bar was also used for the compression test. From the measured wave data, the strain rates at the contact end of input bar/specimen and specimen/output bar were evaluated by one-wave and two-wave calculation (Gray 2000) as shown in Fig. 3. The experimental evidence in this figure supports the possibility of stress state equilibrium at an early stage of deformation of the present foam.

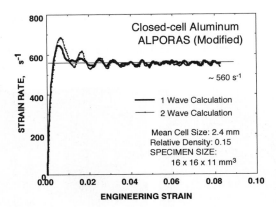

Fig. 3   Variations of strain rate as a function of engineering strain for the contact end of input bar/sample and sample/output bar.

Typical stress-strain curves for the modified foam are shown in Fig. 4. The diagram shows an elastic region at the initial stage, then, followed by a plateau region (with nearly constant flow stress). After the plateau region, the flow stress rapidly increases because the specimen densifies. The stress-strain characteristic for the closed-cell aluminum has already been reported (Gibson & Ashby 1998). As can be seen in this figure, the compressive stress of the present foam obviously exhibits a strain rate sensitivity. Stress-strain relations for the conventional foam(#C) and the modified foam(#M) are shown in Fig. 3 at a dynamic strain rate. The relative density of both samples is identical. It is clearly observed that the plateau stress in the modified foam, however, exhibits higher value than that in a conventional foam. The increase in the plateau stress seems to be due to the increase in the membrane stress of cell walls with the aspect ratio. This experimental evidence suggests that the modified foam obviously has a higher capacity for energy absorption than the conventional foam.

ENERGY ABSORPTION

The absorption energy per unit volume(W) of the modified foam was further evaluated. The absorption energy for a sample can be calculated by integrating the area under the stress-strain curve, namely,

$$W = \int_0^\varepsilon \sigma(\varepsilon)d\varepsilon \tag{1}$$

The average values of absorption energy per unit volume of ALPORAS at a strain of 0.5 for #C and #M are evaluated as 1.30 and 1.72 MJ/m^3, respectively. The value of W in the modified foam, #M, is about 30% higher than that in #C. It is noted that the enhancement of energy absorption can be achieved by the present modification of the structure. The selection of cellular materials for applications such as cycle helmet inner liner, bumper for automobiles or motor cycles is based on energy absorption (Gibson & Ashby 1998). It is noted that the optimization of cellular structures is essentially important for the enhancement of energy absorption in a closed-cell metallic foam.

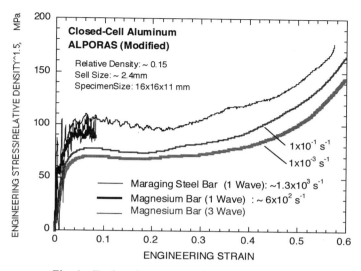

Fig. 4 Engineering stress-strain curves of a modified foam.

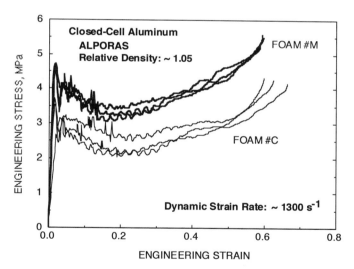

Fig. 5 Effect of the modification in cellular structure on the engineering stress-strain relations.

SUMMARY

The effect of cell size on the compressive behavior in a closed-cell aluminum foam was investigated. From the compressive tests at a dynamic strain rate of 1.1×10^3 s^{-1}, the plateau stress could be increased with decreasing the cell size without increasing the relative density. This increase may be due to the increase in the membrane stress with an aspect ratio of the wall thickness against the edge length. As a result, the absorption energy per unit volume of the modified foam can be increased by ~30% higher than that of a conventional foam.

REFERENCES

Gibson L.J. and Ashby M.F. (1997), *Cellular Solids, Structure and Properties - Second edition*, Cambridge University Press, Cambridge, UK.

Lankford J. Jr. & Dannemann K.A. (1998). Strain Rate Effects in Porous Materials. *Porous and Cellular Materials for Structural Applications, MRS Symposium Proceedings, 521*, 103-108.

Deshpande V. S.& Fleck N. A. (2000). High Strain Rate Compressive behaviour of Aluminium Alloy Foams. *Int. J. Impact Eng.* **24**, 277-298.

Hall I. W., Guden M. and Yu C.-J. (2000). Crushing of Aluminum Closed Cell Foams: Density and Strain Rate Effects. *Scripta Mater.* **43**, 515-521.

Mukai T., Kanahashi H., Miyoshi T., Mabuchi M., Nieh T.G. and Higashi K. (1999). Experimental Study of Energy Absorption in a Close-Celled Aluminum Foam under Dynamic Loading. *Scripta Mater.* **40:8**, 921-927.

Dannemann K. A. & Lankford J. Jr. (2000). High Strain Rate Compression of Closed-Cell Aluminium Foams. *Mater. Sci. Eng. A* **293**, 157-164.

Grenestedt J.L. (1998). Influence of Imperfections on Effective Properties of Cellular Solids, *Porous and Cellular Materials for Structural Applications, MRS Symposium Proceedings, 521*, 3-13.

Sanders W. and Gibson L.J. (1998) Reduction in Young's Modulus of Aluminum Foams due to Cell Wall Curvature and Corrugation, *Porous and Cellular Materials for Structural Applications, MRS Symposium Proceedings, 521*, 53-57.

Miyoshi T, Itoh M, Mukai T., Kanahashi H., Kohzu H., Tanabe S. and Higashi K. (1999). Enhancement of Energy Absorption in a Closed-Cell Aluminum by the Modification of Cellular Structures. *Scripta Mater.* **41:10**, 1055-1060.

Gray G. T. III. (2000). Split-Hopkinson Pressure Bar Testing of Soft Materials, *Mechanical Testing and Evaluatiuon, ASM Handbook,* Vol. 8, 488-496.

DEVELOPMENT OF VERY LOW DECELERATION SHOCK-ABSORBERS

Kazuo ASADA, Yasuhiro TAN, Katsunari OHSONO, Suguru HODE,
Toshihiro MATSUOKA and Shuhei KURI

Mitsubishi Heavy Industries, LTD.

ABSTRACT

We developed a 131 tons weights cask that can transport and storage 69 BWR nuclear spent fuels. A deceleration of the cask at 9m horizontal drop test have to keep below 50G(G= gravity acceleration) in order to assure both structure integrity and seal integrity. Then, 2types of shock-absorbers were developed. Those shock-absorbers are made of woods that have low compression strength and large lock-up strain. Through a quarter(1/4) model 9m horizontal drop test, we proved that a real scale cask with balsa and redwood shock-absorbers have below 40G deceleration .

KEWWORDS

CASK, SHOCK-ABSORBER, 9m DROP TEST, DECELERATION, BALSA WOOD, REDWOOD, FIRPLY-WOOD, CASH –II CODE, 1/4 SCALE MODEL

1. MSF69B CASK

Mitsubishi MSF69B cask, dry type and both transport / storage was newly developed. Property of this cask is as follows. Weight is 131 tons, total length is 6.6m, and maximum diameter is 3.55m. 69 BWR nuclear spent fuel assemblies are inserted into B-AL(boron aluminum) box-type basket. The seal boundary is low-alloy steel vessel and two lids(first lid is inner and second lid is outer). Metallic gasket suits for long time storage. Then, two lids have metallic gasket for seal. At transport, top and bottom shock-absorbers are set on the cask to decrease impact force. This MSF69B cask is shown in Fig. 1.

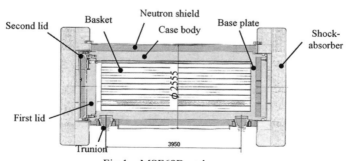

Fig.1 : MSF69B cask

9m drop test simulates a car crash accident at 50km/h. A horizontal drop test is the most severe for basket and metal gasket compared with vertical drop and corner drop. We have to keep the deceleration below 50G at horizontal drop. For this reason, two types shock-absorbers were developed. The first type is composed of balsa wood and redwood. The second type is composed of balsa wood and firply-wood.

2. SHOCK-ABSRBERS
2.1 ASSUMPTIONS

Same shock-absorbers are set to a cask at top and bottom. Mass gravity is the center of cask, and top shock-absorber and bottom shock-absorber have same structures. Then, top and bottom shock-absorbers have same deformation and same resistance.

The structure of the shock-absorber is shown in Fig. 2. The properties of this shock-absorber are assumed as follows.
(1) Outer steel shell thickness is thin. Then, the absorbing energy and impact force of the outer steel shell is adequately small.
(2) Material A is balsa wood in Fig. 2. Its strength at horizontal drop is adequately small.
(3) Material B and C are same woods, but grain direction is different. Material C only can absorb energy, and material B can not absorb energy.

The candidate wood of material C is as follows.
  (a) firply-wood   : σ =15MPa   (b) redwood   : σ =27Mpa
  (c) American cedar : σ =29Mpa   (d)American hiba : σ =39Mpa
  (e) American hemlock spruce : σ =44MPa

Here, σ is compression flow stress which is derived through static test by us, and compression stress-strain relation of red-wood is shown in Fig.3 for example.

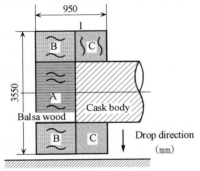

Fig.2 : Shock-absorber

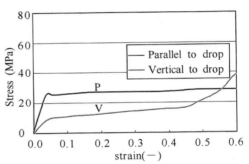

Fig.3 : Stress-strain of redwood

2.2 SIMPLE EQUATION at 9m HORIZONTAL DROP

The deformation of shock-absorber at horizontal drop is shown in Fig.4. The deformation of shock-absorber Y(θ) determines impact force F(θ), absorbing energy E(θ), and deceleration A(θ).

$Y = R_o(1-\cos\theta)$ (1)
$F = 2\sigma R_o L \sin\theta$ (2)
$E = \sigma R_o^2 L(\theta - 1/2\sin 2\theta) = E_o$ (3)
$A = F/M$ (4)

Here,
  R_o = Outer radius of shock-absorber = 1775mm
  L = Total effective length of shock absorber = 795mm
  σ = Compression strength of material C
  E_o = Kinetic energy of cask = MGH = 1.155E7 J
  M = Total weight of cask   =131E3 kg

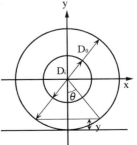

Fig.4 Deformation of shock-abosorber

2.3 SELECTION OF WOOD

The deformation of shock-absorber depending on wood strength is derived from EQ.(1) and EQ(3). This relation is shown in Fig. 5. Here, symbol (//) is that the grain direction of wood is parallel to drop direction, and symbol($\perp$) is that the grain direction of wood is vertical to drop direction. In Fig.5, 667mm is lock-up deformation(at this deformation, the cask body impacts directly rigid ground). The strength of firply-wood($\perp$) shock-absorber is too small to avoid lock-up deformation. The deceleration of cask depending on wood strength is derived from EQ.(3) and EQ.(4).This relation is shown in Fig.6. Deceleration of the fir-ply wood(//) and of the red-wood (//)are respectively 23G and 37G. Then, The firply-wood(//) and redwood(//) are selected for shock-absorber materials.

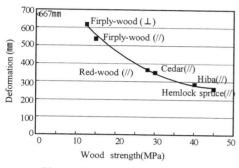 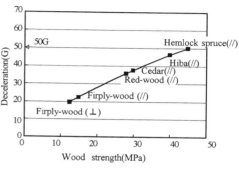

Fig.5 : Deformation of shock-absorber   Fig.6 : Deceleration of cask

In order to get more precise deceleration, we analyze a cask shock-absorber through 3D cask shock-absorber code CASH-II (Asada et al. (1988), Asada et al. (1984)) that was developed in MHI. CASH-II code can apply to any drop direction using non-linear stress-strain wood relation, and can get cask deceleration and shock-absorber deformation. Cask deceleration and shock-absorber deformation through CASH-II code and through simple equation are shown in Table 1.
Stress-strain relation of fir-ply wood have not plane. Then, Simple equation results is different from CASH-II code results.

Table 1 : CASH - II code results

	Deceleration(G)	Deformation(mm)
Firply wood	45.75(23)	356(540)
Red wood	36.25(37)	392(375)

( ) : Simple equation results

3. 9m HORIZONTAL DROP TEST
3.1 1/4 SCALE MODEL

A quarter(1/4) scale cask body model is rigid steel, the diameter is 502mm, the length is 990mm, and the weight is 1.89ton. Length of cask body is not 1/4 scale because of horizontal drop. 1/4 scale shock-absorber is precise scale model, the diameter is 887.5mm, the length is 237.5mm, and the weight is 92kg. Then, total weight of 1/4 scale cask(a cask body and two shock-absorbers) is 2.07ton. This 1/4 scale model is shown in Fig.7.

Two types of shock-absorber are manufactured. One is balsa wood and redwood type. The other is balsa wood and fir-ply wood type.

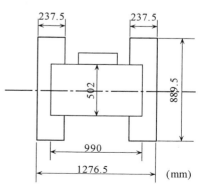

Fig.7 : 1/4 scale model

3.2 DROP TEST

For each shock-absorber, 9m horizontal drop tests are executed twice. Deceleration of cask body (top, center, and bottom) and shock-absorber deformation are measured. The acceleration pick-up is KYOWA strain-gauge type (AS-1000HA), the dynamic amplifier is KYOWA 200kHz type(CVD-230C), and the data-recorder is KYOWA(RTP-670A).

Deceleration of cask body and shock-absorber deformation is effected by oblique drop angle(angle at horizontal drop is 0°).Then, we measured oblique drop angle through delayed time between top acceleration pick-up data and bottom acceleration pick-up data. All deceleration data are passed through low pass filter(400Hz).

Deceleration dynamic data(center and top of cask) for the second experiment of red wood shock-absorber is shown in Fig.8. Scale laws shows that 1/4 scale model deceleration is 4 times of real scale model, and 1/4 scale model time is 1/4 times of real scale model. Then, we find that real model deceleration is 30G and real model duration is 80ms.

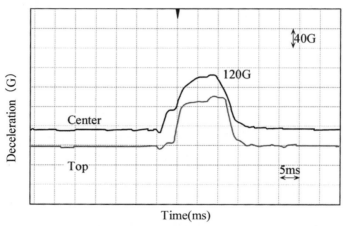

Fig.8 : Deceleration of redwood shock-absorber drop test(2nd)

Deformation of redwood shock-absorber before and after drop test is shown in Fig.9.   A photograph of cask at impact onto 100 tons steel and RC concrete ground is shown in Fig.10. Deformation of shock-absorber is measured using steel gages after drop test.

Before test   After test

Fig.9 : Deformation of redwood shock-absorber

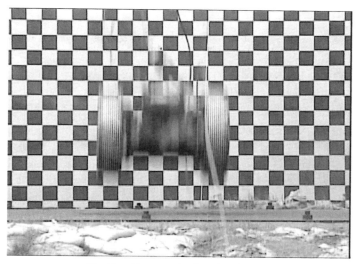

Fig.10 : Photograph of cask at impact

We can not do exact horizontal drop test, but do small angle oblique drop test. Then, horizontal drop deceleration and deformation(θ =0°) is extrapolated by using 2 decelerations and 2 deformations at different small oblique angle.

The decelerations at horizontal drop for firply-wood shock-absorber and redwood shock-absorber are shown in Table 2, and the deformations at horizontal drop for firply-wood shock-absorber and redwood shock-absorber are shown in Table 3. We assume the second order polynominal function depending on oblique angle for deceleration and deformation.

Table 2 : Deceletion at 9m drop test (1/4 scale model)

	Test	Experiment				Cash-II (G)
		Angle θ (°)	Deceleration A(G)	Equation	A at 0° (G)	
Firply wood	①	4.04	107.2	$A=-2.48\theta^2+147.64$	147.64	183
	②	1.66	140.8			
Red wood	①	2.27	105.6	$A=-6.40\theta^2+138.56$	138.56	145
	②	1.91	120.0			

Table 3 : Deformation at 9m drop test (1/4 scale model)

	Test	Experiment				Cash-II (mm)
		Angle θ (°)	Deformaion Y(mm)	Equation	Y at 0° (mm)	
Firply wood	①	4.04	67.3	$Y=0.33\theta^2+61.92$	61.92	89
	②	1.66	62.8			
Red wood	①	2.27	80.1	$Y=1.47\theta^2+87.67$	87.67	98
	②	1.91	82.3			

Then, 1/4 scale model deceleration of firply-wood shock-absorber is 148G(real scale deceleration is 36.9G), and 1/4 scale model deceleration of redwood shock-absorber is 139G(real scale deceleration is 35G). 1/4scale model deformation of firply-wood shock-absorber is 62mm(real scale deformation is 248mm). 1/4 scale model deformation of redwood shock-absorber is 88mm(real scale model deformation is 351mm). All experiment data(deceleration and deformation) is smaller than CASH-II code data.

4. DISCUSSION

The deceleration and deformation of experiments are smaller than those of CASH-II code. This is contradiction. Then, we derived cask velocity V(t) and cask deformation Y(t) by integral of dynamic deceleration data A(t), and cask force F(t). Here, t is time, Vo is impact velocity 13.28m/s, and M is cask mass 2.07E3 kg.

$$V(t) = V_o - \int A(t)dt \quad (5)$$
$$Y(t) = \int V(t)dt \quad (6)$$
$$F(t) = MA(t) \quad (7)$$

In Fig.11, the relation of Force F and deformation Y(EQ(6)) at redwood shock-absorber experiment (second time) is shown. There is difference between this cask deformation 121mm and shock-absorber deformation 82.3mm. This is because that cask body deforms larger than shock-absorber deformation. In Table 4, the energy absorption of 1/4 scale model derived from EQ(5),EQ(6), and EQ(7) is shown. Kinetic energy of 1/4 scale cask is 18.3E4 J. Then, all kinetic energy of 1/4 scale cask is absorbed by shock-absorber deformation energy.

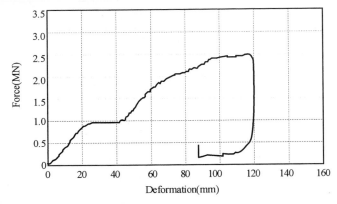

Fig.11 : Relation of force and deformation at red wood shock absorber (2nd)

Table4 : Energy absorbtion of 1/4 scale model

	Deceleration(G)	Deformation (mm)	Force(MN)	Energy(10^4J)
Firply wood ①	107.2	146	2.21	18.6
Firply wood ②	140.8	114	2.9	18.4
Red wood ①	105.6	141	2.17	18.4
Red wood ②	120.0	121	2.46	18.4

5. CONCLUSIONS

We showed that real scale cask with redwood and balsa wood shock-absorber have below 40G deceleration at 9m horizontal drop test through 1/4 scale model experiment.

REFERENCES

Asada K. et al. (1988). Development of Simplified Analysis Code for 9m Drop and 1m Puncture Test for A Radioactive Material Cask. *Proc. Waste Management '88*, Tusco.
Asada K. et al. (1984). Cask Shock Absorber Analysis Code– CASH. *First Symposium of Material Impact Problem,* The Society of Material Science, Kyoto, 113-117.

SYNTHESIS BEHAVIOUR AND HOT SHOCK COMPACTION OF SILICON CARBIDE-CHROMIUM BORIDE COMPOSITES

R. Tomoshige[1], H. Kainou[2] and A. Kato[2]

[1] Research Center for Advances in Impact Engineering, Faculty of Engineering, Sojo University, Kumamoto 860-0082, Japan
[2] Department of Applied Chemistry, Faculty of Engineering, Sojo University, Kumamoto 860-0082, Japan

ABSTRACT

A novel hot shock compaction process that underwater-shock compaction technique were combined with self-propagating high temperature synthesis (SHS) was applied to prepare SiC-CrB_2 system ceramic composites. This process is one-step-process that only a few minutes are required as a processing time from synthesis through compaction of the compounds. The generated shock pressure was estimated at about 10 GPa. This experiment demonstrated that SiC-CrB_2 composites could be densified for short time by the hot shock compaction process. Density and hardness values of the densified composites depended on time window from initiation of SHS reaction. In particular, relative density attained up to 95% or more in time window of 40 seconds. Its hardness value, however, was low comparing with that of monolithic sintered SiC. SEM observations revealed that SiC and CrB_2 phases dispersed homogeneously. In addition, variation in the electric resistivity depended on measurement temperature.

KEYWORDS

Shock Compaction, SHS, Synthesis behaviour, Ceramic Composites, Microstructures, Electric Resistivity, Hardness, Density

INTRODUCTION

Combustion synthesis or self-propagating high temperature synthesis (SHS) has been highlighted as a unique technique, which can be produced readily and shortly various ceramic composites or intermetallic compounds accompanied with intensely exothermic heat. In particular, it is useful for

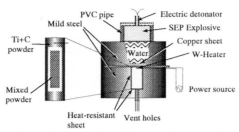

Figure 1: A schematic illustration of hot shock compaction apparatus.

producing high melting point materials because SHS requires no extra energy after the synthesis initiates. However, SHS has a demerit that the synthesized materials include many pores in it. In general, it is known that the pores are formed by both gas generated from volatile in starting materials and rapid volume expansion during the reaction. Thus, post-sintering process must be required for removing the pores.

By the way, some of the authors have been studying about explosively shock compaction for various powders, Tomoshige et al., (1994). The compaction technique can densify shortly difficult-to-consolidate powders. So that, it will be useful to combine SHS reaction with this compaction technique for densification of the porous synthesized materials.

On the other hand, silicon carbide (SiC) is expected to be used as a good thermoelectric material at high temperature because it has semi-conductive properties and indicates high electric conductivity at high temperature. As SiC, however, has also low conductivity at lower temperature range, it is needed by adding some material with good conductivity such as borides into SiC.

In this study, preparation and densification of ceramic composites consisting of silicon carbide (SiC) and chromium boride (CrB_2) have been attempted to do the shock compaction immediately after SHS, which is called hot shock compaction technique. Microstructural observations, density measurements, and hardness tests were performed for the obtained composites. Furthermore, electric resistivity was also measured to search a possibility to be used as thermoelectric materials.

EXPERIMENTAL PROCEDURE

Silicon (Mitsuwa Chemical Industry, Osaka, Japan, mean particle size: 47 μm), graphite (Aldrich Chemical Company, Inc., Milwaukee, WI, USA, ~ 2 μm), chromium (Wako Pure Chemical, Osaka, Japan, 30 μm) and boron powders (Sigma Chemical Co. St. Louis, MO, USA, ~ 2 μm) were used as starting materials. These powders were wet-mixed for one hour with ethyl alcohol according to the following equation.

$$0.25\ Si + 0.25\ C + 0.75\ Cr + 1.5\ B \rightarrow 0.25\ SiC + 0.75 CrB_2 \tag{1}$$

This composition was selected to become eutectic point composition in the SiC-CrB_2 system. The mixture was dried fully at 110 ℃, then it was tapped into a reaction container with a hole of 20 mm in diameter and 40 mm in depth as shown in Figure 1. We have also another reason on adding CrB_2 phase into the reaction system. Silicon carbide has only enthalpy of -69 kJ/mol, Samsonov & Vinitsuki

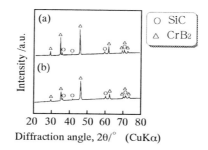

Figure 2 : XRD patterns of the composite obtained by (a) SHS and (b) hot shock compaction

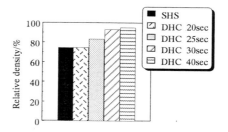

Figure 3 : Relative densities of the composites obtained by SHS reaction and hot shock compaction

(1977), so that, it is needed to assist the synthesis reaction of SiC. Here, the enthalpy is important factor to determine whether the SHS reaction proceeds successfully or not. Accordingly, it will be possible to improve the poor reactivity of SiC by adding chromium and boron powders since synthesis reaction of CrB_2 phase generates high enthalpy of -125 kJ/mole, Samsonov & Vinitsuki (1977). In addition, a chemical oven consisting of powder mixture of titanium (Sumitomo Sitix, Hyougo, Japan, 40 μm) and graphite with a molar ratio of Ti: C=1:1 was used for promoting the reaction of SiC-CrB_2 system. The chemical oven has higher enthalpy of –159 kJ/mole Samsonov & Vinitsuki (1977). Actually, surface temperature of the chemical oven attained 2132 ℃, which was measured by optical pyrometer. The specimen powder mixture was surrounded with the chemical oven as shown in Fig.1. Explosive used was provided from Asahi Kasei Corporation, Tokyo, Japan, which has detonation velocity of 6900 m/sec. About 60-g explosive was detonated after 20 to 40 seconds from initiation of the SHS reaction, which is called time window. In this case, shock pressure was estimated to be about 5 to 10 GPa, for example, Tomoshige R. et al. (1995). The compacted specimens were evaluated by X-ray diffraction experiments (XRD), density and electric resistivity measurements, micro Vickers hardness tests, scanning electron microscopy (SEM) and energy dispersive X-ray spectroscopy (EDS).

RESULTS AND DISCUSSION

XRD experiments were performed for both as-SHSed and shock-compacted specimens (Figure 2(a) and (b)). The shocked samples were prepared to appear a surface cut perpendicular to direction of the shock wave. Consequently, only β-SiC and CrB_2 phases were detected in both the specimens.

Density of the as-SHSed specimen is shown in Figure 3. The specimen indicated low relative density, about 75% of theoretical density (TD). It was considered that the low density attributed to large pores arisen from the volatiles in the starting powders during SHS reaction and rapid volume expansion by intense heat of the exothermic reaction. Meanwhile, densities of the shock compacts produced under some conditions were measured for determining optimum condition in the shock compaction. It is obvious that the density varied with the time window. From this variation, it was considered that both the remaining volatile gas between particles and the specimen softening affected the densification at the shock compaction. However, it will be easy to consolidate the material after the volatile gas is released. Actually, though the compacts obtained at the time window of 20 seconds indicated lower density, the one consolidated at the time window of 40 seconds indicated the highest density among them, which had relative density of 95%TD. Thus, the condition of the time window of 40 seconds was selected as the optimum one in this study.

Figure 4(a) shows a SEM micrograph of polished surface of the shocked specimens. Fig.4 (b), (c) and (d) correspond to X-ray mapping of Cr $K\alpha$, Si $K\alpha$ and C $K\alpha$, respectively, by EDS analyses. According to the microstructures, it was found that the composite was consisted of SiC (dark phase) and CrB_2 (white phase). Furthermore, EDS analyses revealed that chromium silicide phase (gray phase) was also observed, which was not detected in XRD experiments (Fig.2). The silicide twined complexly around CrB_2 phase. Unfortunately, it was hard to get fine eutectic microstructures due to a poor reactivity of the synthesized materials though we expected the microstructures to be formed.

Figure 5 shows micro Vickers hardness values versus the time window. The hardness datum were averaged five values obtained from different portions. The hardness value changed from 7 to 12 GPa, and was about one third that of a hot-pressed monolithic SiC. These results suggested existence of pores located at internal specimen and insufficient interparticle bonding strength. It was also found hardness distribution to be narrow as the time window was long.

Results of electric resistivity are shown in Figure 6. The resistivity was measured by four probe method from room temperature to 800 ℃. As-SHSed and shock-compacted specimens indicated low resistivity from room temperature to 400 ℃. For example, as-SHSed specimen indicates about $0.02 \Omega \cdot cm$ at room temperature, which was lower than $0.13 \Omega \cdot cm$ of monolithic SiC, Samsonov & Vinitsuki (1977). This resulted from addition of CrB_2 with low resistivity of $30 \times 10^{-6} \Omega \cdot cm$, Samsonov & Vinitsuki (1977). In addition, the shock-compacted specimen had lower conductivity than as-SHSed one. This phenomenon might be generated by passing preferentially through conductive phase. Around 400 ℃, the resistivity increased in both specimens. It appeared that electric properties of the boride and existence of high resistivity layer (depletion layer), positioned at SiC grain boundaries generated the high resistivity. While, SiC and CrB_2 have thermal expansion coefficients of 4.7×10^{-6}/deg and 10.5×10^{-6}/deg, respectively, Samsonov & Vinitsuki (1977). Therefore, the difference in thermal expansion coefficient between the particles might cause to section or divide conductive passes in the composite at higher temperature. Then, the resistivity began to decrease around 500 or 600 ℃. It was speculated that the decrease resulted from the semi-conductive properties of SiC or slight densification by heating. It was suggested that the reaction at higher temperature and addition of more SiC to the composites were required to get denser composites with good thermal electric properties.

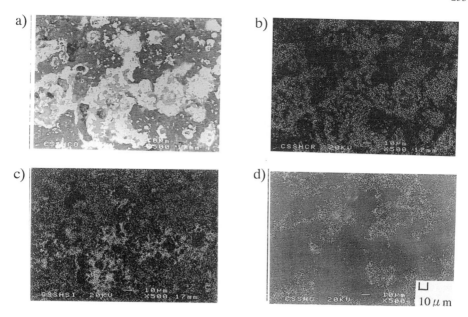

Figure 4 : SEM microscopy and X-ray mapping of the shock-compacted composite

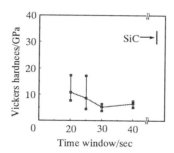

Figure 5 : Micro Vickers hardness values as a function of the time window

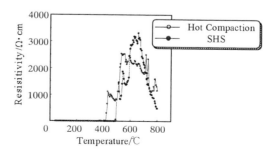

Figure 6: Electric resistivity of the SiC-CrB$_2$ composites with heating

CONCLUSIONS

SiC-CrB$_2$ composites were fabricated shortly by dynamic hot shock compaction. The shock compaction at the time window of 40 seconds made the synthesized materials dense. The compacted composites had relative density of about 95 %. However, its hardness value was lower than that of hot-pressed SiC, which meant that insufficient interparticle bonding between the two phases. Though the formation of the eutectic microstructures was attempted, the composites showed hardly it due to poor reactivity. The synthesized and compacted specimens indicated good electric conductivity at lower temperature regions. However, the resistivity varied significantly with the measurement temperature.

ACKNOMLEDGEMENTS

We express our appreciation to Mr. Susumu Akimaru of Shock wave and Condensed Matter research Center in Kumamoto University for his assistance in shock compaction experiments.

References

Samsonov G.V. and Vinitsuki I.M. (1977). *Handbook on High Melting Point Compounds*, Translated by Nisso-Tsushin-Sya Publisher, Wakayama, Japan (in Japanese).

Tomoshige R, Chiba A, Nishida M, Matsushita T. (1994). Explosive Compaction of Silicon Nitride Powder without Additives Utilizing Underwater-Shock Wave. *Trans. Mat. Res. Soc. Jpn*, , **14A**, 643-46.

Tomoshige R. Kakoki Y. Chiba A. Imamura K. and Matsushita T. (1995) *Metallurgical and Materials Applications of Shock-Wave and High-Strain-Rate Phenomena*, Murr L.E. Staudhammer K.P. and Meyers M.A. eds, Elsevier, Amsterdam

SHOCK ACTIVATION AND REACTION SYNTHESIS IN THE Ti+Si+C AND Ti+AlN SYSTEMS

Jennifer L. Jordan and Naresh N. Thadhani

School of Materials Science and Engineering, Georgia Institute of Technology, Atlanta, GA 30332-0245

ABSTRACT

Shock-activated reaction synthesis is used to study the formation of Ti_3SiC_2 and Ti_2AlN, novel ceramics with metal-like properties (high electrical conductivity and plastic-like deformation). Titanium, silicon carbide, and graphite and titanium and aluminum nitride precursor powders were ball milled and densified using an 80-mm gas gun and double implosion cylinder techniques followed by reaction synthesis heat treatments. The reaction mechanism is being studied using differential thermal analysis to determine the kinetic activation energy and consequently the degree of shock activation necessary for complete reaction synthesis of Ti_3SiC_2 and Ti_2AlN in the solid state.

KEYWORDS

titanium-silicon carbide, titanium-aluminum nitride, shock compaction, materials synthesis, activation energy

1.0 INTRODUCTION

Titanium-silicon ternary carbide (Ti_3SiC_2) and titanium-aluminum nitride (Ti_2AlN) are unique ceramics because they maintain high-temperature properties and wear resistance while also demonstrating metal-like properties including electrical conductivity, thermal conductivity, and easy machinability. Several mechanical and physical properties of these ceramics are listed in Table 1. While in most materials, hardness typically scales with the elastic modulus, it is interesting to note that the ternary ceramics are very stiff but soft materials. They are very incompressible, and Ti_3SiC_2 has a bulk modulus similar to that of TiC, while its elastic modulus and shear modulus are similar to metallic molybdenum. Its hardness, on the other hand, is comparable to that of quenched high carbon steels. Ti_2AlN is also expected to demonstrate similar properties, although some data is unavailable. The unique properties of these ternary ceramics are attributed to their layered structure. The incompressibility and the deformation behavior of these ternary ceramics make them interesting candidates for damage tolerant armor applications.

Table 1: Physical and Mechanical Properties of Ti-based Ternary Ceramics

Property	Ti_3SiC_2	Ti_2AlN
Density[1]	4.5 g/cm^3	4.3 g/cm^3
Young's Modulus[2]	~ 320 GPa	
Bulk Modulus[3]	206 ± 6 GPa	
Shear Modulus[3]	133 GPa	
Vickers Hardness[1,2]	4 GPa	3.5 GPa
Compression Strength (Room temperature)[1,2]	580 ± 20 MPa	471 MPa
Longitudinal Sound Speed[4]	8875 ± 8	
Transverse Sound Speed[4]	5712 ± 21	
Poisson's Ratio[4]	0.180 ± 0.001	
K_{IC} (Vickers)[4] / KIC (Three point bend)[4]	9.4 ± 1.3 / 6.2 ± 0.4	

[1] Barsoum, et al. (1997), [2] Barsoum and El-Raghy (1996), [3] Onodera, et al. (1999), [4] Lis, et al. (1998)

Many attempts have been made to synthesize the ternary carbide Ti_3SiC_2 in bulk form. However, most of these methods typically produce Ti_3SiC_2, in addition to TiC and/or SiC. Barsoum and El-Raghy (1996) employed a process of hot pressing for producing pure, bulk Ti_3SiC_2 starting with a mixture of Ti, SiC, and graphite. A controlled rate of heating, which ensures reaction occurring in the solid state, has been found to be essential for the synthesis of single phase Ti_3SiC_2, Barsoum and El-Raghy (1996). However, if at any stage the heat released from the reaction exceeds the rate of heat dissipation, resulting in a combustion type reaction, then TiC is the preferred phase. Barsoum, et al. (1997) have also synthesized Ti_2AlN using the same method as presented for Ti_3SiC_2 - hot pressing Ti and AlN powders at 1600 °C for 4 hours. Using this method, they have achieved phase pure materials. Similar to Ti_3SiC_2, Ti_2AlN is believed to be formed predominately via a solid state diffusion process.

Shock compression of powders can lead to highly activated states forming defect microstructures and enhancing the solid state reactivity, Thadhani (1993). Thus, thermally activated processes can occur at lower temperatures and significantly shorter times in such a material, during post-shock processing. In prior work, Lee (1997), on the shock modification of powder mixtures, it has been observed that the activation energy for solid state diffusion of C and Ti into TiC_x is lowered by four to six times, which results in TiC_x formation with significantly reduced porosity during post-shock reaction synthesis. Likewise, in Ti-Si powder mixtures, it has been shown, Namjoshi (1999), that prior shock consolidation of the constituent powders allows control of the reaction mechanism during subsequent thermal treatments and, at the same time, accelerates the reaction synthesis process. Thus, Ti-Si powder mixtures can be made to react in the solid state to form products free from defects associated with combustion type chemical reactions.

The objectives of the present work are to investigate the formation of ternary ceramics, the titanium-silicon carbide and titanium-aluminum nitride, in bulk form using shock densification followed by reaction synthesis of the precursor powders. The reaction processes are investigated using differential thermal analysis performed on shock densified samples to determine the reaction mechanisms and kinetics.

2.0 EXPERIMENTAL PROCEDURE

The precursor powders used for shock densification and subsequent reaction synthesis experiments of Ti+SiC+graphite were titanium (Alfa Aesar), silicon carbide (Superior Graphite Company and Performance Ceramics Company), and graphite (Cerac, Inc. and Aldrich Chemical Company). One batch of the precursor powder was prepared by combining Ti and SiC and graphite in the stoichiometric ratio and mixing in a V blender. Another batch was prepared by ball milling the powders for 2 hours in a Spex mill with WC balls weighing approximately 16 grams or roller milling for 6 hours with alumina balls weighing total approximately 330 grams. Graphite was then added to the milled mixture. The final mixing of the powders was performed using a V-blender which was run overnight. The ball to powder weight ratio was the same for both the Spex mill and the roller mill. The precursor powders used for shock densification and reaction synthesis of Ti+AlN were titanium (-325 mesh, Alfa Aesar) and AlN (Alfa Aesar). The powders were mixed and blended in a V blender overnight.

The shock compression experiments are performed using the three capsule plate impact (PI) fixture with the single-stage 80-mm diameter gas gun at Georgia Tech. Additional experiments were conducted a the Energetic Materials Research and Testing Center (EMRTC) in Socorro, NM, using a double-tube cylindrical implosion (CI) design with ANFO and ANFOIL explosives. The powders were pressed in the steel capsules at ~65% of theoretical maximum density (TMD) in the three capsule experiments and ~50-55 % TMD in the implosion cylinders.

Table 2 lists the experimental conditions for each of the powder types. After shock compaction, the consolidated samples were recovered by machining the fixtures and characterized by XRD, which was used to determine the degree of retained residual strain in the unreacted reactants in shock-compressed compacts, and the reaction products formed in the recovered compacts, both before and after post-shock reaction synthesis.

The shock-compressed compacts were subjected to heat treatments in a Lindberg horizontal tube furnace to 1600 °C at varying heating rates and for varying hold times. All heat treatments were performed under conditions of flowing Argon gas. Using a Perkin-Elmer DTA, differential thermal analysis was also used for the heat treatments to determine the activation energies for reactions occurring in the various compacts upon heating at varied rates of 5, 10, 25, 40, 50, and 65 °C/min.

Table 2: Sample contents for shock loading conditions

Sample Contents*	Experimental Conditions*	Experimental Conditions*	Experimental Conditions*
AB Ti + SiC + graphite	PI - 590 m/s, 5 GPa	PI - 870 m/s, 9 GPa	
RM - Ti + SiC + graphite	CI – ANFOIL, 4 GPa		
SM - Ti + SiC + graphite	PI - 590 m/s, 5 GPa	PI - 870 m/s, 9 GPa	PI - 905 m/s, 9 GPa
AB Ti + AlN	CI – ANFOIL, 4 GPa	CI –ANFO, 6 GPa	PI - 905 m/s, 9 GPa

* AB – as blended; RM – Roller milled 6 hours; SM – Spex milled 2 hours; PI – plate impact geometry; CI-cylindrical implosion geometry; pressures calculated using Autodyn-2D (1995) based on reported compressibility characteristics

3.0 RESULTS AND DISCUSSION

3.1 Shock Densified State

Table 3 gives the as shocked states and residual microstrain for the Ti,SiC, and graphite experiments. The recovered shock compressed compacts of all low pressure (< 6 GPa) experiments showed retention of reactants in the Ti + SiC + graphite mixture. XRD line broadening analysis showed residual microstrain ($\varepsilon \approx 10^{-2}$) retained in all of the precursor powders. At higher pressures (~ 9 GPa), the recovered powder compact of the as blended Ti + SiC + graphite powder showed evidence of partial reaction forming TiC_x, while the compact of the Spex milled precursor showed almost complete reaction to TiC_x. For the reaction product observed in the almost completely reacted compact, the TiC has a lattice parameter of 0.430 nm corresponding to a non-stoichiometric TiC_x phase, while that in the partially reacted sample is close to the stoichiometric value (0.433 nm). No reaction was observed in either the low or high pressure experiments in the case of Ti + AlN precursors. The powders are intimately mixed and show no reaction. However, the reactants revealed residual microstrain ($\varepsilon \approx 10^{-2}$), which appeared to somewhat decrease with increase in shock pressure as illustrated in Table 4.

Table 3. Retained strain and as shocked composition for Ti + SiC + graphite

Sample**	Retained Strain	As Shocked	10 °C/min to 1600 °C, Hold 4 hours
As blended	4.7 x 10^{-3}		Ti$_3$SiC$_2$, TiC (0.430)*
Rolling Mill 6 hours	7.2 x 10^{-3}		
Spex Mill 2 hours	2.6 x 10^{-2}		Ti$_3$SiC$_2$, TiC (0.430)*
As blended; P ≈ 5 GPa, PI, 590 m/s		Ti, SiC, graphite	TiC (0.431)*
As blended; P ≈ 9 GPa, PI, 870 m/s		Ti, TiC (0.432)*, graphite	TiC (0.433)*
RM; P ≈ 4 GPa, CI, ANFOIL	1.3 x 10^{-2}	Ti, SiC, graphite	Ti$_3$SiC$_2$, TiC (0.432)*
SM; P ≈ 5 GPa, PI, 590 m/s	2.0 x 10^{-2}	Ti, SiC, graphite	
SM; P ≈ 9 GPa, PI, 870 m/s	2.2 x 10^{-2}	TiC (0.430)*	Ti$_3$SiC$_2$, TiC (0.429)*

* lattice parameter in parenthesis
** RM – Roller milled 6 hours; SM – Spex milled 2 hours; PI – plate impact geometry; CI-cylindrical implosion geometry

Table 4. Retained strain and as shocked composition for Ti + AlN

Sample*	Retained Strain	As Shocked	10 °C/min to 1600 °C, Hold 4 h
As blended	4.7 x 10^{-3}		Ti$_2$AlN, TiN
P ≈ 4 GPa – CI, ANFOIL	3.08 x 10^{-2}	Ti, AlN	Ti$_2$AlN, TiN
P ≈ 6 GPa – CI, ANFO	2.41 x 10^{-2}	Ti, AlN	Ti$_2$AlN, TiN
P ≈ 9 GPa - PI, 905 m/s	1.46 x 10^{-2}	Ti, AlN	TiN, Ti$_3$Al$_2$N$_2$

* PI – plate impact geometry; CI-cylindrical implosion geometry

3.2 Reaction Behavior of Shock Densified Compacts

3.2.1 Titanium-Silicon Carbide, Ti$_3$SiC$_2$

The as blended and as Spex milled powder mixtures and sections of the recovered, shock densified compacts were heat treated in a tube furnace to 1600 °C at 10 °C/minute, with a hold time of four hours, and subsequently characterized by XRD analysis, shown in Table 3. Reaction synthesis of the as blended, shock densified precursor powder compacts at both pressures showed formation of TiC with a lattice parameter the same as that of the stoichiometric compound for the 9 GPa sample and the non-stoichiometric compound for the 5 GPa sample and the as blended sample. No Ti$_3$SiC$_2$ phase was found in either shocked sample. Formation of the stoichiometric TiC phase appears to be due to a self-sustained SHS-type combustion reaction, which can inhibit the formation of the ternary carbide.

Reaction synthesis of the as Spex milled powder and the milled and subsequently shock densified powder compacts showed the formation of Ti$_3$SiC$_2$ and TiC, which has a lattice parameter less than that of the stoichiometric (0.433 nm) value. Thus, while reaction synthesis of the as blended precursors following shock compression reveals a tendency to form stoichiometric TiC and no ternary carbide, the milled and shock densified compacts yield the ternary phase along with TiC following reaction synthesis. Furthermore, the non-stoichiometric TiC$_x$ phase formed during reaction synthesis of the milled, shocked compacts appears to be a TiC – Si solid solution having possibly formed by silicon diffusion into the TiC, which could be an intermediate phase prior to Ti$_3$SiC$_2$ formation. The intimate mixing during milling and the dense-packed highly activated state attained during shock compaction aid the solid-state diffusion of carbon into titanium and subsequently of silicon into TiC$_x$ prior to the formation of the ternary carbide.

To further verify the effect of shock compression on enhanced chemical reactivity and Ti$_3$SiC$_2$ phase formation, reaction kinetic studies were conducted in the DTA at varied heating rate (5, 10, 25,

40, 50, 65 °C/min). At low heating rates, a single broad peak was evident, characteristic of a solid state diffusion reaction. At high heating rates, two peaks were obvious – a solid state diffusion peak and a higher temperature peak from a self-propagating high temperature synthesis reaction (SHS). For SHS reactions, the rate of heat release, following a certain degree of solid state reaction, is faster than the rate of heat dissipation. The resulting retention of heat then triggers a catastrophic combustion type reaction causing all of the remaining reactants to immediately convert to products. Hence, depending on the activation induced by the shock compression process, both the reaction onset and peak temperatures and the degree of reaction by solid state and SHS mechanisms are expected to be manifested by the exotherms observed in the DTA traces. It is interesting to note that while the peak temperature of the solid state exotherm varied from 680 – 1024 °C depending on heating rate and shock pressure, the SHS exotherm was observed to be in the rather narrow range of 1320 – 1468 °C.

The modified Kissinger (1957) method, Boswell (1980), was used to determine the activation energy from the peak temperature of the reaction exotherm obtained from the DTA traces. The results illustrate that the activation energy of the solid state reaction decreases from 80 kJ/mole for the as blended powder to 71 kJ/mole in the as milled, and 58 kJ/mole in the case of the precursor shock compressed using the cylindrical implosion geometry (~ 4 GPa) and 68 kJ/mole for the 9 GPa gas gun sample. At higher shock pressures (9 GPa) a slight increase in activation energy is observed, possibly due to thermal annealing of the defects, although the entire reaction is observed to occur in the solid state. The results indicate that shock compression allows the generation of dense compacts of precursor powders, which can subsequently undergo controlled solid state reactions with accelerated kinetics.

3.2.2 Titanium-aluminum Nitride, Ti_2AlN

Reaction synthesis experiments on Ti + AlN have also been performed using the heat treatment similar to that used for the formation of Ti_3SiC_2 (Table 4). Reaction heat treatment of the 4 GPa samples showed the maximum formation of Ti_2AlN. However, the amount of the Ti_2AlN phase decreased from the 4 GPa to the 6 GPa sample indicating that there might be an optimum pressure window for shock activation. Reaction synthesis of the high pressure (9 GPa) sample showed formation of TiN and $Ti_3Al_2N_2$. The $Ti_3Al_2N_2$ ternary phase has a stacking sequence of ABABACBCBC and is typically known to be formed in a narrow temperature range (1200 – 1300 °C), Schuster and Bauer (1984). This phase does not exhibit the desirable properties of Ti_2AlN.

To further elaborate the effect of shock compression on reaction synthesis and Ti_2AlN formation, reaction kinetic experiments were conducted, to determine the activation energy for reaction, on Ti + AlN powder, in the as blended and as shocked (~ 4 GPa, 6 GPa and 9 GPa) states. The powders were heated in the DTA with varied heating rate (5, 10, 25, 40, 50, 65 °C/min). At all heating rates only a single broad peak is evident, characteristic of a solid state diffusion reaction. The activation energies were determined using the modified Kissinger (1957) method, Boswell (1980). Just like in the case of the Ti_3SiC_2 samples, the as blended mixture shows the highest value of the activation energy (97 kJ/mole), which is observed to be significantly lowered in the sample shock densified at 4 GPa (16 kJ/mole). However, the activation energy is observed to increase in the samples shock densified at higher pressures of 6-9 GPa (58 kJ/mole), possibly due to shock heating and annealing of defects. The overall trend of the effect of shock activation is similar to that observed with the XRD results described above.

4.0 CONCLUSIONS

Shock compression of Ti, SiC, and graphite powders (ball milled for 2 hours) at ~ 4 – 5 GPa showed formation of a dense packed highly activated state of reactants, while compression at a high pressure (~ 9 GPa) resulted in TiC formation in the ball milled compacts. Reaction heat treatment of ball milled and shock densified powder compacts resulted in formation of Ti_3SiC_2 and a TiC phase. The lattice parameter of the TiC phase was different from that of the stoichiometric value, suggesting that the TiC phase may be an intermediate state prior to Ti_3SiC_2 formation. These results along with those of the reaction kinetic studies to determine the activation energies of solid state and combustion type reactions illustrate that shock compression activates powder precursors and promotes the formation of the Ti_3SiC_2 phase.

Preliminary results of shock activated reaction synthesis of Ti + AlN showed the formation of Ti_2AlN. However, the amount of phase formed decreased with increasing shock pressure indicating that there is an optimum window of shock compression pressure in which shock activation can be beneficially used for the formation of Ti_2AlN.

ACKNOWLEDGEMENTS

This work is funded by DOD/ASSERT program through the Army Research Office, contract number DAAG55-98-1-0161.

REFERENCES

Arunajatesan S. and Carim A.H. (1994). Synthesis of Titanium Silicon Carbide. *Journal of the American Ceramic Society* **79:3**, 667-672.

Autodyn-2D. Century Dynamics Incorporated, Oakland, California, 1995.

Barsoum, M.W. and El-Raghy T. (1996). Synthesis and Characterization of a Remarkable Ceramic: Ti_3SiC_2. *Journal of the American Ceramic Society* **79:7**, 1953-1956.

Barsoum M.W., Brodkin D., and El-Raghy T. (1997). Layered Machinable Ceramics for High Temperature Applications. *Scripta Materialia* **36: 5**, 535-541.

Boswell, P.G. (1980). On the Calculation of Activation Energies Using a Modified Kissinger Method. *Journal of Thermal Analysis* **18**, 353 – 358.

Kissinger H.E. (1957). Reaction Kinetics in Differential Thermal Analysis. *Analytical Chemistry* **29: 11**, 1702 – 1706.

Lee J.H. (1997). Synthesis of TiC by Shock-assisted Solid-state Reaction Synthesis. Ph.D. Thesis, Georgia Institute of Technology.

Lis J., Pampuch R., Piekarczyk J., and Stobierski L. (1993). New Ceramics Based on Ti_3SiC_2. *Ceramics International* **19**, 219-222.

Lis J., Rudnik T., and Pampuch R. (1998). Controlled Reactions in SHS-derived Ti-Si-C Materials. *International Journal of Self-Propagating High-Temperature Synthesis* **7: 2**, 189-198.

Namjoshi S. (1999). Reaction Synthesis of Dynamically-Densified Ti-Based Intermetallic and Ceramic Forming Powders. Ph.D. Thesis, Georgia Institute of Technology.

Onodera A., Hirano H., Yuasa T., Guo N.F., and Miyamoto Y. (1999). A Soft but Incompressible Material: Ti_3SiC_2. *International Conference on High Pressure Science and Technology 1999 Abstracts*, 111.

Racault C., Langlais F., and Naslain R. (1994). Solid-state synthesis and characterization of the ternary phase Ti_3SiC_2. *Journal of Materials Science* **29**, 3384-3392.

Schuster J.C. and Bauer J. (1984). The Ternary System Titanium-Aluminum-Nitrogen. *Journal of Solid State Chemistry*, **53**, p. 260 – 265 (1984).

Thadhani N.N. (1993). Shock Induced Chemical Reactions and Synthesis of Materials. *Progress in Materials Science* **37**, 117-226.

HYPERVELOCITY IMPACT CONSOLIDATION OF MECHANICALLY ALLOYED POWDER IN THE Ni-Al-B SYSTEM

K. AYABE[1] and T. OKABE[2]

[1] Graduate Student, Hiroshima Institute of Technology, 2-1-1 Miyake, saeki-ku, Hiroshima 731-5193, Japan
[2] Department of Intelligent Machine Engineering, Faculty of Engineering, Hiroshima Institute of Technology, 2-1-1 Miyake, saeki-ku, Hiroshima 731-5193, Japan

ABSTRACT

Hypervelocity impact consolidation tests were performed at room temperature in expectation of a multiplication effect between mechanical alloying (MA) and combustion synthesis using MA powders composed of Ni-33.0mol%Al-$x$mol%B ($x=0$ and 1.0). The consolidation characteristics were examined as a function of MA milling duration. Some specimens underwent combustion synthesis directly on impact while others did not. The latter subsequently underwent a synthesis reaction by annealing at a comparatively low annealing temperature (near 673K). The density and micro Vickers hardness of the specimens annealed after impact were higher than those not annealed. The density reached a maximum at the optimum milling duration.

KEYWORDS

hypervelocity, impact consolidation, mechanical alloying, combustion synthesis, nickel aluminide, intermetallic compound, boron

INTRODUCTION

Combustion synthesis is a powder synthesis method that uses the heat of a chemical reaction. It is considered to be one of the most useful processes in the production of intermetallic compounds as mentioned by Miura & Mishima (1996). On the other hand, mechanical alloying (MA) is a method by which the constituent elements of an alloy are mixed mechanically and turned into an alloy in a solid state. By MA, the non-equilibrium and metastable phases (amorphous phases, quasi crystals, nano crystals, supersaturated solid solutions) are formed comparatively easily and improved control of sintering can be expected as explained by Hashimoto et al. (1997). In the consolidation of MA powder, it is desirable to maintain non-equilibrium and metastable phases. Therefore, it is necessary to consolidate the MA powder at as low a temperature as possible.

In this study, we performed hypervelocity impact consolidation tests at room temperature in expectation of a multiplication effect between MA and combustion synthesis using MA powders in the Ni-Al-B system as described by Ayabe & Okabe (2001). We examined the possibility that both types of synthesis and the consolidation take place simultaneously. The density and micro Vickers hardness of the consolidated specimens were measured as a function of MA milling duration.

EXPERIMENTAL PROCEDURE

The initial materials were the commercial powders Ni (99.8% purity, -325 mesh particle size), Al (99.9% purity, -100 mesh particle size) and B (99.0% purity, -300 mesh particle size), which were obtained from the Furuuchi Chemical Corporation. They were mixed to Ni-33.0mol%Al-$x$mol%B (x=0 and 1.0) in a glove box with Ar gas atmosphere. These data are summarized in Table 1. The powder mixture of 10.68g was enclosed in 80cc Cr steel (Fe-12.0mass%Cr-2.1mass%C-0.3mass%Si-0.3mass%Mn) vial (inside measurement), together with twenty-seven 10mm diameter and 4.00g Cr steel balls (a ball-to-powder weight ratio of approximately 10) with stearic acid (0.030g) as the lubricant. The powder mixture was mechanically alloyed in Ar gas by a planetary ball mill (the P-5 type made by the Fritsch Corporation) for various milling times (10.8, 21.6, 43.2, 86.4 and 172.8ks) with the number of revolutions of 170rpm (a rotation-to-revolution number ratio of 2). The MA was carried out with the cycle set to rest for 1.8ks every 5.4ks. The characteristics of the MA powder were analyzed by XRD and SEM, and the MA powder was used for the hypervelocity impact

TABLE 1
DATA FOR RAW POWDERS AND CHEMICAL COMPOSITION OF MIXED POWDERS

Element	Purity(%)	Mesh	Ni-33.0mol%Al	Ni-33.0mol%Al-1.0%mol%B
Ni	99.8	-325	Bal.	Bal.
Al	99.9	-100	33.0	33.0
B	99.0	-300	0	1.0

consolidation tests performed by a hypervelocity impact machine made by the Tokyo-koki Seizosho Corporation. The MA powder was packed in a stainless steel (JIS: SUS304) pipe (10mm outside diameter, 1.0mm thickness, 4.0mm length). The impact specimens were prepared by pre-consolidation under uniaxial static compression stress of 600MPa. Then, as illustrated in Figure 1, an impact resisting tool made of tool-steel (JIS: SKS3) was placed on the specimen and the impact test was performed with a bullet weight of 150g and bullet speed of 57m/s. Consolidation characteristics were investigated as a function of MA milling duration, measuring the density and micro Vickers hardness of the specimens.

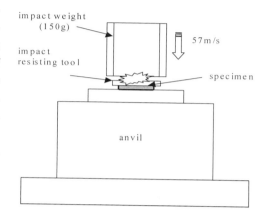

Figure 1: Schematic illustration of hypervelocity impact compression of MA powders.

RESULTS AND DISCUSSION

Characteristics of the MA Powder

The MA process was investigated using SEM and XRD. The SEM micrographs (back scattered images) for various milling times of Ni-33.0mol%Al-1.0mol%B powders mechanically alloyed are shown in Figure 2. The powder particles themselves are flattened and fragmented with a milling time of 10.8ks, though the powders agglomerate partly. Considerable refinement and reduction in particle size are evident with a milling time of 86.4ks. However, the longer milling duration (172.8ks)

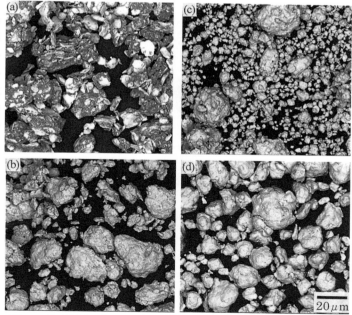

Figure 2: SEM micrographs of Ni-33.0mol%Al-1.0mol%B powders mechanically alloyed for various milling times: (a)10.8ks, (b) 43.2ks, (c)86.4ks, and (d)172.8ks.

causes powder particle coarsening, where serious cold welding takes place and the percentage recovery of powder decreases remarkably. The results without B (Ni-33.0mol%Al) were almost the same as those described above.

The XRD patterns of the MA powders were examined after each MA stage. The results obtained are shown in Figure 3 (with B). These figures show sharp peaks of Ni and Al at the early stage of milling that were later lowered and broadened. However, the peaks of the intermetallic compounds (NiAl, Ni$_3$Al, etc.) were minimal within the range of milling duration in this experiment.

Hypervelocity Impact Consolidation of MA Powder

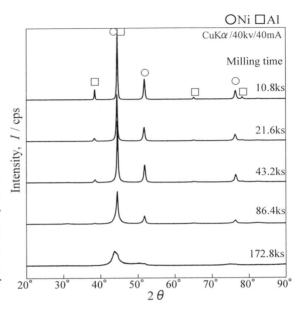

Figure 3: XRD patterns of Ni-33.0mol%Al-1.0mol%B powders mechanically alloyed for various milling times.

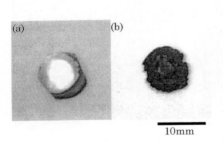

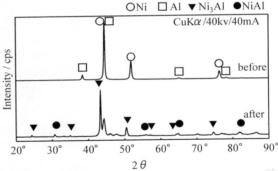

Figure 4: Aspect of the specimen after the impact test: (a) no combustion synthesis and (b) combustion synthesis.

Figure 5: XRD patterns of the specimen before and after the impact test, which indicates combustion synthesis.

When the hypervelocity impact consolidation tests were performed at room temperature, we found that some specimens, especially without B, underwent remarkable combustion synthesis. The specimen shown in Figure 4 is an example (milling time of 21.6 ks). It was quite hot to the touch after the tests and was discolored. It seemed to combustion synthesized, but the relative density was only 79%. The XRD patterns before and after the impact tests are shown in Figure 5. It was found that intermetallic compounds (NiAl and Ni$_3$Al) were formed by the impact. However, this combustion synthesis phenomenon by the impact was infrequent. Specimens that did not undergo combustion synthesis were vacuum annealed after the impact, and the heat treatment effects were examined. The XRD patterns after annealing at 673K are shown in Figure 6. It can be seen from Figure 6 that intermetallic compounds such as NiAl and Ni_3Al are formed by the heat treatment even at this relatively low temperature. This suggests that greater energy and more lattice defects remain stored in the impact specimens. The MA increases the driving force for a change in state as explained by Lü & Lai (1998). Thus, the specimens are considered to have formed a plentiful product (intermetallic compounds) easily by the heat treatment (annealing).

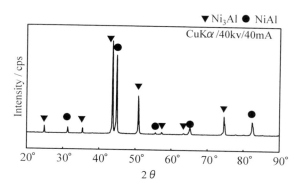

Figure 6: XRD patterns of the specimen after annealing at 673K, where the milling time is 43.2ks.

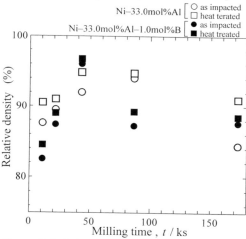

Figure 7: Relation between relative density and milling time for the specimens after the consolidation by impact and after the vacuum annealing.

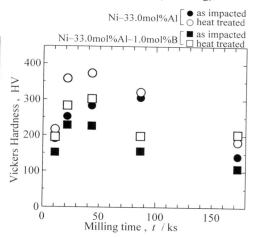

Figure 8: Relation between Vickers hardness and milling time for the specimens after the consolidation by impact and after the vacuum annealing.

Finally, we measured the density and micro Vickers hardness of the specimens after the consolidation by impact and after the vacuum annealing. The results are shown in Figure 7 (relative density) and Figure 8 (micro Vickers hardness) as a function of milling time. From these figures, it can be seen that the density and hardness with annealing are higher than those for impact alone.

There is an optimum milling time for high density. Milling times that are too long or too short do not give good results for density. However, the hardness of the specimens for both impact and annealing does not appear to depend on the milling duration. The effect of B was unclear in this experiment. It is interesting to note that the density of the specimens in which combustion synthesis was induced by the impact was lower than the annealed ones. This is probably because many pores develop when sudden combustion synthesis is induced by impact.

CONCLUSIONS

Hypervelocity impact consolidation tests were performed at room temperature in expectation of a multiplication effect between MA and combustion synthesis. Using MA powders in the Ni-Al-B system, some specimens underwent combustion synthesis directly on impact while others did not. The latter did undergo a synthesis reaction by annealing even though the annealing temperature was comparatively low (673K). The density and micro Vickers hardness of the specimens annealed after the impact were higher than those that were not annealed. There was an optimum (not too short and not too long) milling time for high density. The effect of B on impact consolidation characteristics was unclear in this experiment. More detailed investigation is needed.

ACKNOWLEDGMENTS

This work was supported in part by a Grant-in-Aid from the Ministry of Education, Culture, Sports, Science and Technology, Japan provided to the Academic Frontier Research Center (projected duration 1997-2002) founded at Hiroshima Institute of Technology.

References

Ayabe K. and Okabe T. (2001). Effect of Mechanical Alloying on Combustion Synthesis in the Ni-Al-B System. *J. Japan Inst. Metals* **65:3,** 195-198.
Hashimoto H. Ho Park Y. and Abe T. (1997). Mechanical Alloying. *Materia Japan* **36:10,** 1021-1025.
Lü L. and Lai M. O. (1998). *Mechanical Alloying*, Kluwer Academic Publishers, Massachusetts, USA
Miura S. and Mishima Y. (1996). Estimation of Amount of Liquid Phase During Reaction Synthesis in the Ni-Al System. *Materia Japan* **35:6,** 632-636.

PREPARATION OF Co-Cu METASTABLE BULK ALLOY BY MA AND SHOCK COMPRESSION

Xu FAN[1], Tsutomu MASHIMO[1], Yuyang ZHANG[1] and Akira CHIBA[2]

[1] Shock Wave and Condensed Matter Research Center, Kumamoto University, Kumamoto 860-8555, Japan
[2] Faculty of Engineering, Kumamoto University, Kurokami 2-39-1, Kumamoto 860-8555, Japan

ABSTRACT

Metastable solid solution alloy powders and the bulk bodies in the $Co_xCu_{(100-x)}$ (x=10, 20, 30, 40, 50, 60, 70, 80, and 90) system, which was an almost immiscible system at ambient state, were prepared by mechanical alloying (MA) and shock compression. The MA-treated powders (for 21 hours) showed the X-ray diffraction (XRD) patterns of a single phase of face-centered cubic (FCC) structure, in which the lattice parameter changed with Cu contant almost in accordance with Vegard's law. The XRD patterns of the shock-consolidated bulk bodies did not change much from those of the MA-treated powders. This showed that the metastable solid solution phases were successfully consolidated without decomposition or recrystallization.

KEYWORDS

Mechanical alloying; Shock compression; Metastable bulk alloy; Cobalt; Copper; Magnetic property.

INTRODUCTION

Metastable materials are expected to offer a variety of new properties and applications. Mechanical alloying (MA) has recently been used for nonequilibrium materials processing, including the preparation of amorphous phases, metastable solid solution phases, nanocrystals, high pressure phases. It is important to consolidate the nonequilibrium material powders for evaluations of physical properties, and for industrial applications. By using shock compression, we can consolidate metastable material powders without recrystallization or decomposition due to pulsed short duration and high pressure.

The equilibrium phase diagram of the cobalt (Co)-copper (Cu) system shows virtually no solubility of Co in Cu, and Cu in Co, below 500℃. Many works have been done to produce a metastable solid solution in Co-Cu binary alloy system. Properties of the resulting powders or thin film in this system already have been reported. Gente et al. (1993) produced Co-Cu solid solution alloy powders by MA, and discussed the thermodynamic mechanisms in the process. Yoo (1999), Uimin (1999) and Aizawa et al. (1998) studied the atomic structure and magnetic properties of Co-Cu system.

But, the physical properties, such as the magnetic property of the Co-Cu solid solutions are not well known. The Slater-Pauling carve has not included the data of Co-Cu system. By using the bulk form specimens, the magnetic property can be more precisely measured, and the electrical conductivity or thermal expansion coefficient etc. can be investigated. In this study, the MA treatment and shock-compression recovery experiments were performed to prepare the bulk bodies of metastable solid solution alloy in Co-Cu system.

EXPERIMENT

Starting powders were provided by Rare Metallic Co., Ltd. The Co and Cu powders consisted of irregular particles of 1-2 mm and 325 mesh (<44 mm) in diameter, and the purity of Fe and Cu in catalog were 99.9 wt% and 99.99 wt%, respectively. The MA experiments were carried out by using the planetary micro ball mill (P-7 of Fritsch Co., Ltd.) in an argon atmosphere grove box [4]. A mill capsule with an inner-diameter of 41 mm and a depth of 38 mm and balls with a diameter of 5 mm were used, which were made of silicon nitride (Si_3N_4) and zirconia (Y_2O_3-doped tetragonal ZrO_2), respectively. The starting powder with a weight of 20 grams and 200 zirconia balls were contained into the capsule with a ball-to-powder weight ratio of 4 : 1. The rotation speed of the ball mill was 2840 rpm. The resultant acceleration was estimated to be about 12 g (1g=9.8 m/s^2). The milling was interrupted each 30 min for 35 min to cool mill capsule to avoid the hesting. The milling duration was 21 hours, and small amounts of the material were taken for further analysis after selected milling times.

Shock-compression recovery experiments were conducted using a propellant gun. The MA-treated powder specimens were enclosed in a brass (Cu:Zn=70:30 in wt%) capsule with an inside diameter of 12 mm and with an inside height of 3.0-6.8 mm. The porosities of pellets were 40-54 %. Shock loading was carried out by impacting the capsule with an aluminum alloy (2024Al) flat flyer plate whose thickness was 2.5 mm.

The MA-treated and shock-consolidated specimens were investigated by powder X-ray diffraction (XRD), and Electron Probe Micro Analysis (EPMA). Powder XRD analyses were carried out using monochromatized Fe-Kα radiation with a Rigaku Goniometer. Calibration of the goniometer were performed by measuring diffraction peaks of pure silicon powder mixed in specimen. The lattice parameter was caculated by using Cohen's method, and the extrapolation function $\cos^2 q/\sin q$ was used for extrapolating to zero diffraction angle. The analyses of the Fe and Cu content distribution were carried out for the shock-consolidated bulk bodies using the EPMA apparatus, JXA-8900 of JEOL LTD.

RESULTS AND DISCUSSION

MA-treatment

The X-ray diffraction (XRD) patterns of the starting powder and the MA-treated powders in the $Co_{50}Cu_{50}$ systems are shown in Fig. 1. Even after 3h of milling, the Co peaks have fully disappeared, and the broaden peaks of fcc phase can be observed. Figure 2 shows the changes in lattice parameter of the fcc phase as a function of milling time in the $Co_{20}Cu_{80}$, $Co_{50}Cu_{50}$ and $Co_{80}Cu_{20}$ systems. The lattice parameters of the fcc solid solutions remarkably decreased during the initial stages of MA treatments (MA 3h) and approached the smallest values of about 0.36091, 0.35836 and 0.35632 nm in the $Co_{20}Cu_{80}$, $Co_{50}Cu_{50}$ and $Co_{80}Cu_{20}$ systems, respectively. It was found that the lattice parameter changed with Cu contant almost in accordance with Vegard's law, which showed that true alloying took place.

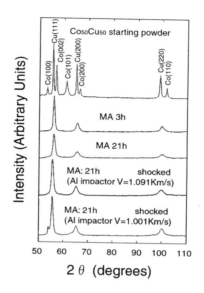

Figure 1: XRD patterns of the starting material, the MA-treated powders and the shock-consolidated bulk bodies in $Co_{50}Cu_{50}$ system

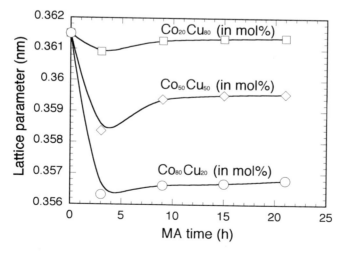

Figure 2: Changes of the lattice parameters as a function of milling time in $Co_{20}Cu_{80}$, $Co_{50}Cu_{50}$ and $Co_{80}Cu_{20}$ system

Shock consolidation

Figure 3 shows a photograph of the shock-consolidated alloy bulk body ($Co_{50}Cu_{50}$, Al impactor V=1.00 km/s) with a diameter of about 12 mm, and a thickness of about 2.5 mm. No large crack could be observed, and the cross sections of bulk bodies show a metallic gloss. The morphology of the polished surfaces of the shock-consolidated bulk bodies appeared almost as a uniform single phase over the entire surface. The EPMA results of the shock-consolidated bulk bodies showed that Co and Cu dispersed well at the submicron level in the bulk bodies.

Figure 3: Photograph of the shock-consolidated bulk body formed at an impact velocity of 1.00 km/s in $Co_{50}Cu_{50}$ system

The XRD pattern of the shock-consolidated bulk body in the $Co_{50}Cu_{50}$ system formed at an impact velocity of 1.00 km/s did not change much from that of the MA-treated powder, as show in Fig. 1. This showed that the metastable solid solution powder was successfully consolidated without decomposition or recrystallization. But, a weak peak of Co (100) was observed in the XRD patterns of those formed at the impact velocities of 1.091 km/s. This may be due to decomposition and/or recrystallization caused by the higher residual temperature after shock compression.

CONCLUSIONS

The bulk bodies of metastable solid solutions alloys over the whole concentration range in the Co-Cu system were prepared by mechanical alloying and shock compression. The lattice parameters of the MA-treared powders were smaller than that of pure Cu, and were larger than that of pure Co, respectively. The X-ray diffraction patterns of the shock-consolidated bulk bodies in a certain low pressure range did not change from those of the MA-treated powders. The bulk alloys of fcc metastable solid solution are expected to show advanced magnetic or electrical properties.

References

Gente C., Oehring M. and Bormann R. (1993). Formation of thermodynamically unstable solid solutions in the Cu-Co system by mechanical alloying. *Physical Review B* **48,** 13244-13252.

Ivchenko V.A., Uimin M.A., Yermakov A.Y., Korobeinikov A.Y. (1999). Atomic structure and magnetic properties of Cu80Co20 nanocrystalline compound produced by mechanical alloying. *Surface Science* **440,** 420-428.

Yoo Y.G., Yang D.S., Yu S.C., Kim W.T., Lee J.M. (1999). Structural and magnetic properties of mechanically alloyed $Co_{20}Cu_{80}$ solid solution. *Journal of Magnetism and Magnetic Materials* **203:1-3,** 193-195.

Zhou Cheng, Aizawa T., Tokumitsu K., Tatsuzawa K. and Kihara J. (1998). Magnetic Properties of Nano-Particulate Cu-Co Alloy on the Route of Bulk Mechanical Alloying. *Materials Scence Forum* **269-272,** 913-918.

EXPLOSIVE WELDING OF BULK METALLIC GLASS PLATE ON CRYSTALLINE TITANIUM PLATE

Y. Kawamura and A. Chiba

Department of Mechanical Engineering and Materials Science, Kumamoto University
2-39-1 Kuro-kami, Kumamoto 860-8555, Japan

ABSTRACT

Recently, a number of amorphous alloys having a wide supercooled liquid region before crystallization and high glass-forming ability have been discovered. Such amorphous alloys can be fabricated directly from the melt in a bulk form with a thickness of ~10 mm at slow cooling rates of the order of 1-100 K/s. The new amorphous alloys are, therefore, called "bulk metallic glasses". The bulk metallic glasses have solved two major problems in amorphous alloys, namely, the limitation of product size and the lack of workability. On the contrary, the problem of welding is not yet adequately solved. For establishment of metallurgical bonding technology of bulk metallic glasses, we have tried to weld $Zr_{55}Al_{10}Ni_5Cu_{30}$ (at%) bulk metallic glass having a wide supercooled liquid region and high glass forming ability by explosive welding. We have succeeded in joining bulk metallic glass plate on polycrystalline titanium metallic plate. No crystallization was observed in the interface. No visible defect was recognized at the interface, showing an achievement of metallurgical bonding of bulk metallic glasses. The successful results obtained in this study are expected to push forward the application of bulk metallic glasses.

KEYWORDS

amorphous alloy, metallic glass, joining, explosive welding, cladding

INTRODUCTION

Amorphous alloys have useful properties such as high strength, high stiffness and good soft-magnetic properties (Masumoto (1994)). The amorphous alloys, however, have a deficiency of weldability as well as a limitation of product size and a lack of workability and machinability. Recently, a number of amorphous alloys having a wide supercooled liquid region before crystallization and high glass-forming ability have been discovered (Masumoto (1994), Inoue (1998)). Such amorphous alloys can be fabricated directly from the melt in a bulk form with a thickness of ~10 mm at slow cooling rates of the order of 1-100 K/s. The new amorphous alloys are, therefore, called "bulk metallic glasses". The bulk metallic glasses exhibit high-strain-rate superplasticity (Kawamura et al. (1999), Kawamura et al. (1997), Kawamura et al. (1998), Kawamura and Inoue (2000)) and excellent workability

(Kawamura et al. (1997)) in the supercooled liquid state. Powder metallurgy processing using the superplasticity, moreover, enables the production of large-scaled bulk metallic glasses with the same tensile strength as the cast bulk and melt-spun ribbon (Kawamura et al. (1995), Kawamura et al. (1997), Kawamura et al. (1996)). These results demonstrate that the bulk metallic glasses can solve two major problems in amorphous alloys, namely, the limitation of product size and the lock of workability. In 1998, the bulk metallic glasses produced by metallic mold casting of melts or consolidation of gas-atomized glass powders have been put on the market as a face material of newly designed golf club heads (Johnson (1999), Inoue (1998)). On the other hand, Kawamura et al. have succeeded in welding of bulk metallic glasses by friction and pulse-current methods (Kawamura and Ohno (2001), Kawamura and Ohno (2001), Kawamura and Ohno (2001)). The problem of welding is, however, not yet adequately solved because the unsuitable for welding of large plates.

The crystallization nose of Time-Temperature-Transition (TTT) diagram of the bulk metallic glasses has been reported to be in the range of 1 to 100 s (Johnson (1999), Inoue (1998)). Explosive welding is expected to be sultable for cladding the bulk metallic glasses. In the explosive welding, there is almost no diffusion of alloying elements between components. Moreover, there is no time for heat transfer to the component metals and no appreciable temperature increase in the metals, resulting in no heat-affected zones. The welding duration of the explosive welding can be controlled below several tens μs (Metals Handbook (1979)). The explosive welding is, therefore, expected to enable welding the bulk metallic glasses without a devitrification. We have tried explosive welding of a $Zr_{55}Al_{10}Ni_5Cu_{30}$ bulk metallic glass that has the highest glass forming ability in Zr-Al-Ni-Cu system. In this paper, we will report a successful result of cladding the $Zr_{55}Al_{10}Ni_5Cu_{30}$ bulk metallic glass plate on crystalline titanium metallic plate by the explosive welding.

EXPERIMENTAL PROCEDURE

$Zr_{55}Al_{10}Ni_5Cu_{30}$ ingots were prepared by arc melting a mixture of the pure elements. The ingots were remelted in an evacuated quartz tube using an induction-heating coil and then injected though a nozzle into a copper mold using high purity argon gas. The dimensions of the obtained bulk metallic glasses were 2.5 mm in thickness, 12 mm in width and 40 mm in length. Specimens with 2 mm in thickness, 10 mm in width and 20 mm in length were prepared by machining the cast bulk metallic plates. The formation of a single glassy phase was confirmed by X-ray diffractometry. Thermal properties were investigated by differential scanning calorimetry (DSC).

Figure 1 shows a schematic illustration of the explosive welding assembly. We used a parallel-plate explosive welding process. The bulk metallic glass plate was mounted on a copper block with 10 mm in thickness, 40 mm in width and 90 mm in length. A pure titanium plate with 3 mm in thickness, 40 mm in width and 90 mm in length was set below an explosive with 30 mm in thickness, 40 mm in

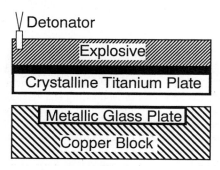

Figure 1 : Schematic illustration of an employed parallel-plate explosive welding process. Explosive cladding assembly before (a), during (b) and after (c) detonation.

width and 90 mm in length. The distance between the bulk metallic glass plate and the titanium plate was about 10 mm. The explosive used in this study was PAVEX consisting mainly of ammonium nitrate, which was provided by Asahi-Kasei Corporation. The detonation speed of the explosive was estimated to be about 2400 m/s.

The joining was evaluated by optical microscopy (OM) and scanning electron microscopy (SEM) of the polished cross-sectional area of the interface in the welded specimens. The glassy structure was investigated by micro-area X-ray diffractometry using Cr$K\alpha$ radiation. The diameter of the X-ray beam was 100 µm. The interface was investigated using EPMA.

RESULTS AND DISCUSSION

Thermal properties of the $Zr_{55}Al_{10}Ni_5Cu_{30}$ bulk metallic glass are listed in Table 1. The melting, glass transition and crystallization temperatures are 1200 K, 653 K and 753 K, respectively. The nose time can be estimated to be 1 to 10 s.

Figure 2 shows the optical micrograph of the polished cross-sectional area of the $Zr_{55}Al_{10}Ni_5Cu_{30}$ bulk metallic glass welded to polycrystalline titanium metal. The bulk metallic glass has a high strength of 1500 MPa, a small Young's modulus of 80 GPa and a large elastic strain of 0.02 at ambient temperature, and deforms inhomogeneously. The inhomogeneous deformation is localized in discrete and thin shear bands because of its non-hardenable nature. No clack was observed in the welded bulk metallic glass plate. There is no visible defect or pore in the interface. A wavy pattern that is typical bond zone morphology in explosive welding was observed in the bond zone (Metals Handbook (1979)). Figure 3 shows the scanning electron micrograph of the polished cross-sectional area of the explosively welded sample. There is no visible defect or pore in the interface. This reveals that metallurgical bonding was achieved successfully by the explosive welding method.

TABLE 1
THERMAL PROPERTIES OF $Zr_{55}Al_{10}Ni_5Cu_{30}$ BULK METALLIC GLASS

T_g	T_x	ΔT_x	T_m
652 K	757 K	105 K	1121K

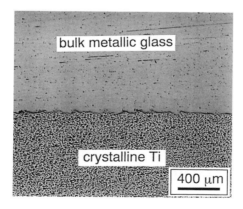

Figure 2 : Optical micrograph of the bonding interface in the $Zr_{55}Al_{10}Ni_5Cu_{30}$ bulk metallic glass plate explosively welded on a crystalline Ti plate.

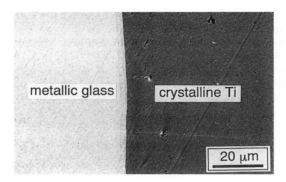

Figure 3 : Scanning electron micrograph of the bonding interface in the $Zr_{55}Al_{10}Ni_5Cu_{30}$ bulk metallic glass plate explosively welded on a crystalline Ti plate.

Figure 4 shows the micro-area X-ray diffraction patterns of the polished cross-section of the interface in the explosively welded specimen. The positions of the X-ray beam are represented in the inset of Fig. 4. The diffraction pattern taken from the bulk metallic plate (S3) consisted a halo pattern that is a typical diffraction pattern of amorphous materials. This result shows an existence of only glassy phase. No detectable amounts of other phases were observed. The diffraction pattern taken just from the interface (S2) exhibited many peaks superimposed on a halo pattern. The diffraction peaks corresponded to those of an α-titanium crystalline phase. Except for the amorphous and α-titanium phases, no detectable amount of any crystalline phases formed by devitrification of the bulk metallic glass were observed. These results reveal that the bulk metallic glass kept the amorphous structure and no devitrification occurred in the interface.

Figure 5 shows EPMA profiles of Ti, Zr and Al elements on the polished cross-section of the interface in the explosively welded sample. There is no diffusion of alloying elements between the $Zr_{55}Al_{10}Ni_5Cu_{30}$ bulk metallic glass plate and pure titanium plate. It has been reported that an amorphous phase was formed in the interface of explosively welded Ti/steel samples and the thickness of the amorphous layer was within 200 nm (Nishida et al. (1995)). This reveals that the shallow interface can be melted and followed by rapid solidification under a higher cooling rate than the ordinary melt spinning method. Moreover, it has been previously reported that the addition of 5 at% Ti to $Zr_{55}Al_{10}Ni_5Cu_{30}$ alloy improves the glass forming ability. The melting and Ti-alloying of the shallow interface between the $Zr_{55}Al_{10}Ni_5Cu_{30}$ bulk metallic glass plate and Ti plate may result in no crystallization during the explosive welding process.

As described above, we have succeeded in cladding of bulk metallic glass plate on crystalline Ti plate by explosive welding. The explosive welding seems to make possible clad a bulk metallic glass to the same bulk metallic glass or another bulk metallic glass. In 2001 it has been reported that bulk metallic glasses were successfully welded to bulk metallic glasses by friction and pulse-current methods. The friction welding is restricted to axis-symmetric components. The pulse-current welding is applicable to only precious components. These methods are unsuitable to clad large bulk metallic glass plates. However, the explosive welding enables to clad bulk metallic glasses to crystalline materials through a large area. The success of explosive welding is encouraging for the future applications of bulk metallic glasses.

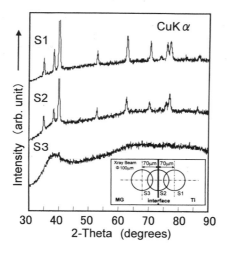

Figure 4 : Micro-focused X-ray diffraction patterns taken from the Ti plate (S1), the interface (S2) and the bulk metallic glass plate (S3) on the polished cross section of the bulk metallic glass plate explosively welded on a crystalline Ti plate. The positions of the X-ray beam are represented in the inset.

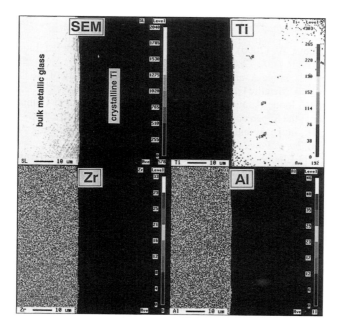

Figure 5 : EPMA profile of Ti, Zr and Al elements on the bonding interface in the $Zr_{55}Al_{10}Ni_5Cu_{30}$ bulk metallic glass plate explosively welded on a crystalline Ti plate. The SEM micrograph of the polished cross-section where the EPMA profiles were taken is shown in the Fig. 5.

CONCLUSIONS

We have succeeded in cladding a $Zr_{55}Al_{10}Ni_5Cu_{30}$ bulk metallic glass plate on a polycrystalline Ti plate by explosive welding. No crystallization was observed in the bonding interface. No visible interface was, moreover, recognized. Given the possibility of welding bulk metallic glasses, one can expect that the bulk metallic glasses will make further progress in diverse applications.

ACKNOWLEGEMENT

We would like to acknowledge Mr. T. Yamamuro of Kumamoto University for his kind assistance with micro-area X-ray diffractmetry.

REFERENCES

Inoue A. (1998), *Bulk Amorphous Alloys, Preparation and Fundamental Characteristics*, Trans Tech Publications, NH, pp.1-37.
Johnson W. L. (1999). Bulk Glass-Forming Metallic Alloys: Science and Technology, *MRS Symp. Proc.* **554**, 311.
Kawamura Y., Nakamura T. and Inoue A. (1999). High-Strain-Rate Superplasticity due to Newtonian Viscous Flow in La55Al25Ni20 Metallic Glass, *Mater. Trans. JIM* **40**, 749.
Kawamura Y., Shibata T., Inoue A. and Masumoto T. (1997). Superplastic Deformation of $Zr_{65}Al_{10}Ni_{10}Cu_{15}$ Metallic Glass, *Scripta Metall. Mater.* **37**, 431.
Kawamura Y., Nakamura T. and Inoue A. (1998). Superplasticity in $Pd_{40}Ni_{40}P_{20}$ Metallic Glass, *Scripta Metall. Mater.* **39**, 301.
Kawamura Y. and Inoue A. (2000). Newtonian Viscosity of Supercooled Liquid in a $Pd_{40}Ni_{40}P_{20}$ Metallic Glass, *Appl. Phys. Lett.*, **77**, 1114.
Kawamura Y., Shibata T., Inoue A. and Masumoto T. (1997). Workability of the Supercooled Liquid in the $Zr_{65}Al_{10}Ni_{10}Cu_{15}$ Bulk Metallic Glass, *Acta Mater.* **46**, 253.
Kawamura Y., Kato H., Inoue A. and Masumoto T. (1995). Full Strength Compacts by Extrusion of Glassy Metal Powder at the Supercooled Liquid State, *Appl. Phys. Lett.* **67**, 2008.
Kawamura Y., Kato H., Inoue A. and Masumoto T. (1997). Fabrication of Bulk Amorphous Alloys by Powder Consolidation, *Int. J. Powder Metallurgy.* **33**, 50.
Kawamura Y., Kato H., Inoue A. and Masumoto T. (1996). Effects of Extrusion Conditions on Mechanical Properties in $Zr_{65}Al_{10}Ni_{10}Cu_{15}$ Glassy Alloy Compacts, *Mater. Sci. Eng.*, **A219**, 39.
Kawamura Y. and Ohno Y. (2001). Metallurgical Bonding of Bulk metallic Glasses, *Mater. Trans.*, **42**, 717.
Kawamura Y. and Ohno Y. (2001). Superplastic Bonding of Bulk Metallic Glasses by Friction, *Scripta. Metall*, **45**, 279.
Kawamura Y. and Ohno Y. (2001). Pulse-Current Welding of Bulk Metallic Glasses, *Scripta. Metall*, **45**, in press.
Masumoto T. (1994). Recent Progress in Amorphous Metallic Materials in Japan, *Mater. Sci. Eng.* **A181/A182**, 8.
Metals Handbook, 9th ed. (1979). Vol. 6, American Society for Metals, pp.467-580.
Nishida M, Chiba A., Honda Y., Hirazumi J. and Horikiri K. (1995). Electron Microscopy Studies of Bonding Interface in Explosively Welded T/Steel Clads, *ISIJ International*, **35**, 217.

JET INITIATION OF HIGH EXPLOSIVE CHARGES BY LEAD COVERS IN DIRECT CONTACT

M. Held
TDW – 86523 Schrobenhausen, Germany

ABSTRACT

The jet initiation behavior with lead covers in contact to normal cast TNT/RDX 35/65 charges with their initiation behavior is observed. The supersonic penetration with drastically reduced bulging effect on the surface to the high explosive charge gives a more sensitive high explosive behavior with reduced build-up distances and shorter delay and initiation times compared to steel covers.

KEYWORDS

Initiation, shaped charge jet initiation, covered high explosives, lead cover, high speed diagnostic, build-up distances, delay times, initiation times, radial breakthroughs.

INTRODUCTION

A lot of tests with jet initiations are performed and published by Held (1968, 1980, 1987a, 1987b, 1989) and by others, where especially Chick & Hatt (1981a, 1981b, 1983) have made a lot of fundamental investigations to this topic. But using flash X-ray as a diagnostic tool the output of threshold values with respect to the distances of the first radial breakthrough and the achieved delay times are only rough achievable compared to the diagnostic with ultra high speed streak and framing cameras (Held, 1993), especially if they are simultaneously achieved (Held, 1990).

Already in a very early investigation Zernow (1955) has found that "retonation" can occur after a buildup distance in the by shaped charge jets initiated covered high explosive charges.

Interesting was the finding by Chick & Hatt (1985) that arrangements with an air gap between a cover plate and a high explosive charges or with slits in the high explosive charge are much more sensitive to directly covered high explosive charges without any air gaps.

The author (Held 1987a, b, c) explains differently this phenomena compared to Chick & Hatt (1985). Chick & Hatt have given a desensitation of the loaded high explosive charge by the earlier arriving shock wave, whereas the author has given the reason to a precompression of the acceptor charge by the bulge of the cover plate and to the load by a ramp wave which has a rise time of a few microseconds compared to the steep rise by direct jet impact, and finally to a larger loaded area which comes from the spall fragments from the exit hole of the barrier.

Different barrier materials are not investigated in the fundamental studies of the jet initiation of high explosive charges. The author (Held 2001) has compared steel with ceramics and lead with two different jet load conditions. Here should be especially presented the results with lead covers under different shaped charge jet load conditions.

2. TEST SETUP

The three different investigated test setups with lead materials as barrier in contact to the cast TNT/RDX 35/65 charge are presented in Fig. 1. It was used the 44 mm standard shaped charge for such tests of the author with wave shaper and a jet tip velocity of 8.3 mm/μs. To see the reproducibility of the tests the first test arrangement I was repeated with a total distance from virtual origin to the high explosive charge of 158 mm and a lead thickness of 42.3 mm. The residual jet velocity v_{jr} for this arrangement can be calculated to 5.8 mm/μs (Held 1991a). In the arrangement II the thicker barrier in the 158 mm distance consumes more jet velocity. Therefore the residual jet velocity is reduced from 5.8 mm/μs to 4.6 mm/μs.

The same lead thickness of 64.9 mm, but arranged at a longer standoff, gives a little more time for stretching of the jet and therefore an increased residual jet velocity of 5.1 mm/μs at the arrangement III.

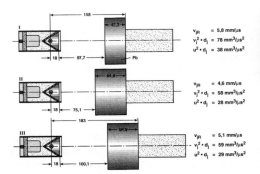

Fig. 1: Jet initiation of lead covered high explosives with different loads

Fig. 2: Test setup of jet initiation with a modified diagnostic arrangement

Fig. 2 shows the from the author typically used test setup with the shaped charge on the left side, the lead barrier more or less in the middle of the picture with the direct attached high explosive charge of 100 mm length and 48 mm diameter. For better observation with the frame and streak camera the barrier and the charge is back-illuminated by an argon balloon. A 45° mirror is arranged on the right side for the observation of the detonation breakthrough with a double flash gap technique (Held 1970).

In all test conditions the jet was continuously stretching. The residual jet velocities v_{jr} can be calculated with the following equation 22 of Held (1991a), where γ is the square root of the target to jet density.

$$v_{jr} = v_{j0}(Z_0 /(P+Z_0))^\gamma \qquad \gamma = \sqrt{\rho_t / \rho_j} \qquad \gamma = \sqrt{11{,}4/8{,}9} = 1{,}132$$

The times of the jet impact on the high explosive charge can be simply calculated with

$$t_a = (Z_0+P)/v_{jr} \tag{3}$$

These times are shorter than the particulation times of the corresponding jet velocities. The diameter of the particulized jet d_{jP} was analyzed from flash X-ray pictures as function of jet velocity some time ago. The mass conservation is used for defining the diameter d_j of the continuously stretching jet.

$$d_j = d_{jP} \cdot \sqrt{t_P/t_{ji}} \tag{4}$$

The calculated values d_j are presented also in table 1. With these values the v^2d loads can be calculated. The cratering velocity is defined by the Bernoulli equation (Held 1991a) with

$$u = v_j/(1+\gamma) \tag{5}$$

This velocity belongs to the penetration in the HE charge with a copper jet which leads to the γ_{HE} value of 0.43 and to $u = 0.7\ v_j$

The v^2d_j values have to be multiplied with 0,49 to achieve the u^2d_j values. All these data are summarized in the Table 1 and are partially listed also in Fig. 1.

TABLE 1

Fall	$P+Z_0$ (mm)	Z_0 (mm)	P (mm)	v_{jr} (mm/µs)	d_{jP} (mm)	t_{jP} (µs)	t_a (µs)	d_j (mm)	$v_{jR}^2 d_j$ (mm³/µs²)	$u^2 d_j$ (mm³/µs²)
I	158	115.7	42.3	5.8	1.4	60	27	2.1	71	34
II	158	93.1	64.9	4.6	1.9	90	35	3.1	64	31
III	183	118.1	64.9	5.1	1.6	75	36	2,3	59	28.5

TEST RESULTS

Case I

$P + Z_0 = 158$ mm $v_{jR} = 5.8$ mm/µs $v^2 \cdot d = 71$ mm³/µs² $u^2 \cdot d = 34$ mm³/µs²

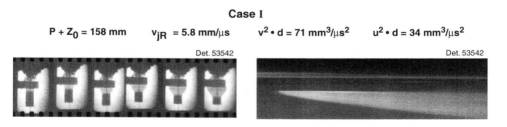

Case II

$P + Z_0 = 158$ mm $v_{jR} = 4.6$ mm/µs $v^2 \cdot d = 64$ mm³/µs² $u^2 \cdot d = 31$ mm³/µs²

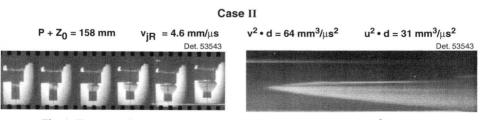

Fig. 3: Frames and streak records of jet initiation with lead covers, 10^6 frames/sec

The achieved frames and simultaneous gained streak records, taken with a rotating mirror camera CORDIN 330 at 10^6 frames/sec, for 2 firings are presented in Fig. 3 where the upper raw shows the 42.3 mm thick lead barrier. In the lower raw a 64.9 mm lead barrier thickness was used. The from the streak records analyzed time/distance plots of the radial breakthroughs of the detonation waves are presented in Fig. 4

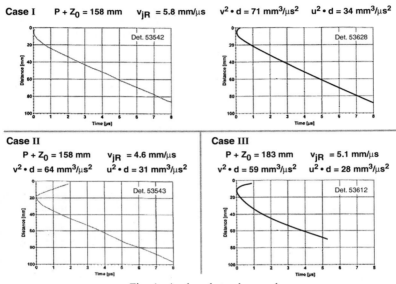

Fig. 4 : Analyzed streak records

The analyzed first radial breakthroughs, resp. buildup distances Δs and measured delay times Δt_{Exp} are summarized in table 2. The buildup distances Δs and delay times Δt_{Exp} are graphically shown as a function of the impacting jet velocities in Fig. 5. The diagram Fig. 6 gives the delay times Δt_{Exp}, the time differences $\Delta\Delta t$ and the initiation times t_i as function of buildup distances Δs. To get the values of the initiation time t_i some times have to be subtracted from the measured delay time Δt_{Exp} (Held, 2001b):
- Perforation time of the lead barrier
- Penetration time into the high explosive charge, at least some small distances before the detonation starts
- Time delay for starting the detonation
- Detonation time from the initiation at the charge axis to the charge surface

Surprising is the fact that this initiation times t_i are typically a little shorter compared to the ideal prediction with an initiation exactly in the axis and not over some area etc.

TABLE 2

Det.-No.	Setup	v_{jR}	Δs	Δt_{Exp}	Δt_{Per}	$\Delta\Delta t$
53542	I	5.833	3.4	15.3	13.145	+2.2
53628	I	5.833	4.5	14.7	13.145	+1.6
53543	II	4.561	17.2	25.9	23.421	+2.5
53612	III	5.056	14.3	25	21.965	+3.0

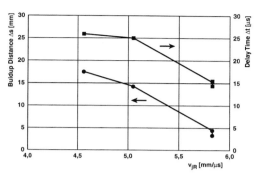
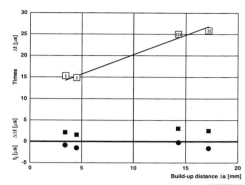

Fig. 5: Build-up distances Δs and delay times Δs

Fig. 6: Delay and initiation times vs build-up distances

CONCLUSION

Up to now are typically used steel barriers in contact to the high explosive charge to reduce the jet velocities or the jet load on the high explosive charge, similar to gap test arrangements. No care was taken on the influence of the barrier material to the jet initiation phenomena. Four tests with lead barriers in contact to the high explosive charge are described which definitely desensitize less the cast high explosive charge compared to steel barriers which have much larger bulging behavior.

REFERENCES

Chick M.C. and Hatt D.J. (1981a), "The Mechanism of Initiation of Composition B by a Metal Jet", 7th Symposium (International) on Detonation, pp. 352-361

Chick M.C. and Hatt D.J. (1981b), "Metal Jet Initiation of Bare and Covered Explosives; Summary of the Mechanism, Empirical Model and Some Applications", Department of Defence, Material Research Laboratories, Melbourne, Victoria, Australia, Report, MRL-R-830

Chick M.C. and Hatt D.J. (1983), "The Initiation of Covered Composition B by a Metal Jet", Propellants, Explosives, Pyrotechnics **8**, 121-126,

Chick M.C. and Hatt D.J. (1985) "A Noval Technique for the Controlled Initiation of Explosives", Journal of Energetic Materials, Vol. **3**, 221-238

Held M (1968), "Initiierung von Sprengstoffen ein vielschichtiges Problem der Detonationsphysik", Explosivstoffe **5**, 2-17

Held M.(1970), "Orthogonal Multi-Streak Recording Technique", Proceeding of the 9th International Congress on High Speed Photography, Denver, CO, 126-129

Held M. (1980), "Initiation of Explosives, A Multi-Layered Problem in Detonation Physics", *Translation of Held M. (1970)* by LLL, Reference 02973

Held M (1987a), "Experiments of Initiation of Covered, but Unconfined High Explosive Charges by Means of Shaped Charge Jets", Propellants, Explosives, Pyrotechnics **12**, 35-40

Held M. (1987b), "Experiments of Initiation of Covered, but Unconfined HE Charges under Different Test Conditions by Shaped Charge Jets", Propellants, Explosives, Pyrotechnics **12**, 97-100

Held M. (1989), "Discussion of the Experimental Findings from the Initiation of Covered but Unconfined High Explosive Charges with Shaped Charge Jets", Propellants, Explosives, Pyrotechnics **12**, 245-249

Held M. (1990), "The advantage of simultaneous streak and framing records in the field of detonics", **19**[th] International Congress on High-Speed Photography and Photonics, SPIE Vol. **1358**, 904-913

Held M.(1991a), "Hydrodynamic Theory of Shaped Charge Jet Penetration", Journal of Explosive and Propellants, R.O.C., **7**, 9-24

Held M. (1991b); "Initiierungsabstand und Detonationsradius", **22**[nd] International Annual Conference of ICT **3**: 1-15

Held M. (1993), "High Speed Photography", Chapter 11 of the book "Tactical Missile Warheads", AIAA Vol. **155**, 609-673

Held M .(2001), "Jet Initiation of Covered High Explosives With Different Materials", Propellants, Explosives, Pyrotechnics, Submitted March 2001

Zernow L., Liebermann I.. and Kronmann S. (1955), "An Explanatory study of the Initiation of Steel-Shielded Composition B by Shaped Charge Jets", Memorandum Report No 944

CURLING OF SQUARE TUBE BY ELECTROMAGNETIC FORMING

S.B. Zhang and H. Negishi

Department of Mechanical Engineering and Intelligent Systems, The University of Electro-Communications, 1-5-1 Chofugaoka, Chofu-shi, Tokyo 182-8585, JAPAN

ABSTRACT

In this paper, curling of an aluminum square tube by an electromagnetic forming is investigated. In free curling, irregularities are produced in the corner surface of the square tube when magnetic pressure reaches in a certain value. When deformation on the side part of the square tube is restricted by a die, it is possible to obtain regular curling profile in the case of short curling length. Simulation of curling square tube is performed by a finite element method and compared with experimental results.

KEYWORDS

Curling, Square tube, Electromagnetic forming, Aluminum, Impulsive pressure, Simulation

INTRUCTION

The square tubes are used for household appliances such as a refrigerator, a washing machine and a microwave oven. Curling of the square tube is generally carried out by use of a roller or a die, which have complex construction, Ping & Negishi (1996). To develop available manufacturing process and reduce manufacturing cost, curling of the aluminum square tube by electromagnetic forming is investigated. Simulation of the curling process is performed by a finite-element method program and the analytical results on curling profile and strain distribution of the square tube agree with experimental ones.

MEASUREMENT OF MAGNETIC FLUX DENSITY

Workpiece

The square tube is shown in Figure 1, which is made by bending and butt-welding of an aluminum plate sheet (JISA1050). The workpieces are annealed at 400°C for 1 hour and have the axial length of

50mm, the wall thickness of 1 mm, the corner radius of 10 mm, and the external width of a.

Equipment

The electromagnetic forming equipment consists of a forming coil with a field shaper, an ignition switch, and an energy-storage condenser bank with capacitance $C=400\mu F$ and maximum charging voltage $V=10kV$.

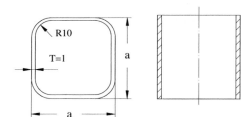

Figure 1: Size of a workpiece

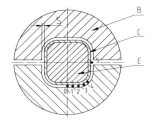

B: Field shaper   C: Workpiece   E: Die
S: Gap(1.5,2,2.5mm)   •0~5: Position of search coil
Figure 2: Cross-section of a forming set showing the position of measuring magnetic flux density

Experimental method

Magnetic flux density in the gap between the field shaper and the external surface of the square tube was measured by use of a search coil having the section area of $1.17mm^2$, 18 turns of copper wire and the axial length of 4mm. The search coil is sited on the axial center of the square tube and its position is shown in Figure 2, in which position 5 indicates a slit of field shaper. When magnetic flux density is measured, the square tube external width a is 48, 47 and 46 mm, respectively, so that the gap S between the workpiece and the field shaper can be varied from 1.5 to 2.5mm.

Experimental results

Magnetic flux density is simply an exponentially decaying function of time. Figures 3-4 show the measured results on the peak value of the first half cycle of magnetic flux density. The magnetic flux density in the gap between the workpiece and the field shaper is almost identical along the gap, and decreases slightly on the slit of the field shaper. The magnetic flux density is reduced as the gap width is increased.

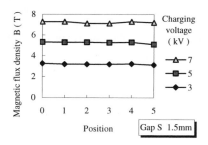

Figure 3: Axial magnetic flux density

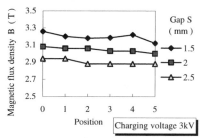

Figure 4: Influence of the gap S between the field shaper and the workpiece on magnetic flux density

CURLING OF SQUARE TUBE

Experimental method

The experimental set-up shown in Figure 5 includes a coil with a field shaper, a square tube workpiece and a die. Curling length H is 4,6mm, respectively. The external width a of the workpiece shown in Figure 1 is 50mm.

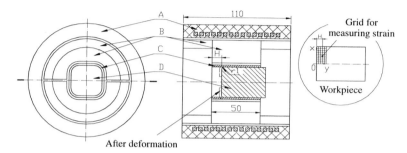

A: Coil   B: Field shaper   C: Workpiece   D: Die   H: Curling length (4,6mm)
Figure 5: Experimental set-up for curling

Experimental results

Profile of deformed square tube

A photograph of the deformed square tube for charging voltage of 8kV is show in Figure 6. The side part of the square tube is deformed and completely contacted with the die, however deformation on the corner is still small in the case of high charging voltage.

Strain distribution

The circumferential strain distribution of the square tube is shown in Figure 7, in which direction of x and y denotes circumferential and axial direction of the square tube, respectively. The strain distribution has irregularity and the compressive strain on the side part near the corner occurs, while the tensile strain on the corner is developed.

H 6mm   V 8kV
Figure 6:   Profile of the deformed square tube

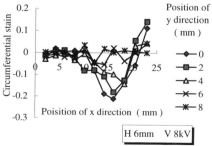

Figure 7:   Circumferential strain distribution of the deformed square tube

Experiment by using restricting die

In free curling, the regular curling profile could be not obtained. For confining deformation of the side part of the square tube, die with taper on the top end shown in Figure 8 is inserted into the square tube. The photographs of the deformed square tube by using the restricting die are shown in Figure 9, when the curling length is 4mm, the regular curling profile is obtained (a), however, when the curling length is 6mm, deformation on the corner is still smaller than that on the side part of the square tube and wrinkles on the corner are developed (b).

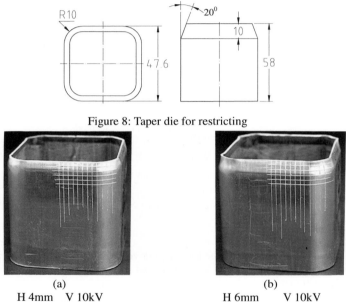

Figure 8: Taper die for restricting

(a) H 4mm   V 10kV   (b) H 6mm   V 10kV
Figure 9: Profile of the deformed square tube by using the restricting die

SIMULATION OF CURLING SQUARE TUBE

Model and analytical method

Simulation of curling square tube is performed by a finite element analysis program. A quarter of the square tube is considered in the analysis because of its symmetry. An element is a bilinear shell element with four nodes. The square tube is regarded as the isotropic material following von-Mises Yield condition. The Newton-Raphson method and the updated Lagrangian formulation are used as the solution methods of the non-linear equation, and the Newmark-beta of implicit solution time-integration method is used for the analysis of the dynamic forming process. Except in curling region, the square tube being deformed is supported by a rigid body with 0.3 friction coefficient. The friction model between the square tube and the rigid body is Coulumb friction model.

Material properties

The material properties used in the analysis are that the young's modulus is 70.3GPa, the poisson's ratio is 0.28, and the mass density is $2.71 \times 10^3 kg/m^3$. The flow stress-strain of the square tube was represented by Eqn. 1, Negishi, Murata, Suzuki & Maeda (1980).

$$\sigma = 132\varepsilon^{0.356} + 1.2(\dot{\varepsilon}/\dot{\varepsilon}_0)^{0.198} \qquad \text{MPa} \qquad (1)$$

where σ is flow stress, ε is strain, $\dot{\varepsilon}$ is strain rate, and $\dot{\varepsilon}_0$ is material constant ($\dot{\varepsilon}_0 = 10^{-4}\text{sec}^{-1}$).

Magnetic pressure

Applied magnetic pressure to the square tube is calculated by Eqn. 2, Suzuki (1993).

$$P = \frac{B^2}{2\mu_0}\left(1 - \exp\left(-\frac{2T}{\delta}\right)\right) \qquad (2)$$

where P is magnetic pressure, μ_0 is permeability, B is magnetic flux density, T is thickness of the square tube and δ is skin depth ($\delta = \sqrt{2\rho/\omega\mu_0}$, ρ is electric resistance coefficient of workpiece material and ω is angular frequency of magnetic flux density). The magnetic pressure calculated from Eqn. 2 by using the measured data of magnetic flux density is shown in Figure 10.

When deforming, the magnetic pressure loading to the square tube is reduced with increasing of the displacement of the tube wall. According to the experimental result of magnetic flux density, magnetic pressure is

$$P(t) = \frac{1}{(1+ks)^4} P_0(t) \qquad (3)$$

where s is displacement, k is experimental coefficient (0.05 in the present report) and $P_0(t)$ is the magnetic pressure shown in Figure 10.

For considering the effect of the axial distribution of magnetic flux density on curling deformation, the axial distribution of magnetic flux density is measured by use of the search coil mentioned previously, and the result is shown in Figure 11, in which position 0 and 25mm denotes the center and the end of the side part of the square tube in the axial direction, respectively. In simulation, the axial distribution of magnetic flux density is simplified to curve 2 showing that magnetic flux density decreases about 13.5 per cent when the axial position is varied from 15 to 25mm.

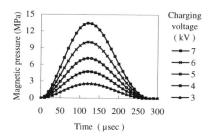

Figure 10: Magnetic pressure vs time

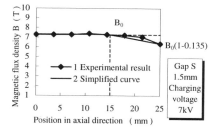

Figure 11: Axial distribution of magnetic flux density

Results of simulation

In free curling, the profile and the circumferential strain distribution of the deformed square tube are shown in Figures 12 and 13 respectively. The analytical results agree well with experimental ones.

In curling by using a restricting die, the profiles of the deformed square tube are shown in Figure 14.

Regular curling profile is obtained in curling length of 4mm, and deformation on the corner is smaller than that on the side part of the square tube in curling length of 6mm. However the wrinkle on the corner in experimental result is not developed.

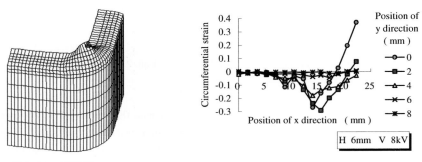

H 6mm V 8kV
Figure 12: Profile of the deformed square tube

Figure 13: Circumferential strain distribution of the deformed square tube

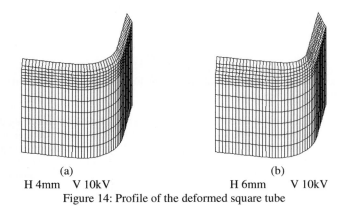

(a) H 4mm V 10kV (b) H 6mm V 10kV
Figure 14: Profile of the deformed square tube

CONCLUSIONS

Curling of the square tube is investigated through the experiment and simulation of FEM, the conclusions are summarized as follows.
1. The magnetic flux density in the gap between the field shaper and the square tube is almost identical along the gap and reduced as the gap is increased.
2. The curling profile and the strain distribution in free forming have irregularities.
3. When deformation of the square tube is restricted by a die, the regular curling profile of the square tube can be obtained in the case of short curling length.
4. Simulation results of curling profile and circumferential strain distribution agree with experimental ones.

References

Ping Z.Z. and Negishi H. (1996). Electromagnetic Curling of Square Tube. *The Proceedings of the 47 th Japanese Joint Conference for the Technology of Plasticity*, 337-338

Negishi H., Murata M., Suzuki H. and Maeda T. (1980). An Analysis of Forming Process for Tube Expansion, *Journal of the Japan Society for Technology of Plasticity* **21:234**, 728-734.

Suzuki H. (1993). *High Energy Rate Forming*, Corona Publishing CO., LTD., Tokyo, JAPAN

PERFORATION OF HIGH STRENGTH FABRIC SYSTEM BY VARYING SHAPED PROJECTILES: SINGLE-PLY SYSTEM

C.H. Cheong, V.B.C. Tan and C. T. Lim

Impact Mechanics Laboratory,
Department of Mechanical Engineering
National University of Singapore
10 Kent Ridge Crescent, Singapore 119260

ABSTRACT

This paper investigates the ballistic limits, energy absorption behaviour and the mechanisms that lead to perforation in Twaron® CT 716 plain-woven, single-ply fabric by different shaped projectiles. The projectile shapes tested are flat head, hemispherical head, ogival head (CRH 2.5) and conical head (half angle of 30°).

Results show that while the amount of energy absorbed by the fabric is quantitatively different for all four projectiles, they show similar trends – energy absorbed increases with impact velocity up to a critical impact velocity before it starts to decrease. The energy absorption capability of the Twaron® fabric is explained by considering how impact energy is converted to strain energy and kinetic energy of the fabric. Different projectile shapes were also found to perforate the fabric through different mechanisms - yarn rupture, fibrillation, failure by friction, and bowing.

KEYWORDS

Ballistic limit, high strength fabric, perforation, impact, projectile shapes, energy absorption, fibrillation, rupture, friction, bowing.

INTRODUCTION

The high impact resistance of high strength fabrics is attributed to the high strength polymeric fibres used (Laible, 1980). The aramid fibres in the Twaron® CT family of fabric consist of highly aligned macromolecules strongly bonded to one another resulting in excellent tenacity and modulus, making them ideal for ballistic applications. The lightness of this material adds to their popularity. In order to achieve optimal impact protection capabilities, a clear understanding of the dynamic mechanical response of the fabric is essential. However, this is often complicated by multiple stress wave interactions at yarn crossovers making the problem almost analytically intractable. Further, impact outcome differs for different shaped projectiles and different impact velocities. This paper evaluates the ballistic performance of Twaron® CT 716 when impacted by four different projectile shapes - hemispherical head, flat head, ogival head (CRH 2.5) and conical head (half angle of 30°). The ballistic limits, energy absorption characteristics and failure mechanisms of the fabric are examined. Impact tests are conducted by subjecting fabric specimens to projectile impacts up to 600 m/s. Each specimen has an impact area of 120mm (clamped) by 118mm (free).

BALLISTIC LIMIT

The ballistic limit is the velocity at which 50% of the impacts result in complete penetrations and 50% in partial penetration according to the Protection Ballistic Limit or PBL (Laible, 1980). In short, it is a statistical measure of the velocity at which penetration just occurs.

Ballistic limits for all projectile shapes are estimated from experimental data. The values obtained are 163 m/s, 103 m/s, 76 m/s and 60 m/s for hemispherical, flat, ogival and conical projectiles respectively.

Despite its more streamlined profile, the hemispherical projectile exhibits a higher ballistic limit compared to the less streamlined flat head projectile. This phenomenon is attributed to the cutting action facilitated by the peripheral edge of the flat head.

IMPACT ENERGY ABSORPTION

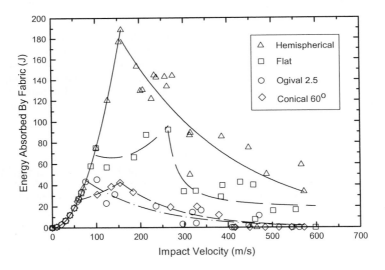

Figure 1: Energy absorption curve for all projectile shapes

The energy expended by the projectiles in perforating the fabric specimens is regarded as the energy absorbed by the fabric. It is found by subtracting the residual energy of the projectile from its initial impact energy. In the event when no perforation occurs, the energy absorbed by the fabric is taken as equal to the initial impact energy. The energy absorption curves shown in Figure 1 consist of a sub-ballistic limit impact regime A, low velocity perforation regime B and high velocity perforation regime C as illustrated in Figure 2.

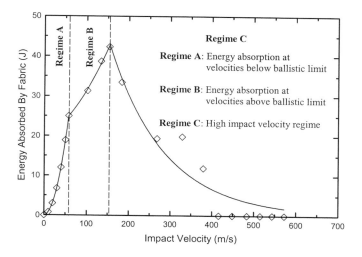

Figure 2: Typical energy absorption graph for a conical head projectile

Sub-ballistic Limit Impacts

This portion of the energy absorption curve spans from zero impact velocity to the ballistic limit of the projectile concerned. All projectiles exhibit the same trend in this regime since energy absorbed equals impact energy, i.e.,

$$E_{abs} = \frac{1}{2} m V_{impact}^2 \qquad (1)$$

where E_{abs} is the energy absorbed by the fabric, V_{impact} is the impact velocity and m is the projectile mass.

On impact, the fabric is pulled in towards the impact point and transverse deflection resembles a pyramid (Adanur, 1995). The impact energy is dissipated through kinetic energy due to transverse deflection of the fabric, stretching of yarns resulting from transverse deflection and in-plane deformation (that resembles a square pyramid (Adanur, 1995)) and overcoming yarn-to-yarn friction. It was observed that yarns are ravelled at the free edges and the specimens are creased. The extent of creasing and ravelling increased with impact velocity, suggesting an increase in transverse deformation and yarn-to-yarn sliding. Hence, all modes of energy dissipation - kinetic energy, strain energy of the fabric and frictional losses - increase with impact velocity.

Low Velocity Perforation

Regime B shows the emergence of differences in the effects of the various projectile shapes. This portion of the curve ranges from the ballistic limit to the velocity that corresponds to maximum energy absorption. This regime is only evident and more pronounced for certain projectile shapes. Within this regime, the impact energy absorbed by the fabric continues to rise with impact velocity but at a lower rate than sub-ballistic limit impacts. The lower rate of energy absorption is attributed to the occurrence of perforation resulting in earlier unloading of stresses in the fabric. Specimens continue to show extensive creases and are especially severe for the hemispherical head projectiles. This suggests the dominance of energy dissipation through stretching of the fabric. The perforation is generally smaller than the projectile diameter except in the case of the flat head projectile. This phenomenon implies significant friction between the projectile and fabric surfaces, which contributes an additional mode of impact energy dissipation.

High Velocity Perforation

The drop in the energy absorbed by the fabric marks the start of the high velocity perforation regime. Fabric specimens perforated at such high impact velocities showed markedly more localised damage than low velocity perforations. Creasing is less obvious and confined to small regions of the fabric. There is also less damage at the clamped edges. Ravelling of yarns at the free edges of the fabric is considerably lessened. It is concluded that at high impact velocities, only part of the fabric is deflected and stretching of the fabric is reduced. At high velocities, the fabric is no longer effective in absorbing the impact energy as the fabric is perforated on impact so that there is no time for the impact energy to be dissipated away from the impact point to the rest of the fabric. The amount of fabric affected during perforation decreases with increasing impact velocity. Other researchers (Cunniff, 1992, Shim et al., 1995 and Yong et al., 1999) also arrived at these conclusions.

All projectile types show a drastic drop in energy absorbed after the velocity that corresponds to maximum energy absorption. Comparison of the energy absorbed in this regime for all the projectiles show the same trend as in previous regimes. That is, hemispherical head projectile facilitates the highest energy absorption followed by flat head, ogival and conical head projectiles. The streamlined profiles of sharp projectiles enable them to slip through the fabric between adjacent yarns, keeping yarn breakage to a minimum whereas the blunt projectiles defeat the target mainly through stretching and breaking the yarns. Contribution by the kinetic energy of the deforming fabric to impact energy absorption is most substantial in this regime.

FABRIC FAILURE MECHANISMS

Rupturing of Yarns

Rupturing or severing of aramid yarns has been associated with the breaking of primary or covalent bonds of macromolecular chains (Shim et al., 1995). With the exception of flat head projectile, the ends of the yarns ruptured by all other projectile types appear messy and disorderly. Unruly yarn pullout is especially prominent for the hemispherical head, which suggests that the yarns are broken mainly by head-on impact. On the contrary, conical and ogival projectile impacts cause the least yarn pullout, which implies that they wedged through the fabric. The ends of the yarns severed by the flat head projectile are observed to be more orderly and even. Close scrutiny of the impacted specimen illustrated in Figure 3 suggests that the fabric is sheared by the circumferential edge of the projectile when it presses against the fabric upon impact. Such a cutting action helps explain the orderly and even ends of the broken yarns.

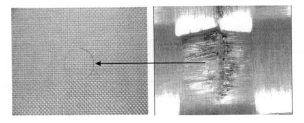

Figure 3: Evidence of shear in the fabric impacted by a flat head projectile

Fibrillation

Fibrillation (Laible, 1980) refers to the splitting of a fibre along its length and is achieved by breaking the weaker secondary bonds that hold the individual macromolecules together. It is promoted by actions that facilitate the breaking of secondary bonds, such as abrasive action of a projectile sliding across the fabric, especially in a direction that is perpendicular to the length of the fibre. Fibrillation of fibres is more severe for projectiles that have angled nose profiles such as that of the conical and flat head projectiles.

Friction

Friction tests were performed and results indicate that the static coefficient of friction is 0.31 in the weft direction and 0.20 in the warp direction of the fabric. Although the coefficient of friction is considerably small, there is strong evidence of frictional effects from abrasion. Small patches of fibre breakage were seen at the yarn crossovers near the impact region. Frictional effects are most prominent at low velocities but diminish at higher velocities. At high velocities, material at the impact point is broken on contact and more yarns are severed, hence, friction from projectiles squeezing through the perforation is less significant. This phenomenon however is less distinct with the sharper projectiles. Scanning Electron Microscopy (SEM) shows that the failure mechanisms associated with friction include flattening of fibres, fibrillation and rupture.

Flattening of fibres occurs when projectiles press directly onto the fibres. Flattening is most severe for the hemispherical head projectile but least severe for the ogival and conical head projectiles.

Fibrillation is more pronounced for conical projectiles because the angled edge adjoining the cone to the shank of the projectile scrapes against the fibres. Flat head projectiles also possess sharp edges, however, flattening of the fibres is as prominent as fibrillation for such projectiles since the flat surface presses directly onto the fibres upon impact.

Bowing

Bowing (Adanur, 1995) refers to the phenomenon where the warp yarns become non-orthogonal to the weft yarns. Bowing can be caused by two mechanisms. At the region of impact, yarns are pushed aside of yarns during a projectile's passage through the fabric. Away from the impact point, severe deformation causes the strained yarns to displace from crossover points of the weave.

Bowing is almost non-existent in specimens impacted by the flat head projectile because penetration is mainly due to shearing and also because of the constant cross-section of the projectile. Conversely, bowing is most substantial for the hemispherical head. Although its round and blunt profile does not enable it to slip through small perforations by pushing aside surrounding yarns, stresses developed in the fabric are much higher and more extensive prior to yarn breakage.

The ogival and conical head projectiles do not cause large stresses because of their narrower and sharper profiles. Instead, bowing is mainly through the pushing aside of yarns as they penetrate the fabric, which is less intense compared to the hemispherical head. Between the ogival and conical head projectiles, the ogival head projectile causes more bowing because its nose profile is larger in cross-section nearer the tip.

At higher velocities, bowing is more localised for all projectile types and its significance is reduced for the blunt projectiles since more yarns are broken.

CONCLUSION

The ballistic limits for the hemispherical, flat, ogival and conical heads are found to be 163 m/s, 103 m/s, 76 m/s and 60 m/s respectively. The energy absorbed by the fabric specimens generally rises with impact velocity up to a certain critical velocity before it drops steeply and then decreases to an almost constant value at high velocities. The critical velocity signifies the impact velocity at which the yarns are broken at the impact point before impact energy can be transmitted to the entire fabric, thus causing a sudden drop in the energy absorbed. Sharper projectile shapes result in less energy absorbed than flat and hemispherical head projectiles. Impact energy is dissipated and absorbed by the fabric as kinetic energy due to deflection and strain energy due to deformation as well as frictional losses through yarn-to-yarn friction and friction between the projectile and fabric. Of the four projectile types tested, fabric specimens absorbed the most impact energy when the projectile is hemispherical and the least energy when they are ogival or conical.

Failure mechanisms that lead to perforation are rupturing of yarns, fibrillation, friction and bowing. Hemispherical projectiles were found to perforate the fabric specimens mainly by stretching the yarns to rupture while flat projectiles are able to shear the yarns because of their angled edges in addition to rupturing them. Conical and ogival projectiles are sharper and bowing is a significant mode of fabric perforation for these projectiles.

Overall, the close-knit relationship between modes of energy dissipation and failure mechanisms for different projectile geometry determine the ballistic limits and the level of energy absorbed in the high strength fabric.

REFERENCES

Adanur S. (1995), *Wellington Sears Handbook of Industrial Textiles,* Technomic Publishing Company, Inc.

Cunniff P.M. (1992). An Analysis of the System Effects in Woven Fabrics Under Ballistic Impact. *Textile Res. J.* 62:9, 495-509.

Laible R.C. (1980), *Ballistic Materials and Penetration Mechanics. Method and Phenomena: Their Applications in Science and Technology*, Elseiver.

Shim V.P.W., Tan V.B.C. and Tay T.E. (1995). Modelling Deformation and Damage Characteristics of Woven Fabric Under Small Projectile Impact. *Int. J. Impact Engng* 16:4, 585-605.

Yong S.Y., Shim V.P.W. and Lim C.T. (1999). An Experimental Study of Penetration of Woven Fabric By Projectile Impact. 3rd *Int. Sym. on Impact Engng* Sect. 12, 559-565.

PERFORATION OF HIGH STRENGTH FABRIC SYSTEM BY VARYING SHAPED PROJECTILES: DOUBLE-PLY SYSTEM

C.H. Cheong, C. T. Lim and V.B.C. Tan

Impact Mechanics Laboratory,
Department of Mechanical Engineering
National University of Singapore
10 Kent Ridge Crescent, Singapore 119260

ABSTRACT

Work done previously on single-ply system is now extended to double-ply system. Experiments were carried out to investigate the impact phenomenon on a 2-ply system that consists of Twaron® CT 716 fabric and projectiles of the following nose shapes: hemisphere, flat, ogive (CRH 2.5) and cone (half angle of 30°). Results obtained revealed that the increase in energy absorption does not necessarily double when the ply-number is increased to two. In fact, the ratio of energy absorbed in the 2-ply system to that of the 1-ply system varies with impact velocity. The maximum energy absorption ratio of 2-ply to 1-ply system for all projectile types scatters about an average of two. The amount of deviation from the ratio two depends on the mechanisms that lead to perforation. Failure mechanisms of a 2-ply system are similar to those of a 1-ply system, but the degree of damage between the front ply and distal ply differs for certain projectile shapes.

KEYWORDS

Failure mechanisms, high strength fabric, 2-ply system, perforation, projectile shape, front ply, back ply.

INTRODUCTION

The resistance of high-strength fabrics to ballistic impact is highly dependent on the nose profile of the projectile. Thus, it is not entirely sound to analyse the impact strength of such fabrics in isolation from projectile geometry. It had been established that impact energy is absorbed by the fabric through the following modes - kinetic energy of the deforming fabric, stretching of yarns, friction between yarns and that between projectile and fabric surfaces. Contribution to impact energy absorption by each mode varies in proportion as velocity changes. The energy absorption curve is broadly classified into a low velocity regime characterised by extensive creasing in the impacted specimen and a high velocity regime that exhibits limited creasing. Following the work done on a single-ply system, double-ply system is now studied. An additional ply of fabric would undoubtedly improve the impact energy absorption capacity but the increment may not necessarily double. Nonetheless, the ratio of the maximum energy absorbed for 2-ply to 1-ply system averages to 2.1. The failure mechanisms associated with a 2-ply system are very much similar to a 1-ply system except for some variations in intensity between the front and back plies.

EXPERIMENTAL SETUP

The fabric used is Twaron® CT716. Fabric samples are clamped between two ends spaced 118mm apart. The target area measures 120mm by 118mm. The nose profiles used are hemisphere, flat, ogive of CRH 2.5 and cone of half angle 30°. Impact velocities ranges from 0 to 600 m/s.

BALLISTIC LIMIT

The ballistic limits or V50 velocities for all projectile shapes show clear increment over the 1-ply system except for the flat head projectile. The values of which for hemispherical, flat, ogival and conical heads are 246 m/s, 124 m/s, 118 m/s and 115 m/s respectively. In comparison with the corresponding V50 velocities of 163 m/s, 103 m/s, 76 m/s and 60 m/s in the single-ply system, it is noticeable that the flat head projectile exhibits the least change. The small disparity is attributed to the peripheral edge of the flat head projectile, which shears the fabric yarns across their thickness rather than stretching them along the longitudinal direction where the strength is highest. Therefore, the additional strength provided by the second ply of fabric is not optimally realized.

IMPACT ENERGY ABSORPTION

The energy absorption curves are similar to that of a 1-ply system. Effect of the second ply becomes conspicuous only after the sub-ballistic regime of the 1-ply system as shown in Figure 1.

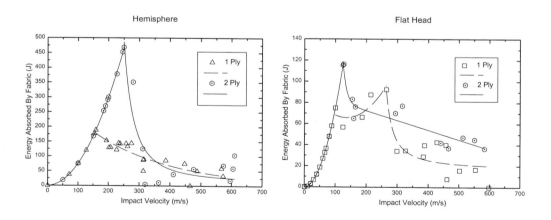

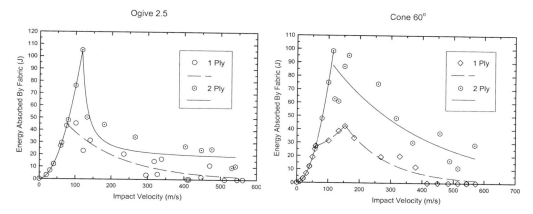

Figure 1: Energy absorption curves for 2-ply and 1-ply systems

More broken yarns and larger out-of-plane transverse deformation indicated by an increased number of ravelled yarns at the unclamped edges give rise to higher impact energy absorption (Montgomery et al., 1982 also arrived at similar results). Even so, having an extra ply of fabric is not significantly beneficial at velocities in the sub-ballistic limit regime and at the upper end of the high velocity regime where stretching of the fabric is highly localised near the impact point limiting impact energy absorption. However, improvement is more substantial at velocities that correspond to maximum impact energy absorption. In some cases, the improvement can be more than double as illustrated in Figure 2.

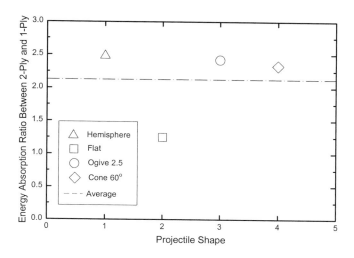

Figure 2: Ratio of maximum impact energy absorption between 2-ply and 1-ply systems

Interestingly, the ratio of the maximum impact energy absorbed by the 2-ply system to that by 1-ply system is scattered about an average value of two and the extent of deviation depends on projectile geometry. Nose profiles which facilitate superior penetration mechanisms give lower ratios. For instance, the penetration mechanism associated with hemispherical head projectiles primarily entails stretching the yarns to failure. Thus, the reinforcement provided by the additional ply of fabric is clearly effective and the ratio is the highest at 2.5. On the other hand, the cutting action of flat head projectiles easily perforates the 2-ply system, thus giving the lowest ratio of 1.25. Between these two extreme cases are the ogival and conical head projectiles. Both projectile types possess the same narrow nose profile which is highly effective in slipping through the fabric between yarns, hence there is close similarity in their ratios.

FABRIC FAILURE MECHANISMS

The mechanisms that lead to perforation in a 2-ply system show similarities to the 1-ply system. Differences between the two systems are in the intensity of the various failure modes discussed. In summary, the failure modes are yarn rupture, fibrillation, friction and bowing.

Yarn Rupture

Yarn rupture for different projectile shapes is characterised by how disorderly the yarns appear after breakage. This aspect is the same for both types of systems. However, the back ply in a 2-ply system generally shows fewer number of broken yarns compared to the front ply. This suggests that the two plies of fabrics behave like discrete layers which sequentially reduce the impact energy of the projectile. This phenomenon is especially distinct with hemispherical head projectiles at low velocities. Near the ballistic limit, the projectile was found sandwiched between the two plies. Figure 3 shows that there are no yarns broken in the back ply whereas approximately 20 yarns are broken in the front ply.

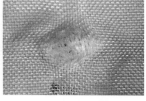

Front Ply   Back Ply

Figure 3: Disparity in yarn breakage between front and back plies of a specimen impacted by a hemispherical head projectile

Fibrillation

The extent of the splitting of fibres along their lengths remains similar to a 1-ply system. There are also no noticeable differences between the front and back plies in a 2-ply system.

Friction

Frictional effects in the form of patches of broken fibres due to abrasion between projectile and fabric surfaces are most prominent in the front ply. The pullout of yarns during perforation results in broken yarns from the front ply crossing into the plane of the back ply of fabric, thus providing a protective interface that alleviates projectile-fabric abrasion in the back ply. Figure 4 is a sample illustration of less intense fibre damage by friction in the back ply for all projectile types.

Front Ply   Back Ply

Figure 4: Fibre breakage in the front and back ply specimens impacted by a conical head projectile

Bowing

With impact energy considerably reduced after perforating the front ply, perforation of the back ply via slipping through the fabric between fabric yarns becomes more dominant. This effect is evident even in specimens impacted by flat head projectiles whose constant cross section does not facilitate pushing aside of yarns as shown in Figure 5.

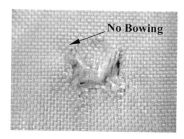 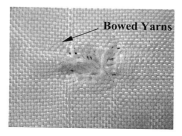

Front Ply   Back Ply

Figure 5: Bowing in back ply of fabric impacted by flat head projectile

Bowing in the back ply diminishes at higher impact velocities.

CONCLUSION

A multi-ply fabric armour system undoubtedly increases impact energy absorption, however the reinforcement that results depends on the nose geometry of the projectile. The degree of improvement varies with impact velocity. Nevertheless, the ratio of maximum impact energy absorption by a 2-ply system to that by a 1-ply system for all projectile types appear scattered about an average of two. Projectiles that facilitate superior penetration mechanisms give lower ratios and vice versa. Changes in the ballistic limits are clearly distinct for all projectile types. The least change is witnessed in the flat head projectile since this nose shape shears the fabric yarns across their thickness so that the additional yarn strength from the second ply of fabric is not optimally realised. Failure mechanisms remain similar but intensity of damage differs between the front and back plies. Yarn rupture and frictional effects are less intense in the back ply. Reduced impact energy after perforation of the front ply causes more bowing in the back ply of specimens at low velocities. This is so even for the flat head projectile whose projectile geometry does not facilitate bowing as established in the single-ply system.

REFERENCES

Cunniff P.M. (1992). An Analysis of the System Effects in Woven Fabrics Under Ballistic Impact. *Textile Res. J.* 62:9, 495-509.

Montgomery T. G., Grady P. L. and Tomasino C. (1982). The Effects of Projectile Geometry on the Performance of Ballistic Fabrics. *Textile Res. J.* 52:7, 442-450.

EVALUATION OF IMPACT PERFORATION CHARACTERISTICS OF LAMINATED COMPOSITES USING A PUNCH-LOADING HOPKINSON BAR

T. Yokoyama

Department of Mechanical Engineering, Okayama University of Science
Okayama 700-0005, Japan

ABSTRACT

The impact perforation characteristics of three different laminated composites are determined using a punch-loading Hopkinson bar. Square cross-ply CFRP laminates, plain-weave CFRP laminates and GFRP laminates each having three different numbers of plies are tested at room temperature. The test results for the maximum punch loads and the perforation energies at various displacement rates are presented. The effects of thickness variation and the types of reinforcing fiber and lamina on the perforation characteristics are examined. It is shown that except for cross-ply CFRP laminates, the maximum punch load/thickness increases, and the perforation energy/thickness decreases with increasing displacement rate. Also, the thickness effects are significant in quasi-static punch shear tests.

KEYWORDS

Impact punch shear test, Cross-ply CFRP, Plain-weave CFRP, Plain-weave GFRP, Hopkinson bar, Perforation energy, Displacement rate, Thickness effect

INTRODUCTION

Laminated composite materials have been extensively used in aerospace and other applications. These materials offer definite advantages over conventional materials in terms of their high specific stiffness and strength. Their impact properties are a major concern and have been mainly examined using the split Hopkinson bar. Harding (1979) first determined the dynamic through-thickness shear strength of woven-roving composites using a punch-loading Hopkinson bar. Subsequently, Harding and Welsh (1983) measured the dynamic stress-strain curves for CFRP and GFRP using a tensile split Hopkinson bar. Werner and Dharan (1986) obtained the high strain rate response of graphite/epoxy in interlaminar and transverse shear using a modified Hopkinson bar. Harding and Li (1992) characterized the rate-dependent interlaminar shear strength of CFRP and GFRP using a tensile Hopkinson bar in conjunction with double-lap specimens. Staab and Gilat (1995) determined the dynamic stress-strain curves for glass/epoxy laminates with a direct tension Hopkinson bar. Riendeau and Nemes (1996) studied the rate-dependent behavior of graphite/epoxy in transverse shear using a punch shear version of the split Hopkinson bar. However, the impact resistance of composite materials to penetration and/or perforation has not been fully investigated. The objective of the present work is to evaluate the impact perforation characteristics of three kinds of laminated composites using a punch-loading Hopkinson bar. Three different kinds of laminated composites each having three different numbers of plies are tested at room temperature. Comparative tests at low and intermediate displacement rates are performed on the same designs of composite specimens in an Instron testing machine. The test results for the maximum punch (or perforation) loads and the perforation energies at different displacement rates are presented.

TEST COMPOSITES AND SPECIMEN GEOMETRY

The test composites were 40 mm square cross-ply CFRP laminates, plain-weave CFRP laminates, and GFRP laminates. The mechanical properties of the laminae are given in Table 1. The types of reinforcing fiber and resin matrix used for the laminates are listed in Table 2.

TABLE 1
MECHANICAL PROPERTIES OF LAMINAE USED FOR LAMINATED COMPOSITES

Property	CFRP (Unidirectional)		CFRP (Plain - weave)	GFRP (Plain - weave)
	0°	90°		
E (GPa)	137	8.4	63	33
σ_B (MPa)	2548	67	568	755
τ_{ILSS} (MPa)	89	30	59	—
δ (%)	1.8	0.8	0.9	2.3
ρ (kg/m³)	1,570		1,550	2,170

TABLE 2
TYPES OF REINFORCING FIBER AND RESIN MATRIX

	CFRP (Unidirectional)	CFRP (Plain - weave)	GFRP (Plain - weave)
Fiber	T700S (Torayca)	T300B-3000 (Torayca)	E-Glass
Matrix	Epoxy #2500 (130℃)*	Epoxy #2500 (130℃)	Epoxy #2500 (130℃)
Volume fraction V_f	0.67	0.60	0.45

* Values in parentheses denote the hardening temperature

TEST PROCEDURE

Punch Shear Version of Split Hopkinson Bar

A schematic drawing of a punch shear version of the split Hopkinson bar apparatus is shown in Figure 1. The apparatus consists principally of a striker bar, an input (or punch) bar, an output (or die) tube and a recording system. A composite specimen is placed between the input bar and the output tube on a specimen holder. The radial clearance between the input bar and the output tube is 0.85mm. When the input bar is impacted with the striker bar fired through a gun barrel by compressed air released from a pressure tank, a compressive strain pulse is generated and propagated along the input bar toward the specimen. When the strain pulse reaches the specimen, part of the pulse is reflected back into the input bar due to the impedance mismatch at the bar/specimen interface, and part of the pulse is transmitted through the specimen to the output tube. The strain pulses are measured by two sets of semiconductor strain gages with a gage length of 2 mm. The strain gage signals are digitized and recorded using a 10-bit digital storage oscilloscope at a sampling rate of 1 μs/word. The digitized data are then transferred to a 32-bit microcomputer for data processing.

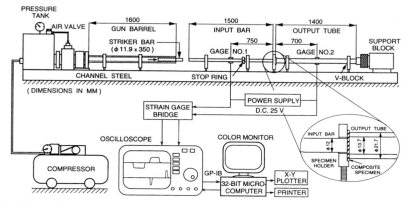

Figure 1: Schematic of punch shear version of split Hopkinson bar apparatus and recording system

Method of Data Analysis

By application of elementary one-dimensional elastic wave theory, the applied punch loads P and the displacements u at each face of the input bar and the output tube are given by

$$P_1(t) = A_1 E_1 \{\varepsilon_i(t) + \varepsilon_r(t)\}, \quad P_2(t) = A_2 E_2 \varepsilon_t(t) \tag{1}$$

$$u_1(t) = c_1 \int_0^t \{\varepsilon_i(t') - \varepsilon_r(t')\} \, dt', \quad u_2(t) = c_2 \int_0^t \varepsilon_t(t') \, dt' \tag{2}$$

where A is the cross-sectional area, E Young's modulus, c the longitudinal elastic wave velocity and subscript 1 and 2 refer to the input bar and the output tube respectively. Here, ε_i, ε_r and ε_t are the incident, reflected and transmitted strain pulses, and t is the time from the start of the pulse. The average punch load $P(t)$ applied to the specimen, the displacement $\delta(t)$, and the displacement rate $\dot{\delta}(t)$ between both surfaces of the specimen are determined as

$$P(t) = \frac{1}{2}(P_1(t) + P_2(t)) = \frac{1}{2}[A_1 E_1 \{\varepsilon_i(t) + \varepsilon_r(t)\} + A_2 E_2 \varepsilon_t(t)] \tag{3}$$

$$\delta(t) = u_1(t) - u_2(t) = \int_0^t [c_1 \{\varepsilon_i(t') - \varepsilon_r(t')\} - c_2 \varepsilon_t(t')] dt' \tag{4}$$

$$\dot{\delta}(t) = c_1 \{\varepsilon_i(t) - \varepsilon_r(t)\} - c_2 \varepsilon_t(t) \tag{5}$$

When $P_1(t) = P_2(t)$ is assumed, the average punch load reduces to

$$P(t) = A_2 E_2 \varepsilon_t(t) \tag{6}$$

Eliminating time t between Eqns. 4, 5 and 6 leads to the punch load and displacement rate versus displacement relations for the specimen.

RESULTS AND DISCUSSION

Quasi-Static Perforation Tests

Perforation tests at quasi-static and intermediate rates were performed in the Instron testing machine using a specially designed punch shear fixture at two different crosshead velocities of 10 mm/min and 100 mm/min. At least five tests were conducted on each laminated composite.

Impact Perforation Tests

A number of impact perforation tests were carried out using the punch shear version of the split Hopkinson bar. Typical strain gage records from impact punch shear tests on the cross-ply CFRP laminate are shown in Figure 2. The upper trace from gage No.1 gives the incident and reflected strain pulses, and the lower trace from gage No.2 gives the strain pulse transmitted through the specimen. Figure 3 shows the resulting impact punch load and displacement rate vs. displacement for the cross-ply CFRP laminate. The displacement rate is almost uniform, except for the initial and final stages of the test, and the peak punch load corresponds to the plugging formation.

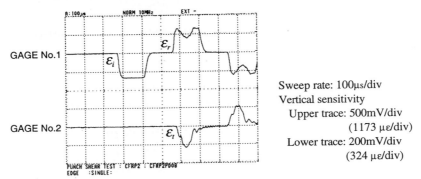

Figure 2: Oscilloscope records from impact punch shear test on cross-ply CFRP laminate

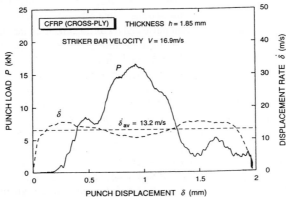

Figure 3: Impact punch load and displacement rate vs. displacement for cross-ply CFRP laminate

In an effort to examine the effect of displacement rate on the perforation characteristics, the impact and quasi-static punch load-displacement curves for the cross-ply CFRP laminates are compared in

Figure 4. It is observed that the maximum punch load and the displacement at the maximum punch load decrease with increasing displacement rate. The reason why quasi-static and intermediate rate punch loads after the plugging do not drop to zero is due to the friction between the specimen and both the plug and the punch rod.

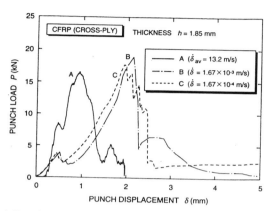

Figure 4: Punch load-displacement curves for cross-ply CFRP laminates at different displacement rates

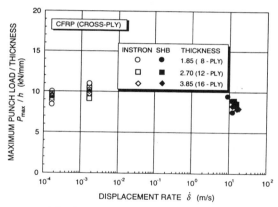

Figure 5: Maximum punch load/thickness vs. displacement rate for cross-ply CFRP laminates

The maximum punch load/thickness for the cross-ply CFRP laminated composites with different numbers of plies are plotted against the displacement rate in Figure 5. The perforation energy is obtained from the area under the punch load-displacement curve up to the displacement at the peak punch load. Figure 6 shows the perforation energy/thickness versus displacement rate for the cross-ply CFRP laminates. The variation of the perforation energy/thickness at low displacement rates is much greater than that at high displacement rates. The perforation energy/thickness decreases with increasing displacement rate. The rate dependence of the maximum punch load and the energy absorption is attributed to the rate dependence of the resin or epoxy in the laminated composites. The perforation characteristics data on the plain-weave CFRP laminates and GFRP laminates are omitted owing to the limited space.

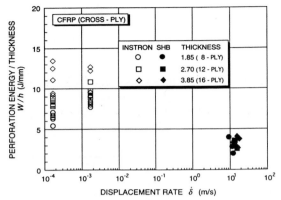

Figure 6: Perforation energy/thickness vs. displacement rate for cross-ply CFRP laminates

SUMMARY AND CONCLUSIONS

The impact resistance of three different laminated composites has been studied using the punch-loading Hopkinson bar. The effects of displacement rate up to the order of $\dot{\delta}$ = 10 m/s and the types of reinforcing fiber and lamina on the maximum punch loads and perforation energies were investigated in detail. The main findings of the present work are summarized as follows:
(1) The rate dependence of the perforation characteristics varies, depending on the types of reinforcing fiber and lamina used for composites.
(2) The maximum punch load/thickness slightly increases and the perforation energy/thickness creases with increasing displacement rate for the plain-weave CFRP and GFRP laminates.
(3) With respect to the maximum punch load/thickness and the perforation energy/thickness, regardless of the displacement rate, the three laminated composites could be ranked as: cross-ply CFRP > plain-weave GFRP > plain- weave CFRP.
(4) The thickness of laminated composites has a more significant effect on the quasi-static perforation characteristics than on the impact perforation characteristics.

References

Harding, J. (1979). The High-Speed Punching of Woven-Roving Glass-Reinforced Composites. *Proc. 2nd Oxford Conf. Inst. Phys. Conf. Ser.* No.47, 318 - 330.
Harding, J. and Welsh, L. (1983). A Tensile Testing Technique for Fibre-Reinforced Composites at Impact Rates of Strain. *J. Mater. Sci.* **18**, 1810 - 1826.
Harding, J. and Li, Y. L. (1992). Determination of Interlaminar Shear Strength for Glass/Epoxy and Carbon/Epoxy Laminates at Impact Rates of Strain. *Comp. Sci. and Technol.* **45**, 161-171.
Riendeau, S. and Nemes, J. A. (1996). Dynamic Punch Shear Behavior of AS4/3501-6. *J. Comp. Mater.* **30:13**, 1494-1512.
Staab, G. H. and Gilat, A. (1995). High Strain Rate Response of Angle-Ply Glass/Epoxy Laminates. *J. Comp. Mater.* **29:10**, 1308-1320.
Werner, S. M. and Dharan, C. K. H. (1986). The Dynamic Response of Graphite Fiber-Epoxy Laminates at High Shear Strain Rates. *J. Comp. Mater.* **20**, 365-374.

DEFORMATION MECHANISM OF WHA COMPOSITE BLOCK IMPACTING ON RIGID ANVIL

Z.G..Wei Y.C. Li J.R.Li X.Z. Hu and S.S. Hu

Department of Modern Mechanics, University of Science and Technology of China, Hefei 230027, Anhui, P.R.China

ABSTRACT

The deformation mechanism of a tungsten heavy alloys (WHA) blocks impacting at normal incidence a rigid target is investigated numerically. The simulations are performed with the Lagrange hydrocode ABAQUS implemented with Johnson-Cook constitutive models. The material of the blocks is modeled as a mixture of Fe-Ni-W grains interspersed periodically among the W-grains. Four different distributions are considered and the deformation is assumed to be locally adiabatic. The shape and orientation effects of the weak matrix on the deformation mechanism are investigated and discussed. The simulation results show that the shape and orientation of the matrix have significant influence on the initiation, development of adiabatic shear bands of the WHA penetrators. For a fixed volume percentage of Fe-Ni-W particles, different distributions result in different deformation modes and patterns of shear bands.

KEY WORKS: Tungsten heavy alloy penetrator, impact, adiabatic shear band

INTRODUCTION

Experiments have shown that the depleted uranium penetrators penetrate deeper than the tungsten rods. Magness and Farrand (1990) explained this difference by suggesting that shear bands near the penetrator nose develop sooner in depleted uranium (DU) than in tungsten heavy alloys (WHA), leading to the earlier failure and "self-sharpening" effects for depleted uranium penetrators. Recently, it was found that the pre-twisted WHA penetrators show better ballistic performance than that of un-twisted ones, Wei et al. (1998). It was found that the channel diameter in target for pre-twisted WHA was thinner than that of the un-twisted WHA and the ballistic limit velocity of pre-twisted WHA was less than that of un-twisted WHA. The reason can also be attributed to the "self-sharpening" effect due to the microstructure influence of the penetrator material, Wei et al. (2000). This motivated us to numerically study the influence of shape and orientation of microstructure of tungsten particles/matrix on the initiation and development of adiabatic shear bands in tungsten composites during impact. A penetration problem in general will involve three-dimensional deformations of the WHA. Here we study, for simplicity, the impact of composite WHA penetrators normal to rigid target under plane strain deformation condition. The method used in this study is similar to that used by Batra et al.(1995,1998). The detailed investigations of microstructure influence on the adiabatic shear localization and on ballistic performance of composite WHA penetrators were given elsewhere, Wei (2001). It is can be seen from this paper that a possible way to improve the ballistic performance of WHA penetrators in the future is to design and fabricate the penetrators at micro/meso-scale level.

FORMULATION OF THE PROBLEM

We analyze thermalmechanical deformations of a WHA penetrator impacting at normal incidence a rigid target by considering the following material configurations for the penetrator's material: pure tungsten, Fe-Ni-W matrix, for comparison purpose, and WHA alloys, in which undeformed or artificially elongated Fe-Ni-W particles are periodically distributed in W particles. No separation across the interface between two different material phases is allowed. The deformation is assumed to be locally adiabatic (heat conduction is neglected) and the following Johnson-Cook relation (1983) is used.

$$\sigma = (A + B\varepsilon^n)\left(1 + C \ln \frac{\dot{\varepsilon}}{\dot{\varepsilon}_0}\right)[1 - (T^*)^m] \quad (1)$$

where $T^* = (T - T_r)/(T_m - T_r)$, A, B, C, n, m are material parameters, T_r, T_m are room and melting temperature, respectively. $\dot{\varepsilon}_0$ is reference strain rate.

In order to obtain the parameters of the pure tungsten and Fe-Ni-W matrix phase, we refer to Zhou (1993), who obtained the following constitutive relation, which differs from the Johnson-Cook relation only in the dependence of the flow stress upon the temperature rise, i.e. the different temperature softening coefficients $1 - (T^*)^m$ in Eqs.(1) and $1 - \beta[(T/T_R)^q - 1]$ in Eqs.(2).

$$\sigma = (A + B\varepsilon^n)\left(1 + C \ln \frac{\dot{\varepsilon}}{\dot{\varepsilon}_0}\right)\left\{1 - \beta\left[\left(\frac{T}{T_R}\right)^q - 1\right]\right\} \quad (2)$$

We fit Johnson-Cook by least square method using the dada provided by Zhou(1993) listed below. For Tungsten, we have $m = 0.536$ and for Fe-Ni-W, $m = 0.638$. Other parameters in Johnson-Cook model are the same as that given by Zhou. From Fig.1 it can be found that the fitted curves based on Johnson-Cook model are in good agrement with Zhou et al. model over long range, except the tungsten at low temperature(for example, from 400K to 700K). For simplicity, we implement the Johnson-Cook model to the ABAQUS code.

Tungsten: $A = 730 MPa$, $B = 562 MPa$, $n = 0.0751$ $C = 0.02878$, $q = 0.15$, $\beta = 2.4$
$G = 155 GPa$, $T_r = 293K$ $\rho = 19300 kg/m^3$ $c = 138 J/kg.K$ $\dot{\varepsilon}_0 = 1.355 \times 10^{-7} s^{-1}$

Fe-Ni-W: $A = 150 MPa$, $B = 546 MPa$, $n = 0.208$ $C = 0.0838$, $q = 0.2$, $\beta = 2.4$
$G = 98.84 GPa$, $T_r = 293K$ $\rho = 9200 kg/m^3$ $c = 382 J/kg.K$ $\dot{\varepsilon}_0 = 6.67 \times 10^{-7} s^{-1}$

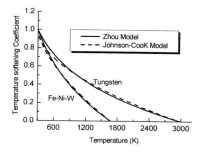

Figure 1: The comparison between the Zhou et al. model and Johnson-Cook model

Initially, all the blocks (length of 30mm, half width of 5mm and thickness of 2mm) considered are moving with a uniform speed of 250 m/s in the direction normal to the plane surface of the rigid target. The effect of heat conduction, failure criterion, the friction between the rigid target and the block surface are not considered here. The problem is solved by the explicit finite element code ABAQUS, and the 6-noded wedge elements with one-point integration rule are used. In order to minimize the CPU time, only half of the problems were run for all the blocks according to their symmetry. The finite element mesh had 120 uniform elements in vertical direction and 40 elements in horizontal direction. To calculate the temperature rise due to plastic deformation under approximately adiabatic condition a heat conversion coefficient of 0.9 is assumed. Six cases are considered in this computational study. Two of them for pure tungsten and Fe-Ni-W matrix, for comparison purposes, and four cases for composite blocks. For composite cases, the weak matrix elements, each of them is consisted of 8 elements, are periodically distributed in the tungsten elements. For convenience, the overall volume fraction of the matrix is selected as 20%, slightly higher than that for conventional 93%wt WHA, and is held constant. The inclination angle of the elongated weak matrix in the blocks is just 45^0 clockwise or counterclockwise to the horizontal level. All weak matrix groups have the same volume. The detailed arrangement is assigned as follows. For case 1 and case 2, the elongated and inclined matrix is located on the outer half of the block near the periphery with inclined 45^0, as shown in Fig.3 (a) (left for case 1 and right for case 2). However, the orientations of weak matrix are different for case1 and case 2. The distribution of case 1 approximately presents the microstructure of the pre-twisted WHA to some degree, Wei (1998,2000). For case 3 and case 4, the elongated and inclined matrix is disposed in the inner of the block with inclined 45^0, as shown in Fig.4 (a) (left for case 3 and right for case 4). The difference of matrix orientations between case 3 and case 4 is evident. The rest of the matrix groups in above cases are squares without any inclination.

RESULTS AND DISCUSSION
Homogenous Materials

Fig.2 depicts the deformed meshes of pure tungsten (left) and Fe-Ni-W (right) blocks at the same time. It is can be seen that there is a large plastic deformation at the impact end of the specimen, and the block particles outside of the mushroomed region do not undergo much plastic deformation. It should be mentioned that the deformation is roughly uniform, neither the pure tungsten nor the matrix shows the localized deformation.

Figure 2: Initial and deformed configurations for pure W (left) and Fe-Ni-W (right)

Composite Materials
Cases 1 and 2

The arrangements of these weak matrix elements in cases1 and case2 in the reference configuration are combined together and shown in Fig.3(a). Two typical graphic views of the evolution process of the deformed meshes for each of the two blocks are presented in Fig.3 (b)(c). The results shown here contrast dramatically with those of homogenous pure tungsten and Fe-Ni-W blocks just mentioned above. The most noticeable characteristics is that there are several areas of intense deformation initiating from the weak matrix and propagating toward the bottom of the block. A scrutiny of the equivalent plastic strain rate distribution and temperature field, not shown herein, reveals the emerging localized deformation band.

However, some differences also appear between case 1 and case 2. For the same amount of overall deformation, shear strain concentration is much severer, and the shear band extension is larger in case1 than that in case 2, indicating the strong influence of the orientation of matrix on the initiation and propagation of shear band. It is hardly discern the localized deformation within the block in case 2 except near the periphery. It is interesting to note that the region near the periphery in case 2 has a pair of shear bands in each site, where a matrix group resides. One is along the direction of the inclined and elongated matrix and another is cross the former one with a certain angle along the direction of local maximum shear stress. However, it is should be emphasized that the deformation slip on the periphery in case1 is much higher than case 2.

Since the specimen geometry, loading conditions and percentage of the volumes of the weak matrix are identical for case 1 and case 2, the differences in deformation patterns of the two materials must be due to the different orientation of the weak matrix.

In case 1, the direction of the elongated matrix is approximately parallel to the direction of local maximum shear direction through all the deformation process, so the retardation effect of the hard tungsten on the shear band formation becomes lessened and the tendency to adiabatic shear localization increases. Shear localization begins in the soft phase and propagates along a favorable path, and eventually causes the "self-sharpening" effects under certain stress situations, for example, in penetration process. However, in the cases studied here, due to the lack of sustained stress concentration, the localized deformation can not proceed anymore, so it is "frozen" in the blocks.

Contrast to case1, in the case 2 the direction of the elongated matrix is approximately parallel to the direction of local maximum shear direction in advance at small deformation stage. As the deformation proceed, due to the rotation of the material in the mushroom region, hard grains located in the paths of shear band impede the development of shear band. The tendency of the shear deformation becomes lessened. However, since the significant shear stress concentration in the transition region between the mushroom and straight body, the shear band can form along another direction, where the maximum shear stress appears. This reasoning is consistent with the numerical results.

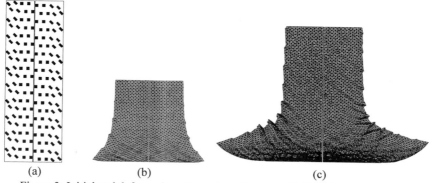

Figure 3: Initial and deformed configurations for case 1 (left) and case 2 (right)

Cases 3 and 4

The arrangements of these weak matrix elements in the reference configuration are shown in Fig.4 (a), called case 3 and case 4, respectively. The deformation time in Fig.4(b) is the same as that in Fig.3(c). The deformation patterns in Fig.4 are similar to those in Fig.3. However, some differences really exist in the appearance, profiles and general patterns of the deformed meshes between them. For example, the deformation slip on the periphery in case 3 and case 4 is the same, this phenomenon is evidently contrast to the difference between case1 and case 2, indicating the arrangement of the weak matrix has

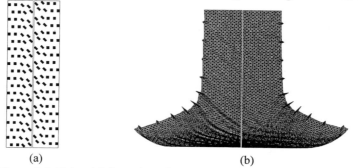

Figure 4: Initial and deformed configurations for case3 (left) and case 4 (right)

strong influence on the development of the shear band. It is can be seen that pronounced shear band can form easier and earlier in the case1, where the inclined and elongated matrix is located on the outer half of the block, than case3, where the inclined matrix on the inner half of the block

It should be mentioned that the mesh used herein is not fine enough to capture the dimensions of real shear band. Therefore, the shape and length of the localized deformation band in above graphs will only be rough estimates.

CONCLUSIONS

1. Due to the inhomogeneities of the morphology of the microstructure, the localized deformation always occurs easier and earlier in the composite than in pure tungsten and matrix.
2. The orientation of elongated weak matrix has significant influence on the deformation pattern, especially on the susceptivity of adiabatic shear localization.

Acknowledgement—This work is supported by the Chinese National Science Foundation (project No. 10002017) and the National Defense Key Laboratory (Project No. OOJS75.1.3.ZK0103)

References

Batra R.C, Peng Z. (1995). Development of shear bands in dynamic plane strain compression of depleted uranium and tungsten blocks, *Int. J. Impact. Engng.* **16:3**, 375-395

Batra R.C. Wilson N.M. (1998). Adiabatic shear bands in plane strain deformation of a WHA. *International Journal of Plasticity* **14:1-3**, 43-60

Johnson G.R. Cook W.H. (1983) A constitutive model and data for metals subjected to large strain rates and high temperatures. *Proceedings of the Seventh International Symposium on Ballistics*, The Hague, The Netherlands, 541-548.

Magness Jr L.S. Farrand T.G.(1990). Deformation Behavior and its Relationship to the Penetration Performance of High-Density KE Penetrator Materials. in *Proc. 1990 Army Science Conference,* Durham, NC, 149-164.

Zhou M.(1993). Dynamic shear localization in a tungsten heavy alloy and ductile rupture in a spheroidized 1045 steel, Doctoral dissertation, Brown University, Providence, RI,USA.

Zhou M. Needleman A. and Clifton R.J.(1994). Finite element simulations of dynamic shear localization. *J. Mech. Phys. Solids* **42**: 423-458

Zhou M.(1998). The growth of shear bands in composite microstructures, *International Journal of Plasticity* **14**, 733-754.

Wei Z.G. Hu S.S. Li Y.C. et al.(1998a) Dynamic properties and ballistic performance of pre-torqued tungsten heavy allays, In C.V.Niekerk ed. *Proceeding of the 17th International Symposium on Ballistics.* South Africa, The South Africa Ballistics Organization, **3**,391-398

Wei Z.G. Yu J.L. et al.(1998b) Adiabatic shear localization of pre-twisted tungsten heavy alloys under dynamic compression/shear loading.*3th International Symposium on Impact Engineering*, Singapore.

Wei Z.G. Yu J.L. Hu S.S. et al.(2000). Microstructure influence on adiabatic shear localization of tungsten heavy alloys, *International Journal on Impact Engineering* **24**,747-758.

Wei Z.G (2001). Dynamic behavior of tungsten heavy alloys and ballistic performance of WHA penetrators. Doctoral Dissertation, University of Science and Technology of China. P.R.China.

A STUDY ON IMPACT FATIGUE BEHAVIOR OF STRUCTURE STEELS UNDER REPEATED TENSILE LOADS
- INVESTIGATION FOR A HIGH STRENGTH STEEL -

Ken'ichi KITAURA and Hiroo OKADA

Graduate School of Engineering, Osaka Prefecture University
Sakai-shi, Osaka 599-8531, JAPAN

ABSTRACT

In this study, to obtain the basic data on impact fatigue behavior of structure steels, low- and medium-cycle impact fatigue tests are carried out on several kinds of specimens of a high strength steel of 590 MPa class, *i.e.*, smooth specimens and notched specimens under repeated tensile loads. The features of impact fatigue behavior of structure steels are clarified through discussions for experimental results.

KEYWORDS

Impact fatigue, Structure steels, Fracture behavior, Low- and medium-cycle, Repeated tensile load, High strength steel, Smooth specimen, Notched Specimen, Cyclic creep, Permanent strain rate

INTRODUCTION

This study deals with impact fatigue behavior of structure steels under low- and medium-cycle tensile loads. First, two types of testing machines are produced for investigating impact fatigue behavior of steels under repeated tensile loads. Next, impact fatigue tests are carried out on several kinds of specimens of a high strength steel of 590 MPa class, *i.e.*, smooth specimens and notched specimens under repeated tensile loads. Finally, the features of impact fatigue behavior of structure steels are clarified through discussions for experimental results.

EXPERIMENTAL APPARATUS AND METHOD

In figures 1 and 2, two types of testing machines are shown for investigating impact fatigue behavior of steels under repeated tensile loads. Fig. 1 shows arrangement of low-cycle impact fatigue testing machine (spring-weight type). In this figure, ① represents a bar held rigidly at the top end and carrying a specimen ⑦ and stop at lower end onto which a moving weight (hammer) ⑤ by spring ③ can be forged. Fig. 2 shows arrangement of high-cycle testing machine (revolution-hammer type). In this figure, ① represents a bar held rigidly at the top end and carrying a specimen ⑦ and stop at lower end onto which a revolution ball (hammer) ⑤ by motor ⑧ can be forged.

By using these testing machines, low- and medium-cycle impact fatigue tests are carried out on several kinds of specimens of a high strength steel of 590MPa class, *i.e.*, smooth specimens and notched specimens under repeated tensile loads. In Fig. 3, shape and dimension of specimens are shown. Mechanical properties and the chemical composition of materials are given in tables 1 and 2.

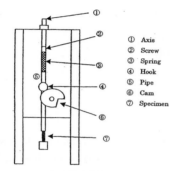

Figure 1: Experimental apparatus of low-cycle testing machine (Spring-weight type)

① Axis
② Screw
③ Spring
④ Hook
⑤ Pipe
⑥ Cam
⑦ Specimen

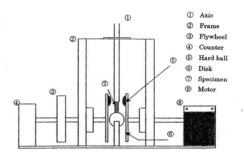

Figure 2: Experimental apparatus of high-cycle testing machine (Revolution-hammer type)

① Axis
② Frame
③ Flywheel
④ Counter
⑤ Hard ball
⑥ Disk
⑦ Specimen
⑧ Motor

TABLE 1 CHEMICAL COMPOSITION (%)

Material	C	Si	Mn	P	S
HT600	0.09	0.26	1.7	0.017	0.003

TABLE 2 MECHANICAL PROPERTIES

Material	K_t	Ultimate stress MPa	Reduction of area %	Fracture strain
A	1.0	646.8	66.8	1.10
B	1.7	923.2	51.6	0.726

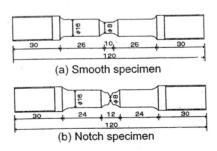

(a) Smooth specimen

(b) Notch specimen

Figure 3: Shape and dimension of specimens

EXPERIMENTAL RESULTS AND DISCUSSIONS

Experimental Results and Discussions for Spring-Weight Type Tests
Deformation Behavior

The process of fracture of smooth and notched specimens having small K_t-values is creep type as well as that of mild steel specimens [Kitaura, and et al. (1990)]. The process of fracture of notched specimens having large K_t-values is local creep type in the case of high stress level and is crack type in the case of low stress level. Moreover, permanent strain ε is evaluated by using the change of the diameter at the minimum cross-section of the specimen in the following

$$\varepsilon = 2 \cdot \ln(d_0/d) \qquad (1)$$

where d is diameter measured in the N cycle and d_0 is initial diameter of the specimen.

Variation of Impact Stress Range Ratio

The typical cyclic behavior of stress range measured by two strain gages attached on the bar ① is shown in Fig. 5. In almost range, stress range ratio is constant.

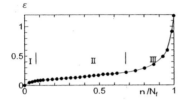
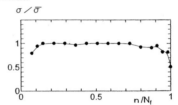

Figure 4: Cyclic behavior of permanent strain ε Figure 5: Cyclic behavior of stress ratio $\sigma/\bar{\sigma}$

Variation of Permanent Strain Rate

Pursuing permanent strain ε, it is found that there exist three types of cyclic behavior of ε depending on above mentioned types (A), (B) and (C) of the process of fracture and the stable stage where permanent strain rate $\Delta \dot{\varepsilon}_c$ is constant in each type occupies most of fatigue life. The relation between $\Delta \dot{\varepsilon}_c$ and $\bar{\sigma}$ is shown in Fig. 6, and it can be expressed by

$$\bar{\sigma}/\sigma_B = S(\Delta \dot{\varepsilon}_c/\varepsilon_f)^\beta \qquad (\Delta \dot{\varepsilon}_c = \Delta \varepsilon_c/\bar{T}) \qquad (2)$$

where S and β are experimental constants. σ_B and ε_f are the ultimate strength and the fracture ductility of the material under the static tensile test. $\bar{T}$ is the duration of maximum stress under the impact fatigue test.

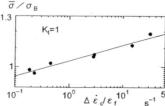

Figure 6: Relation between $\bar{\sigma}/\sigma_B$ and $\Delta \dot{\varepsilon}_c/\varepsilon_f$

318

Relationship between Deformation Behavior and Fatigue Life

Next, the relation between permanent strain rate $\Delta \varepsilon_c$ and the cycle number N_{us} at the starting point of the unstable stage is given for the case of the two creep type (A) and (B) in Fig. 7. At the same time, it is found that these relationships for the case of the crack type (C) are different from above relations because the value of $\Delta \varepsilon_c$ is fairly smaller than that for the creep types. And the relation between N_{us} and impact fatigue life N_f is given in Fig. 8. From both figures, the relatoin between $\Delta \varepsilon_c$ and N_f can be expressed in the following

$$(\Delta \varepsilon_c / \varepsilon_f) N_f^m = C \tag{3}$$

where m and C are material constants

Relationship Between Fatigue Strength and Fatigue Life

Next, the relation between impact tensile stress $\bar{\sigma}$ and impact fatigue life N_f for the case of the two creep type (A) and (B) are given in Fig. 9. The relation between impact stress range $\bar{\sigma}$ and fatigue life N_f for the smooth specimen is expressed by

$$\bar{\sigma} (N_f \cdot \bar{T})^n = D \tag{4}$$

where n and D are experimental constants.

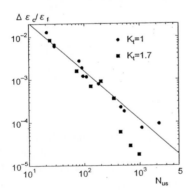

Figure 7: Relation between $\Delta \varepsilon_c / \varepsilon_f$ and N_{us}

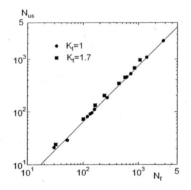

Figure 8: Relation between N_{us} and N_f

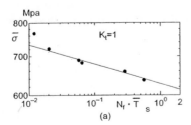

(a)

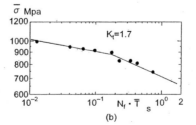

(b)

Figure 9: Relation between $\bar{\sigma}$ and $N_f \cdot \bar{T}$

Moreover, the strength parameter D in equation (4) can be estimated by using the ultimate strength σ_B and the reduction in area Φ in the following form [Kitaura, and et al. (1990)]:

$$D = (0.8 + 0.0016\ \Phi)\ \sigma_B \qquad (5)$$

This estimation shows comparatively good agreement with the experimental results as shown in Fig. 10.

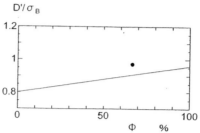

Figure 10: Relation between D/σ_B and Φ

Experimental Results and Discussions for Revolution-hammer Type Tests

Variation of Impact Stress Range

The typical cyclic behavior of stress ratio for the revolution-hammer type testing machine is shown in Fig. 11. It is found that the behavior is similar to that for spring-weight type testing machine. That is, in almost range, stress ratio is constant.

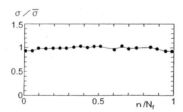

Figure 11: Cyclic behavior of stress ratio $\sigma/\bar{\sigma}$

Relationship Between Fatigue Strength and Fatigue Life

Next, the relation between impact tensile stress σ and impact fatigue life N_f for the cases of low- and medium-cycle tests are given in Fig. 12. Relations between $\bar{\sigma}$ and $N_f \cdot \bar{T}$ are also given in Fig. 13.

It is seen from these figures that the relation between σ and impact fatigue life N_f for the case of the spring-weight type testing machine is different from that for the case of the revolution-hammer type testing machine. On the other hand, the relation between $\bar{\sigma}$ and $N_f \cdot \bar{T}$ is almost same to each other for cases of smooth specimens and notched specimen with the small stress concentration factor.

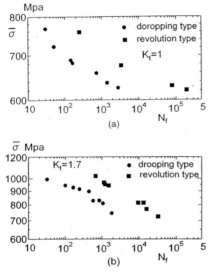
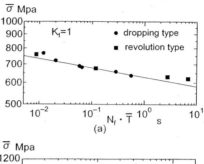
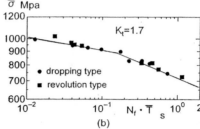

Figure 12: Relation between $\bar{\sigma}$ and N_f

Figure 13: Relation between $\bar{\sigma}$ and $N_f \cdot \bar{T}$

CONCLUSIONS

This study dealt with impact fatigue behavior of structure steels under low- and medium-cycle tensile loads. First, two types of testing machines were produced for investigating impact fatigue behavior of steels under repeated tensile loads. Next, impact fatigue tests were carried out on several kinds of specimens of a high strength steel of 590 MPa class, *i.e.*, smooth specimens and notched specimens under repeated tensile loads. Finally, the features of impact fatigue behavior of structure steels were clarified through discussions for experimental results.

References

K. Kitaura, Y. Okamura, and H. Okada (1990), On the Effects of Static Stress and Pre-strain on Low-cycle Impact Fatigue of Steels under Axial Load, Journal of the Kansai Society of Naval Architects, Japan, **214**, 187-195 (in Japanese).

CORRELATION OF DYNAMIC BEHAVIOR WITH MICROSTRUCTURAL-BIAS IN TWO-PHASE TiB$_2$+AL$_2$O$_3$ CERAMICS

[a]Andrew Keller, [b]Greg Kennedy, [b]Louis Ferranti, [a]Min Zhou, and [b]Naresh Thadhani;

[a]*Woodruff School of Mechanical Engineering,* [b]*School of Materials Science and Engineering, Georgia Institute of Technology, Atlanta, GA 30332-0405*

ABSTRACT

The dynamic compressive strength and microscopic failure behavior of TiB$_2$/Al$_2$O$_3$ ceramic composites with a range of microstructural morphologies and size scales are analyzed. A split Hopkinson pressure bar (SHPB) is used to achieve loading rates of the order of 400 s^{-1}. The dynamic compressive strength of the materials is found to be between 4.3 and 5.3 GPa, indicating a strong dependence of strength on microstructure. Microstructures with finer phases have higher strength levels. The dynamic strength levels are approximately 27% higher than the values of 3-4 GPa measured at quasi-static loading rates for these materials. These strength levels are also higher than the strength levels of monolithic TiB$_2$ and Al$_2$O$_3$ under similar dynamic conditions. A soft recovery mechanism in the experimental configuration allows the specimens to be subjected to loading under a single, well-defined stress pulse. Scanning electron microscopy (SEM) and energy dispersive spectrometry (EDS) indicate that failure associated with the Al$_2$O$_3$ phase is transgranular cleavage in all microstructures. On the other hand, failure associated with the TiB$_2$ phase is a combination of transgranular cleavage and intergranular debonding and varies with the microstructures. Quantitative image analysis shows that the measured compressive strength of the materials directly correlates with the fraction of TiB$_2$-rich areas on fracture surfaces.

KEYWORDS: dynamic failure resistance, strength, ceramic composites, Hopkinson bar, impact, spall strength

INTRODUCTION

Cceramics possess many desirable properties which contribute to their increasing use in areas previously dominated by metals and metallic alloys. Examples of such applications include cutting tools, drill bits, wear parts, sensors, structural and electronic components, electrodes, biomechanical devices, lightweight armor, and gas turbine components. Ceramics are well-suited for such applications due to their excellent mechanical properties at high temperatures, high strength, excellent chemical stability, and creep, wear, oxidation and impact. Unfortunately, ceramics are also characterized by a brittle nature which can potentially lead to sudden and catastrophic failure. One method to reduce brittleness and enhance failure resistance is the development of ceramic composites. For example, Niihara et al. (1991) reported that a 5% population of SiC nanoparticles increases the tensile strength of Si_3N_4 from 350 MPa to 1 GPa and improves its fracture toughness from 3.25 $MPa\sqrt{m}$ to 4.7 $MPa\sqrt{m}$. Although microstructure-induced, size-dependent toughening mechanisms at the micro and nano levels are demonstrated approaches for property enhancement, so far such effects have not been well quantified. In order to develop more advanced materials, it is necessary to characterize the influences of phase morphology, phase length scale, and interfacial behavior on fracture toughness. Recently, two-phase ceramics of titanium diboride/alumina (TiB_2/Al_2O_3) with a range of phase sizes and phase morphologies have been developed (Logan, 1992a,b). These materials have shown a wide range of fracture toughness values and some of the values are higher than those of both constituents produced separately in bulk. These materials provide an opportunity to study the correlation between microstructure and mechanical behavior and are analyzed in this research.

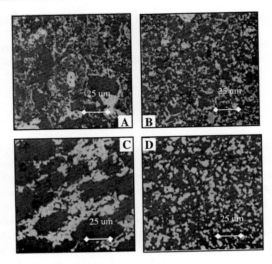

Fig. 1 Microstructures of Materials Analyzed (TiB_2 white, Al_2O_3 dark)

MATERIALS

The materials are produced through self-propagating high temperature synthesis (SHS) and mechanical mixing (MM) of powder followed by hot pressing (Logan 1992b; Carney 1997). Each material has a nominal composition of 70% Al_2O_3 and 30% TiB_2 by weight. Figure 1 shows the microstructures of the four Al_2O_3/TiB_2 composites analyzed. To quantify the different phase morphologies and sizes, the linear intercept length (LIL) (Underwood, 1970) for the two phases are measured on digital images of the microstructures. Kennedy et al. (2000) analyzed the dependence of the Hugoniot Elastic Limit (HEL) and spall strength on microstructure. Using a cohesive finite element method (CFEM), Zhai et al. (1999, 2000) have quantified the effect of microstructure on energy release rate for the materials analyzed here.

EXPERIMENTS

In order to characterize the response of the materials to dynamic compressive loading, a split Hopkinson pressure bar (SHPB) apparatus as illustrated in Fig. 2 is used. This experimental apparatus permits time-resolved analyses of material response to transient compressive loading. Soft recovery of specimens is achieved through the use of a momentum trapping technique developed by Nemat-Nasser et al. (1991). This mechanism allows specimens to be recovered after loading under a single stress pulse, eliminating any unintended reloading. The specimen fragments were collected for postmortem analysis and characterization of microscopic failure. Two strain gauges are mounted on opposite faces of the specimen to measure its strain and strain rate. The specimen stress is calculated using signals from strain gauges mounted on the incident and transmission bars, see Follansbee (1985). A pair of tungsten carbide platens is placed between the specimen and the bars to prevent the bar ends from being indented. Gradual loading is achieved through the use of a copper disc between the striker bar and the incident bar, as in Nemat-Nasser et al. (1991). More details of the experiments are given in Keller (2000).

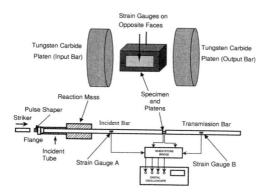

Fig. 2 A Schematic Illustration of the Experimental Configuration

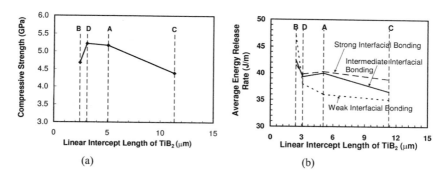

Fig. 3 Correlation between Failure resistance and Linear Intercept Length,
(a) measured compressive, (b) calculated average energy release rate

COMPRESSIVE STRENGTH

Microstructures A and D show significantly higher failure resistance than microstructures B and C. The superior levels of failure resistance are reflected in their higher average failure stresses than those of B and C. Figure 3(a) shows the correlation between the compressive strength and the average linear intercept length. Since material B has a much higher level of porosity than the other materials, attention is given only to materials A, C, and D. A clear trend is seen. Higher strengths coincide with smaller LIL. This can be phenomenologically explained. Microstructures with smaller LIL values present a more heterogeneous media for propagating cracks, inducing more tortuous cracks paths and, therefore, increasing energy dissipation for crack growth. This finding is consistent with the results of the explicit fracture modeling of Zhai et al. (2000) and Zhai (2000) using the same microstructures. The micromechanical quantification of the fracture resistance of the four microstructures accounts for the actual phase distributions and arbitrary fracture patterns. A comparison of the experimental results and computed results are given in Fig. 3(b). The calculated average energy release rate is plotted as a function of the LIL. Note that in the modeling, porosity in microstructure B is not considered, therefore, only results from microstructures A, C, and D should be used to assess the effect of the LIL. Three different levels of TiB_2/Al_2O_3 interphase bonding strength are considered. Clearly, a consistent trend is observed in both the experiments and the calculations, regardless of the level of interphase bonding strength.

FRACTOGRAPHY

The fragments of fractured specimens are recovered and the morphologies of the fracture surfaces are analyzed. Coinciding energy dispersive spectrometry (EDS) maps showing the distributions of Al_2O_3-rich and TiB_2-rich areas are taken for each region analyzed. The simultaneous use of these phase maps and fractographs greatly facilitates the identification of phases on fracture surfaces and the intergranular or intragranular nature of the fracture process. The fractographs and the corresponding EDS maps for microstructure A are shown in Fig. 4.

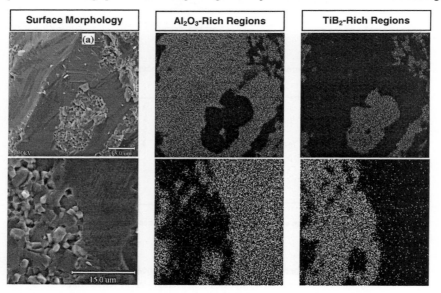

Fig. 4 SEM Images of Fracture Surfaces after Experiment, Microstructure A

The failure surfaces show large regions of Al_2O_3 surrounded by a continuous network of TiB_2. The Al_2O_3 areas are smooth, indicating transgranular cleavage. Failure involving the TiB_2 phase appears to be primarily intergranular pullout. Approximately 74% of the fracture surfaces is cleavage planes within the Al_2O_3 and 26% are associated with the TiB_2. Note that the volume fraction of Al_2O_3 in this material is 72.6%. The fracture surfaces for microstructure B are dominated by small regions of transgranular cleavage of the Al_2O_3 phase. Failure involving the TiB_2 phase occurred primarily by intergranular pullout. TiB_2 particles are relatively evenly distributed on the fracture surfaces, suggest a fracture process which is not significantly affected by the presence of the particles. Numerous pores were observed in the Al_2O_3 phase, consistent with the measured low density (95% of theoretical value) of this material. Approximately 84% of the fracture surfaces is Al_2O_3-rich, significantly higher than the 72.6% volume fraction of Al_2O_3 in this material. Since Al_2O_3 has a lower strength and lower fracture energy compared with TiB_2, this higher ratio of fracture surfaces in Al_2O_3 appears to be the primary reason for the observed low strength reported earlier. A primarily transgranular mode of fracture inside Al_2O_3 is also observed for microstructure C. However, failure involving the TiB_2 phase occurred through a combination of transgranular cleavage and intergranular pullout. The fact that cleavage in Al_2O_3 occurs over multiple grains contributes to its lower failure resistance, due to the fact that the effectiveness of the TiB_2 as a reinforcing phase is decreased. Approximately 77% of the fracture surfaces is Al_2O_3-rich, higher than the 72.6% volume fraction of Al_2O_3 in the material. The failure in microstructure D is unique in that transgranular cleavage is the primary fracture mechanism for both Al_2O_3 and TiB_2. Although failure involving the TiB_2 phase occurred through a combination of transgranular cleavage and intergranular pullout, the dominant mechanism is cleavage. This is in sharp contrast to what is observed for the other microstructures. It also appears that the homogeneous distribution of TiB_2 reinforcement inhibits intergranular separation, thus forcing cracks to go through the stronger TiB_2 phase and enhancing failure resistance. Approximately 71% of the fracture surfaces is Al_2O_3-rich, lower than the 72.6% volume fraction of Al_2O_3 in this material. This is the only microstructure that shows a TiB_2-rich fracture surface fraction higher than its corresponding volume fraction in the material. This bias toward the TiB_2 points to an unusual shift of failure into the stronger phase and can be associated with the significantly higher strength. Clearly, increasing fracture surface areas in Al_2O_3 corresponds to lower strength. Microstructure D shows the lowest fraction of Al_2O_3 fracture surfaces (72.6%) among all materials analyzed. This is also the only microstructure that shows an Al_2O_3 surface fraction lower than the volume fraction of this phase. This indicates a preference for failure in the TiB_2 phase, leading to the higher strength of this material.

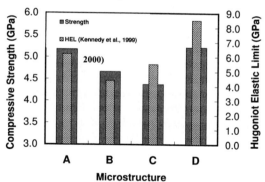

Fig. 5 A Summary of compressive strength and HEL

DISCUSSION

The analysis using four TiB_2/Al_2O_3 ceramic composites with different microstructural morphologies and size scales has yielded dynamic compressive strength between 4.3 and 5.3 GPa, indicating a strong dependence of strength on microstructure. Microstructures with finer phases as indicated by smaller linear intercept length values have higher strength levels. This result follows the general relationship between grain size and strength (Nordgen and Melader, 1988). Also, low density has a significant negative influence on strength. The dynamic strength levels are approximately 27% higher than the values of 3-4 GPa measured at quasi-static loading rates. Scanning electron microscopy (SEM) and energy dispersive spectrometry (EDS) indicate that failure associated with the Al_2O_3 phase is transgranular cleavage in all microstructures. On the other hand, failure associated with the TiB_2 phase is a combination of transgranular cleavage and intergranular debonding and varies with the microstructures.

The fraction of TiB_2-rich fracture surface areas directly correlates with the measured compressive strength of the materials. The size distributions of the fragments indicate no direct correlation between strength and average fragment size for the materials under the conditions analyzed. The addition of titanium diboride reinforcement clearly improves the failure resistance of the materials. Hot-pressed monolithic alumina has a reported compressive strength of 3.00 GPa under quasi-static loading conditions. At a strain rate of 10^3 s^{-1}, compressive strength is between 3.70-4.00 GPa, (Logan, 1992b). At a strain rate of 412 s^{-1}, the weakest of the four materials tested (material C), has a compressive strength of 4.39 GPa, significantly higher than that for the monolithic matierial. The four materials also have higher strength levels than that reported for hot pressed titanium diboride under quasistatic loading conditions.

A comparison of the compressive strength and the HEL of the four materials is given in Fig. 5. The HEL values for these materials are obtained by Kenndy et al. (2000). There is a strong correlation between these two measures, with microstructure D showing the highest resistance followed by microstructure A. This correlation confirms the findings on the influence of phase size and failure mechanism described in this paper.

ACKNOWLEDGEMENT

Support from grant DAAG55-98-1-0454 from the Army Research Office and CAREER grant CMS9984289 from the National Science Foundation is gratefully acknowledged. We are grateful to Dr. K. V. Logan for providing the test materials and for helpful discussions.

REFERENCES

Anderson, C.E., Morris, B.L., (1992). The Ballistic Performance of Confined Al_2O_3 Ceramic Tiles, *International Journal of Impact Engineering*, 12(2), 167-187;

Carney, A.F., (1997). The Effect of Microstructure on the Mechanical Properties of a Titanium Diboride/Alumina Composite, Masters Thesis, Georgia Institute of Technology, Atlanta, GA;

Follansbee, P., (1985). The Hopkinson Bar, *Metals Handbook*, 8, 198-203, Amer. Soc. Metals, Metals Park, Ohio;

Keller, A. R., (2000). An Experimental Analysis of the Dynamic Failure Resistance of TiB_2/Al_2O_3 Composites, M.S. Thesis, Georgia Institute of Technology;

Kennedy, G.B., Keller, A.R., Russell, R.R., Ferranti, L., Zhai, J., Zhou, M., Thadhani, N., (2000). Dynamic Mechanical Properties of Microstructurally Biased Two-Phase $TiB_2+Al_2O_3$ Ceramics, *manuscript in preparation*;

Logan, K.V., (1992a). Shapes Refractory Products and Method of Making the Same, U.S. Patent # 5141900;

Logan, K.V., (1992b). Elastic-Plastic Behavior of TiB_2/Al_2O_3 Produced by SHS, Ph.D. Dissertation, Georgia Institute of Technology;

Nemat-Nasser, S., Isaacs, J.B., Starrett, J.E., (1991). Hopkinson Techniques for Dynamic Recovery Experiments, *Proc. R. Soc. Lond.*, 435, 371-391

Niihara, K., Nakahira, A., Sekino, T., (1991). New Nanocomposite Structural Ceramics, *Material Research Society Symposium Proceedings*, 286, 405-412;

Nordgen, A., Melander, A., (1988). Influence of Porosity on Strength of WC-10%Co Cemented Carbide, *Powder Metallurgy*, 31(3), 189-200;

Underwood, E.E., (1970). *Quantitative Stereology*, Addison Wesley Publishing Company;

Zhai, J. and Zhou, M., (1999). Micromechanical Modeling of Mixed-mode Crack Growth in Ceramic Composites, *Mixed Mode Crack Behavior*, ASTM STP 1359, 174-200, K. Miller and D. L. McDowell, eds.;

Zhai, J. and Zhou, M., (2000). Finite Element Analysis of Micromechanical Failure Modes in Heterogeneous Brittle Solids, *International Journal of Fracture*, 101, 161-180.

COMPUTATIONAL ANALYSIS OF OBLIQUE PROJECTILE IMPACT ON HIGH STRENGTH FABRIC

V.B.C. Tan, V.P.W. Shim and N. K. Lee

Impact Mechanics Laboratory,
Department of Mechanical Engineering
National University of Singapore
10 Kent Ridge Crescent, Singapore 119260

ABSTRACT

This paper presents the results of computational simulation of oblique projectile impact and perforation of woven aramid fabric. The fabric is modelled as a network of pin-jointed viscoelastic springs with strain rate dependent failure strains. Effects of crimping within the fabric are also accounted for. The model is verified by its good agreement with ballistic tests involving normal impact of fabric samples in terms of fabric deformation and energy absorption characteristics. For oblique impacts, the model predicted that the energy absorbed by the fabric during perforation shows similar trends for low energy impact regardless of angle of obliquity. The energy absorbed by the fabric increases linearly with impact energy in the low impact energy regime up to a critical impact energy when the fabric deformation becomes localised and energy absorption drops. This critical impact energy was found to increase with angle of obliquity. Energy absorbed by the fabric as strain energy is almost constant for the low impact energy regime but starts to decrease when the critical impact energy is exceeded. Fabric kinetic energy is lower than strain energy just above the ballistic limit but becomes the major form of energy dissipation as impact energy increases.

KEYWORDS

Fabric armour, aramid fibre, oblique impact, ballistic

INTRODUCTION

Twaron, a woven fabric made from high strength aramid fibres, possesses high impact resistance. These fabrics are widely used in flexible armour applications because of their high strength to weight ratio while maintaining good drapability. The ballistic performance of high strength fabric is still not well understood and most papers present results on normal impacts only. This paper discusses the response of plain-woven Twaron™ T713 fabric under both normal and oblique impacts based on results from computational simulation. The fabric is represented by a model described by Shim et al. (1995). The simulation is validated by comparison with experimental data from normal impact tests.

COMPUTATIONAL MODELLING AND SIMULATION

Computational simulations were performed for both normal and oblique impacts. Results of ballistic tests for normal impacts were used for verification. In both simulations and impact tests, the fabric specimens had dimensions of 120 mm by 110 mm and were clamped along the 120 mm edges. The spherical steel projectile had a mass of 7.0g and radius of 6mm. It was modelled as a smooth rigid sphere in the simulation. The impact velocity ranged from 100 m/s to 430 m/s.

The woven fabric is modelled as a network of nodes, pin-jointed to one another by one-dimensional viscoelastic elements. Each node corresponds to a crossover point of the orthogonal yarns in the weave. They are assigned a mass such that the overall areal density of the target is preserved.

Twaron is a woven fabric made from aramid fibres, i.e. PPTA (p-phenyleneterephthalamide) fibres, which exhibit significant strain rate sensitivity. The fibres can be considered to be made entirely of highly aligned PPTA molecular chains. Based on the morphology of PPTA fibres, Termonia and Smith (1988) represented the fibres by chains of nodes connected in series by strong intramolecular bonds. Each molecular chain interacts with neighbouring molecular chains through weak but numerous intermolecular bonds. When the fibres are stretched, both primary (intramolecular) and secondary (intermolecular) bonds extend. At high strain rates, the degree of intermolecular slippage decreases; hence, it is mainly the primary bonds that extend. At low strain rates, extension of the weaker secondary bonds becomes significant.

The constitutive response of Twaron yarns (Shim et al., 2001) can be represented as a three-element linear viscoelastic model shown in Fig. 1. When the spring-dashpot system is subjected to a load, both the springs will extend but the extension of spring K_2 will be restricted by the Newtonian dashpot at high strain rates. In effect, the extension of spring K_2 represents intermolecular slippage being restrained at high strain rates and the extension of the system is mainly due to that of spring K_1 representing the primary bonds.

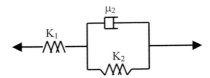

Fig. 1 Three-element viscoelastic solid constitutive model

At low strain rates, failure of fibres results primarily from the breakage of weaker secondary bonds whereas at high strain rates, this intermolecular slippage is restricted, causing fibres to fail from breakage of the primary bonds (chain scission). Such strain rate dependent failure is achieved in the three-element model by assigning a failure strain to spring K_1 ($\varepsilon_{1,max}$) and a failure strain to spring K_2 ($\varepsilon_{2,max}$). Although the failure strains of primary and secondary are constant, the total strain at failure varies with strain rate, giving rise to rate sensitive failure strains.

Crimp describes the degree of undulation (wave like appearance) of yarns due to the weave architecture. A fraction of the total strain is due to the straightening of crimped yarns and not the extension of the yarns. This crimp effect is included in the numerical analysis by assuming that a fraction of the total strain does not give rise to stresses.

Based on the theoretical strengths of highly oriented PPTA fibres, K_1 and μ_2 are assigned values of 160 GPa and 35 MPas respectively. The value of K_2 is obtained from the equation 1 where K_{static} is the stiffness of the fibres under quasi-static loads (80 GPa).

$$K_2 = (K_1 K_{static})/(K_1 - K_{static}) \qquad (1)$$

The value of $\varepsilon_{1,max}$ is estimated to be 5% while $\varepsilon_{2,max}$ is obtained from equation 2 so that the elongation at break under static load equals the static failure strain $\varepsilon_{static, max}$, (3.3%).

$$\varepsilon_{2,max} = (K_1 - K_{static}) \varepsilon_{static,max} / K_1 \qquad (2)$$

NORMAL IMPACT

To check the validity of the simulation, numerical results are compared to test results of normal impacts. The energy absorbed by the fabric specimens for the range of impact energies tested is shown in Fig. 2. The energy absorbed by the fabric refers to the drop in kinetic energy of the projectile after it perforates the fabric specimen. Both the numerical and experimental results show that in the low impact energy regime, the energy absorbed increases almost linearly with impact energy up to a critical impact energy, at which energy absorbed by the fabric is maximum; beyond this impact energy, the energy absorbed decreases. The decrease in energy absorbed is sharper than that predicted by the simulation. At high impact energies (i.e. impact energies greater than the critical impact energy), the energy absorbed is almost constant.

When a projectile impacts a fabric, the impacted yarns (primary yarns) are pushed outwards directly by the projectile and material is pulled in towards the impact point. As the impact progresses, the transverse wave front and the in-plane stress waves travel further away from the impact point to orthogonal yarns via the crossover points in a zigzag manner, thereby causing more fibre elements to stretch and more fibre particles to be pulled towards the impact point. This causes a pyramid shaped deflection centered at the impact point. Such a deflected shape is also predicted in the simulation.

Figure 3 shows the predicted fabric deformation at the instant prior to perforation. At low impact energies, stress waves and transverse waves have enough time to propagate away from the impact point before perforation occurs, resulting in deformation of the entire fabric. This gives rise to high levels of energy absorption. Although the model does not permit ravelling of the yarns, it shows that there is significant pull-in of the free edges, which corresponds to the larger degree of ravelling at low velocities as shown in Fig. 3. At low velocities, the fabric exhibits a high degree of stretching corresponding to the widespread creasing observed in test specimens illustrated in Fig. 3. The numerical simulation also shows some tearing at the clamped edges as observed in test specimens perforated at low velocity.

At high impact energies, the contact time between the fabric and projectile is short. Therefore the transverse deflection does not have enough time to travel outwards to the edges before the failure strain of the fibre elements at the impact point is exceeded. As a result, the degree of deformation is localized around the impact point. This means that the stretching of the fibre elements and their transverse deflection are much less. Hence, energy absorbed in the form of strain and kinetic energy is reduced. Figure 3 shows that at a high impact energy, the free edges are pulled in to a lesser extent, which corresponds to the smaller degree of raveling observed in test specimens. The degree of stretching is more localized at high velocity, which matches the less extensive creasing observed in test specimens.

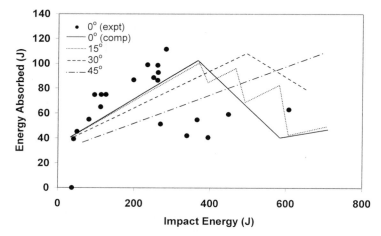

Fig. 2 Variation of energy absorbed with impact energy
impact angle, θ = 0° to 45°

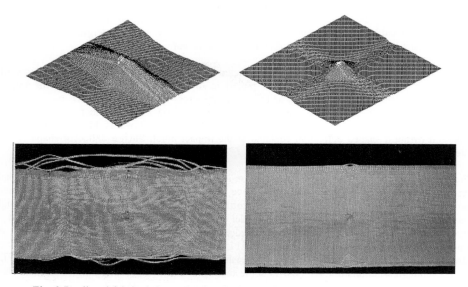

Fig. 3 Predicted fabric deformation just before perforation and test specimens after perforation for normal impact. (left 200J, right 560J)

OBLIQUE IMPACT

Figure 3 also shows the predicted energy absorption characteristics of Twaron fabric specimens for impacts at different angles of obliquity (θ = 0°, 15°, 30° and 45° to the normal). The target size, boundary conditions, projectile mass and geometry remain unchanged. The energy absorbed by the fabric initially increases with impact energy for all angles of obliquity up to a value of about 100J before it starts to drop. The only exception is when the impact angle is 45°, where the energy absorbed shows no apparent decrease. The decrease in energy may occur at an impact energy above the range of impact energies in the simulations. It is also predicted that the rate of increase in energy absorbed with impact energy decreases with angle of obliquity.

The extent of fabric deformation is a good indication of the amount of energy absorbed by the fabric because it shows the amount of material affected by the impact and hence the amount of fabric involved in absorbing impact energy. Prevorsek et al. (1989) reported that the energy absorbed by the fabric is proportional to the volume within its deflected surface. As fabric deflection is not possible without some stretching of the fabric, the size of the deflection is an indication of the amount fabric being strained. The extent of the deflection also gives an indication of how much of the fabric is put into transverse motion. The kinetic energy of the deflected yarns constitutes an important component of energy absorbed by the fabric.

Figure 4 shows the division of energy absorbed by the fabric via strain and kinetic energy for impact at 30° obliquity just prior to perforation. Strain energy constitutes a larger portion of the energy absorbed compared to kinetic energy for impacts just above the ballistic limit. Beyond this range, the energy absorbed becomes predominantly that of kinetic energy of the fabric. At the critical impact energy there is a drop in both strain and kinetic energy of the fabric. The strain and kinetic energy of the fabric at other impact angles show similar trends.

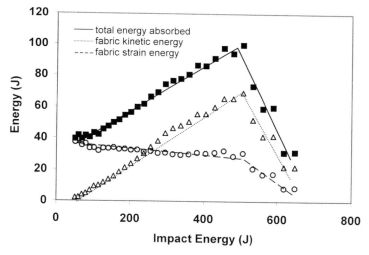

Fig. 4 Variation of fabric energy with impact energy for θ = 30°

Below the critical impact energy, the strain energy remains fairly constant as the impact energy increases because the entire fabric specimen is deflected regardless of impact energy in this regime. However, the kinetic energy absorbed increases with impact energy causing the total energy absorbed by the fabric to increase. This increase in kinetic energy is due to more momentum being transferred to the deflected fabric with higher impact velocity.

Figure 5 shows fabric deformation at various angles of obliquity just before perforation for an impact energy of 560J. The degree of deformation and stretching changes from a localized region around the impact point for normal impact to a fully developed transverse deflection for oblique impact at 45°. It can be seen from Fig. 3 that this impact energy exceeds the critical impact energy for all impact angles except 45°. This clearly demonstrates that with increasing obliquity, the fabric is able to disperse the impact energy of the projectile more effectively. Both the kinetic and strain energy absorbed by the fabric increase with obliquity. This is because at high obliquity, the transverse wave fronts have more time to propagate to the clamped edges before the impacted fiber elements fail. This results in extensive transverse deflection, which accounts for the high kinetic energy and high degree of stretching and deformation giving rise to the large fabric strain energy. It should also be noted that at high angles of obliquity, the point of perforation occurs farther away from the point of impact. At low obliquity, it can be seen that the amount of deformation and extent of stretching is very localized around the impact region because fiber elements near the impact point fail before the transverse wave fronts can propagate to the edges.

Fig. 5 Fabric deformation just before perforation at 560J impact energy

CONCLUSION

The impact perforation of Twaron fabric has been investigated through computational simulation. The simulation models the fabric as a network of nodes connected together by three-element viscoelastic springs that mimic the mechanical behaviour of PPTA fibres. Results of the simulation were verified by experimental results for normal impacts on fabric samples. The model predicts that the energy absorbed by the fabric has two distinct regimes regardless of impact angle. The low impact energy regime is described by increases in energy absorbed with impact energy. The rate of increase of energy absorbed with impact energy decreases with impact angle. Within this regime, fabric samples are fully deflected and the edges of the fabric are pulled in towards the impact point significantly. The maximum energy absorbed by the fabric is not sensitive angle of obliquity. Transition from the low impact energy regime to the high impact energy regime is marked by a drop in energy absorbed by the fabric. Within the high impact energy regime, fabric deformation is localised around the point of perforation. Energy is absorbed primarily in the form of kinetic energy of the fabric rather than strain energy of the fabric.

REFERENCES

Prevorsek D.C., Kwon Y.D., Harpell G.A. and Li H.L. (1989). Spectra Composite Armour: Dynamics of Absorbing the Kinetic Energy of Ballistic Projectiles. *34th International SAMPE Symposium*, 1780-1789.

Shim V.P.W., Lim C.T. and Foo K.J. (2001). Dynamic Mechanical Properties of Fabric Armour. *Int. J. Impact Engng* 25:1, 1-15.

Shim V.P.W., Tan V.B.C. and Tay T.E. (1995). Modelling Deformation and Damage Characteristics of Woven Fabric Under Small Projectile Impact. *Int. J. Impact Engng* 16:4, 585-605.

Termonia Y. and Smith P. (1988). A Theoretical Approach to the Calculation of the Maximum Tensile Strength of Polymer Fibers. *High Modulus Polymers*, Marcel Dekker Inc., 321-362.

ANALYSIS OF DYNAMIC THERMAL STRESSES IN MODERATELY THICK SHELLS OF REVOLUTION OF FUNCTIONALLY GRADED MATERIAL WITH TEMPERATURE-DEPENDENT PROPERTIES

E. Inamura[1], S. Takezono[2], K. Tao[2] and T. Kawasaki[3]

[1]Department of Mechanical Engineering, Tokyo Metropolitan College of Technology,
Higashi-ohi, Shinagawa-ku 140-0011, Japan
[2]Department of Mechanical Engineering, Toyohashi University of Technology,
Tempaku-cho, Toyohashi 441-8580, Japan
[3]KEYENCE Corporation,
Meitoh-hondohri, Meitoh-ku, Nagoya 465-0087, Japan

ABSTRACT

This paper is concerned with the numerical analysis of dynamic thermal stress and deformation in moderately thick shells of revolution made of functionally graded material with temperature-dependent properties by using the finite element method. The temperature distribution through the shell thickness is expressed by a curve of high order, and thick shell elements are utilized for the analysis. As numerical examples, two kinds of shells composed of SUS304 and PSZ subjected to thermal loads due to heat generation in the shell bodies are treated; one is a cylindrical shell and another is an axisymmetric shell having a parabolic meridian. In comparison with the results from the characteristic method, a good agreement is obtained. It is also found that temperature distributions, stress distributions and deformations are significantly influenced by temperature-dependent properties of the materials.

KEYWORDS

Structural Analysis, Functionally Graded Material, Temperature-Dependent Properties, Dynamic Thermal Stress, Thick Shells, Elasticity, FEM

INTRODUCTION

Structures made of functionally graded material(FGM) have been actively developed under severe thermal conditions owing to their excellent mechanical and thermal properties. It is well known that the material properties are dependent on the temperature, so the influence of temperature dependence appears greatly where large temperature variation occurs. For dynamic thermal stress analyses of structural components and/or mechanical components of FGM, some investigations have been carried out considering temperature-dependent material properties. These investigations, however, are almost limited to the simple geometries such as plates, hollow cylinders and hollow spheres(e.g. Sumi & Sugano(1997) and Sumi et al.(1996),(1998)). The authors proposed the numerical method for the dynamic thermal stress problem of moderately thick shells of revolution made of FGM subjected to thermal loading due to fluid or heat generation, not considering temperature-dependent properties of the materials (Takezono et al.(2000)). On the other hand, the quasi-static thermal stress problem considering temperature-dependent material properties was analyzed by using the finite element method (Inamura et al.(2001)). Then in the present paper, an analytical method for the dynamic thermal stress problem is developed by using this numerical method. Material parameters are given by functions of the shell thickness, and each coefficient of basic equations related to material properties is determined from numerical integration through shell thickness. In order to improve the accuracy of the solutions, the temperature distribution through the thickness is supposed to be a curve of high order. Then using the obtained temperature distribution, the stresses and deformations are derived from the thermal stress equations.

As numerical examples, functionally graded axisymmetric shells composed of stainless steel(SUS304) and partially stabilized Zirconium (PSZ) subjected to impulsive heat generation due to electromagnetic radiation are analyzed. Firstly, we analyze the long cylindrical shell subjected to inner surface heating. The obtained results are compared with ones obtained by characteristic method (Sumi et al.(1998)), and the accuracy of the solutions is evaluated. Secondly, we analyze the parabolic shells of revolution subjected to outer surface heating with cosine distribution along the meridian. The results are compared with those from the theory not considering the temperature dependence of material properties.

FUNDAMENTAL EQUATIONS

The parametric expressions of a middle surface of shells of revolution may be written as $r = r(s)$ and $z = z(s)$, where s is the meridional length along the middle surface from a boundary, and r and z are coordinates of the middle surface as shown in Figure 1.

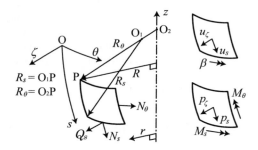

Figure 1: Coordinates and notations

Heat Conduction Equations

If the shell is subjected to axisymmetric heating, the equation of heat conduction at a point in the shell body, whose material properties vary along the meridional and thickness directions, is given in the orthogonal curvilinear coordinates (s, θ, ζ) as follows;

$$\rho c \frac{\partial T}{\partial t} - \frac{1}{RL_s L_\theta}\left[\frac{\partial}{\partial s}\left(R\lambda \frac{L_\theta}{L_s}\frac{\partial T}{\partial s}\right) + R\frac{\partial}{\partial \zeta}\left(\lambda L_s L_\theta \frac{\partial T}{\partial \zeta}\right)\right] - \eta = 0 \qquad (1)$$

where $L_s = 1 + \zeta/R_s$, $L_\theta = 1 + \zeta/R_\theta$, R_s and R_θ are radii of curvature, T is the temperature at (s, ζ, t), c is the specific heat, ρ is the mass density, λ is the coefficient of thermal conductivity, η is the heat generation per unit volume and per unit time, and t is time. c, ρ and λ are functions of T.

Making use of the shape function matrix $[N_T]$, which connects the temperature in the internal parts of the element with the nodal temperature vector $\{\phi\}$, and employing the finite element method based on the Galerkin method, we obtain the following finite element equation of heat conduction considering heat transfer on the shell surface,

$$[K^e]\{\phi\} + [C^e]\frac{\partial\{\phi\}}{\partial t} = \{F^e\} \qquad (2)$$

$$\begin{aligned}
{[K^e]} &= \int_{V^e} \lambda\left[\frac{1}{L_s^2}\frac{\partial[N_T]^T}{\partial s}\frac{\partial[N_T]}{\partial s} + \frac{\partial[N_T]^T}{\partial \zeta}\frac{\partial[N_T]}{\partial \zeta}\right] dV + \int_{S^e} K[N_T]^T[N_T]\, dS \\
{[C^e]} &= \int_{V^e} \rho c[N_T]^T[N_T]\, dV, \quad \{F^e\} = \int_{V^e} \eta[N_T]^T\, dV + \int_{S^e} K\Theta[N_T]^T\, dS
\end{aligned} \qquad (3)$$

In the above equations, V^e and S^e are volume and boundary area of the element, respectively. K is the coefficient of heat transfer and Θ is the ambient fluid temperature of the shell.

To improve the accuracy of the solutions, the temperature distribution through the shell thickness is expressed by a curve of p-th degree with respect to temperature and its derivatives on the middle surface as follows, and the value of p is decided in consideration of the convergence of the solutions.

$$T(s, \zeta, t) = T|_{\zeta=0} + \zeta\left.\frac{\partial T}{\partial \zeta}\right|_{\zeta=0} + \frac{\zeta^2}{2}\left.\frac{\partial^2 T}{\partial \zeta^2}\right|_{\zeta=0} + \cdots + \frac{\zeta^p}{p!}\left.\frac{\partial^p T}{\partial \zeta^p}\right|_{\zeta=0} \qquad (4)$$

Equations of Motion

Using the obtained temperature distribution, the stresses and deformations are derived from the thermal stress equations. The equations of motion for the element are derived by the principle of virtual work, and thick shell elements are used in the analysis. In the present paper the stress component σ_ζ, normal to the middle surface, can be assumed to be neglected, and the membrane forces and the resultant moments per unit length shown in Figure 1 are obtained as follows;

$$\{N_s, N_\theta, Q_s, M_s, M_\theta\} = \int_{-h/2}^{h/2} \{\sigma_s L_\theta, \sigma_\theta L_s, \sigma_{s\zeta} L_\theta, \sigma_s L_\theta \zeta, \sigma_\theta L_s \zeta\}\, d\zeta \qquad (5)$$

NUMERICAL METHOD

The derivatives with respect to time in the heat conduction equation are treated by the Crank-Nicolson method. λ, c and ρ at each time step are initially evaluated with the temperature at the previous time step. The heat conduction equations are solved by using these initially evaluated values, and the temperature distribution is obtained. Again, new values of λ, c and ρ are evaluated with the obtained temperature, and by using these values modified temperature distribution is calculated. These procedures are continued until the temperature distribution is converged enough. Then by using the obtained temperature, Young's modulus E, Poisson's ratio ν and the coefficient of linear thermal expansion α are evaluated, and the deformations and the resultant forces are obtained from equations of motion. In order to calculate the inertia term in equations of motion the Houbolt method is used. For the integration along the meridian and through the shell thickness, the Gauss-Legendre quadrature and the Simpson's 1/3 rule are applied, respectively.

NUMERICAL EXAMPLE

As numerical examples, two geometrical types of functionally graded shells composed of SUS304 and partially stabilized Zirconia (PSZ) subjected to impulsive thermal loads due to electromagnetic wave are treated. The irradiated surface and the opposite surface consist of PSZ only and SUS304 only, respectively, and the volumetric ratio of both materials varies lineally through shell thickness. Material constants E, ν and α are evaluated by the Mori-Tanaka theorem (Mura 1987) that is available for particle reinforced composites, and λ, c and ρ are assumed to be lineally dependent on the volume fraction (Sugano et al.(1993)). Material constants of SUS304 and PSZ except for ν and ρ are given as a function of temperature as shown in Figure 2. ν and ρ vary so slightly with the temperature that they are supposed to be constant. ν is 0.3 for both materials and ρ of SUS304 and ρ of PSZ are $8.00 \times 10^3 \text{kg/m}^3$ and $6.07 \times 10^3 \text{kg/m}^3$, respectively. In the following two numerical examples shell surfaces are assumed to be adiabatic, and the reference and initial temperatures are both 300K.

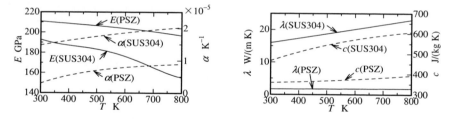

Figure 2: Temperature-dependence of SUS304 and PSZ (Sugano et al. (1993))

Each element has two nodes and in the Gaussian quadrature two integration points along the meridian are used. The division number through shell thickness is chosen to be 128. The increment of time Δt is selected as $0.1 \mu s$ in Example 1 and $2 \mu s$ in Example 2. Also in both examples p in Eqn. 4 is chosen to be 3. These values are decided in consideration of convergency of the solutions, capacity of the computer and computing time.

Example 1: Long cylindrical shell

A long cylindrical shell subjected to axisymmetric heat generation irradiated by electromagnetic wave on the inner surface is treated (Figure 3). In this figure t_0 is the irradiation time of electromagnetic energy. ρ_0 and c_0 are a reference density and a reference specific heat, respectively, and the material constants of PSZ at the reference temperature T_0 are used for these parameters. The obtained results are compared with solutions by the characteristic method (Sumi et al.(1998)) based on three dimensional theory. In the characteristic method the division number through the shell thickness is chosen to be 500, and strain component in the axial direction ε_z is supposed to be zero.

Figure 4 shows the variations of temperature and stress σ_θ with time on the inner, middle and outer surfaces of the shell. The solutions from the present theory are in good agreement with the results from the characteristic method.

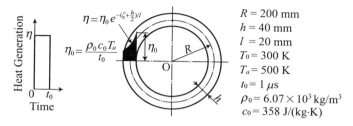

Figure 3: Example 1

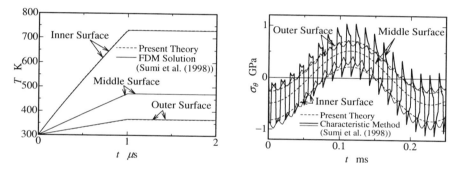

Figure 4: Variations of temperature T and stress σ_θ with time

Example 2: Parabolic shell of revolution

The simply supported axisymmetrical parabolic shell subjected to irradiation by electromagnetic wave on outer surface is analyzed (Figure 5). The heat generation η due to irradiation varies along meridian and through thickness as shown in this figure. η_0 is given as $\eta_0 = \rho_0 c_0 T_a/t_0$, and ρ_0, c_0, t_0 are the same as example 1. At the boundary (Point B), displacement in the axial direction of the shell is zero. The

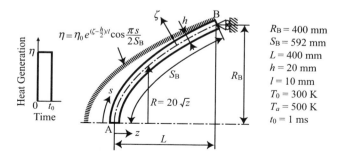

Figure 5: Example 2

results considering temperature dependent properties of the material are compared with ones ignoring temperature-dependency and using material constants at fixed temperature 300K or 500K. The number of nodal points along the meridian is 101.

Figure 6 shows the variations of temperature, displacement u_ζ and stress σ_θ with time at the vertex of the shell (Point A), where σ_s is equal to σ_θ. The temperature and stresses considering temperature-dependent material properties are almost similar to the results using material properties at 500K, but are different from ones at 300K. On the other hand displacements from this theory are different from the any results using material properties at 300K and 500K. So the analysis considering temperature-dependent material properties is very important to obtain more accurate solutions.

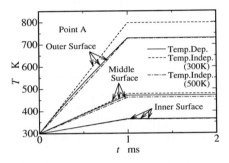
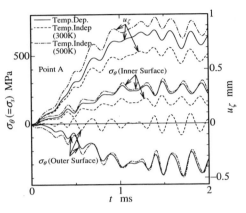

Figure 6: Variations of temperature T, stress σ_θ (= σ_s) and displacement u_ζ with time at point A

REFERENCES

Houbolt J.C. (1950). Journal of Aeronautical Science. *A Recurrence Matrix Solution for the Dynamic Response of Elastic Aircraft* **17**, 540-550.

Inamura E., Takezono S. and Tao K. (2001). Transactions of the Japan Society of Mechanical Engineers. *Analysis of Transient Thermal Stresses in Moderetely Thick Shells of Revolution of Functinally Graded Material with Temperature-Dependent Properties* **A 67:653**, 142-149 (in Japanese).

Mura T. (1987). *Micromechanics of Defects in Solids*, Martinus Nijhoff.

Sugano Y. Kataoka S and Tanaka K. (1993). Transactions of the Japan Society of Mechanical Engineers. *Analysis of Transient Thermal Stresses in a Hollow Circular Cylinder of Functionally Graded Material with Temperature-Dependent Material Properties* **A 59:562**, 1505-1513 (in Japanese).

Sumi N., Monna H. and Sugano Y. (1996). Transactions of the Japan Society of Mechanical Engineers. *Dynamic Thermal Stresses in a Composite Hollow Circular Cylinder with an Interlayer of Functionally Graded Material* **A 62:597**, 1189-1196 (in Japanese).

Sumi N. and Sugano Y. (1997). Journal of Thermal Stresses. *Thermally induced stress waves in functionally graded materials with temperature-dependent material properties* **20:3**, 281-294.

Sumi N., Monna H. and Sugano Y. (1998). Transactions of the Japan Society of Mechanical Engineers. *Dynamic Thermal Stresses in a Functionally Graded Material Subjected to Electromagnetic Radiation* **A 64:618**, 333-338 (in Japanese).

Takezono S., Tao K., Inamura E. and Ozawa Y. (2000). Transactions of the Japan Society of Mechanical Engineers. *Thermal Stress and Deformation in Moderately Thick Shells of Revolution of Functionally Graded Material under Thermal Impulsive Loading* **A 66:645**, 1060-1067 (in Japanese).

IMPACT EROSION ON INTERFACE BETWEEN SOLID AND LIQUID METALS

M. Futakawa[1], H. Kogawa[1], Y. Midorikawa[2], R. Hino[1], H. Date[3] and H. Takeishi[2]

[1]Japan Atomic Energy Research Institute
Tokai-mura, Naka-gun, Ibaraki-ken, 319-1195, Japan
[2]Chiba Institute of Technology
2-17-1, Tsudanuma, 275-8588, Japan
Tohoku Gakuin University
1-13-1, Tagajyo, 985-8537, Japan

ABSTRACT

JAERI is carrying out R&Ds for constructing the facility of spallation neutron source which may bring us innovative science fields. A high power proton beam will be injected into the liquid mercury target for the neutron yield. The mercury vessel will be subjected to the pressure waves generated by rapid thermal expansion. The pressure waves propagate from the liquid mercury into the vessel solid metal, and vice versa. The pressure waves may induce erosion on the interface between the vessel solid metal and the liquid mercury under certain loading conditions, e.g. impact. In order to investigate the impact erosion damage due to the pressure wave, we have carried out impact experiments on the mercury by using a modified conventional split Hopkinson pressure bar apparatus. The chamber for the mercury consists of the input, the out put bars and a collar. Disk specimens with 16 mm diameter and 5 mm thickness ; A6061, 316ss, Inconel 600 and Maraging steel, were set on the ends of both bars contacting the mercury. Many pits by the impact erosion were observed on the specimen surface at 40 and 80 MPa compressive pressure, which is sufficiently lower than the yield stress of each material. The ranking order of damage due to the impact erosion is A6061>316ss≅Inconel600>Maraging steel. In particular, A6061 was fractured completely after imposed 100 impacts.

KEYWORDS

Impact, Erosion, Pressure wave, Liquid metal, split Hopkinson pressure bar, Solid metal, Pit

INTRODUCTION

The Japan Atomic Energy Research Institute (JAERI) is conducting the research and development concerning the Neutron Science Project, aiming at the expansion of basic research such as neutron structural biology and material science using a high intensity neutron beam and nuclear technology in the 21st century including actinide transmutation with ADS (Accelerator Driven System)[Planning

Division for Neutron Science (1999)]. In this project, a MW-scale target which can produce the high intensity neutron beam by spallation reaction between proton beam and target materials is planed to be constructed.

The liquid mercury target system for the MW-scale target is developed in JAERI as taking into account advantages from the viewpoint of the self-circulating heat removal and the neutron yield. Figure 1 shows a schematic drawing of the liquid mercury target structure. The proton beam with 1 µs pulse duration is injected into the mercury through the beam window at 25 Hz. The moment the proton beam hits the target vessel, stress waves are imposed on the beam window and pressure waves generate in the mercury by the thermal shock due to the proton beam injection. The pressure waves travel in the mercury to the vessel wall and vice versa. The stress waves excited by the pressure wave propagate in the vessel wall. As a result, the dynamic stress distribution in the vessel becomes very complicated. To understand the dynamic behavior due to the interaction between the liquid mercury and the solid vessel, has been carried out the international ASTE collaboration using AGS (the Alternating Gradient Synchrotron)[Futakawa(2000a)], in which the proton beam (24 GeV, proton intensity:10^{12} per a pulse, pulse duration: 40 ns) was injected into the cylinder-type target vessel filled with the mercury. The stress wave propagating behavior in the vessel wall was measured using a laser Doppler technique. The measured results were compared with calculated results using a liquid-elastic solid model for the mercury. The measured dynamic strain response of the vessel wall damped more slowly as compared with the calculated results and the superimposed high frequency components in the calculation was not observed in the experiment. The discrepancy between experimental and calculated results may be attributed to the modeling for the liquid mercury, because the liquid is likely to behave nonlinearly by viscosity and/or cavitation under certain negative pressure. It is deduced that negative pressures might be induced along the interface between the vessel wall and the liquid mercury because of the difference of inertia effect between them [Ishikura(2001)]. Provided that the negative pressure cause cavitations in mercury and the following collapse of the cavitations generates microjets to form pits on the surface of the vessel wall, the erosion damage accumulate on the surface of the vessel wall. It hazards the structural integrity and the life of the vessel.

In order to examine the impact damage on the interface between the liquid mercury and the solid metals, the plane-strain-wave incident experiment was carried out by using modified Hopkinson bar impact technique [Futakawa(2000b)]. The chamber for the mercury consists of an input, an out put bars and a collar. Disk specimens; A6061, 316ss, Inconel 600 and Maraging steel, were set on the ends of both bars in contact with the mercury. The relationship between impact erosion damage and the material properties will be discussed.

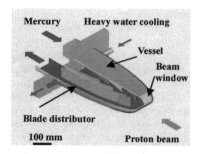

Fig. 1 Mercury target vessel

EXPERIMENT

The experimental apparatus for the stress-wave incident into the mercury, using the a modified conventional split Hopkinson pressure bar (SHB), consists of a striker of 500 mm length, the input bar of 1500 mm length, the output bar of 1500 mm length and an air gun to inject the striker to the input bar. The diameter of the bars is 16 mm. The detailed description of the apparatus is seen elsewhere [Futakawa(2000b)]. The chamber filled with the mercury is formed by the collar and the ends of the input and output bars. The specimens of 5 mm thickness and 16 mm diameter are screwed on the end of both bars. The chemical compositions and mechanical properties of the specimens made of A6061, 316ss, Inconel 600 and Maraging steel are shown in Tables 1 and 2. The distance between both specimens is set to be 5 mm. The impact velocities of the striker are 5.7 m/s and 2.0 m/s.

As assuming that a uniaxial homogeneous stress distribution is given in the axial direction of the specimen, the strain ε , the strain rate $\dot{\varepsilon}$ and the stress σ are obtained from the measured reflected strain ε_r and transmitted strain ε_t, as follows:

$$\varepsilon(t) = \frac{-2c_0}{L}\int_0^t \varepsilon_r(t')dt' \quad (1)$$

$$\dot{\varepsilon}(t) = \frac{-2c_0}{L}\varepsilon_r(t) \quad (2)$$

$$\sigma(t) = \frac{EA}{A_s}\varepsilon_t(t), \quad (3)$$

the pressure p(t), and volumetric strain in mercury $\Delta v/v$, therefore, are estimated from

$$\sigma(t) = p(t), \quad \varepsilon(t) = \Delta v(t)/v. \quad (4)$$

Table 1 Chemical compositions of specimens

(%)	C	Si	Mn	P	S	Ni	Cr	Mo	Fe
316ss	0.05	0.49	1.30	0.03	0.02	10.3	16.9	2.07	Bal

(%)	Si	Fe	Cu	Mn	Mg	Cr	Zn	Ti	Al
A6061	0.74	0.20	0.23	0.08	1.00	0.05	0.02	0.02	Bal

(%)	C	Mn	Fe	S	Si	Cu	Ni	Cr
Inconel600	0.07	0.36	9.28	0.002	0.21	0.11	73.30	15.83

(%)	Ni	Co	Mo	Al	Ti	Fe
Maraging steel	18.0	9.0	5.0	0.9	0.1	Bal

Table 2 Hardness H_V and yield stress

	316ss	A6061	Inconel600	Maraging steel
H_V	211	129	215	310
Yield stress(MPa)	204	283	356	1910

RESULTS

Material, Impact Number and Velocity Dependency

The strain responses measured on the input and output bars are illustrated in Fig.2. The amplitude of the strain wave traveled through the output bar becomes smaller than that of the incident strain wave injected through the input bar into mercury. That is because the acoustic impedance of mercury chamber was affected by the boundary conditions between the chamber and the input and output bars [Futakawa (2000b)].

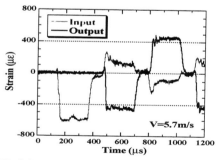

Fig.2 Strain time response on output/input bar

Figure 3 shows the surface of the specimens after imposed 100 impacts with the velocity of 5.7 m/s, which is estimated to generate the macroscopic compressive pressure of about 80 MPa in the mercury. Many pits are observed on the surface and the degradation due to the impact erosion is dependent on the kinds of materials. In particular, A6061 specimen shattered into some fragments up to 100 impacts as shown in Figure 3. To quantitatively evaluate the impact erosion damage, image analyses were applied to the surface. Figure 4 shows the relationship between the number of eroded pits and the eroded area after 10 impacts. The damages evaluated using the number of pits and the eroded area are not clearly distinguished between the specimens of the output and input bars. The materials dependency on the eroded damage exhibits the same tendency between the number of pits and the eroded area. The most sever damage was observed in A6061, and then the ranking order of damage is A6061> Inconel 600≅316ss>Maraging steel.

Figure 5 shows the relationship between the number of impacts with 5.7 m/s and the eroded area of the specimens fixed on the output bar. In the same figure, the results at Vi=2.7 m/s, corresponding to about 40 MPa macroscopic compressive pressure, in the 316ss only are plotted. The impact erosion damage evaluated by the number of pits and the eroded area increases with the impact velocity and the number of impacts regardless of the kinds of materials.

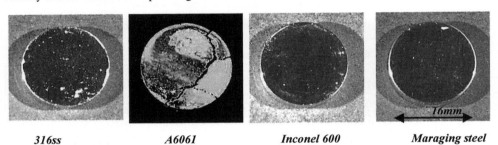

Fig. 3 Eroded surface of specimens after imposing 100 impacts with 5.7 m/s

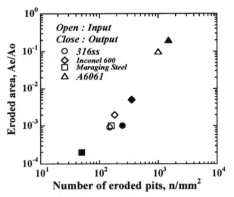

Fig. 4 Realtionship between number and area of eroded pits

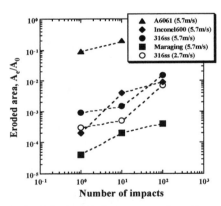

Fig.5 Relationship between number of impacts and eroded area

Feature of Eroded Pits

Figure 6 shows the micrograph around pits in 316ss at the various number of impacts with 5.7 m/s. The pits were formed not only at the same position to become larger but also at different zone to increase the number of the pits, as increasing the number of impacts. It is recognized from Figs. 3, 4 and 6 that the size and number of the pits of Maraging steel is the smallest among them. In A6061, many cracks propagate and link among pits to finally introduce failure, as shown in Fig.7. Figure 8 shows many slip lines formed around the edge of the pit due to plastic deformation in 316ss. Such slip lines were observed in Inconel 600 and Maraging steel, as well. Figure 9 shows SEM micrographs of the bottom of the pits. In the case of Inconel 600, striations were observed like typical fatigue degradation. The number of striations is about 10 that is almost equal to the number of impacts. In the case of A6061 and 316ss, some cracks are observed nearby the pit bottom.

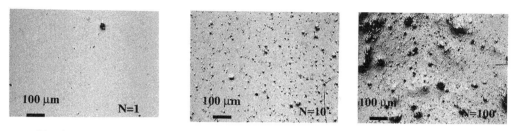

Fig. 6 Micrograph eroded surface in 316ss at various number of impacts with 5.7 m/s.

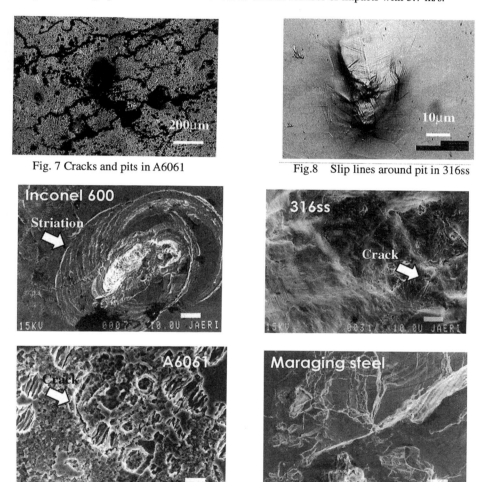

Fig. 7 Cracks and pits in A6061

Fig.8 Slip lines around pit in 316ss

Fig. 9 SEM micrograph nearby bottom of pits

DISCUSSION

The ranking order of the degradation due to impact erosion, A6061> Inconel 600≅316ss>Maraging steel, corresponds to that of the hardness H_V rather than the σ_y as recognized in Table 2. The impact load may produce the cavitation in the mercury along the interface between the specimen surface and mercury. The collapse of the cavitation bubble imposes high compressive pressures on microzone of the surface due to the microjet injection and cumulative damages cause pits like fatigue[Mori(1996)]]. In the case of the steels which have the hardness ranging from 100 H_V to 400 H_V, the fatigue strengths examined on the uniaxial tensile, bending and twisting fatigue tests are associated with the hardness and linearly increase with the hardness [JSME(1977)]. Additionally, Heymann reported that the erosion resistance defined as mass loss due to liquid impact erosion increases in proportion to $H_V^{5/2}$ [Heymann(1970)]]. These trends are recognized in our results on the liquid metal impact erosion by using the SHB technique. The hardness of non-damaged area hardly changed even after impacts because the microscopic pressure generated in mercury is not higher than the yield stresses of all materials used in the experiments. The hardness around the pits, however, got to be higher and many slip lines were observed as shown in Fig.8. It is deduced that the locally loaded pressures to make pits on the surface in contact with the mercury are higher than the yield stress. Aluminum alloys are reported to be completely embrittled by liquid mercury around room temperature [Price(1994)]. LME (liquid metal embrittlement) influences the pitting formation. The nickel base alloys become sensitive to LME in mercury that deteriorates the fatigue life. LME seems to assist the pitting formation caused by the impact erosion.

SUMMARY AND REMARKS

In order to examine the impact erosion damage on the liquid/solid metals interface, which might be produced on the inner wall of the target vessel filled with the mercury, the plane strain wave incident experiments were carried out by using the Hopkinson bar impact technique. The specimens made of 4-kind metals were installed to the input/output bars as being in contact with the mercury. Many pits due to erosion were observed on the specimen surface under impact loading with 40 and 80 MPa microscopic compressive pressures, which are sufficiently lower than the yield stress of each material. The ranking order of damage due to the impact erosion is A6061>316ss≅Inconel600>Maraging steel, which is the same order as the hardness of materials. In particular, A6061 was fractured completely after imposed 100 impacts. The impact erosion damage should be taken into account from the viewpoint of the structural integrity of the mercury target vessel.

REFERENCES

Futakawa M., Kikuchi K., Conrad H., Stechemesser H. (2000a), Pressure and stress waves in a spallation neutron source mercury target generated by high-power proton pulses, Nucl. Inst. Meth. in Phys. Res. A, 439pp.1-7.
Futakawa M., Kogawa H. and Hino R. (2000b), Measu rement of dynamic response of liquid metal subjected to uniaxial strain wave, J. Phys. IV France 10Pr9-237-242.
Heymann F. J. (1970), Toward quantitatiove prediction of liquid impact erosion, ASTM STP 474pp.212-248.
Ishikura S.(2001), Private communication.
JSME (1977), Mechanical Engineers' Handbook, pp.4-34-35.
Mori H., Hattori S., Okada T., Mizushima K. (1996), An approach to studying cavitation bubble collapse pressure and erosion, Kiron-A, Vol. 62, No.602, pp.2326-2332.
Planning Division for Neutron Science (1999), Proceeding of the 3rd workshop on neutron science project –Science and technology in the 21st century opened by intense spallation neutron source-, JAERI-Conf 99-003.
Price C. E.(1994), Contrasting LME in aluminum and nickel alloys with overtones to SCC, Corrosion 94, Paper No. 128, 1994.

FATIGUE STRENGTH OF CARBON STEELS UNDER REPEATED IMPACT TENSION

A. Chatani[1], A. Hojo[1], H. Tachiya[1], N. Yamamoto[2] and S. Ishikawa[3]

[1] Department of Mechanical Systems Engineering, Kanazawa University
2-40-20 Kodatsuno, Kanazawa-shi 920-8667, Japan
[2] Yamazaki Mazak Co. Ltd.
1 Oguchinorifune, Ooguchi-cho, Niwa-gun, Aichi-ken 480-0197, Japan
[3] Graduate School of Natural Science and Technology, Kanazawa University
2-40-20 Kodatsuno, Kanazawa-shi 920-8667, Japan

ABSTRACT

In order to clarify the impact fatigue strength, it is required to know the behaviour of slipped grains having close relation with fatigue fracture. So the repeated impact tensile experiments of carbon steels (S45C and SS400) were carried out and the slip bands in ferrite grains were observed by optical microscope. Main results obtained in case of S45C are as follows. It was confirmed that the impact fatigue strength is lower than the ordinary non-impact one, and the number of slipped grains under the repeated impact is less than that under the ordinary non-impact. In addition, strain hardening due to the repeated impact is also lower than that due to the ordinary non-impact.

KEY WORDS

Impact fatigue, Fatigue strength, Slip bands, Repeated impact tension, Stress waveform, Carbon steel

INTRODUCTION

Impact fatigue strengths of carbon steel have been obtained by many investigators, e.g. Tanaka et al (1992), Maekawa(1994), Yang et al(1994), Yang and Zhou(1994) and Chatani et al (1998). In addition, the overview of impact fatigue fracture was introduced by Yu et al(1999). From these results we can see that the impact fatigue strength depends on the stress waveform applied to the material, but that in high cycle range is often less than the ordinary fatigue strength. On the other hand, it is well known that cracking developed from slip bands causes fracture. For the reason, crack behaviours under repeated impact have been also studied by, e.g. Johnson et al (1990) and the others. Such a process as cracking, however, under the repetitions of impact stress has been little observed concerning slip bands initiation except a few of observations of Zhang et al (1999). Hence, it is required to clarify microscopic conditions of impact fatigue mechanism. Thus, the present paper shows the fatigue strength of steels S45C and SS400 under repeated impact tension and ordinary pulsating tension for comparison, and describes the initiation of slip bands in grains observed microscopically by replicas.

EXPERIMENTAL PROCEDURE

An outline of the originally built-up apparatus for repeated impact tension is shown in Figure1. Running mechanism of the apparatus is as follows. Slider 9 reciprocates with the rotation of disk 7 driven by motor 6. When slider hook 10 fixed at the slider strikes pipe hook 11 fixed at pipe 4 and coated with leathers at its both sides, the pipe guided along circular bar 3 is accelerated and strikes flange 2 near the right end of the bar. While the compressive stress is produced at the impact end of the pipe,the tensile stress is also produced in the bar and propagates in its left direction. It results in the application of impact tensile stress to specimen 1. The specimen is linked to base 5 and screwed into the left end of the bar 3. The tensile stress of the specimen disappears on arrival of the stress wave reflected from the other free end of the pipe. The duration of the stress and its waveform depend on the length L of the pipe, and so they are variable to an extent. L is 600 mm in the present experiment. The stress waveform is measured by strain gages 12 mounted at equal intervals (a,b) from the end of the bar. Main components (pipe,bar and son) of the present apparatus are made of steels. The central cut surfaces of the specimen as shown in the Figure1 are electropolished and nital etched for microscopic observations.

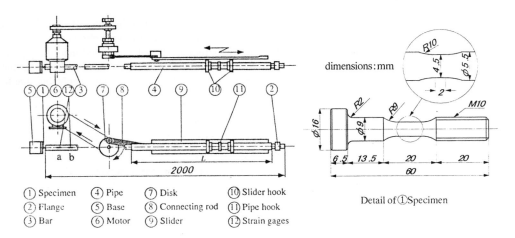

① Specimen   ④ Pipe   ⑦ Disk   ⑩ Slider hook
② Flange   ⑤ Base   ⑧ Connecting rod   ⑪ Pipe hook
③ Bar   ⑥ Motor ⑨ Slider   ⑫ Strain gages

Detail of ①Specimen

Figure 1: Experimental apparatus and specimen

Figure 2 shows triangular stress waveforms applied to the specimen. As is in Figure 2(a), the wave form is measured by the two-point strain measurement method based on the elementary theory of stress wave propagation developed by Lundberg and Henchoz (1977). The waveform of specimen in Figure 2(b) shows that corresponding to the former. Since the both are in good agreement, measured stress σ_{max} by the method of Figure 2(a).was used hereafter. The duration from loading to unloading is about 0.5 ms. The rate of repetitions of such an impact stress is 4 - 4.5 cycles/s. For

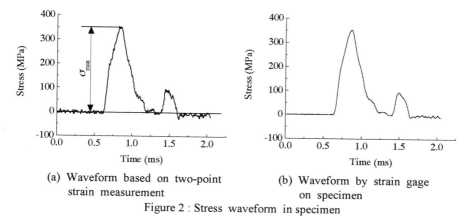

(a) Waveform based on two-point strain measurement

(b) Waveform by strain gage on specimen

Figure 2 : Stress waveform in specimen

TABLE 1
CHEMICAL COMPOSITION AND MECHANICAL PROPERTIES

	Chemical Composition (wt%)					Yield Point (MPa)	Tensile Strength (MPa)
	C	Si	Mn	P	S		
S45C	0.43	0.21	0.61	0.011	0.006	367	638
SS400	0.11	0.17	0.48	0.031	0.028	304	434

comparison, the other fatigue tester of loop dynamometer type giving ordinary pulsating tension (non-impact) to a specimen was used. The rate of repetitions is about 10 cycles/s.

Table 1 shows the chemical composition and mechanical properties of the present materials S45C and SS400 which are annealed at 850℃ and 900℃ respectively in vacuum. In order to find slip bands in ferrite grains, the surface of specimen was observed microscopically by replicas at appropriate cycles of stress, and to show slip bands conditions, the following coefficient K was defined, that is, K (Slipped grains ratio) = (Number of slipped grains of designated grains) / (Number of designated grains). The number of grains designated prior to experiment is about 100.

EXPERIMENTAL RESULT AND DISCUSSION

Case for S45C

Figure 3 is S-N curves in the range of stress cycles to fracture of the order of 10^5 - 10^6. From the result, it is confirmed that impact fatigue strength is less than non-impact one, though there is scatter. An example showing the change of surface conditions with stress cycles is Figure 4, where N is the number of cycles at surface observation, N_f is the number of cycles to fracture. We see some slipped grains, e.g. in the upper region and the other in the figure (b). And we also see that the number of slip bands in the grain increases with increasing number of stress cycles, or the slip bands become broad in comparison with the figure (b) and (c). Figure 5 shows the relation between slipped grains ratio K and cycles ratio N/N_f. As a whole, the both cases of the impact and the non-impact indicate that slipped grains under the impact is less than those under the non-impact for the same stress cycles except for

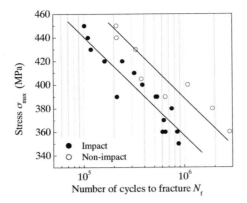

Figure 3 : S-N curves (S45C)

(a) $N/N_f = 0$, $K = 0$ (b) $N/N_f = 0.80$, $K = 0.40$ (c) $N/N_f = 1.0$, $K = 1.0$

Figure 4 : Micrograph of surface conditions under repeated impact (σ_{max}= 400 MPa , S45C)

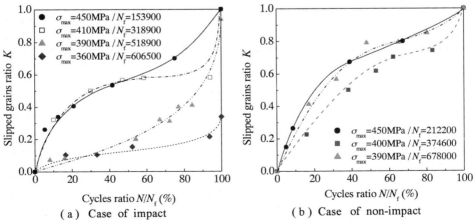

(a) Case of impact (b) Case of non-impact

Figure 5 : Tendency of increase of slipped grains

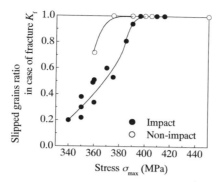

Figure 6 : Slipped grains ratio in case of fracture

the case of $K=1.0$. Such an inclination is shown in Figure 6, where K_f denotes K in case of fracture. And the figure shows that slip bands do not appear in some of grains in spite of fracture caused by a crack propagation. In the present experiment, microcrack of the order of a grain size could not be sufficiently observed, but sometimes it was found in the neighborhood of 50-70% of fracture life.

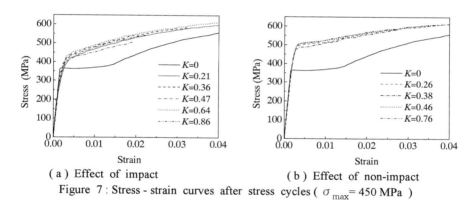

(a) Effect of impact (b) Effect of non-impact

Figure 7 : Stress - strain curves after stress cycles (σ_{max} = 450 MPa)

In order to clarify the differences between the impact fatigue and the non-impact one, strain hardening concerning slip bands of the both was measured. Figure 7 shows the stress-strain curves after impact fatigue and non-impact one with the same stress 450 MPa each. Strain hardening under impact fatigue is lower than that under non-impact one. This fact corresponds to the result of Figure 5 showing less slipped grains. While this fact may seem contradictory, the fracture life is considered to depend on crack initiation and its propagation predominantly from less slipped grains.

Case for SS400

Figure 8 shows *S-N* curve under the repeated impact tension and corresponding slipped grains ratio. We see that K value for SS400 approaches 1.0 more early than that for S45C, in comparison with Figure 5 (a) and 8 (b). In other words, ferrite grains of SS400 deform more easily than that of S45C, because ferrite grains of S45C are restricted by hard pearlite structure. The tendency, however, of the increase of slipped grains under the low stress level is similar to that of S45C. And so the initiation of microcrack was like a case of S45C.

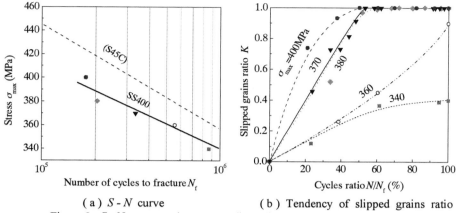

(a) S-N curve (b) Tendency of slipped grains ratio
Figure 8 : S-N curve and corresponding slipped grains ratio for SS400

CONCLUSIONS

The repeated impact tensile experiments of carbon steels (S45C and SS400) were carried out and the slip bands in ferrite grains were observed by optical microscope. Main results obtained in case of S45C are as follows. 1) It was confirmed that the impact fatigue strength is lower than the ordinary non-impact one in the range of stress cycles to fracture of the order of 10^5-10^6. 2) The number of slipped grains under the repeated impact is less than that under the ordinary non-impact. 3) Strain hardening due to the repeated impact is also lower than that due to the ordinary non-impact.

REFERENCES

Chatani A., Hojo A. and Tachiya H. (1998). Slip Bands and Fatigue Life of a Steel under Repeated Impact Tension. *Key Engineering Materials Vols.***145-149**, 329-332.
Johnson P., Zhang X. P. and Pluvinage G.(1990). Crack Growth Rate in Impact Fatigue and in Programmed Variable Amplitude Loading Fatigue. *Engineering Fracture Mechanics* **37:3**, 519-525.
Lundberg B. and Henchoz A. (1977). Analysis of Elastic Waves from Two-Point Strain Measurement. *Experimental Mechanics. June*, 213-218.
Maekawa I. (1994). Behaviour of Materials Subjected to an Impact Force and New Evaluation Method for Impact Strength. *Nuclear Engineering and Design* **150**, 315-322.
Tanaka T. Kinoshita K. and Nakayama H. (1992). Effect of Loading Time on High-Cycle Range Impact Fatigue Strength and Impact Crack Growth Rate. *JSME International Journal* **35:1**, 108-116.
Yang P., Liao X.,Zhu J. and Zhou H. (1994). High Strain-Rate Low Cycle Impact Fatigue of a Medium-Carbon Alloy Steel. *International Journal of Fatigue* **16:5**, 327-330.
Yang P. and Zhou H. (1994). Low-Cycle Impact Fatigue of Mild Steel and Austenitic Stainless-Steel. *International Journal of Fatigue* **16:8**, 567-570.
Yu J., Liaw P. K. and Huang M. (1999). The Impact-Fatigue Fracture of Metallic Materials. *JOM April*, 15-18.
Zhang M., Yang P., Tan Y. and Liu Y. (1999). An observation of Crack Initiation and Early Crack Growth under Impact Fatigue Loading. *Materials Science & Engineering* **A271**, 390-394.

EXPERIMENTAL AND NUMERICAL STUDY ON HIGH VELOCITY IMPACT AGAINST STEEL PLATE

S. Yoshie[1] and T. Usui[2]

[1] Research & Development Department, Power Plant Division, Kawasaki Heavy Industries, LTD.
2-6-5 Minamisuna, Koto-ku, Tokyo 136-8588 JAPAN
[2] Department of Applied Physics, The National Defense Academy
1-10-2 Hashirimizu, Yokosuka-shi 239-0811 JAPAN

ABSTRACT

It is necessary to develop the material that considered a meteorite collision for constructing space station and so forth. The experiments at the impact velocity in the order of 1 km/s have been executed as the first stage for accumulating data, which quantitatively shows the perforation or penetration behavior, using stainless steel as a target whose dynamic property is already known. In the tests, the target plate shows the ductile crater formation as a milk crown at impact, and plugging failure due to shear. Initial strain wave near perforated point is independently distinguished and shifts to residual strain. The duration time of initial strain wave can correspond to the time that a projectile needs to perforate through a target. Also, accuracy of numerical simulation has been checked with the experimental results of both perforation and penetration behavior. And the authors showed that analyses was able to simulate the experiments by appropriately expressing the strain energy up to uniform elongation every strain rate according to high tensile test results of material.

KEYWORDS

Impact, Perforation, Penetration, High strain rate, Plugging fracture, Dynamic material property

INTRODUCTION

To check the impact behavior between a tank and flying objects, the author carried out the experiments for aiming at large deformation of shell plate rather than perforation and showed the differences of the behavior depending on the restriction conditions, Yoshie (1993). Also, such mechanism of dynamic deformation was clarified including numerical results, Yoshie (1999). In this paper, perforation and penetration behavior were discussed through both experimental and numerical approach. The condition that a projectile is buried in the target material without a perforation is defined as a penetration in this paper.

EXPERIMENTS

Experimental Method and Measuring Way

M1 Rifle was used to materialize the impact velocity range in the experiments. Flat-nosed copper projectile, which was 7.8 mm in diameter, 15 mm in length and 6.3g in average weight, was chosen. The velocities at impact and after perforation were measured as average by setting a pair of electron foils back and forth of a target. For perforation and penetration, two kinds of impact velocity were prepared by adjusting the explosive charge. Stainless steel type 316 was chosen as before. Thickness of 10 mm was predicted for perforation and penetration. The rear side of target (outside of 150×150 mm) was rigidly supported. Appearance of the target is shown in Figure 1, left. Plastic foil gages of 5 mm in length were attached around impact point. Location of strain gages is shown in Figure 1, right. Sampling time of the counter using electron foils was 0.1 μ s and 0.5 μ s was set for strain history.

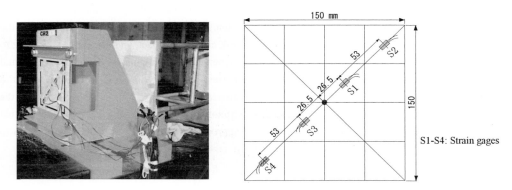

Figure 1 : Setting of target plate & location of strain gage

Experimental Results and Discussions

The case of impact velocity of 0.8424 km/s for example of perforation and the cases of 0.6682 km/s for penetration are shown in this paper. In the case of perforation, plugging fracture was presented and the velocity after perforation was 0.1344 km/s. Because the weight of projectile is not maintained after penetration, it is difficult to estimate distortional energy of the target using velocity before and after perforation. Correlation between the rupture signals of electron foils and strain history is shown in Figure 2. It is understood the strain shifts to residual level of about 1000 μ strain.

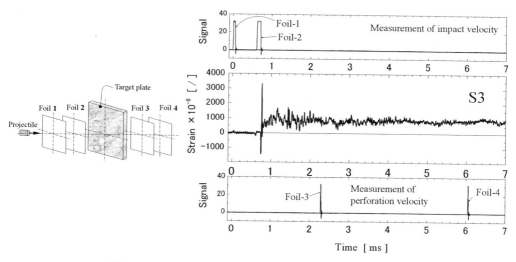

Figure 2 : Signal for velocity and strain history in experiment
Test No.CH203 (Impact velocity : 0.8424 km/s)

Appearance of the target after perforation is shown in Figure 3. The actual impact point became about 10 mm of oblique lower part from frame center. After impact the outside part of the projectile adhered to the inside and impact side of perforation hole with friction and melting. The hole diameter was 13 mm in average and had a tendency to spread to about ϕ 14 mm at the rear side of impact. In the case of aiming at the center of S1 and S2 of the same target with the velocity of 0.6682 km/s, it presented penetration as shown in Figure 3.

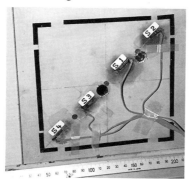

Center : Perforation by impact
velocity of 0.8424 km/s
Right : Penetration by impact
velocity of 0.6682 km/s

Figure 3 : Appearance of target plate after test

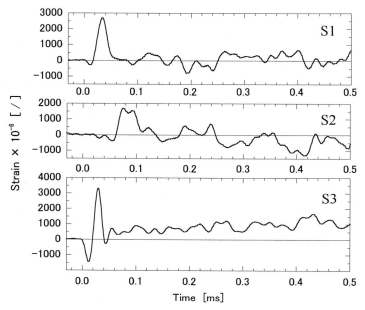

Figure 4 : Distribution of strain history in experiment
Test No.CH203 (Impact velocity : 0.8424 km/s)

Distribution of the strain history is shown in Figure 4. At S1 or S3, the initial strain wave is distinguished as a triangular wave with about 3000 μ strain of peak. Because the impact point was near the S3 gauge, the peak value of the initial strain wave is bigger than S1. The compression wave of bending due to milk crown style transformation proceeds to the tension wave in S3 and is more obvious than S1. The initial strain wave can be cancelled after a projectile perforates, the duration time of about 50 μ s in the initial wave of S3 corresponds the time for perforation. It is supported by the analytical results.

NUMERICAL APPROACH

Analytical Method

Code AUTODYN2D, Birnbaum, Cowler and Hayhurst (1996), was used. The dynamic property of target was realized by logarithmic interpolation every strain rates and strain up to the uniform elongation. Material data were referred to high tensile test results of stainless steel 316 type, Albertini and Montagnani (1989). Failure condition was defined as the uniform elongation. In the case of strain rate of 6000 s^{-1} or more, the trend was kept from the condition of the strain rate of before and was exterpolated. The physical property of copper of the projectile was referred to LA-4167-MS.

Analytical Results and Comparison with Experiments

Perforation behavior, when the impact velocity of 0.8667 km/s measured in experiment was set, is shown in Figure 5. The state like milk crown at impact, the situation before perforation at 42 μ s (Perforation time was about 50 μ s in experiment) and also the situation of the plugging failure are explained in this figure. Regarding the initial strain wave at S1, the comparison with experiment is shown in Figure 6. It can be reproduced almost with the 2D analysis.

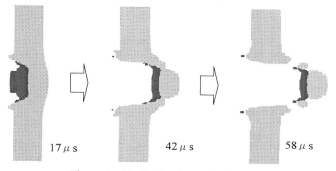

Figure 5 : Perforation by analysis
(Impact velocity: 0.8424 km/s)

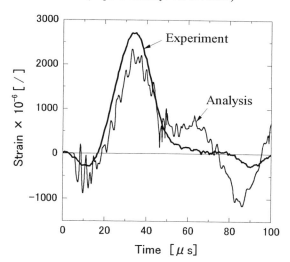

Figure 6 : Comparison between experiment and analysis on strain history
(Impact velocity: 0.8424 km/s)

Histories of the projectile velocity and of the target distortional energy are compared in Figure 7 at each impact velocity. The velocity after perforation was calculated largely than the experiment and the analysis shows perforation even when it was penetrated in the experiment. So the analyses is of conservative in this sense. If the impact velocity that penetrates was calculated in the analysis, it was

0.585 km/s and there are about 10% of margin. However, in the case that the decrease of uniform elongation exceeds the increase of flow stress in strain rate hardening, such as carbon steel, Yoshie (1997), neglecting the dynamic material property leads to unconservative evaluation. So the sufficient attention is necessary. If the distortional energy of the target is roughly calculated by using the crater volume and material strain energy up to uniform elongation of 274 MJ/m^3 at a strain rate in the order of 1000 s^{-1}, it becomes about 0.42 kJ. Such value is also shown in analytical results in Figure 7. Hence it becomes an important to know the dynamic strain energy of the material.

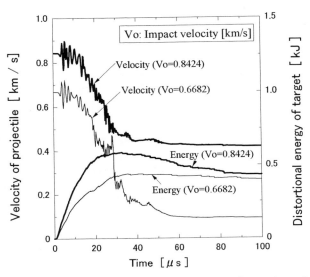

Figure 7 : Projectile velocity & distortional energy of target in analyses

CONCLUSIONS

Perforation and penetration behavior at impact velocity in the order of 1 km/s was discussed. In the experiment, it is shown that the initial strain wave was independently generated in case of perforation, which is related to perforation time and so on. When the dynamic material property is expressed in a code as authors proposed, numerical approach could simulate well the experimental results.

REFERENCES

Albertini C. and Montagnani M. (1989). "Dynamic Material Properties Laboratory," JRC.
Birnbaum N.K., Cowler M.S. and Hayhurst C.J. (1996). 2nd ISIE.
Yoshie S., Iwasaki M., Makino M., Watanabe H. and Kennedy J.M. (1993). 12th SMiRT.
Yoshie S., Albertini C. and Mentani Y. (1997).   J.Soc. Material Science, Japan, **46** : **11**, 1286-1292.
Yoshie S. (1999). 4th SILOS.

DEFORMATION AND FAILURE MECHNISM OF THIN-PLATE UNDER COMBINED LASER HEATING AND PRE-STRESS

Z.G.Wei Y.C. Li Z.P. Tang

Department of Modern Mechanics, University of Science and Technology
of China, Hefei,230027 Anhui P.R.China

ABSTRACT

A new analysis method has been developed to study the transient response of a thermal-viscoplastic thin-plate made of 4340 steel and 2024-T351 aluminum subjected to constant plane-stress loading and to rapid laser heating. A continuum statistic damage and a thermodynamic damage model are implemented into the constitutive equation of materials. This model can be used to determine parameters related to the critical rupture initiation of materials subjected to deposited laser energy, loading, and duration of heating. The threshold time of heating and threshold temperature of material for laser thermal heating coupled with the tensile stresses causing the structural failure are calculated. The results show that there exist three types of deformation and failure modes in structure under laser heating and pre-stress: rapid rupture, no rupture and delayed rupture according to the duration of the laser heating. In general, the higher the pre-stress level and the higher the laser power density, the shorter the rupture time of the material. The pre-stress level can significantly effect the damage evolution, while the influence of laser power density on damage is relatively minor.

KEY WORDS

Laser heating, deformation and failure mechanism, thermal viscoplastic material, structure instability

INTRODUCTION

The interaction of high power laser beam with materials/structures has been a subject of considerable interest during the last decades. Recently, the emphasis has been shifted to developing a fundamental understanding of the interaction of low and middle power continuous wave laser with pre-stressed structures, for example, Zhang et al. (1995).

The dominant effect of a low and middle-energy laser weapon on the target is thermal. The transfer of energy from the laser to the target causes a succession of events: thermal-softening of material, melting and the evaporation of the target area depending on laser power density, the absorption of the target as well as thermal characteristics of the target material. When the thermal stress equal to or greater than a certain condition, the material will be softened or cracks occur in the laser irradiation zone, which may induce the macro-catastrophe of the material or structure which are subjected to force. In practice, during laser heating, bulging may occur in structures such as vessel with inner pressure, and the effect of geometric curvature should be considered in the deformation process. Due to the complexity of the problem, finite element calculation is needed. For simplicity, only thin plate subject the impingement of a heat source and plane stress is considered in this paper. This work focuses attention on the fundamental deformation and failure mechanism of thermoplastic material and structure induced by a heat source. Based on two types of continuum damage model, the deformation, damage and fracture mechanism of 4340 steel and 2024T-3 aluminum plate are determined.

ANALYTICAL MODEL

Damage Evolution Equation

Let us define a damage variable $D = (A - A_S)/A$, where, A and A_S are gross and effective areas of cross section, respectively. Two types of damage models are used in this paper.

Model 1. Continuum statistic damage model. From the conservation law of microcracks in phase-space, Bai et al. (1991) obtained a statistic damage model for ideal cracks. The formula deduced by Bai et al. also can be used for void type damage, Li et al. (1999). For obtaining damage equation, void growth equation is needed. An essential result for void growth problem is offered by Rice-Tracey's analysis (1969), which derives the rate of growth of a spherical cavity in a perfectly plastic infinite body as a function of the accumulated plastic strain rate $\dot{\varepsilon}$ and the triaxiality ratio σ_m/σ_{eq}. If nucleation rate of void is negligible, combining the results of Bai et al. (1991), and Rice et al. (1969), we can obtain the following damage evolution equation:

$$\partial D/\partial t = 0.576 \exp\left(\frac{3\sigma_m}{2\sigma_{eq}}\right) \dot{\varepsilon} D \qquad (1)$$

Eq. (1) is the same as that derived by Lemaitre (1996).

Model 2. Thermodynamic damage model. Chandrakanth et al. (1995) considered a damage evolution equation for 2024-T3 using an internal variable theory in the framework of thermodynamics. The damage evolution law has been verified with experimental results given by Chow et al. (1987). For simplification, the damage formula can be written as follows:

$$\dot{D} = -\frac{\sigma_{eq}^2}{2ES_0(1+D)^2} f\left(\frac{\sigma_m}{\sigma_{eq}}\right)\dot{\varepsilon}_{eq} \quad S_0 = \frac{\sigma_y^2}{2E(1-D)^2}\frac{dD}{d\varepsilon} \qquad (2)$$

where $E = 67 GPa$, $dD/d\varepsilon = 0.6$, $\sigma_y = 330 MPa$

Constitutive Equation with Damage

Using strain equivalence principle (Lemmatize, 1996), the Johnson-Cook with damage can be written as:

$$\sigma = (1-D)(A+B\varepsilon^n)\left(1+C\ln\frac{\dot{\varepsilon}}{\dot{\varepsilon}_0}\right)[1-(T^*)^m] \qquad (3)$$

where $T^* = (T-T_r)/(T_m-T_r)$, A, B, C, n, m are material perimeters, T_r, T_m are room temperature (293°K) and melting temperature, respectively. $\dot{\varepsilon}_0$ is reference strain rate. If the damage is negligible, the above equation can be deduced to the Johnson-Cook model (1983). The materials selected for this research are 4340 steel and 2024 aluminum, Johnson and Cook (1983). For T2024 Aluminum

$$A = 265 MPa, \quad B = 426 MPa, \quad n = 0.34 \quad C = 0.015, \quad m = 1.0$$
$$T_m = 775K \quad \rho = 2770 kg/m^3 \quad c = 875 J/kg.K \quad \dot{\varepsilon}_0 = 1 s^{-1}$$

Relationship between Stress and Strain at Constant Loading

True strain ε can be defined as: $d\varepsilon = dL/L$, then $\varepsilon = \ln(L/L_0)$. Initial force: $P_0 = \sigma_0.S_0$, contemporary force $P = \sigma.S$. Assuming $P_0 = P$ during deformation process and the contribution of the damage to volume is negligible, and the volume of material keeps constant, then:

$$\sigma = \sigma_0 (S_0/S) = \sigma_0 (L/L_0) = \sigma_0 e^{\varepsilon} \quad (4)$$

Temperature Rise in Thin-Plate Material

Assuming 25% of the energy is absorbed by the material, from the point of energy conversation, we have (see Appendix):

$$dT/dt = 0.25 I/\rho ch \quad \text{or} \quad T = T_r + \alpha t \quad (5)$$

Where, ρ, c T_r α are the density of mass, the specific heat, initial temperature and temperature rise rate, respectively. The relationship between temperature rise rate and laser power density I for 2024-T351 aluminum is shown in Table 1. In calculation, the thickness of the plate is constant ($h = 2.5mm$)

TABLE 1
RELATIONSHIP BETWEEN α AND LASER POWER DENSITY

Laser Power Density (W/cm^2)	100	500	1000	5000	10000
α (K/s) for T2024 aluminum	41	206	410	2060	4100

RESULTS AND DISSCUSION

Giving the initial conditions (Initial damage $D_0 = 10^{-5}$, Threshold Damage $D_s = 0.3$, at which the material can be considered as fully fractured), combining Eq. (1)(2)(3)(4) and (5), the deformation and failure mechanism can be determined. It was found that the damage in statistic damage model (*model 1*) has very little influence on the deformation process except at very late stage. So the main attention will be paid to the thermodynamic damage evolution for T2024 aluminum material.

Strain Rate-Time Relationship

For most of the process, as illustrated in Fig.1 (a)(b), the material deforms very slowly. Once the strain reaches a certain criterion (for example, a critical strain), the strain rate begins to rise dramatically. These phenomena are attributed to geometric instability which means the structure can not afford further loading.

Damage is very sensitive to pre-stress level, as shown in Fig.1 (a), but is insensitive to laser power density, as shown in Fig. (b). From Fig.1 (a), it can be seen that the higher the pre-stress level, the larger the difference between the predicted results with damage and that without damage.

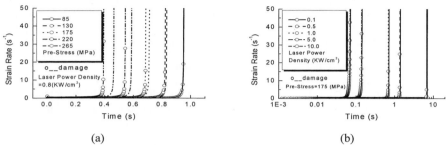

(a) (b)
Figure 1 (a)(b): Comparison of strain rate *versus* time for the cases, with and without damage

Effect of Laser Power Density

Fig.2 (a)(b) show the relationship between failure time (t) and laser power density (I). It is can be seen that the higher the laser power density, the lower the temperature at failure. The $log(t)$ is inversely proportional to $log(I)$.

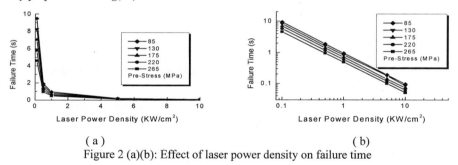

(a) (b)
Figure 2 (a)(b): Effect of laser power density on failure time

Effect of Pre-Stress

The failure time and material temperature at failure, As illustrated in Fig.3 (a) and Fig.3 (b), respectively, decreases significantly as the value of pre-stress level increases.

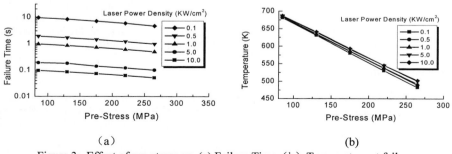

（a） (b)
Figure 3: Effect of pre-stress on (a) Failure Time （b） Temperature at failure

Effect of Duration of Laser Heating

Fig.4 (a), (b) show the effect of the duration of laser heating on the deformation mechanism at pre-press level $P = 175 MPa$ and laser power density $I = 1KW/cm^2$. (There exists a corresponding threshold strain ε_c. In this paper, the prescribed ε_c are 0.01, 0.02, 0.04, 0.05 and 0.1). The temperature keeps constant during calculation when the criterion is met. It can be seen that the duration of laser heating plays important role in the deformation and failure mechanism. The results show that three types of deformation mode may occur according to the prescribed strain: 1. Rapid rupture, as $\varepsilon_c = 0.1$ shown in Fig.4 (a)(b); 2. No rupture, as $\varepsilon_c = 0.01$, $\varepsilon_c = 0.02$ shown in Fig.4(a)(b); and 3. Delayed rupture, in this case, the magnitude of the strain rate decreases quickly at the beginning of the ceasing heating, then stay at some level during the rest of the process and then rise rapidly, as $\varepsilon_c = 0.05$, shown in Fig.4(a). Fig.4(c) shows the comparison between the two cases, with damage and without damage. It is can be seen that the deformation process with damage develops faster than that without damage. For instance, at $\varepsilon_c = 0.05$, the instability time are approximately $1.2s$ and $4.6s$, respectively, for the model with damage and without damage (the time difference is $3.4s$); at $\varepsilon_c = 0.04$, the instability time is approximately $2.2s$ for the model with damage, however the instability time for the model without damage is more than $10s$ (no plotted).

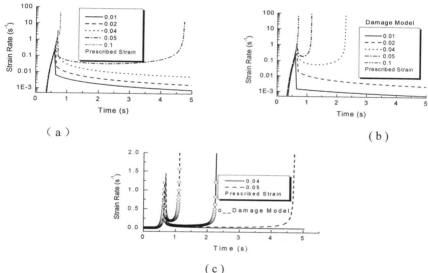

Figure 4: Strain rate-time relationship (a) without damage, (b) with damage, and (c) their comparison

CONCLUSION

1. Thermal instability, damage softness and geometry softness play important role in deformation and failure mechanism of structures imposed by laser heating combined with external loading or pre-stress.
2. The higher the pre-stress and the higher the laser power density, the faster the instability of structure and the shorter the failure time.
3. There are three types of deformation modes: rapid ruptures, no rupture and delayed rupture according to the duration of laser heating.

Acknowledgement—This work is supported by the Chinese National Science Foundation (project No. 10002017) and the National Defense Key Laboratory (Project. OOJS75.1.3.ZK0103)

References

Bai Y.L. Ke F.J. Xia M.F. (1991). Formulation of statistical evolution of microcracks in solids. *ACTA Mechanica Sinica* **7:1**, 59-66
Carslaw H.S. Jaeger J.C. (1959). *Conduction of Heat in Solids*, Oxford University Press, UK
Chandrakanth S. Pandey P.C. (1995) An isotropic damage model for ductile material. *Engineering Fracture Mechanics* **50:4**, 457-465
Chow C.L. Wang J. (1987) A anisotropic theory of continuum damage mechanics for ductile fracture. *Engineering Fracture Mechanics* **27**, 547-558.
Johnson G.R. Cook W.H. (1983) A constitutive model and data for metals subjected to large strain rates and high temperatures. *Proceedings of the Seventh International Symposium on Ballistics*, The Hague, The Netherlands, 541-548.
Lemaitre J. *A Course on Damage Mechanics*, (1996) Springer-Verlag, Berlin, Heidelberg.
Li Y.C. Li D.H. Wei Z.G. et al.(1999) Research on the deformation, damage and fracture rules of circular tubes under inside-explosive loading, Acta Mechanica Sinica **31:4**, 442-449.
Rice.J.R. Tracey D.M.(1969) On the ductile enlargement of voids in triaxial stress fields. *J. Mech. Phys. Solids* **17**, 210-217.
Zhang.N. Liu.C.L. Sun C.W.(1995) The thermolcoupling effect of CW COIL beam on composites. 1-7, *26 th AIAA Plasmadynamics and Lasers Conference*, June 19-22,/ San Diego, CA, USA.

APPENDIX Temperature Rise in Thin-plate Material

Let us consider the case that the heat flux F_0 generated by laser heating imposes on front surface of a thin-plate. Heat conduction obeys the Fourier equation. h, z are the thickness of the plate and the distance from the front surface of the plate, respectively. k, ρ, c are the thermal resistance of the material, the density of mass and the specific heat of the material, respectively, and remain constant throughout the process. T is the temperature at any point in the plate. Melting and vaporization phenomena are not considered here. From the point of energy conversation, the energy absorbed by plate is fully converted to heat energy which cause the increase of the temperature. The temperature field in thin-plate with imposed heat flux is listed below, for example, Carslaw et al. (1959).

$$T(z,t) = \frac{F_0 k}{KL}t + \frac{F_0 L}{K}\left\{\frac{3(L-z)^2 - L^2}{6L^2} - \frac{2}{\pi^2}\sum_{n=1}^{\infty}\frac{(-1)^n}{n^2}e^{-kn^2\pi^2 t/L^2}\cos\left[\frac{n\pi(L-z)}{L}\right]\right\} \quad (A1)$$

$T(0,t)$ and $T(L,t)$ are the temperature at the front and rear of the plate, respectively. Let us define average temperature $T_a = [T(0,t) + T(L,t)]/2$, temperature difference $\Delta = T(0,t) - T(L,t)$, and relative temperature difference $\delta = \Delta/T_a$, Assuming 25% of the energy are absorbed by the plate during heating, from the point of energy conversation, we have: $dT/dt = 0.25 I/\rho c h$ or $T = T_r + \alpha t$. Fig.5 shows the distributions of T_a and δ for 2024-T351 aluminum. The thickness of plate in Fig.5 (a)(b) varies from *1* to *5mm*, while $I=1KW/cm^2$; The range of laser power density in Fig.5 (a)(b) is *1-5 KW/cm²* and at constant thickness of plate $h = 2.5mm$. It is shown that, in most of the cases, δ is very little (within 10%), so the average stress method used in text is reasonable.

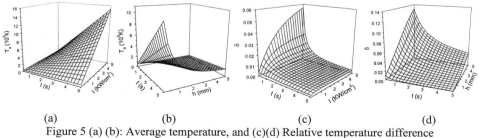

(a)   (b)   (c)   (d)

Figure 5 (a) (b): Average temperature, and (c)(d) Relative temperature difference

HIGH STRAIN RATE DEFORMATION BEHAVIOR OF AL-MG ALLOYS

T. Masuda[1], T. Kobayashi[2] and H. Toda[2]

[1]Graduate Student, Graduate School of Toyohashi University of Technology
[2]Department of Production Systems Engineering, Toyohashi University of Technology,
Toyohashi-city, AICHI, 441-8580, Japan

ABSTRACT

The aim of the present study is to determine the accurate measuring method of the stress-strain curve in impact tensile tests and to investigate the impact response of commercial Al-Mg series alloys at wide stain rate range. Impact tensile tests are carried out using a split-Hopkinson bar apparatus and servo-hydraulic impact testing machine. In order to record true load-time and strain-time histories avoiding the effects of oscillation, strain gages are attached onto specimen surfaces with two different gage sections to measure load directly in the case of the servo-hydraulic impact testing. On the other hand, an impact absorber is used to eliminate the superimposed oscillations for the split-Hopkinson bar apparatus. The strain rate sensitivity of the 0.2% proof stress, the ultimate tensile strength and n value decreases slightly with increase of strain rate up to approximately $10^2 s^{-1}$. However, they increase with strain rate exceeding approximately $10^2 s^{-1}$. The elongation and reduction of area show linear increases with increasing strain rate. Although fracture surfaces mainly exhibit the shear type dimple pattern under low strain rates, ordinary equiaxed dimple fracture is observed under the high strain rates.

KEYWORDS

Hopkinson bar, strain rate dependency, Al-Mg series alloys, impact tensile test, SEM, fracture surface, dimple, impact stress-strain curve

INTRODUCTION

It is a pressing need that properties of light metals should be measured and evaluated under dynamic loading condition, because they have been began to be used widely for the civil engineering, constructions and transportations. The Al-Mg series alloy has superior mechanical properties such as a high strength/weight ratio, good corrosion resistance and deformability, it is considered for many advanced applications where the structural components are subjected to dynamic loading. In order to optimize deformation and fracture performance of this alloy under the high strain rates, it is necessary to understand the dynamic deformation mechanisms. On the other hand, since various impact test machines such as the split-Hopkinson bar apparatus, the servo-hydraulic impact testing machine and etc. have been developed owing to the above-mentioned social demands, dynamic mechanical properties became to be measured at high strain rate in the well-equipped laboratories. However, measurement and evaluation of the stress-strain curves under impact loading conditions have not been well established due to difficulties in the measurement such as superimposed oscillation mainly caused by inertial effect.

It is recognized that materials often respond in different ways to high strain rate loading as compared to low strain rate loading. As strain rate is increased from static to dynamic, it results in an increase of the strength with increasing strain rate. Further, the flow stress of a material depends not only on the strain and strain rate but also on its microstructure especially at the dislocation level. In order to combine the micromechanisms of deformation and macroscopic constitutive relationships, there is a need to understand the microstructural changes which take place during deformation. The aim of the present study is to determine the accurate measuring method of the stress-strain curve in impact tensile tests and to investigate the impact response of commercial Al-Mg series alloys at wide stain rate range.

MATERIALS AND EXPERIMENTS

Two kinds of work hardening aluminum alloys used in this investigation are 5052-H112 and 5083-H112 alloys. The chemical compositions of used samples are shown in Table 1. Geometry and nominal dimensions of the tensile specimens are given in Fig. 1 and Fig. 2. All specimens used in static tensile, dynamic servo-hydraulic tensile and dynamic split-Hopkinson bar tests were sampled in longitudinal orientation and have the same gage length and diameter.

TABLE 1
CHEMICAL COMPOSITIONS OF SAMPLES USED

(mass%)

	Si	Fe	Cu	Mn	Mg	Cr	Zn	Ti	Al
5052-H112	0.09	0.25	0.03	0.04	2.50	0.19	0.01	0.01	bal.
5083-H112	0.14	0.20	0.03	0.65	4.64	0.11	-	0.02	bal.

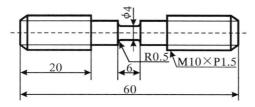

Figure 1: Geometry of a specimen in the static and dynamic tensile tests.

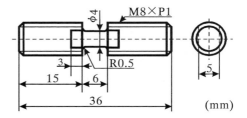

Figure 2: Geometry of a specimen in the split-Hopkinson bar test.

Static tensile experiments were carried out using the Instron testing machine. The intermediate rate tensile tests were conducted using a servo-hydraulic impact test machine with a capacity of 49kN. In order to record the true load-time and strain-time history avoiding the effects of oscillation, strain gages were attached onto specimen surfaces of the two parallel bodies shown in Fig. 1 to measure load directly. The loading velocity was varied from 0.01 to 12m/s. The split-Hopkinson bar test was conducted for strain rate exceeding approximately $10^3 s^{-1}$. The loading velocity was 30m/s. In order to record the true strain-time history eliminating the superimposed oscillations, an impact absorber was used between the striker and the input bar for the split-Hopkinson bar apparatus. As a result, accurate stress-strain curves under impact loading conditions have been recorded. These data obtained from a stress-strain curve were compared as a function of strain rate. Also, fracture surfaces of the tested specimens were observed with a scanning electron microscope.

RESULTS AND DISCUSSION

Stress-Strain Curves over a Wide Range of Strain Rates

Figures 3 and 4 shows stress-strain curves of the 5052-H112 and 5083-H112 alloys at four strain rates (4×10^{-4}, 10^{-1}, 1×10^3 and $3 \times 10^3 s^{-1}$). In all of the cases, Young's modulus is almost constant at all strain rates. 0.2% proof stress and ultimate tensile strength slightly changes with increase strain rate, while elongation showed rapid increase with increasing strain rate. Stress-strain curves showed a three-step process in which there were an initial elastic deformation region, uniform plastic deformation region until the maximum stress, then unstable deformation to the failure. In the static loading, the fracture occurred immediately when the stress reached the maximum stress. However, the flow stress decreased gradually at the high strain rate after the maximum stress. In the case of the dynamic tensile test using the split-Hopkinson bar apparatus, the stress-strain curves could be recorded exactly from the initial elastic deformation to the final fracture as compared to the other techniques.

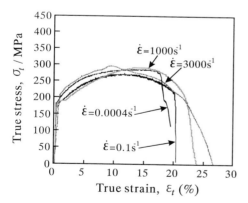

Figure 3: Typical stress-strain curves of the 5052-H112 alloy at various strain rates.

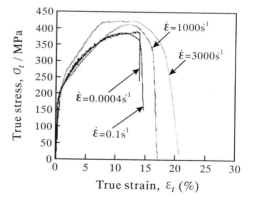

Figure 4: Typical stress-strain curves of the 5083-H112 alloy at various strain rates.

Strain Rate Dependency of Tensile Properties

Figures 5 and 6 show the 0.2% proof stress and ultimate tensile strength as a function of strain rate in the 5052-H112 and 5083-H112 alloys. The strain rate sensitivity of the 0.2% proof stress and the ultimate tensile strength decreased slightly with increase of strain rate up to approximately $10^2 s^{-1}$.

However, the 0.2% proof stress and the ultimate tensile strength increase with increase strain rate exceeding approximately $10^2 s^{-1}$. Tanimura, *et al.* reported that the aluminum alloys containing magnesium solute atoms showed the negative strain rate dependency at intermediate strain rate. In this study, the negative strain rate dependency showed at strain rate region between 10^{-4} and $10^2 s^{-1}$. The 5083-H112 alloy shows higher strength than the 5052-H112 alloy. Under the same strain rate range, the 5083-H112 alloy showed a larger strain rate dependency compared with the 5052-H112 alloy. While the 5052-H112 alloy showed an increase in the ultimate tensile strength of approximately 31MPa in the strain rate range of 4×10^{-4} to $3 \times 10^3 s^{-1}$, that of the 5083-H112 alloy was approximately 43MPa.

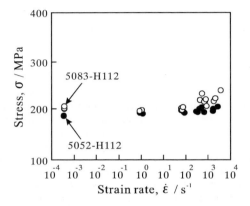

Figure 5: Variation of 0.2% proof stress with nominal strain rate.

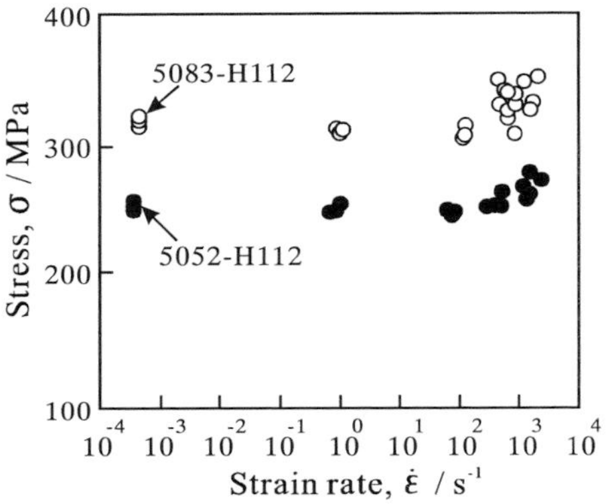

Figure 6: Variation of ultimate tensile strength with nominal strain rate.

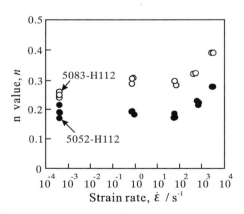

Figure 7: Variation of strain hardening exponent with nominal strain rate.

Figure 7 presents the variation of strain hardening exponent with the strain rate. The strain hardening exponent showed a similar strain rate dependency with the 0.2% proof stress and ultimate tensile strength. However, the increase in the strain hardening exponent was kept constant up to a strain rate of approximately $10^2 s^{-1}$. The 5083-H112 alloy contains approximately twice magnesium element as compared to the 5052-H112 alloy. However, strain rate dependency was similar.

Figures 8 and 9 present the variation of elongation and reduction of area with strain rate. The elongation and reduction of area increase linearly with increasing strain rate. Although the strain rate dependency in elongation and reduction of area is negligible up to the strain rate of approximately $10^0 s^{-1}$, the degree of rate sensitivity appears to increase rapidly for the higher strain rates.

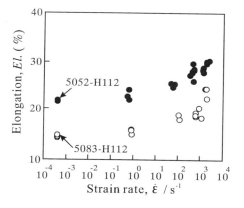

Figure 8: Variation of elongation with nominal strain rate.

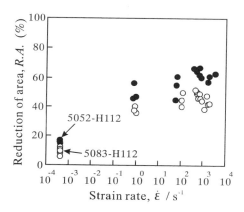

Figure 9: Variation of reduction of area with nominal strain rate.

Effect of the Strain Rate on Fracture Surface and Structure

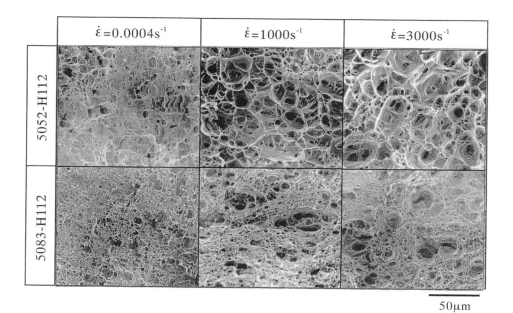

Figure 10: SEM micrographs of fracture surfaces at various strain rates.

Figure10 shows the SEM micrographs of fracture surfaces of 5052-H112 and 5083-H112 alloys at various strain rates. Both Al-Mg series alloys provide the flat surfaces at static loading conditions. However, the fracture occurred by a combination of two different dimple sizes at higher strain rates, one of which is approximately 10μm and the other is several micrometer in the case of 5083-H112 alloy having higher magnesium content. The fracture surface of low magnesium 5083-H112 alloy is mainly dominated by dimples of approximately 20μm in diameter. Although fracture surfaces mainly exhibited the shear type dimple pattern under low strain rates, ordinary equiaxed dimple fracture surfaces were observed under high strain rates. The specimen was largely necked at high strain rates. In both materials, the dimple size and depth increased with increase strain rate.

CONCLUSIONS

Using a split-Hopkinson bar apparatus and servo-hydraulic impact testing machine, accurate measuring methods of the stress-strain curve under impact loading conditions were determined and the effect of strain rate on mechanical properties in 5052-H112 and 5083-H112 aluminum alloys were examined over a wide range of strain rate. The following conclusions could be drawn.

1. Strain gages were attached onto specimen surfaces with two different gage sections to measure load directly in the case of the servo-hydraulic impact testing. An impact absorber was sandwiched between the striker and the input bar for the split-Hopkinson bar apparatus. As a result, accurate stress-strain curves under impact loading conditions have been recorded.
2. The strain rate sensitivity of the 0.2% proof stress, the ultimate tensile strength and n value decreased slightly with increase of strain rate up to approximately $10^2 s^{-1}$. However, they increased with increasing strain rate exceeding approximately $10^2 s^{-1}$.
3. The elongation and reduction of area increased rapidly with increase of strain rate, and these dependencies were linear. Although strain rate dependencies in the elongation and reduction of area were negligible up to the strain rate of approximately $10^0 s^{-1}$, the degree of rate sensitivity appears to increase rapidly for the higher strain rates.
4. Although fracture surfaces mainly exhibited the shear type dimple pattern under low strain rates, ordinary equiaxed dimple fracture surfaces were observed under high strain rates. The specimen was largely necked at high strain rates. In both materials, the dimple size and depth increased with increasing strain rate.

REFERENCES

Lindholm U. S. (1964). Some Experiments with the Split Hopkinson Pressure Bar. *Journal of Mechanics Physics Solids* **12**, 317-335.

Mukai T., Higashi K., Tuchida S. and Tanimura S. (1993). Influence of Strain Rate on Tensile Properties in Some Commercial Aluminum Alloys. *Journal of Japan Institute of Light Metals* **43:5**, 252-257.

Nicholas T. (1981). Tensile Testing of Materials at High Rates of Strain. *Experimental Mechanics* **21**, 177-185.

Sugiura N., Kobayashi T., Yamamoto I., Nishido S. and Hayashi K. (1995). Strain Rate Dependency of Impact Tensile Properties in an AC4CH-T6 Aluminum Casting Alloy. *Journal of Japan Institute of Light Metals* **45:11**, 633-637.

Sun Z. M., Kobayashi T., Fukumasu H, Yamamoto I. and Shibue K. (1998). Tensile Properties and Fracture Toughness of a Ti-45Al-1.6Mn Alloy at Loading Velocities of up to 12m/s. *Metallurgical and Materials Transaction A* **29A**, 263-277.

Yokoyama T. (1996). Impact Tension and Compression Testing of Ductile Cast Iron with Split Hopkinson Bar. *Journal of the Society of Materials Science, Japan* **45:7**, 785-791.

A VISCO-PLASTIC ANISOTROPIC MODEL FOR THE IMPACT DEFORMATION OF CRUSHABLE FOAM

V.P.W. Shim and L.M. Yang

Impact Mechanics Laboratory
Department of Mechanical Engineering
National University of Singapore
10 Kent Ridge Crescent, Singapore 119260

ABASTRACT

This study focuses on the rate-dependent constitutive behaviour of crushable foam. Two rigid polyurethane foams of different density are subjected to static and impact deformation to investigate their response. Experimental results indicate that the behaviour of polyurethane foam is both orthotropic and rate dependent. An anisotropic visco-plastic model is developed to characterize the observed behaviour. By employing the results of previous microstructural analyses, the effects of foam density on the response are also incorporated. Comparison of theoretical results with experimental data shows that the model is able to describe the visco-plastic behaviour of foams.

KEYWORDS

Impact, crushable foam, visco-plastic, anisotropic, foam model

INTRODUCTION

Cellular materials such as polyurethane foams are employed in mitigating impact damage by limiting the force that can be transmitted through them. For efficient usage of such foams, an understanding of their response to loading is essential, particularly how the global stiffness and strength are governed by the nature of the constituent material and the relative density of the foam. Gibson and Ashby (1988) have undertaken a comprehensive study of foams and documented their findings in their well-cited treatise. Attempts to describe the rate-dependent plastic behaviour of foams have also been made, based on the assumption that the material is a homogeneous continuum (Neilson et al, 1995; Puso and Govindjee, 1995; Zhang, 1997). The use of idealized cell geometries (Gibson and Ashby, 1988) facilitate modelling of the plastic and anisotropic behavior in foams. Shim and Yang proposed a cell model that accounts for the observed anisotropic response of deformation localization for compression in the foam rise direction and uniformly-distributed deformation for compression in the transverse direction.

Although often deemed more detailed, characterization of the anisotropic visco-plastic behavior of foam based on cell microstructure is very complicated. Frequently, what is required for engineering applications, is a macroscopic description of the observed responses. In this effort, the mechanical behavior of rigid polyurethane foams is examined from the viewpoint of a homogeneous continuum and constitutive equations which incorporate orthotropy and foam density are derived.

EXPERIMENTS

Rigid polyurethane foams of two densities – 27.2 kg/m^3 ($=\rho_{oL}$) and 44.8 kg/m^3 ($=\rho_{oD}$) – were studied and are denoted as L-foam and D-foam, respectively. The density of polyurethane is 1,300 kg/m^3 ($=\rho_s$) (Orchon, 1990). Microscopic examinations of the microstructure show that the foams comprise primarily open cells that have an orthotropic geometry.

The stress-strain curves for compression in the foam rise and transverse directions for both foams were determined. Uniaxial loading at strain rates ranging from 10^{-4}/s to10^{-3}/s was applied using an INSTRON universal testing machine. The tests showed that there was negligible difference in the results between the two strain rates. Table 1 presents values of the measured material parameters corresponding to quasi-static loading. Dynamic uniaxial compression tests were also conducted on foam specimens by means of a drop tester and the corresponding stress-strain curves derived from the output of an accelerometer attached to the impacting mass. The global strain rate imposed on the specimen is $\dot{\varepsilon}(t) = \frac{v_0}{H} + \frac{1}{H}\int_0^t a dt$, where v_0 is the initial velocity of the drop mass, a is its acceleration, H the initial height of the specimen and t denotes time. The average strain rate achieved was about 100/s. The experimental stress-strain curves are shown in Figs. 1 to 4. It was found that:

(1) For static compression in the foam rise direction, plastic deformation initiates and localizes within a narrow band. This zone of severe deformation then propagates outwards, accompanied by the generation of an approximately constant stress. However, for compression in the transverse direction, deformation is uniformly distributed and the associated stress increases with deformation.
(2) During compression in both directions, deformation perpendicular to the loading direction is negligible.
(3) The behavior of the material is rate-dependent.
(4) For impact loading in the rise direction, material failure tends to be catastrophic and brittle, whereas for compression in the transverse direction failure is comparatively more gradual and ductile. Consequently, compression in the transverse direction yields a larger energy absorption capacity.

TABLE 1 PROPERTIES OF POLYURETHANE FOAMS

Density (kg/m^3)	Direction	Modulus (MPa)	Yield stress (kPa)
27.2 (L-foam)	Foam Rise	2.9	125
	Transverse	1.4	70
44.8 (D-foam)	Foam Rise	7.7	324
	Transverse	3.8	160

ANISOTROPIC VISCO-PLASTIC MODELLING

The orthotropic nature of the foam cells causes the global mechanical behaviour of the material to also be orthotropic. Characterization of orthotropic elastic properties of materials has been mentioned in detail by many authors (e.g. Mal and Singh, 1991). The relationship between Young's modulus and relative density for cellular materials has also been discussed by Gibson and Ashby (1988). In contrast, the focus of this investigation is on modelling the visco-plasticity of foam after yielding using a continuum approach.

The Bingham model shown in Fig. 5 is one of the simplest visco-plastic idealisations, where Y is the static yield stress and η is a viscosity parameter. The response of the model is described by:

$$\sigma = Y + \eta \dot{\varepsilon} \qquad (1)$$

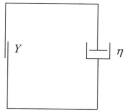

Figure 5: Bingham model

where σ and $\dot{\varepsilon}$ are the stress and strain rate respectively; overstress is defined by $\sigma - Y$. Past test data has show that the effects of strain rate on overstress can be described by a logarithmic function (Campbell, 1973). It is proposed that the relationship described by Eqn. (1) be extended to a three-dimensional logarithmic form and applied to foam material:

$$\sigma_{ij} - \sigma_{ij}^0 = \eta \ln\left(\frac{\dot{\varepsilon}_{ij}}{\dot{\varepsilon}_0}\right) \qquad (2)$$

where σ_{ij} denotes the current stress, σ_{ij}^0 is the static flow stress and $\dot{\varepsilon}_0$ is a reference strain rate (defined as $\dot{\varepsilon}_0 = 10^{-4}/s$ in the present study). For an orthotropic material, σ_{ij}^0 can be expressed in terms of a plastic potential φ, which depends on the orientation of the axes with respect to the foam rise direction; the following form is proposed:

$$\varphi = Y_{ij}\varepsilon_{ij}^P + c_1\sin^2(\theta^i)\left(\frac{1}{1-I^P} + \ln(1-I^P)\right) - \frac{\langle \varepsilon^a - \varepsilon^{rd}\rangle}{\varepsilon^a - \varepsilon^{rd}} E_d \cos^2\left(\frac{\theta^i + \theta^j}{2}\right)\left[\varepsilon^a + (1-\varepsilon^{rd})\ln(1-\varepsilon^a)\right] \qquad (3)$$

The static flow stress is thus:

$$\sigma_{ij}^0 = \frac{\partial \varphi}{\partial \varepsilon_{ij}^P} = Y_{ij} + c_1\sin^2(\theta^i)\frac{I^P \delta_{ij}}{(1-I^P)^2} + E_d \cos^2\left(\frac{\theta^i + \theta^j}{2}\right)\frac{\langle \varepsilon^a - \varepsilon^{rd}\rangle}{1-\varepsilon^a}\frac{\varepsilon_{ij}}{\varepsilon^a} \qquad (4)$$

where Y_{ij} is the initial yield stress, ε_{ij}^P denotes plastic strain, θ^i is the angle between the X_i-axis and the foam rise direction, $I^P = \varepsilon_{ii}^P$ is the first invariant of the plastic strain, $\varepsilon^a = \sqrt{\varepsilon_{ij}\varepsilon_{ij}}$ is introduced to quantify the strain magnitude and c_1 and E_d are material parameters. The singularity function $\langle \varepsilon^a - \varepsilon^{rd}\rangle$ is defined such that $\langle \varepsilon^a - \varepsilon^{rd}\rangle = \varepsilon^a - \varepsilon^{rd}$ when $\varepsilon^a - \varepsilon^{rd} > 0$ and zero otherwise, and ε^{rd} is the strain at the onset of densification for compression in the foam rise direction. The second term, which increases in significance as θ^i tends to $\pi/2$, describes strain-hardening of the material, while the third term, which increases in significance as θ^i tends to 0, defines behaviour during densification. A microstructural analysis (Shim and Yang) has shown that $\varepsilon^{rd} = 1 - 6\sqrt{\frac{1}{\omega(6+4\omega)}\frac{\rho_0}{\rho_s}}$, where the global foam density is denoted by ρ_0, the density of the solid material is ρ_s, ω is a parameter defining the aspect ratio of a typical cell and is approximately 1.4 for the foams studied. The preceding expression yields values of $\varepsilon_{rd} = 0.78$ for L-foam and $\varepsilon_{rd} = 0.72$ for D-foam.

For orthotropic materials that are symmetric about the X_3-axis (taken as the foam rise direction), the initial static yield stress Y_{ij} is assumed to be described by a Hill-type yield function (Hill, 1950):

$$2f(Y_{ij}) = a_1(Y_{11} - Y_{33})^2 + a_1(Y_{22} - Y_{33})^2 + a_2(Y_{11} - Y_{22})^2 + 2a_3(Y_{13}^2 + Y_{23}^2) + 2(a_1 + 2a_2)Y_{12}^2 = 1 \quad (5)$$

where a_i (i=1, 2, 3) are material parameters that depend on the values of the yield stresses.

In general, the viscosity parameter η in Eqn. (2) also varies with θ'. A linear relationship comprising two constants η_0 and η_1, is used to describe this dependence:

$$\eta(\theta') = \eta_0 + \eta_1 \theta' \quad (6)$$

For greater utility, the effects of foam density on material behaviour are also incorporated. Analyses based on cell microstructure models (e.g. Gibson & Ashby, 1988 and Shim & Yang) have shown that the plastic flow stress is proportional to $(\rho_0/\rho_s)^2$ for uniaxial compression in the foam rise direction, and proportional to $(\rho_0/\rho_s)^{3/2}$ for compression in the transverse direction. Shim & Yang have further demonstrated that sensitivity to foam density during densification in the foam rise direction (i.e. the third term in the RHS of Eqn. (4)) is proportional to $(\rho_0/\rho_s)^{1/2}$. Thus, the plastic behaviour of foam under static loading, as described by Eqn. (4), can be re-written as:

$$\sigma_{ij}^0 = \left(Y_{ij}' + c_1' \sin^2(\theta') \frac{I^P \delta_{ij}}{(1-I^P)^2}\right)\left(\frac{\rho_0}{\rho_s}\right)^{m(\theta')} + E_d' \cos^2\left(\frac{\theta^i + \theta^j}{2}\right)\frac{\langle \varepsilon^a - \varepsilon^{rd}\rangle \varepsilon_{ij}}{1-\varepsilon^a} \cdot \frac{1}{\varepsilon^a}\left(\frac{\rho_0}{\rho_s}\right)^{1/2} \quad (7)$$

where $Y_{ij} = Y_{ij}'\left(\frac{\rho_0}{\rho_s}\right)^{m(\theta')}$, $c_1 = c_1'\left(\frac{\rho_0}{\rho_s}\right)^{m(\theta')}$ and $E_d = E_d'\left(\frac{\rho_0}{\rho_s}\right)^{1/2}$, and Y_{ij}', c_1' and E_d' are independent of foam density. A linear relationship is assumed for the function $m(\theta')$:

$$m(\theta') = m_1 + m_2 \theta' \quad (8)$$

where m_1 and m_2 are constants. Substitution of the conditions that $m(0) = 2$ and $m(\pi/2) = 3/2$ results in $m_1 = 2$ and $m_2 = -1/\pi$. If the static behaviour of L-foam (density ρ_{0L}) is used as a reference, Eqn (7) becomes:

$$\sigma_{ij}^0 = \left(Y_{ij}^L + c_1^L \sin^2(\theta') \frac{I^P \delta_{ij}}{(1-I^P)^2}\right)\left(\frac{\rho_0}{\rho_{0L}}\right)^{2-\theta/\pi} + E_d^L \frac{\langle \varepsilon^a - \varepsilon^{rd}\rangle \varepsilon_{ij}}{\varepsilon^a(1-\varepsilon^a)} \cos^2\left(\frac{\theta^i + \theta^j}{2}\right)\left(\frac{\rho_0}{\rho_{0L}}\right)^{1/2} \quad (9)$$

Substituting Eqn. (9) into Eqn. (2) and considering Eqn. (5), equations for the visco-plastic response of foam under compression can be obtained. Taking the X_3-axis to define the foam rise direction, compression in this direction is described by:

$$\sigma_{33} = Y_{33}^L \left(\frac{\rho_0}{\rho_{0L}}\right)^2 + \eta_0 \ln(10^4 \dot{\varepsilon}_{33}) + \langle \varepsilon_{33} - \varepsilon^{rd}\rangle E_d^L \frac{1}{1-\varepsilon_{33}}\left(\frac{\rho_0}{\rho_{0L}}\right)^{1/2} \quad (10)$$

For compression in the transverse direction (X_1-axis), the constitutive relationship is:

$$\sigma_{11} = \left(Y_{11}^L + \frac{c_1^L \varepsilon_{11}^P}{(1-\varepsilon_{11}^P)^2}\right)\left(\frac{\rho_0}{\rho_{0L}}\right)^{3/2} + \left(\eta_0 + \frac{\pi \eta_1}{2}\right)\ln(10^4 \dot{\varepsilon}_{11}) \quad (11)$$

There are six material parameters in Eqns. (10) and (11); these are determined from test data relating to the static and dynamic compression of L-foam. From the values in Table 1, Y_{11}^L =70kPa and Y_{33}^L =125kPa. By noting the densification response of L-foam for compression in the rise direction, E_{rd} is found to be 400kPa. Based on fitting dynamic test results for L-foam in the rise direction, η_0 =4.34kPas. Data derived from static compression in the transverse direction yields the parameter c_1^L =25kPa. A fit of Eqn. (11) with results for dynamic compression in the transverse direction gives η_1 = -1.86kPas. Table 2 summarises the values derived for the parameters.

A comparison of the fitted curves with measured data for L-foam is shown in Figs. 1-4. Substitution of the parameter values in Table 2 into Eqns. (10) and (11) facilitates prediction of the static and dynamic compressive behaviour of D-foam, which is more than 50% denser. The predicted curves and corresponding test data are also shown in Figs. 1-4. These indicate that the proposed model has good potential to describe the anisotropic visco-plastic behaviour of foam of different densities.

TABLE 2. MODEL PARAMETERS OBTAINED FROM DATA FITTING.

Y_{33}^L (kPa)	Y_{11}^L (kPa)	E_d^L (kPa)	η_0 (kPas)	η_1 (kPas)	c_1^L (kPa)
125	70	400	4.34	-1.86	25

CONCLUDING REMARKS

Two rigid polyurethane foams of different density were subjected to static and dynamic compressive gross deformation. Experimental results indicate that the foam behavior is orthotropic and visco-plastic. In the foam rise direction, crushing initiates via localized plastic deformation, which subsequently propagates outwards from the initial zones of failure. This is accompanied by the generation of an approximately constant stress. However, for compression in the transverse direction, deformation is uniformly distributed and the associated stress increases monotonously with deformation. Based on these experimental observations, an orthotropic viscoplastic model is established. Firstly, static response is defined via a combination of two constitutive descriptions, one primarily for post-yield strain-hardening and the other for densification. This is coupled with a Bingham-type rheological model to account for rate-sensitivity. Comparisons between responses predicted by the proposed model and experimental data exhibit good agreement.

It is noted that the ranges of values for the viscosity and relative density parameters are not large; 1.42 kPa s $\leq \eta(\theta') \leq 4.34$ kPa s and $1.5 \leq m(\theta') \leq 2.0$ respectively. Consequently, it is expected that the use of linear relationships to approximate both $m(\theta')$ and $\eta(\theta')$ would not result in significant errors. Also, the unknown parameters were derived from only two values of θ' (i.e. 0 and $\pi/2$). Experimental tests at other values of θ' (e.g. $\pi/4$) should be undertaken to further verify validity of the model and its potential to describe foam behaviour for loading in any direction; i.e. the assumed functions of θ' could be refined. Nevertheless, the present model establishes a basis for the full description of rate-sensitive behaviour of anisotropic foams of different densities.

REFERENCES

Campbell, J.D. (1973). Dynamic plasticity. *Mat. Sci. Engrg*, **12**, p3.
Gibson, L.J. and Ashby, M.F. (1988). *Cellular Solids Structure & Properties*, Pergamon Press.
Hill, R. (1950). *The Mathematical Theory of Plasticity*, Oxford University Press, London.
Mal, A.K. and Singh, S.J. (1991). *Deformation of Elastic Solids*, Prentice-Hall Inc., New Jersey.
Neilson, M. K., R. D. Krieg, H. L. Schreyer (1995). A constitutive theory for rigid polyurethane foam. *Polymer Engineering and Science*, **35**, pp. 387-394.

Orchon, S. (1990). Polyurethanes Thermoset, in *Handbook of Plastic Materials and Technology*, ed. I. I. Rubin. A Wiley-Interscience Publication, John Wiley & Sons Inc., pp. 511-524.

Puso, M. A., S. Govindjee (1995). A phenomenological constitutive model for polymeric foam. *Mechanics of Plastics and Plastic Components*; American Society of Mechanical Engineers, Materials Division (Publication) MD, **68**, pp. 159-176.

Shim, V.P.W. and Yang, L.M. (to be published). Characterizing the Anisotropic Plastic Response of Crushable Foam by a Three-Dimensional Microstructural Cell Model.

Zhang, J., Z. Lin, A. Wang, N. Kikuchi, V.V. Li, A.F. Yee (1997). Constitutive modeling and material characterization of polymeric foams. *Journal of Engineering Materials and Technology*, 119, p. 284.

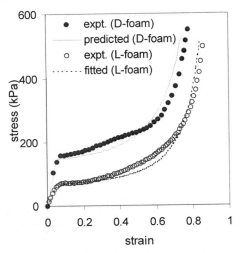

Figure 1: Stress-strain curves for static compression in the transverse direction.

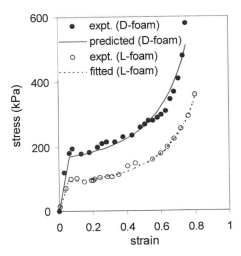

Figure 2: Stress-strain curves for dynamic compression in the transverse direction.

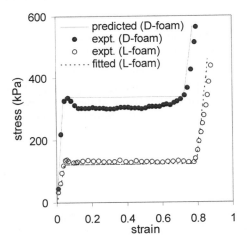

Figure 3: Stress-strain curves for static compression in the foam rise direction.

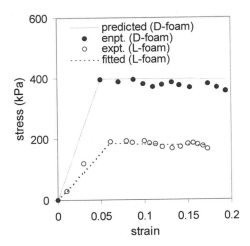

Figure 4: Strain-strain curves for dynamic compression in the foam rise direction.

DEFORMATION AND PERFORATION OF WATER-FILLED AND EMPTY ALUMINUM TUBES BY A SPHERICAL STEEL PROJECTILE: EXPERIMENTAL STUDY

M. Nishida[1], K. Tanaka[1] and M. Ito[2]

[1]Department of Mechanical Engineering, Nagoya Institute of Technology,
Gokiso-cho, Showa-ku, Nagoya 466-8555, JAPAN
[2]Graduate Student, Nagoya Institute of Technology

ABSTRACT

The dynamic behavior of an empty tube and a water-filled tube made of A6063 aluminum alloy subjected to spherical projectile impact was investigated experimentally. Their plastic deformation, crack development, and plug formation were observed in the velocity range from about 80m/s to about 230m/s near their ballistic limit. Also, the plastic deformation of tube in the axial and meridional directions, and the mass and thickness of the plug were measured. It was found that the filling of water significantly affects the ballistic limit, plugging process, and plastic deformation of the tube. The plastic deformation of the water-filled tube was more localized near the impact point and was smaller than that of the empty tube in both of axial and meridional directions. The ballistic limit of the water-filled tube was approximately 20 percent smaller than that of the empty tube. The diameter of the plug of the water-filled tube was slightly larger than that of the empty tube.

KEYWORDS

Perforation, Plugging, Ballistic limit, Water-filled tube, Aluminum tube, Plastic deformation, Impact strength, Projectile impact

INTRODUCTION

In order to prevent any contents from releasing accidentally, thin walled vessels and tubes, such as fuel tank and launch vehicle, should have sufficient strength against any failure, fracture, and cracking caused by the impact of foreign objects. Then, the prediction for the perforation and penetration process and the design criteria of such structures against the projectile impact have become a crucial issue. A numerous experimental, numerical, and theoretical analyses have been made on the fracture and deformation of thin plates and tubes impacted by a projectile for a long time. However, there are some differences in the perforation phenomena between the fluid-filled tube and empty tube (Ma &

Stronge (1985), Corbett et al. (1996)). Some attempts have so far been made at vessels and tubes filled with liquid (Anderson et al.(1999)), pressurized-gas (Jones & Birch (1996), Lambert & Schneider (1997)) and the combination of liquid and gas (Poe & Rucher (1993)).

In the present paper, the perforation mechanism of water-filled tubes and empty tubes impacted by a steel sphere are studied experimentally. Particularly, the mass of plugs and the plastic deformation of tubes after impact are inspected in detail and the influence of the filled water on the ballistic limit is discussed.

EXPERIMENTAL APPARATUS

The schematic of the apparatus for impact experiments is shown in Figure 1. A tube made of A6063-H18 aluminum alloy having an outer diameter of 50 mm and a nominal thickness of 1 mm is fully clamped across a span of 250 mm. The basic mechanical properties of A6063-H18 aluminum alloy used in the present experiment are as follows: Young's modulus is 70 GPa, ultimate strength is 165 MPa, 0.2 % proof stress is 100 MPa and elongation is 20 %. A steel sphere, for bearing use, of diameter 8 mm and mass 2.1 g is used for the projectile. The thickness of tubes is determined so that the perforation through one side of the tube can be achieved by using the present experimental apparatus. The ratio of the tube thickness to the projectile diameter is approximately 0.1. Tap water at atmospheric pressure and room temperature is used as a containing liquid paying attention that any air bubbles do not go into the tube. The projectile embedded in a sabot of 50 mm in length, 29.8 mm in diameter is accelerated using the pressurized nitrogen gas gun of 30 mm in barrel diameter and it is separated from the sabot by the sabot-stopper placed at the muzzle of the gun. The sabots made of polyethylene resin are about 30 g in mass. The tube, sabot-stopper and gun barrel are aligned so that the projectile can impinge on the center of a tube perpendicularly. The impact velocity of projectile is determined from the interruption signals of two light beams (FX-10: SUNX Ltd., Japan) that are parallel and spaced at 40 mm.

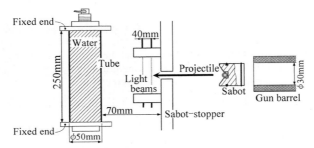

Figure 1: Experimental apparatus for projectile impact on the tube

EXPERIMENTAL RESULTS AND DISCUSSIONS

Observation of Tubes after Impact and Ballistic Limit

The impact experiments are carried out in the velocity range from 80 m/s to 230 m/s. Under the condition of the present experiments, the plastic deformation and perforation are observed only in the vicinity of the impact point, not around the fixed ends of tube. Therefore, an attention is focused on the vicinity of the impact point.

The evolutions of plastic deformation of the empty tube produced by the steel sphere impact are shown in Figure 2. Figure 2(a) shows an indentation of the tube at the velocity less than the ballistic limit. Here the ballistic limit means the critical impact velocity of perforation defined as the complete piercing of one side of the tube by the projectile. At relatively low velocities, the indentation in the region surrounding the impact point becomes deeper and more extensive with increasing the impact velocity, where both the stretching and bending effects contribute to this deformation. At the impact velocity of approximately 150 m/s, a cracking occurs in the tube. In Figure 2(b), it is observed that the brim of the plug is formed clearly by a circumferential crack. At all velocities under the ballistic limit, the projectile rebounds from the surface of a tube without embedding. At velocities exceeding about 160 m/s, the projectile passes through one side of the tube.

In Figure 3(a)-(d), the plastic deformations of the water-filled tubes are shown. General feature of the observation in Figure 3 is almost same as in Figure 2. Also, it is found from Figure 4 that the ballistic limit of the water-filled tube is about 20 percent smaller than that of the empty tube and great effect of the filled water on the ballistic limit can be seen clearly. The present result is opposite to the results of Ma & Strange (1985) who showed that the steel tubes of outer diameter 51 mm, length 250 mm and thicknesses 1.2 mm, 2.1 mm and 3.3 mm were stiffened by the filled water and the ballistic limit was increased markedly. It is considered that the diversity between the results of two investigations is mainly attributed to the material properties of tubes.

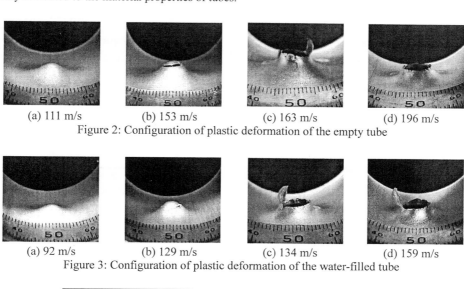

(a) 111 m/s (b) 153 m/s (c) 163 m/s (d) 196 m/s
Figure 2: Configuration of plastic deformation of the empty tube

(a) 92 m/s (b) 129 m/s (c) 134 m/s (d) 159 m/s
Figure 3: Configuration of plastic deformation of the water-filled tube

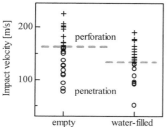

Figure 4: Ballistic limit of the empty tube and water-filled tube

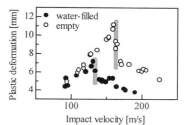

Figure 5: Maximum plastic deformation of the empty tube and water-filled tube

Maximum Plastic Deformation of Tubes

Figure 5 shows the maximum plastic deformation of the tube subjected to the sphere impact. With increasing the impact velocity, the maximum plastic deformation of both the water-filled and empty tubes increases gradually below the ballistic limit but it decreases suddenly at the ballistic limit and decreases further over the ballistic limit. In consequence, there appears a peak in the maximum plastic deformation just below the ballistic limit, in both cases of the water-filled and empty tubes. This tendency about the peak agrees well with the results of Levy & Goldsmith (1984) who dealt with the perforation and penetration of a steel sphere into the thin plates of aluminum 2024-O. For impact velocities larger than the ballistic limit, there appears a pronounced difference in the maximum plastic deformations between the water-filled and empty tubes. The maximum plastic deformation for the water-filled tube is about 40 percent smaller than that for the empty tube.

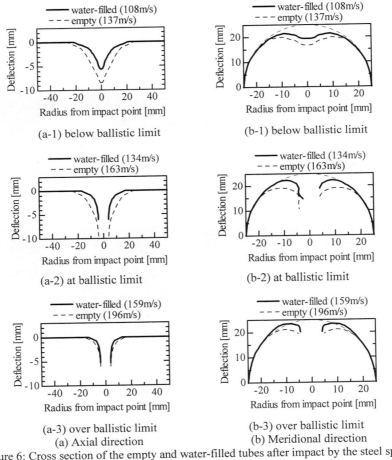

Figure 6: Cross section of the empty and water-filled tubes after impact by the steel sphere

Change of the Cross Section of Tubes after Impact

The deformed shapes of the cross section of tubes after impact are shown in Figure 6 for the axial and meridional directions. Since the ballistic limit is different between the empty tube and water-filled tube, the comparison is made between the results at the impact velocities that are 20 percent larger and smaller than each ballistic limit. It is to be noted that the different length scales are employed between the ordinate and the abscissa in these graphs. The deformation of the water-filled tube is more localized around the immediate vicinity of the impact point in both of the axial and meridional directions and the tube bulges partially in the meridional direction over the ballistic limit. Therefore, it is evident that the plastic deformation mode is greatly affected by the filling of water.

Mass of Plug

Under the condition of the present experiments, it is found that the plug has a shape of flat hemisphere but its perimeter deviates a little from a complete circle and its edge is somewhat jagged in the case of the empty tube. Then, the variation of the mass, not diameter, is compared in Figure 7. Here, the impact velocity is normalized by the each velocity of ballistic limit for the water-filled tube and empty tube.

As the impact velocity increases, the mass of plugs increases in both cases of the water-filled and empty tubes. The plug masses of the water-filled tube are about 5-10 percent larger than those of the empty tube. Under the condition of the present experiments, incomplete separations of the plug from the tube are often found and these plugs remain attaching partly at the intersection of the rim of the perforation hole and a meridian passing through the center of the perforation hole, as shown in Figure 2(c), Figure 3(c) and Figure 3(d).

Ma & Stronge (1985) discussed about the in-plane stretch of tubes and plug formation by measuring the final thickness of plugs. They reported that the thickest cross section of a plug was at its center and the thickness quickly decreased with leaving away from the center of the plug, which is in agreement with the results of Goldsmith & Finnegan (1971). However, the change of plug thickness in the present experiment is less than the results of Goldsmith & Finnegan (1971) or Ma & Stronge (1985), as shown in Figure 8.

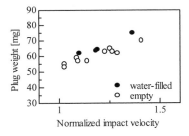

Figure 7: Masses of plugs of the empty tube and water-filled tube

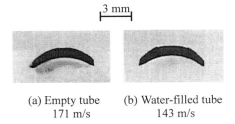

(a) Empty tube   (b) Water-filled tube
171 m/s   143 m/s

Figure 8: Photograph of the cross section of plug

MECHANISM OF PLUGGING

In Figure 9, the photographs of the deformed cross section of both tubes are shown. In the axial direction, the thickness of the tube near the crack is slightly decreased in both cases of the water-filled tube and empty tube. On the other hand, in the meridional direction, the thickness of the empty tube near the crack is slightly decreased, but clear change in the thickness can not be seen for the water-filled tube.

From the above observations, the plugging mechanism of the tube filled with water can be deduced. In the case of perforation of a flat plate, the thickness of plate is first reduced owing to a stretching. At approximately the same time as a thinning, a shearing occurs at the perimeter of the contact surface between the projectile and plate. Finally, an initial crack is generated near the contact perimeter by the combination of thinning and shearing. When a circular tube is filled with water, the effect of the stretching on a plug formation decreases and the shearing becomes effective, since the filling of water restrains from the stretching of the tube especially in the meridional direction. It is considered that the filled water increases the percentage of shear stress in comparison with tensional stress, which affects the formation of a plug.

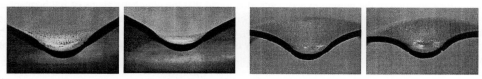

Empty tube (149 m/s) Water-filled tube (131 m/s)   Empty tube (153 m/s) Water-filled tube (121 m/s)
(a) Axial direction   (b) Meridional direction
Figure 9: Photograph of the cross section for the empty and water-filled tubes

CONCLUSIONS

It is evident that the perforation behavior of tubes is greatly affected by the filled water. Under the condition of the present experiment, the critical velocity to generate perforation in a tube is decreased by the filled water. Also, for the water-filled tube, the plastic deformation is more localized near the impact point and is smaller than that of the empty tube. The mass of plugs removed from the water-filled tube is larger than that from the empty tube. In consequence, it is deduced that the filled water increases the percentage of shearing in comparison with stretching which affects the formation of plug.

References

Anderson C.E., Sharron T.R., Walker J.D. and Freitas C.J. (1999). Simulation and Analysis of a 23-mm HEI Projectile Hydrodynamic Ram Experiment. *Int. J. Impact Engng.* **22**, 981-997.

Corbett G.G., Reid S.R. and Johnson W. (1996). Impact Loading of Plates and Shells by Free-Flying Projectiles: a Review. *Int. J. Impact Engng.* **18: 2**, 141-230.

Goldsmith W. and Finnegan S.A. (1971). Penetration and Perforation Processes in Metal Targets at and above Ballistic Velocities. *Int. J. Mech. Sci.* **13**, 843-866.

Jones N. and Birch, R.S. (1996). Influence of Internal Pressure on the Impact Behavior of Steel Pipelines. *Trans. ASME J. Pressure Vessel Tech.* **118**, 464-471.

Lambert M. and Schneider E. (1997). Hypervelocity Impacts on Gas Filled Pressure Vessels. *Int. J. Impact Engng.* **20**, 491-498.

Levy N. and Goldsmith W. (1984). Normal Impact and Perforation of Thin Plates by Hemispherically-Tipped Projectiles-II. Experimental results. *Int. J. Impact Engng.* **2: 4**, 299-324.

Ma X. and Stronge W.J. (1985). Spherical Missile Impact and Perforation of Filled Steel Tubes. *Int. J. Impact Engng.* **3: 1**, 1-16.

Neilson A.J., Howa W.D. and Garton G.P. (1987). Impact Resistance of Mild Steel Pipes: an Experimental Study, AEE Winfrith AEEW-R2125.

Poe R.F. and Rucker M.A. (1993). Evaluation of Pressurized Vessels Following Hypervelocity Particle Impact. *Proc. First European Conference on Space Debris*, 441-446.

GENERATION-PHASE SIMULATION OF DYNAMIC INTERFACIAL FRACTURE UNDER IMPACT LOADING

T. Nishioka[1], Q.H. Hu[1] and T. Fujimoto[1]

[1] Department of Ocean Mechanical Engineering,
Kobe University of Mercantile Marine,
5-1-1 Fukae Minamimachi, Higashinada-ku, Kobe 658-0022, JAPAN

ABSTRACT

In this paper, first, the experimental results of the dynamic interfacial fracture phenomenon in a bimaterial specimen are presented including the caustic patterns in narrow band shapes for transonically propagating cracks. Next, a moving finite element method is developed to accurately simulate the transonic interfacial crack propagation. Using the experimentally measured histories of crack propagation, the generation-phase simulation is carried out. The dynamic J integral and the separated dynamic J integrals are evaluated to investigate the mechanism of fracture energy supply to the propagating interfacial crack tip. Furthermore, the reaction forces and displacements of the supporting and loading points are estimated. The strain energy density distributions and profiles of the crack face during the crack propagation are visualized. From the numerical results for the separated dynamic J integrals and the dynamic J integral, it is found that the energy release rate is not zero even for the transonically propagating crack tip.

KEYWORDS

Interfacial fracture mechanics, Bimaterial, Transonic crack propagation, Intersonic crack propagation, Dynamic J integral, Separated dynamic J integral, Caustics, Contact/non-contact condition, Impact fracture, Moving finite element Method, Path independent integral

INTRODUCTION

The establishment of dynamic interfacial fracture mechanics has received much attention due to broad developments and utilization of composite materials and jointed materials. Recently the transonic (or intersonic) crack propagation has attracted many researchers' interests. Yang et al. (1991) suggested that the interfacial cracks could move faster than the lower Rayleigh wave velocity of a bimaterial system by their analytical study. This activated the study of high-speed interfacial fracture phenomena. In their experimental studies, some researchers observed the interfacial cracks propagating faster than the lower shear wave velocity under certain loading condition (Lambros & Rosakis, 1995, Nishioka et al., 2000). However, the mechanism of transonic crack propagation phenomenon in interfacial fracture has not been fully clarified.

Recently, Nishioka and Yasin (1999) have proposed the concepts of the separated dynamic J integrals and the separated dynamic energy release rates which have the meaning of the rate of energy flowing into the interfacial crack tip from individual material sides. Thus, the dynamic J integral (Nishioka and Atluri, 1983) can be evaluated by the sum of the separated dynamic J integrals. The dynamic J integral and the separate dynamic J integrals provided one of the possibilities for clarifying the mechanism of the origin of the crack propagating energy (Nishioka and Yasin, 2000).

In previous studies (Nishioka et al. 2000), the dynamic interfacial fracture phenomena in bimaterial specimens of aluminum alloy and epoxy resin, were recorded by an ultrahigh-speed camera. The method of transmitted caustics was used to evaluate the energy release rate. The caustic patterns in narrow band shapes for transonically propagating cracks were firstly reported in literature.

In this paper, first, the experimental results are briefly summarized. Next, a moving finite element method is developed to accurately simulate very high-velocity interfacial crack propagation such as transonic crack velocity. Using the experimentally measured histories of crack propagation, the generation-phase simulations are carried out. Using the path independent separated dynamic J integrals, the mechanism of fracture energy supply to the propagating interfacial crack tip is investigated. Also, the displacements and reaction forces of the loading point and supporting points are presented to analyze the status of specimen during the crack propagation. Furthermore, the Mach shock waves emanated from transonically propagating crack tips are visualized by strain energy density. And the profiles of the crack surfaces during the crack propagation are visualized to detect the crack face contact.

EXPERIMENTAL METHODS

The shadow optical method of caustics (Manogg, 1964) is an important experimental technique in fracture mechanics studies. It is sensitive to stress gradients and therefore is an appropriate tool for quantifying stress concentration problems. Shadow optical images of test specimens under loading in general are characterized by simple geometric patterns, which can easily be evaluated. The method of caustics has been extended by many researchers (Theocaris & Gdoutos, 1972, Rosakis et al.,1984, Nishioka & Kittaka, 1990) to different conditions of loading and material behavior in static as well as dynamic situations.

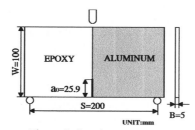

Figure 1: Specimen geometry

In this experiment, a bimaterial plate made of epoxy and aluminum alloy (AL5052) with an interfacial crack is employed. The relevant dimensions are shown in Fig. 1. The width, height and thickness of the specimen are 220, 100 and 5 mm, respectively, while the span between two supporting points is

200mm. The initial crack length is measured as 25.9mm. A dropping hammer (14.7Kg) was impacted on the specimen at the center of the upside span with the velocity of 7.0m/s. The material properties are listed in Table 1. It is seen from the Young's moduli in the table that the aluminum alloy is much stiffer than the epoxy. Thus, dilatational wave velocity C_d, shear wave velocity C_s and Rayleigh wave velocity C_R in the aluminum alloy are much higher than those in the epoxy.

TABLE 1
MATERIAL PROPERTIES

Material	E (GPa)	ν	ρ (kg/m³)	C_d (m/s)	C_s (m/s)	C_R (m/s)
Epoxy	2.3	0.38	1170	1515.7	843.9	792.1
Aluminum alloy	72	0.3	2680	5344.5	3214.5	2977.1

Using an ultrahigh-speed camera, high-speed photographs of the caustic patterns were taken by a laser caustic method which can be quickly synchronized to the onset of brittle fast fracture. The ultrahigh-speed camera is capable of taking 80 high-speed photographs with the maximum framing rate of two million frames per second. The high-speed photographs for dynamic interfacial fracture in the specimen are shown in Fig. 2. The framing rate of 340000 frames/s was used. Narrow band shapes of caustic pattern can be observed in these photographs (see photographs 67 to 74).

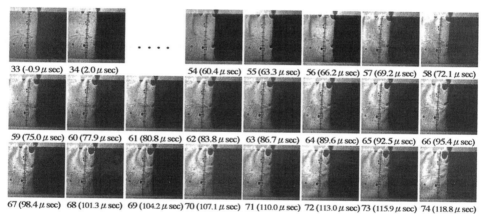

Figure 2: High-speed photographs of impact interfacial fracture

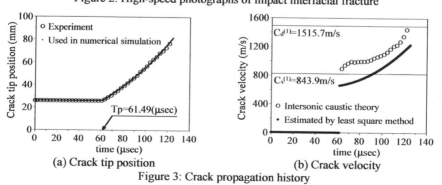

(a) Crack tip position (b) Crack velocity

Figure 3: Crack propagation history

The history of crack propagation is shown in Fig.3 (a, b). The fracture initiation time T_p was 61.49μs. There are two sets of data in both figures. One was evaluated by the crack tip position in the caustic pattern using the method of caustics for intersonically propagating crack (Shen and Nishioka, 2001). Other is calculated by using the least-squares method and will be adopted in the following numerical simulation. Then the crack velocities were also evaluated in two ways. One was obtained by the derivative of the least square fitted curve. Other is approximately estimated from the shapes of the caustic patterns using the theory of caustics developed by Shen and Nishioka (2001). The crack was stationary until the fracture initiation time T_p. After T_p, the crack velocity increased along with the crack propagation and exceeded the shear wave velocity C_s of the epoxy.

DYNAMIC J INTEGRAL AND SEPARATED DYNAMIC J INTEGRALS

The well-known Eshelby-Rice static J integral has been widely used in static fracture mechanics. For dynamic fracture mechanics, Nishioka and Atluri (1983) derived the path-independent dynamic J integral for a homogenous material as

$$J'_k = \lim_{\Gamma_\varepsilon \to 0} \int_{\Gamma_\varepsilon} [(W+K)n_k - t_i u_{i,k}]dS = \lim_{\Gamma_\varepsilon \to 0} \left\{ \int_{\Gamma+\Gamma_c} [(W+K)n_k - t_i u_{i,k}]dS + \int_{V_\Gamma-V_\varepsilon} [\rho\ddot{u}_i u_{i,k} - \rho\dot{u}_i \dot{u}_{i,k}]dV \right\}, \quad (1)$$

where W and K are the strain and kinetic energy densities. Γ_ε denotes the near field integral path, while Γ and Γ_c are the far field path and the crack face integral path, respectively (see Fig. 4(a)). The crack axis components of the dynamic J-integral $J'_k{}^0$ can be obtained by the coordinate transformation: $J'_k{}^0 = \alpha_{kl} J'_l$, where α_{kl} is the coordinate transformation tensor. The tangential component of the dynamic J integral $J'_1{}^0$ has the physical significance of the dynamic energy release rate G. Thus we have

$$J'^0_1 = J' = G. \quad (2)$$

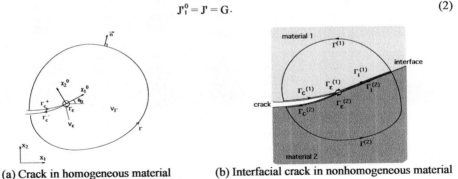

(a) Crack in homogeneous material   (b) Interfacial crack in nonhomogeneous material
Figure 4: Definition of integral paths

Considering a nonhomogeneous plate with a dynamically propagating interfacial crack as shown in Fig. 4(b), recently Nishioka and Yasin (1999) have derived the separated dynamic J integrals as

$$J'^{(m)}_k = \lim_{\Gamma^{(m)}_\varepsilon \to 0} \int_{\Gamma_\varepsilon} [(W+K)n_k - t_i u_{i,k}]dS = \lim_{\Gamma^{(m)}_\varepsilon \to 0} \left\{ \int_{\Gamma^{(m)}+\Gamma^{(m)}_c+\Gamma^{(m)}_I} [(W+K)n_k - t_i u_{i,k}]dS + \int_{V^{(m)}_\Gamma - V^{(m)}_\varepsilon} [\rho\ddot{u}_i u_{i,k} - \rho\dot{u}_i \dot{u}_{i,k}]dV \right\}, \quad (3)$$

$$(m=1,2)$$

where $\Gamma^{(m)}_I$ (m=1,2) are the integral paths along the interface in the sides of the material 1 and 2. Also the crack axis components can be obtained by applying the coordinate transformation: $J'^{0(m)}_k = \alpha_{kl} J'^{(m)}_l$.

The separated dynamic J integrals also have the physical significance of the separated energy release rates $G^{(m)}$ (m=1,2) which are the energy flow rates from material m (m=1,2) into the propagating interfacial crack tip per unit crack extension. Thus, we have the following relations:

$$J'^{0(m)}_1 = G^{(m)} = J'^{(m)}_1 \cos\theta_0 + J'^{(m)}_2 \sin\theta_0, \qquad (m=1,2). \tag{4}$$

Furthermore, the dynamic J integral and the energy release rate can be obtained by the sum of the separated dynamic J integrals:

$$J'^0_1 = J'^{0(1)}_1 + J'^{0(2)}_1 = G = G^{(1)} + G^{(2)}. \tag{5}$$

GENERATION-PHASE SIMULATION

A moving finite element method was used to accurately simulate transonic interfacial crack propagation. By using this method, the meshes around the crack tip are refined to assure the enough precision. The mesh pattern and the integral paths are plotted in Fig.5. The fine central part of the meshes around the crack tip moves along with the crack propagation.

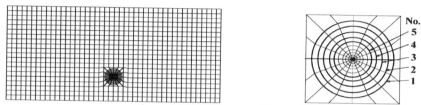

Figure 5: Mesh pattern and integral paths

The result of numerical simulation of the dynamic J integral is shown in Fig. 6. The energy release rates for the stationary crack were evaluated by using the size of caustic patterns and plotted in the same figure to compare with the numerical ones. We can see that these two curves agree well with each other. At the fracture initiation point, the value of J'^0_1 decreases drastically. Then during the transonic crack propagation period, the dynamic J integral, which presents the energy release rate, keeps non-zero values. Thus the non-zero energy flow drives the crack to propagate with transonic velocity. The separated dynamic J integrals are shown in Fig. 7. The superscript (m), m=1,2 means the material side. Thus, $J'^{0(1)}_1$ is the energy release rate from epoxy side while $J'^{0(2)}_1$ is that from aluminum side. It is concluded that the more compliant material epoxy provides much more fracture energy than the stiffer material aluminum does.

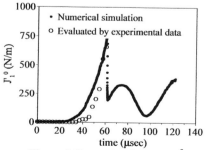

Figure 6: Dynamic J integral J'^0_1

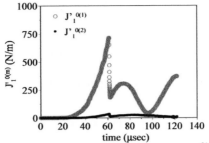

Figure 7: Separated dynamic J integral $J'^{0(m)}_1$

During the impact fracture simulation, the contact/non-contact algorithm was used for the loading and supporting points. It means that the supports are fixed in their positions, when the specimen has the trends to move downward, they will provide supporting forces, otherwise the supporting points on the specimen will be separated from the supports. For the loading point, the hammer keeps the velocity of 7m/s, if the loading point on the specimen moves faster than this speed, it will be separated from the hammer. The variations of the displacements and reaction forces of the loading point and supporting points are plotted in Figs 8 and 9. It can be concluded from Fig.8 and Fig.9 that the specimen is always pushed by the impacting hammer. The load becomes the maximum value of 17.6 kN at about 65μs, and the supporting point for the aluminum side provides most of the reaction forces while the epoxy supporting point has fairly small forces. Comparing to the displacements of the loading point, the supporting points move very slightly.

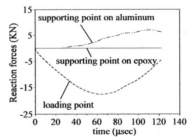

Figure 8: Reaction forces on loading and supporting points

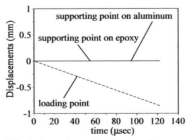

Figure 9: Displacements on loading and supporting points

Figures 10(a) and10(b) show the distributions of strain energy density for the subsonic and transonic crack propagation phases. The high concentrations of the strain energy density can be found around the crack tip, in addition to near the loading and supporting points on the aluminum side. It is seen that there are higher intensities on aluminum side near the loading and supporting points because the aluminum is subjected to more impact force and provides more supporting force. However, around the crack tip, the higher concentration occurs on the epoxy side due to the larger energy flow into the crack tip from the more compliant material epoxy.

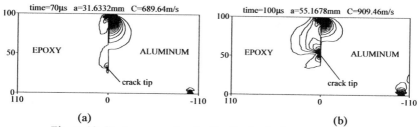

Figure 10: Strain energy density distribution in the entire specimen

The strain energy density distributions near the crack tip at time 110μs and 120μs are shown in Figs. 11(a) and 11(b), respectively. Both are in transonic propagation period. In these regimes, since the shear wave is continuously emanated from the dynamically propagating crack tip to the epoxy side, it is trapped and concentrated along a singular line which is considered as the front of shock wave (Mach shock wave). In the present case, we can see the trajectory of Mach shock wave in almost horizontal direction in the epoxy side.

Furthermore, by observing the profiles of crack face as shown in Fig.12, it can be proved that there is no crack-surface contact occurring even under transonic crack propagation. Thus in present case, we can conclude the energy source for the transonic fracture cannot be attributed to a crack face contact mechanism.

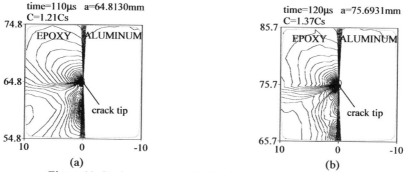

Figure 11: Strain energy density distribution around the crack tip

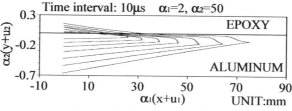

Figure 12: Crack face profiles

CONCLUSIONS

Using the experimental data for the transonic interfacial fracture, the generation phase simulation of this phenomenon was carried out. It was found that the dynamic J integral values are not zero even for the transonic fracture phase. The non-zero dynamic integral values or equivalently non-zero energy release rates are the driving energies or forces for the transonic fracture phenomenon. Furthermore, most of the fracture energy is provided from the more compliant material epoxy. Since in the present case the crack-face contact did not occur at all, the energy source for the transonic fracture cannot be attributed to any crack face contact mechanism.

Acknowledgement: This study was supported by the Natural Science Grant from Mitsubishi Foundation.

REFERENCES

Lambros J. and Rosakis A.J. (1995). Shear Dominated Transonic Crack Growth in Bimaterial-part I: Experimental Observations. *J. Mech. Phys. Solids* **43**, 169-188.

Manogg P. (1964). Anwendung der Schattenoptik zur Untersuchung des Zerreißvorgangs von Plattern, *Dissertation, University of Freiburg*.

Nishioka T. and Atluri S.N. (1983) Path-independent Integrals, Energy Release Rates, and General Solutions of Near-tip Fields in Mixed-mode Dynamic Fracture Mechanics. *Eng. Fract. Mech.* **18**, 1-22.

Nishioka T. and Kittaka H. (1990). A Theory of Caustics for Mixed-mode Fast Running Cracks. *Engng. Fract. Mech.* **36**, 987-998.

Nishioka T., Miyauchi H., Fujimoto T. and Sakakura K. (2000) Measurement of Energy Release Rates for Dynamic Interfacial Fracture Phenomena. *Proceedings of the 9th International Congress on Experimental Mechanics*, 367-370.

Nishioka T. and Yasin A. (2000). Numerical Simulations of Dynamic Interfacial Fracture Phenomena, *Advances on Computational Engineering and Sciences*, (S.N. Atluri and F.W. Brust, editors), **I**, Tech Science Press, 1047-1052.

Nishioka, T., Tokudome, H. and Kinoshita, M. (2001). Dynamic Fracture-Path Prediction in Impact Fracture Phenomena Using Moving Finite Element Method Based on Delaunay Automatic Mesh Generation. *Int. J. Solids Struc.* **38-31**, 5273-5301

Nishioka T. and Yasin A. (1999). The Dynamic J Integral, Separated Dynamic J Integrals and Moving Finite Element Simulations, for Subsonic, Transonic and Supersonic Interfacial Crack Propagation. *JSME International Journal* **A-42**, 25-39.

Rosakis A.J., Duffy J. and Freund L.B. (1984). The Determination of Dynamic Fracture Toughness of AISI 4340 Steel by the Shadow Spot Method. *J. Mech. Phys. Solids* **32**, 443-460.

Shen S.H. and Nishioka T. (2000) Method of Caustics for Intersonically Propagating Interfacial Crack, *Eng. Fract. Mech*, (in preparation).

Theocaris P.S. and Gdoutos E. (1972). An Optical Method for Determining Opening-mode and Edge Sliding -mode Stress Intensity Factors. *J. Appl. Mech.* **39**, 91-97.

Yang W., Suo Z. and Shih C.F. (1991). Mechanics of Dynamic Debonding. *Proc. R. Soc. Lon.* **A433**, 679-697.

Impact Engineering and Application
Akira Chiba, Shinji Tanimura and Kazuyuki Hokamoto (Eds)
©2001 Elsevier Science Ltd. All rights reserved.

DYNAMIC ELASTO VISCO-PLASTIC CRACK PROPAGATION ANALYSIS USING MOVING FINITE ELEMENT METHOD

T. Fujimoto[1] and T. Nishioka[1]

[1]Department of Ocean Mechanical Engineering, Kobe University of Mercantile Marine
Hyogo 658-0022, Japan

ABSTRACT

In crush accidents, fast fracture phenomena often occur with large deformation of structures. However, the mechanism of fast crack propagation under dynamic large deformation has not been fully clarified. Recently, the authors reported that the T* integral is effective to describe elasto visco-plastic fracture. In this study, we develop a moving finite element method to simulate fast crack propagation phenomena under dynamic large deformation. Numerical simulations are carried out for cases of various crack propagation velocities. The path independent T* integral, crack opening and stress distributions are calculated by this moving finite element method. Based on these numerical results, the T* integral demonstrates its ability to clarify the mechanism of fast crack propagation in elasto visco-plastic materials.

KEYWORDS

Moving Finite Element Method, Elasto Visco-Plastic Fracture, Fast Crack Propagation, Large Deformation, Fracture Mechanics Parameter, T* integral,

INTRODUCTION

The plasticity, viscosity and inertial properties of a material are important factors in determining its behavior under dynamic large deformation. When a material is subjected to this type of deformation, crack propagation phenomena often occur. To measure and predict these fracture phenomena requires the formulation of a dynamic elasto visco-plastic fracture mechanics. Recently, many researchers have used the cohesive force model (Xu & Needleman 1994) and the nodal release method (Kobayashi, Mall, Urabe & Emery 1978) to simulate this type of fracture. The authors have also used the nodal release method to analyze fracture behavior under dynamic large deformation (Fujimoto, Akashi & Nishioka 2000). This work showed that the T* integral (Atluri, Nishioka & Nakagaki 1984) can be used as a valid fracture mechanics parameter for elasto visco-plastic fracture. Excellent path independence of the T* integral was maintained in this type of fracture.

In the nodal release method, it is assumed that the crack propagation effect can be expressed by the decrease of holding-back force, instead of modeling the true shape near the propagating crack tip. For fracture under dynamic large deformation, the calculation of the crack tip parameter depends on the strain rate of the near field. Therefore, the assumption used to formulate the nodal release method may cause numerical errors in the crack tip parameter in the case of fracture under dynamic large deformation.

In this study, the moving finite element method (Nishioka 1994) is applied to the problem of fracture under dynamic large deformation. This method allows accurate evaluation of the fracture mechanics parameter by use of moving elements. The true shape near a propagating crack tip is modeled by these moving elements. In this method, mesh regeneration is required to continue the finite element calculation. Delaunay triangulation (Taniguchi 1992) is used for this mesh regeneration with crack propagation.

In this FEM, the updated Lagrangian form for large deformation theory (Tomita et al. 1988) is used to formulate the stress-strain rate relation. The plasticity and viscosity of the material are expressed by the Malvern type constitutive equation (Malvern 1951). Fast crack propagation phenomena are simulated using this moving finite element method. In the numerical results of the moving finite element calculations, the path independence of the T* integral, the crack opening and stress distributions are evaluated and discussed.

THE ELASTO VISCO-PLASTIC CONSTITUTIVE EQUATION

In this study, the Malvern type constitutive equation (Malvern 1951) is used to express the plasticity and viscosity of the material. In this constitutive equation, the dependence of the stress-strain relation on strain rate is formulated based on the quasi-static relation, which is expressed in the following form:

$$\bar{\varepsilon} = \frac{\sigma_f}{E} + \left(\frac{\sigma_f}{F}\right)^{\frac{1}{n}}, \qquad (1)$$

where $\bar{\varepsilon}$ and σ_f are the equivalent strain and the static flow stress, respectively. The parameters E, F and n are the Young's modulus, the reference stress and the hardening exponent, respectively.

The relation between the equivalent stress rate $\dot{\bar{\sigma}}$ and the equivalent strain rate $\dot{\bar{\varepsilon}}$ is shown as:

$$\dot{\bar{\varepsilon}} = \frac{\dot{\bar{\sigma}}}{E} + D\left(\frac{\bar{\sigma}}{\sigma_f} - 1\right), \qquad (2)$$

where $\bar{\sigma}$ is the equivalent stress, and D is the visco-plastic coefficient. A dot above a variable (˙) denotes its time derivative. The material properties are shown in Table 1.

TABLE 1
MATERIAL PROPERTIES

	Symbol	Unit	Value
Young's Modulus	E	GPa	210
Poisson Ratio	ν	-	0.3
Mass Density	ρ	kg/m³	7.9
Reference Stress	F	MPa	300
Hardening Exponent	n	-	0.2
Visco-plastic Coefficient	D	1/sec	100

MOVING FINITE ELEMENT METHOD FOR FRACTURE UNDER DYNAMIC LARGE DEFORMATION

Various types of moving finite element methods were proposed by Nishioka and Atluri (Nishioka 1994) to simulate dynamic crack propagation. This numerical technique has been shown to give accurate solutions for a number of complex crack propagation phenomena, for example kinked crack growth (Nishioka, Tokudome & Kinoshita 1999) and dynamic crack bifurcation (Nishioka et al. 2001). In this study, the moving finite element method is applied to the problem of fracture propagation under dynamic large deformation.

Figure 1 outlines the concept of the moving finite element method. The consistent group of finite elements (moving elements) is located so as to reproduce the valid boundary conditions near the propagating crack tip. The use of a fine mesh subdivision for the moving elements allows accurate estimation of the stress singularity behavior near the crack tip. At each time step the moving elements translate with the propagating crack tip, and the other elements surrounding the moving elements are regenerated automatically by Delaunay triangulation (Taniguchi 1992). For dynamic elastic problems, the moving finite element method based on Delaunay triangulation was developed by Nishioka, Tokudome & Kinoshita (1999), and Nishioka et al. (2001).

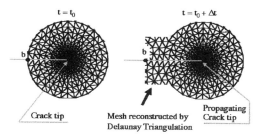

Figure 1: Moving finite element procedures for a propagating crack

In this method, the solution fields in the previous mesh ($t = t_0$) are mapped onto those in the present mesh ($t = t_0 + \Delta t$). Quantities at new nodal points and new integral points are calculated by finite element interpolation. The displacements, velocities, accelerations and some additional visco-plastic parameters are required in dynamic elasto visco-plastic problems

The finite element equation for dynamic large deformation is formulated based on the large deformation theory (Tomita et al. 1988). The relation between the Jaumann rate of the Kirchhoff stress $\dot{S}_{ij}$ and the deformation rate d_{kl} is expressed according to the updated Lagrangian form. This relation has been obtained as

$$\dot{S}_{ij} = L'^{tan}_{ijkl} d_{kl} - \frac{\dot{\bar{\varepsilon}}^{vp}_n}{1+\xi} P_{ij}, \qquad (3.a)$$

$$L'^{tan}_{ijkl} = \frac{H_s}{H_s+3G}\left[D^e_{ijkl} + \frac{3G}{H_s}\left\{\frac{1}{3}\delta_{ij}\delta_{kl}\frac{2G(1+\nu)}{1-2\nu} + \frac{3G\sigma'_{ij}\sigma'_{kl}}{\bar{\sigma}^2}\right\}\right] - \frac{1}{h}\frac{\xi}{1+\xi}P_{ij}P_{kl}, \quad \xi = \left(\Delta t\, h\frac{\partial \dot{\bar{\varepsilon}}^{vp}}{\partial \bar{\sigma}}\right), \quad (3.b, c)$$

$$h = p_{kl}P_{kl} - \frac{\partial \bar{\sigma}}{\partial \bar{\varepsilon}^{vp}}, \qquad P_{ij} = D^e_{ijkl}p_{kl}, \qquad p_{ij} = \frac{3\sigma'_{ij}}{2\bar{\sigma}}, \qquad H = \frac{\bar{\sigma}}{\dot{\bar{\varepsilon}}^{vp}} \text{ and } H_s = \frac{\bar{\sigma}}{\omega\dot{\bar{\varepsilon}}^{vp}}, \qquad (3.d, e, f, g, h)$$

where $\bar{\varepsilon}^{vp}$, D^e_{ijkl}, G and σ'_{ij} are the equivalent visco-plastic strain, the elastic stiffness tensor, the shear modulus and the deviatoric stress tensor, respectively, and Δt is the time increment per step.

For the Malvern type constitutive equation, the terms $\dfrac{\partial \bar{\sigma}}{\partial \bar{\varepsilon}^{vp}}$ and $\dfrac{\partial \bar{\dot{\varepsilon}}^{vp}}{\partial \bar{\sigma}}$ are expressed by the following equations (Fujimoto, Akashi & Nishioka 2000),

$$\frac{\partial \bar{\sigma}}{\partial \bar{\varepsilon}^{vp}} = \frac{\partial \bar{\sigma}}{\partial \sigma_f} \frac{\partial \sigma_f}{\partial \bar{\varepsilon}^{vp}} = \left(\frac{1}{D}\bar{\dot{\varepsilon}}^{vp}+1\right) nF(\bar{\varepsilon}^{vp})^{n-1} \quad \text{and} \quad \frac{\partial \bar{\dot{\varepsilon}}^{vp}}{\partial \bar{\sigma}} = \frac{D}{\sigma_f} . \qquad (4.a, b)$$

By introducing these constitutive equations into the principle of virtual work, the finite element equation is formulated. The Newmark β method is used as the time integral scheme in this study.

FRACTURE MECHANICS PARAMETER T* INTEGRAL

In elasto visco-plastic fracture, the combined effects of the plasticity, viscosity and inertia of a material determine its deformation behavior. The conventional J integral (Rice, 1968) cannot be used to evaluate elasto visco-plastic fracture, because the J integral does not include irreversible energy effects and inertial effects. In this study, the T* integral is used to evaluate this type of fracture. This integral was first proposed by Atluri, Nishioka & Nakagaki (1984), and its use as a visco-plastic fracture parameter has been discussed in the literature (Nishioka, Kobashi and Atluri 1988). The excellent path independence of the T* integral for dynamic elasto visco-plastic fracture was shown in a previous paper by the authors (Fujimoto, Akashi & Nishioka 2000).

The expression of the T* integral is shown as

$$T^*_k = \int_{\Gamma_\varepsilon} [(W+K)n_k - t_i u_{i,k}]dS = \int_{\Gamma+\Gamma_c} [(W+K)n_k - t_i u_{i,k}]dS + \int_{V_\Gamma - V_\varepsilon} [\rho \ddot{u}_i u_{i,k} - \rho \dot{u}_i \dot{u}_{i,k} + \sigma_{ij}\varepsilon_{ij,k} - W_{,k}]dV, \quad (5)$$

where W and K are the stress working density and the kinetic energy density, respectively; n_k, t_i and ρ are the direction cosines, the traction and the mass density, respectively; and Γ_ε, Γ and Γ_c denote a near field path, far field path and crack surface path, respectively. V_Γ is the region surrounded by Γ, while V_ε is the region surrounded by Γ_ε. The physical meaning of the T* integral under steady state is equivalent to the energy flow to the domain V_ε near the crack tip.

NUMERICAL MODEL

Crack propagation phenomena for single-edge-cracked specimens are simulated in this study. The geometry of the specimen is described in Fig.2. The initial width of the specimen W_0, the initial length $2L_0$, and the initial crack length a_0 are set as 60 mm, 30 mm and 15 mm, respectively.

The specimen is deformed by displacing the edges of the specimen in the opposite directions along the x_2 axis, as shown in Fig.2. The deformation rate is applied such that the normalized displacement rate is $\dot{u}/L_0 = 1000$ [1/sec]. The crack propagation velocities C is assumed to be $C=\alpha C_s$ ($\alpha = 0.1, 0.2, 0.3$), where Cs is the shear wave velocity of material (Cs = 3198 m/s). The time increment Δt used in the finite element method is 1.0 μs.

In order to estimate T* integral for propagating crack tip, the moving near filed path model and the extending near filed path model are proposed by Nishioka & Yagami (1988). Here, the moving path model is used as the near-field path model. The diameters of the far field path Γ and the near field path $\Gamma\varepsilon$ are $\Gamma = 6.0\sim14.0$ mm and $\Gamma\varepsilon = 2.0$ mm, respectively. In the initial finite element subdivision, the total number of nodes and elements are 2000 and 3778, respectively.

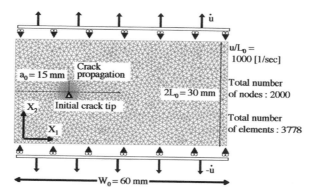

Figure 2: Numerical model for crack propagation

NUMERICAL RESULTS AND DISCUSSION

Figure 3 shows the stress σ_{yy} distributions during the crack propagation process. Stress values become almost zero near the crack propagated zone. On the other hand, wake zone is created by occurrences of the plastic deformation. By high crack velocity, the crack propagation is caused without enough crack opening. At the beginning of crack propagation (see the figure at a=15mm), the higher stress concentration is observed near the crack tip, and is of higher magnitude than during the subsequent crack propagation. After the start of crack propagation, the stress concentration near the propagating crack tip decreases rapidly.

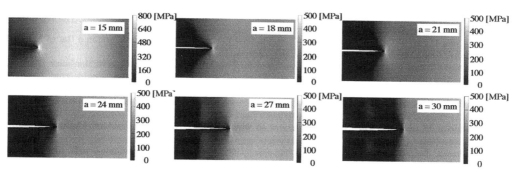

Figure 3: Stress σ_{yy} distributions during the crack propagation process (C=0.1Cs)

The values of the T* integral for each far field path are shown in Fig.4. Excellent path independence can be seen for each deformation stage. Thus, the T* integral can be used as the elasto visco-plastic fracture parameter.

For various crack propagation velocities, the histories of T* integral are shown in Fig.5. For C=0.1Cs, the values of T* integral are estimated as non-zero values. However, the T* values in the case of C=0.2Cs or 0.3Cs are almost zero. These results indicate no supply of fracture energy to crack tip. The limit of crack propagation velocity can be estimated from these results.

HYPER-VELOCITY IMPACT TEST

Experimental Apparatus

All experiments were carried out at the Institute of Laser Engineering (ILE) in Osaka University. We used a neodymium (Nd) glass laser system called "Gekko MII" at ILE. The laser had a single shot energy up to 30J at the 1053nm wavelength. The laser pulse duration was about 1ns at full width on half height of the maximum intensity (FWHM). The focal spot was set as the diameter of about 800μm.

Figure 1 shows the schematic view of the laser acceleration. The laser is focused onto the front surface of an aluminum (Al) foil under vacuum. The irradiation surface of the high intense laser is ablated and the plasma jet is blown off at high speed from the Al surface. By the reaction of the jet, the high pressure occurs and propagates in the foil as the shock wave. The free surface of the Al foil is accelerated by the shock. Reflecting at the free surface, the shock wave turns into the rarefaction wave. The remaining Al foil under the irradiation surface gains momentum, and its part was lunched as a flyer with hyper-velocity.

Using the same setup of laser and Al foils, the flyer velocity measurement experiments were performed. Figure 2 shows the relation between the laser energy and the flyer velocity. The maximum flyer velocity was about 8.6km/s for the laser shot with 30J energy in this setup.

The Experimental setup for the hyper-velocity impact experiments is shown in Figure 3. The Al flyer

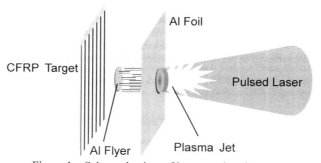

Figure 1 Schematic view of laser acceleration

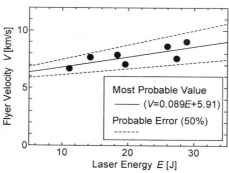

Figure 2 Relation between laser energy and flyer velocity

model is used as the near-field path model. The diameters of the far field path Γ and the near field path $\Gamma\varepsilon$ are $\Gamma = 6.0 \sim 14.0$ mm and $\Gamma\varepsilon = 2.0$ mm, respectively. In the initial finite element subdivision, the total number of nodes and elements are 2000 and 3778, respectively.

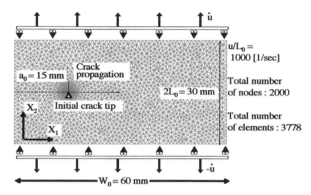

Figure 2: Numerical model for crack propagation

NUMERICAL RESULTS AND DISCUSSION

Figure 3 shows the stress σ_{yy} distributions during the crack propagation process. Stress values become almost zero near the crack propagated zone. On the other hand, wake zone is created by occurrences of the plastic deformation. By high crack velocity, the crack propagation is caused without enough crack opening. At the beginning of crack propagation (see the figure at a=15mm), the higher stress concentration is observed near the crack tip, and is of higher magnitude than during the subsequent crack propagation. After the start of crack propagation, the stress concentration near the propagating crack tip decreases rapidly.

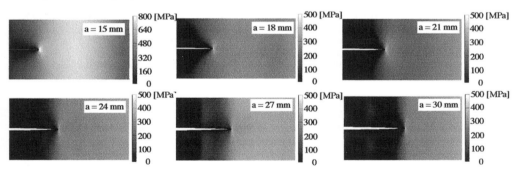

Figure 3: Stress σ_{yy} distributions during the crack propagation process (C=0.1Cs)

The values of the T* integral for each far field path are shown in Fig.4. Excellent path independence can be seen for each deformation stage. Thus, the T* integral can be used as the elasto visco-plastic fracture parameter.

For various crack propagation velocities, the histories of T* integral are shown in Fig.5. For C=0.1Cs, the values of T* integral are estimated as non-zero values. However, the T* values in the case of C=0.2Cs or 0.3Cs are almost zero. These results indicate no supply of fracture energy to crack tip. The limit of crack propagation velocity can be estimated from these results.

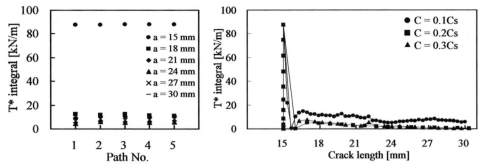

Figure 4: Path independence of T* integral

Figure 5: Variation of T* integral for elasto visco-plastic fracture

CONCLUSION

A moving finite element method for dynamic elasto visco-plastic fracture is developed. The T* integral has excellent path independence. The limit of crack propagation velocity can be estimated by the application of T* integral.

ACKNOWLEGMENT

This work is supported by the Grant-in-Aid for Scientific Research from the Ministry of Education (No.12450051).

REFERENCE

Atluri S.N., Nishioka T., and Nakagaki M., (1984). Incremental Path-Independent Integrals in Inelastic and Dynamic Fracture Mechanics, *Eng. Fract. Mech.*, **20, 2**, 209-244.

Fujimoto T., Akashi T., and Nishioka T. (2000). Crack-tip Deformation Behavior under Nonlinear and Dynamic Large Deformation of Metallic Materials, *Advances on Computational Engineering and Sciences*, (S.N. Atluri and F.W. Brust, editors), **II**, Tech Science Press, Palmdale, CA, USA, 1860-1865.

Kobayashi A.S., Mall S., Urabe Y. and Emery A.F., (1978). Fracture Dynamic Analysis of Crack Arrest Test Specimen, *Numerical Methods in Fracture Mechanics* (Eds. Luxmoore, A.R. and Owen, D.R.J.), Univ. College, Swansea, 709-720.

Malvern L.E. (1951). The Propagation of Longitudinal Waves of Plastic Deformation in a Bar of Material Exhibiting a Strain-Rate Effect, *ASME J. Appl. Mech.*, **18**, 203-208.

Nishioka T. and Yagami H., (1988). Invariance of the Path Independent T* integral in Nonlinear Dynamic Fracture Mechanics, with Respect to the Shape of a Finite Process Zone, *Engineering Fracture Mechanics*, **31, 3**, 481-491.

Nishioka T., Kobashi M. and Atluri S.N., (1988). Computational Studies on Path Independent Integrals for Non-linear Dynamic Crack Problems, *Computational Mechanics*, **3**,331-342.

Nishioka T., (1994). The State of the Art in Computational Dynamic Fracture Mechanics, *JSME International Journal*, Ser. **A**, **37, 4**, 313-333.

Nishioka T., Tokudome K. and Kinoshita M. (1999). Mixed-Phase Simulation with Fracture-Path Prediction Mode Using Moving Finite Element Method Based on Delaunay Automatic Mesh Generation, *Trans. JSME*, **65**, Ser. **A**, 597-604.

Nishioka T., Tchouikov S., Furutuka J. and Fujimoto T., (2001). Fracture Path Prediction Simulations of Dynamic Fracture Phenomena, *Material Science Research International, Special Tech. Publ.*-1, 169-176.

Rice J.R. (1968). A Path Independent Integral and Approximate Analysis of Strain Concentration by Notches and Cracks, *J. Appl. Mech.*, **35**, 379-386.

Taniguchi T., (1992). *Automatic Mesh Generation for FEM: Use of Delaunay Triangulation*, Morikita Publ. (in Japanese).

Tomita Y., Shindou A., Asada S. and Goto H. (1988). Deformation Behavior of a Strain Rate Sensitive Block under Plane Strain Tension, *Trans. JSME*, **54**, Ser. **A**, 1124-1130.

Xu X.-P. and Needleman A., (1994). Numerical Simulations of Fast Crack Growth in Brittle Solids, J. Mech. Phys. Solids, 42, 9, 1397-1434.

DEFORMATION AND FRACTURE OF CFRP UNDER HYPER-VELOCITY IMPACT USING LASER-ACCELERATED FLYER

Y. Yamauchi, M. Nakano, K. Kishida, N. Ozaki, T. Kasai, Y.Sasatani,
H. Amaki, K. Kadono, S. Ikai, K. Nishigaki and K. A. Tanaka

Department of Precision Science & Technology, Faculty of Engineering,
Osaka University, 2-1 Yamada-Oka, Suita, Osaka 565-0871, Japan

ABSTRACT

We performed hyper-velocity impact tests that laser-accelerated aluminum flyer collided to carbon fiber reinforced plastics (CFRP) target. We succeeded in observing the deformation and fracture behavior of the CFRP target with a high-speed framing camera. After the impact experiments, we investigated damages of the CFRP target with an optical microscope and a scanning electron microscope (SEM). As these results, we proposed a series of hyper-velocity impact fracture mechanisms.

KEYWORDS

Fracture, Shock Waves, Hyper-velocity Impact, CFRP, Laser, Orbital Debris, Spallation, Delamination

INTRODUCTION

Hyper-velocity impact by the collision of orbital debris becomes a problem for future space activities of the International Space Station (ISS). Since the formation of damage or fracture in materials under impact loading differs extremely from that under static loading, it is necessary to perform the hyper-velocity impact test that simulated the collision of orbital debris.

In recent years, the flyer acceleration technique using laser attracts attention in many fields with development of high intense laser. [K. A. Tanaka, et al. (2000), N. Ozaki, et al. (2001)] The pulsed laser can concentrate energy in both spatial and temporal regions. Using the shock wave induced by irradiation of the laser, we can accelerate flyer to the velocity exceeding 30 km/s, but the mass of flyer is 10^{-5}g order. Although the orbital debris impact of 1g order cannot be simulated by use of laser-accelerated flyer, it is possible to obtain the fundamental data to clarify the fracture mechanisms of materials under hyper-velocity impact loading.

HYPER-VELOCITY IMPACT TEST

Experimental Apparatus

All experiments were carried out at the Institute of Laser Engineering (ILE) in Osaka University. We used a neodymium (Nd) glass laser system called "Gekko MII" at ILE. The laser had a single shot energy up to 30J at the 1053nm wavelength. The laser pulse duration was about 1ns at full width on half height of the maximum intensity (FWHM). The focal spot was set as the diameter of about 800μm.

Figure 1 shows the schematic view of the laser acceleration. The laser is focused onto the front surface of an aluminum (Al) foil under vacuum. The irradiation surface of the high intense laser is ablated and the plasma jet is blown off at high speed from the Al surface. By the reaction of the jet, the high pressure occurs and propagates in the foil as the shock wave. The free surface of the Al foil is accelerated by the shock. Reflecting at the free surface, the shock wave turns into the rarefaction wave. The remaining Al foil under the irradiation surface gains momentum, and its part was lunched as a flyer with hyper-velocity.

Using the same setup of laser and Al foils, the flyer velocity measurement experiments were performed. Figure 2 shows the relation between the laser energy and the flyer velocity. The maximum flyer velocity was about 8.6km/s for the laser shot with 30J energy in this setup.

The Experimental setup for the hyper-velocity impact experiments is shown in Figure 3. The Al flyer

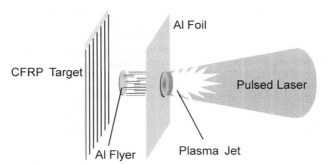

Figure 1   Schematic view of laser acceleration

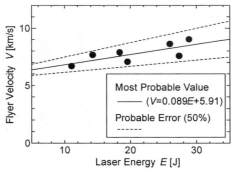

Figure 2   Relation between laser energy and flyer velocity

accelerated by the laser irradiation collided with a CFRP (Carbon Fiber Reinforced Plastics) target after the 100μm flight. Unidirectional CFRP laminates was used for the target. Its size was 20 x 10mm² and 450 μm thickness. The mechanical properties of the target material are shown in Table 1. Deformation and fracture of the target were observed in the perpendicular direction to the flight path using the high-speed flaming camera (Ultra NAC FS501).

Table 1 Mechanical properties of CFRP target

Density ρ [kg/m³]	Young's Moduli		Shear Modulus G_{12} [GPa]	Poisson's Ratio ν_{12}
	E_{11} [GPa]	E_{22} [GPa]		
1.36×10^3	139.5	11.5	4.7	0.29

High-speed Flaming Images

Figure 4 shows the position of the Al foil and the target in the flaming image before the laser shot. Laser was irradiated from the right-hand-side of the image. The Al foil with a thickness of 12.5μm was on the left side of an aperture disk. The CFRP target was spaced in 100μm from the Al foil. Figure 5 shows the series of flaming images obtained by the hyper-velocity impact test for laser energy 26.4J. As the result of the flyer velocity measurement experiments, the impact velocity of flyer was estimated about 8.3 km/s. The exposure time of one frame was 10ns and the time interval between frames was

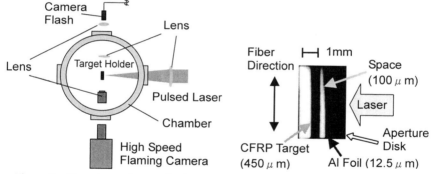

Figure 3 Experimental setup for hyper-velocity impact test

Figure 4 Position of Al foil and target in the flaming image before laser shot

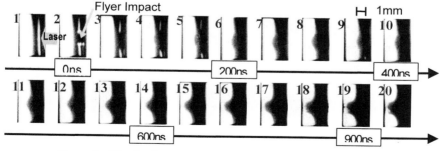

Figure 5 Flaming images of hyper-velocity impact on CFRP target
(Laser energy 26.4J, Flyer velocity 8.3km/s)

40ns, but it was changed 50ns after the 14th frame. In the 2nd frame, the flyer was launched out by laser irradiation and then collided on the target. In the 3rd frame or subsequent ones, the deformation of the target was shown.

Fractography

The damages in the CFRP target after the experiment were observed in detail using an optical microscope and a scanning electron microscope (SEM). Figure 6 shows the damages in the CFRP target for the flyer velocity 8.3km/s. The hyper-velocity impact generated a crater on the collision surface of the target. Its diameter was almost the same as that of the Al flyer and the laser focal spot size. The damage on the back surface changed with laser irradiation energy, i.e., the impact velocity of the flyer. At the low impact velocity (∼7km/s), the damage was not clear. In the case of the flyer impact velocity about 7∼8km/s, two cracks along the fiber direction were seen. The interval distance of these cracks was about 800μm. When the flyer impact velocity was exceeding 8km/s, the two cracks along fibers were not only seen, but the delamination was found between those cracks over 15mm length as shown in the side view image. Moreover, it found that carbon fibers were broken in the central part. Since the broken fibers supported each other at the center, the two split parts of the delamination layer rose up.

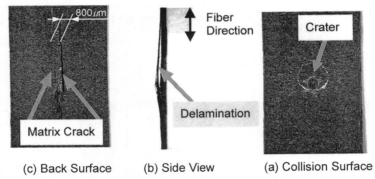

Figure 6 Images of damages in CFRP target after experiment with optical microscope (Laser energy 26.4J, Flyer velocity 8.3km/s)

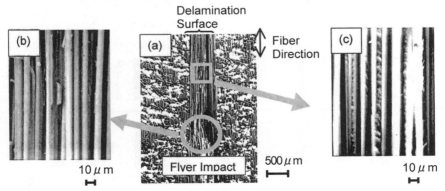

Figure 7 SEM images of delamination surface on CFRP target (Laser energy 26.4J, Flyer velocity 8.3km/s)

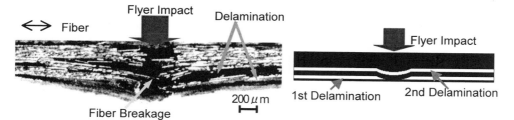

Figure 8 Section view of CFRP target with optical microscope
(Laser energy 27.6J, Flyer velocity 8.5km/s)

Figure 9 Schematic view of interlaminar fracture in CFRP target

To observe the delamination surface with SEM, the delamination layer of the target was removed. Figure 7 shows the typical images of the delamination surface. The circle in the center image shows the crater position on the collision surface. The left image is the delamination surface in this circle. The fiber surfaces were exposed and fracture has occurred mainly at the fiber/matrix interface. The surfaces of epoxy matrix were comparatively smooth. These were the typical mode I fracture surfaces with tensile opening. The right image is the delamination surface at the distance 1mm from the circle. Zigzag patterns like teeth of a saw were found at the fracture surface of epoxy matrix. These characteristic patterns are often seen on the fracture surfaces with shear (mode II or mixed-mode I/II).

In order to observe the target sections, after embedding the targets at an epoxy resin, they cut near the collision part along with the fibers. After grinding the cutting planes, we observed them with an optical microscope. One of the photographs is shown in Figure 8. Two or more delaminations between layers were observed in all the targets. Moreover, it turned out that the delaminations have occurred at almost the same position among the targets. The schematic view of the fracture observed in the targets is shown in Figure 9. The CFRP layer between the two delaminations bent and crushed the lower delamination. Therefore, it was presumed that the lowest delamination occurred first, and then, the second one generated between inside layers.

DISCUSSION

Based on the high-speed flaming image and the damage observation after the experiments, the hyper-velocity impact fracture mechanism of CFRP was considered. The following fracture model was proposed and shown in Figure 10.

(1) When the hyper-velocity flyer collides with a CFRP target, the compressive shock wave is generated. Propagating into the target, it is scattered and decreased.
(2) The compressive wave reflects at the back surface of the target and turns into tensile wave. This tensile wave generates spalling cracks with mode I type at the position near the back surface.
(3) If the compressive shock wave produced by flyer impact is large enough, the reflective tensile wave causes secondary or more spallations and several spalling cracks is generated in the inside part of the target.
(4) When deformation of the target is large, spall layers move with different velocity and shear stress occurs at the target. By this shear stress, the spalling cracks propagate along fibers with mode II or mixed-mode I/II type.
(5) Carbon fibers of the spall layer are kinked and broken by tension. The strain energy stored in the spall layer is released and the delaminations grow further.

If the laser irradiation energy is lower, the above damages cannot extend to the fiber breakage.

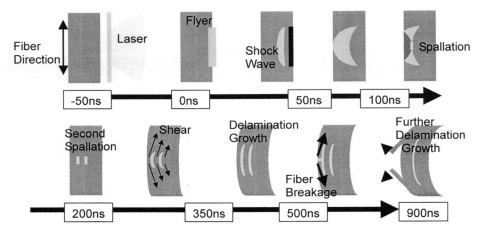

Figure 10 Schematic diagram of fracture mechanism under hyper velocity impact loading

CONCLUSIONS

We performed hyper-velocity impact tests that laser-accelerated aluminum flyer collided to carbon fiber reinforced plastics (CFRP) target. Based on the high-speed flaming image and the damage evaluation after the experiments, the hyper-velocity impact fracture mechanism of CFRP was proposed as follows.

1. On the collision surface, the impact generates compressive waves and a crater with almost the same diameter of the Al flyer.
2. Under the back surface, spallations are caused by reflected tensile waves and the similar surfaces of the crack-opening mode I fracture are created.
3. The spalling cracks propagate along the direction of carbon fibers and produce the fracture surfaces of mode II or mixed-mode I/II.
4. At the center of the spalling layer, carbon fibers are kinked and broken by tension.

For the lower laser energy, we observed that the above damages could not extend to carbon fiber breakage.

ACKNOWLEDGEMENTS

We would like to express our appreciation to Dr. M. Yoshida, Dr. Y. Kitagawa and Dr. R. Kodama for their interest in this research and fruitful discussions. We sincerely thank the laser operation group, especially Mr. K. Suzuki and K. Sawai, for their engineering support. This study is carried out as a part of "Ground Research Announcement for Space Utilization" promoted by Japan Space Forum.

REFERENCES

K. A. Tanaka, et al. (2000). Multi-layered flyer accelerated by laser induced shock waves. *Physics of Plasmas*, **7:2**, 676-680,
N. Ozaki, et al. (2001). Planar Shock Wave Generated by Uniform Irradiation from Two Overlapped Partially Coherent Laser Beams. *Journal of Applied Physics* **89:5**, 2571-2575.

CRACK-TIP STRESS FIELD MEASURED BY INFRARED THERMOGRAPHY AND FRACTURE STRENGTH IN EXPLOSION CLAD PLATE

I. Oda[1], Y. Tanaka[2], A. Masuki[3] and T. Izuma[4]

[1]Dept. of Mech. Eng. and Mater. Sci., Kumamoto University,
Kurokami 2-39-1, Kumamoto 860-8555, Japan
[2]Dept. of Mech. and Elec. Eng., Yatsushiro National College of Technology,
Hirayama-shinmachi 2627, Yatsushiro 866-8501, Japan
[3]JEOL Ltd., Musashino 3-1-2, Akishima 196-8558, Japan
[4]Asahi Chemical Ind. Co., Ltd., Tokyo, Japan

ABSTRACT

Explosion clad plates composed of copper and mild steel were dealt with as a typical example of a bonded dissimilar material. Tensile tests were carried out using rectangular plate specimens extracted from clad plate. An artificial through-the-thickness edge crack was made in each specimen. The cracks were made both close and perpendicular to the explosion interface. The stress field near the crack, the stress intensity factor, the crack opening displacement and fracture strength under tensile loading perpendicular to the crack plane were examined experimentally and by elastic-plastic finite element analysis. The stress field and the stress intensity factor were investigated using an infrared stress imaging system. The effects of material inhomogeneity, residual stresses and the change of material characteristics on the stress field, stress intensity factor, the deformation behaviour and the fracture strength were investigated. The existence of a lower strength material ahead of the bonded interface was found to increase the stress intensity. Conversely higher strength material ahead of the bonded interface was found to decrease the stress intensity factor. The brittle fracture strength of the inhomogeneous specimens can be evaluated using the stress intensity factor and linear fracture mechanics criterion.

KEYWORDS

Bonded Dissimilar Plate, Explosion Welding, Crack-Tip Stress Field, Stress Intensity Factor, Infrared Thermography, Fracture Strength, Fracture Mechanics

INTRODUCTION

Bonded dissimilar plates can offer improved corrosion and heat resistance properties over standard materials. Such properties often make them the preferred choice for chemical apparatus and pressure vessels. The failures of such apparatus and vessels can lead to serious accidents. Therefore it is highly important to examine the deformation and fracture characteristics of dissimilar materials. Bonded dissimilar plates exhibit inherent material inhomogeneity, residual stresses and material characteristics different to those of the base materials. The effect of material inhomogeneity in dissimilar plates on deformation and fracture has been examined analytically [Sato et al. (1983), Vijayakumar & Cormack (1983)] and experimentally [Nishioka et al.(1993)]. A study on J-integral analysis of dissimilar plates has also been performed [Kikuchi & Kobayashi(1994)]. The fracture toughness [Oda et al. (1982)] and the fracture strength [Oda et al. (1985)] of stainless clad-steel have been studied. However, the result of a study, which examines totally the effects of material inhomogeneity, residual stresses and the change of material characteristics on deformation and fracture has not yet been published.

In this paper, an explosion-clad plate is dealt with as a typical example of a bonded dissimilar plate. The clad plate considered is composed of C1100 copper and SS400 mild steel, which have considerably different mechanical properties. The two-dimensional deformation behaviour in the vicinity of the crack and close to the bonded interface, and the fracture strength were examined experimentally and by finite element analysis. The effects of material inhomogeneity, residual stresses and the hardened zone on the deformation and the fracture strength were investigated.

EXPERIMENT

It was recognized by micro hardness test on a vertical section of the clad plate that a remarkably hardened zone developed in the vicinity of the bonded interface. Tensile tests were carried out on rectangular specimens 254 mm in length, 28mm wide and 4mm thick extracted from the clad plate. The surface plane of the specimen is perpendicular to the plane of the bonded interface. Figure 1 shows the shape and dimensions of the central portion of the specimen. An artificial through-the-thickness edge notch was made in each specimen. The notch length used throughout was 10mm. The notch tip portion, 3mm long, was prepared by electrical wire discharge machining with a wire of 0.03mm in diameter. Hereafter, this notch will be referred to as the crack for convenience.

Four types of specimen were used:
- A homogeneous specimen, type SHS, and an inhomogeneous specimen, type SIS, with crack tips in the SS400 mild steel side.
- A homogeneous specimen, type CHS, and an inhomogeneous specimen, type CIS, with crack tips in the C1100 copper side.

Following cracking the residual stress, σ_{ry}, in the specimen is redistributed. The residual stress parallel to the bonded interface was measured by the stress relaxation method using 1mm gauge length bi-axial strain gauges. Figure 2 shows the distribution of residual stress along the extension of the

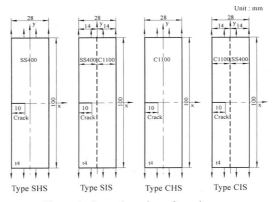

Figure 1: Central portion of specimens

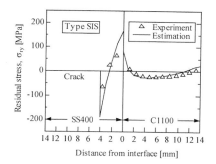

Figure 2: Distribution of residual stress in Type SIS

crack line in specimen type SIS. The curve in figure 2 was evaluated considering the equilibrium condition of the resultant force and the resultant moment produced by the residual stresses. The residual stress can be seen to be compressive near the crack tip, tensile in the bonded region and compressive in the plate edge region opposite the crack.

Tensile tests were performed at temperatures ranging from −180 to 20°C. The tensile load applied was uniform and in the direction perpendicular to the plane of the crack. The applied load and the crack mouth opening displacement were measured. The temperature was monitored throughout using a copper-constantan thermocouple welded to the specimen under test.

Based on Kelvin's equation the following relationship can be derived for adiabatic elastic deformation [Jordan & Sandor (1978)]:

$$\Delta T = -kT\Delta\sigma \qquad (1)$$

where ΔT is the temperature change, k is thermo-elastic modulus, T is absolute temperature of elastic body, and $\Delta\sigma$ is the change in sum of the principal stresses.

A repeated remote elastic tensile load varying from zero to 19.3MPa was applied to some of the specimens. The sum of the principal stresses, $\sigma_x + \sigma_y$, was obtained by consideration of the temperature change, ΔT. The thermo-elastic modulus, k, of the crack-tip material was used in this calculation.

For the case of two-dimensional Mode I loading, the following equation describing the relationship between the increment of the stress intensity factor, ΔK_I, and the corresponding difference in the sum of the principal stresses can be derived [Shiratori et al. (1989)] for points along the extension of the crack line ($\theta = 0$):

$$\Delta K_I = \sqrt{\frac{\pi r}{2}} \Delta(\sigma_x + \sigma_y) \qquad (2)$$

where r is the distance from the crack tip.

An infrared measuring system was used to measure the values of ΔT. Values for $\Delta(\sigma_x + \sigma_y)$ were then calculated by substituting the values of ΔT into Eqn. 1. By further substitution into Eqn. 2 values of ΔK_I were determined.

ANALYSIS

A plane stress analysis was performed to investigate the deformation around the crack tip of a specimen under a uni-axial tensile load in the y-direction. An elastic-plastic finite element analysis with an incremental theory of plasticity was used. The material was assumed to harden according to the following power law relation between stress and strain [Swift (1952)]:

$$\bar{\sigma} = C(\alpha + \bar{\varepsilon})^n \tag{3}$$

where $\bar{\sigma}$ (MPa) is the equivalent stress, $\bar{\varepsilon}$ is the equivalent plastic strain and C, α and n are constants. The material constants for the base materials were obtained by performing tensile tests on small specimens of C1100 and SS400. C1100 was found to have much lower strength and somewhat higher elongation percentage than SS400. By performing a hardness test and then considering the relationship between strength and hardness, the material constants for the hardened region produced by the bonding process were obtained. For the inhomogeneous specimens, residual stresses were also considered in the analysis.

RESULTS AND DISCUSSION

Figure 3 shows the distribution of the sum of the principal stresses, $\sigma_x + \sigma_y$, around the crack tips in the four types of specimen under repeated remote stress. These distributions were measured using the infrared measuring system.

The values of $\Delta(\sigma_x + \sigma_y)$ in the inhomogeneous specimen, type SIS, were observed to be higher than that of the homogeneous specimen, type SHS, despite the fact that the crack tip material is the same. This phenomenon can be attributed to the existence of C1100, which has lower strength than the crack tip material and lies ahead of the bonding interface in the inhomogeneous specimen. Conversely, for specimens CIS and CHS, the higher strength material ahead of the bonded interface in the inhomogeneous specimen causes the values of $\Delta(\sigma_x + \sigma_y)$ to be lower than those in the homogeneous specimen.

For the case where the remote stress is at a maximum (19.3 MPa) the ΔK values were calculated by substituting the results shown in Figure 3 into Eqn. 2 (see TABLE 1). The differences between the

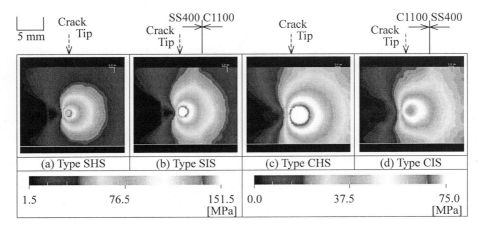

Figure 3: Distribution of sum of principal stresses around crack tip measured by infrared thermography. (Repeated load σ_{max} = 19.3 MPa, σ_{min} = 0).

TABLE 1
STRESS INTENSITY FACTORS EVALUATED BY INFRARED
STRESS MEASURSING SYSTEM (When σ = 19.3 MPa)

Type	SHS	SIS	CHS	CIS
ΔK [MPa·m$^{1/2}$]	5.50	6.45	5.50	4.00

homogeneous and inhomogeneous specimens can be clearly seen.

Figure 4 shows the relationship between fracture strength σ_{nf} and temperature. The value σ_{nf} is the critical value of the net stress at the moment of macroscopic unstable fracture initiation. The symbol (*) indicates that ductile fracture occurs at the crack tip. At very low temperatures, low stress brittle fracture takes place in type SHS although this does not occur in type CHS.

At low temperature, the fracture strength σ_{nf} of the inhomogeneous specimen type SIS was found to be slightly lower than that of the homogeneous specimen type SHS. At higher temperature, the fracture strength of the inhomogeneous type SIS specimen was found to be lower than that of homogeneous type SHS. At low temperatures, when brittle fracture occurs at the crack tip, the crack tip stress field can be assumed to be dominated by the fracture toughness as calculated by linear-elastic fracture mechanics. Therefore, the fracture strength of the inhomogeneous specimen type SIS, $(\sigma_{nf})_{SIS}$, can be expressed in terms of the fracture strength of the homogeneous specimen type SHS, $(\sigma_{nf})_{SHS}$, as follows:

$$(\sigma_{nf})_{SIS} = (F_{SHS} / F_{SIS}) (\sigma_{nf})_{SHS} \qquad (4)$$

where, F_{SHS} and F_{SIS} are the correction factors based on stress intensity factors of specimen types SHS and SIS, respectively. The ratio of the correction factors is the same as that of the stress intensity factors. Substituting the values of $(\sigma_{nf})_{SHS}$ and the stress intensity factors as shown in TABLE 1 allows an estimation for $(\sigma_{nf})_{SIS}$ to be made for the low temperature case.

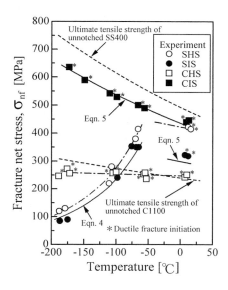

Figure 4: Relationship between apparent fracture net stress and temperature

The σ_{nf} value of the inhomogeneous specimen type CIS is much higher than that of the homogeneous specimen type CHS at all test temperatures. Furthermore, the σ_{nf} value for type CIS is the highest of all the specimens across all the test temperatures. Consequently, in a real world application the use of a inhomogeneous plate like type CIS would mean increased fracture strength. For ductile fracture in type CIS, where plastic deformation occurs at the crack tip, the fracture strength may be dominated by the strength of the cross sectional area ahead of the crack tip. In such a case, σ_{nf} for the inhomogeneous specimens can be expressed as follows:

$$\sigma_{nf} = (\sigma_{uC} A_C + \sigma_{uS} A_S) / (A_C + A_S) \qquad (5)$$

where σ_{uC} and σ_{uS} represent the ultimate tensile strength of the C1100 and SS400 parent materials and A_C and A_S are the areas of the dissimilar materials ahead of the crack tip in the inhomogeneous specimens. The estimated values of $(\sigma_{nf})_{CIS}$ and $(\sigma_{nf})_{SIS}$ as calculated by Eqn. 5, are found to be approximately the same as the values found experimentally.

CONCLUSIONS

1. The effects of a dissimilar material in a clad plate on the crack tip stress field can be evaluated by an infrared stress measuring system.
2. Lower strength material ahead of the bonded interface facilitates growth of the stress near the crack tip in an inhomogeneous specimen. Conversely, higher strength material ahead of the bonded interface restrains the stress near the crack tip.
3. At low temperature, when macroscopic brittle fracture occurs at the crack tip, the fracture strength of inhomogeneous specimens can be estimated by considering that of the homogeneous specimen whose crack tip material is the same as the inhomogeneous one and the relevant ratio of the stress intensity factors as obtained by the infrared stress measuring system.
4. For ductile fracture, where plastic deformation occurs near the crack tip, the fracture strength of the inhomogeneous specimens can be estimated by dividing the strength of the cross-sectional area ahead of the crack tip proportionately according to the ratio of the dissimilar materials in the inhomogeneous specimens.

REFERENCES

Jordan E. H. and Sandor B. I. (1978). Stress Analysis from Temperature Data. *J. Testing and Evaluation* **6**, 325-331.
Kikuchi M. and Kobayashi M. (1994). Two-Dimensional Crack Growth Simulation across the Fusion Line. *Transaction JSME* **60**, 25-30 (in Japanese).
Nishioka T. et al. (1993). Measurements of the Near-Tip Deformation in Inhomogeneous Elastic-Plastic Fracture Specimens Using the Moire Interferometry. *Transaction of JSME* **59**, 558-565 (in Japanese).
Oda I. et al. (1982). Fracture Toughness of Explosive Stainless-Clad HT80 High Strength Steel. *Transaction of Japan Welding Society* **13**, 45-50.
Oda I. et al. (1985). Fracture Strength of Explosion Clad Steel with Surface Notch. *Transaction of Japan Welding Society* **16**, 55-60.
Sato K. et al. (1983). Crack Tip Plastic Deformation of Notched Plates with Mechanical Heterogeneity. *Journal of Japan Welding Society* **52**, 154-161(in Japanese).
Shiratori M. et al. (1989). Measurements of Stress Intensity Factors by Infrared Video System. *Transaction of Japan Society of Mechanical Engineering* **55**, 159-164 (in Japanese).
Swift H. W. (1952). Plastic Instability under Plane Stress. *J. Mech. Phys. Solids* **1**, 1.
Vijayakumar S. and Cormack D. E. (1983). Stress Behavior in the Vicinity of a Crack Approaching a Bimaterial Interface. *Engineering Fracture Mechanics* **17**, 313-321.

DEFORMATION NEAR A CRACK CLOSE TO INTERFACE IN EXPLOSION COPPER-CLAD MILD STEEL PLATE

I. Oda[1], K. Shiraishi[2] and M. Yamamoto[1]

[1] Department of Mechanical Engineering and Materials Science, Kumamoto University,
Kurokami 2-39-1, Kumamoto 860-8555, Japan
[2] Department of Aerospace Systems Engineering, Sojo University,
Ikeda 4-22-1, Kumamoto 860-0082, Japan

ABSTRACT

Bonded dissimilar plates are often used in chemical apparatus and pressure vessels. Failures in such apparatus and vessels can lead to serious accidents. In this paper, explosion clad copper-mild steel plate is dealt with as a typical example of a bonded dissimilar material. Tensile tests were carried out on rectangular specimens extracted from the clad plate. A through-the-thickness edge crack was made in each specimen in the vicinity of and perpendicular to the explosive interface. The effect of material inhomogeneity, residual stresses and the hardened zone on the deformation near the cracks was examined both experimentally and by finite element analysis. Lower strength material ahead of the bonded interface and tensile residual stresses near the bonded interface were found to increase the crack opening displacement. Higher strength material ahead of the bonded interface, compressive residual stresses and the hardened zone near the bonded interface were found to decrease the crack opening displacement.

KEYWORDS

Explosion Welding, Bonded Dissimilar Plate, Material Inhomogeneity, Hardened Zone, Residual Stress, Fracture Mechanics, Crack Opening Displacement, Finite Element Method

INTRODUCTION

Bonded dissimilar plates offer enhanced corrosion and heat resistance properties over standard materials. They are often used in chemical apparatus, pressure vessels, structural transition joints and chemical tankers. The mechanical failure of such structures can lead to serious accidents and it is therefore of great importance to examine the deformation and fracture behavior of dissimilar materials.

Dissimilar materials inherently have inhomogeneous material properties. Furthermore, the bonding process causes residual stresses and altered material characteristics. The effects of material inhomogeneity have been investigated analytically (Vijayakumar et al., 1983) and experimentally (Nishioka et al., 1993). The J integral for dissimilar plates has also been investigated analytically (Kikuchi et al., 1994). However, a study, which examines totally the effects of material inhomogeneity, residual stresses and the change of material characteristics on deformation and fracture, has not been published.

The research presented herein is a basic study of the deformation and fracture of a bonded dissimilar material. An explosion clad plate composed of C1100 copper and SS400 mild steel was dealt with as a typical example of a bonded dissimilar plate. The two-dimensional deformation behavior in the vicinity of a crack close to the bonded interface was examined. The effects of material inhomogeneity, residual stresses and the hardened zone on strain, the plastic zone near the crack and the crack opening displacement were examined both experimentally and by elastic-plastic finite element analysis.

EXPERIMENTS

Figure 1 shows the results of micro hardness tests on vertical sections of the clad plate. The hardened zone produced in the vicinity of the bonded interface is clearly visible.

Uniaxial tensile tests were carried out on small test specimens of both base materials in order to obtain their mechanical properties. Material constants for the region hardened by the bonding process were obtained based on the results of the hardness tests and the relation between strength and hardness.

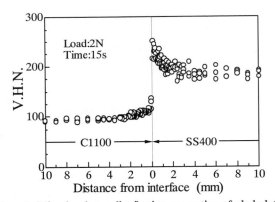

Figure 1: Microhardness distribution on section of clad plate

Tensile tests were carried out using rectangular specimens extracted from the clad plate. The geometry of the central portion of the specimens is shown in Figure 2. Each specimen contains an artificially made through-the-thickness edge crack perpendicular to the bonded interface. The crack tip, 3mm in length, was made by wire-electric-discharge machining with 0.03mm diameter wire. In the homogeneous specimens, Type CH, and inhomogeneous specimens, Type CI, cracks were made such that their tips lay in the copper C1100 region. In the homogeneous specimens, Type SH, and the inhomogeneous specimens, Type SI, cracks were made such that their tips lay in the SS400 mild steel region.

The residual stress distribution parallel to the bonded interface σ_{ry} was obtained by the stress relaxation method using 1mm gauge length biaxial strain gauges. Figure 3 shows the distribution of residual stress σ_{ry} in the material ahead of the crack in specimen Type CI. The residual stress σ_{ry} is compressive near the crack tip, tensile in the bonded region and compressive in the region opposite the crack.

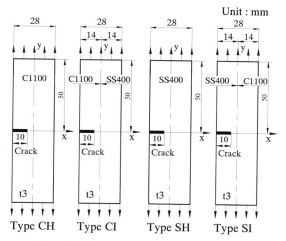

Figure 2 : Central portion of test specimen

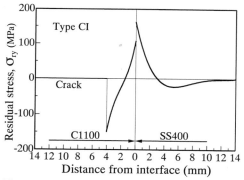

Figure 3 : Distribution of residual stress σ_{ry} in Type CI

A uniformly tensile load perpendicular to the crack plane was applied to the specimen. The load and the crack opening displacement were measured.

ANALYSIS

The deformation near the tip of a crack in a plane stress field, which is under uniaxial tension in the y direction, was analyzed. An elastic-plastic finite element analysis with an incremental theory of plasticity was performed. The material was assumed to work harden according to the following power law relation between stress and strain which was suggested by Swift:

$$\overline{\sigma} = C(\alpha + \overline{\varepsilon}_p)^n \tag{1}$$

where $\overline{\sigma}$ (MPa) is the equivalent stress, $\overline{\varepsilon}_p$ is the equivalent plastic strain and C, α, and n are material constants. Four models as shown in TABLE 1 were used in the analysis of the deformation in each of the inhomogeneous specimens. Different combinations of the three influential factors were considered in the analysis to clarify the effects of the individual factors on the deformation behavior.

TABLE 1
FOUR MODELS USED IN FINITE ELEMENT ANALYSIS

	Factors considered		
	Inhomogeneity	residual stress	hardened zone
Model-M	O		
Model-MR	O	O	
Model-MH	O		O
Model-MHR	O	O	O

RESULTS AND DISCUSSION

Figure 4 shows the relationship between crack opening displacement, V_g, as measured by a clip gauge and the applied net stress. The analytical results are also shown in the figure. The analytical and experimental results correspond well and a trend can be seen. The V_g values for the inhomogeneous specimen, Type CI, is lower than that of the homogeneous specimen, Type CH, even though the material at the crack tip is the same in both types. This phenomenon is caused by a combination of the hardened zone produced by bonding, the residual stresses and the higher strength SS400 steel ahead of the crack tip in the inhomogeneous specimen. The V_g values for the inhomogeneous specimen, Type SI, is higher than that of the homogeneous specimen, Type SH, even though the material at the crack tip is the same in both types. This phenomenon is mainly caused by the existence of the lower strength C1100 copper ahead of the bonded interface in the inhomogeneous specimen.

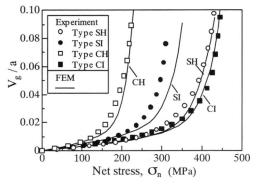

Figure 4 : Relationship between crack mouth opening displacement and applied net stress

The effects of material inhomogeneity, residual stress and the hardened zone on the crack opening displacement δ at 0.25mm from the crack tip in the inhomogeneous specimens were investigated using F.E.A.. At the low applied stress level, as shown in Figure 5(a) the δ value of model MR is lower than that of model M. The δ value of model MHR is also lower than that of model MH. These phenomena are caused by the compressive residual stress near the crack tip. As shown in figures 5(a) and (b), the δ value of model MH is lower than that of model M and the δ value of model MHR is lower than that of model MR. These phenomena show the effect of the hardened zone. At low applied stress levels, the effects of both the residual stress and the hardened zone on the crack opening displacement are obvious. However, at high applied stress levels the effect of residual stress is not clear whilst the effect of the hardened zone is remarkable.

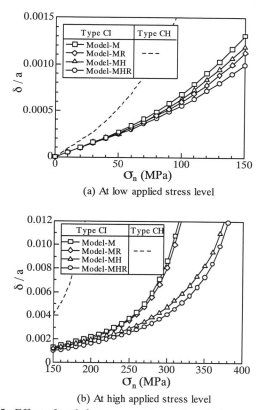

(a) At low applied stress level

(b) At high applied stress level

Figure 5 : Effect of each factor on crack opening displacement in Type CI

Figure 6 shows the F.E.A. results for the distribution of ε_y, the strain in the direction of the tensile load, on the prolongation of the crack at the applied net stress of 300MPa in the inhomogeneous specimen, Type CI. The distribution of ε_y in each model is discontinuous at the bonded interface. This reflects the inhibitory action of the higher strength material SS400 against the growth of strain. The effects of the compressive residual stress near the crack tip and the tensile residual stress at the bonded interface also can be seen. As shown in Fig.6, the strain, ε_y, is restrained from growing by the hardened zone.

The effect of each factor on the distribution of the equivalent plastic strain near the crack tip in the inhomogeneous specimens was also obtained. The behavior of the strain ε_y and the plastic zone closely correspond to the behavior of the crack opening displacement.

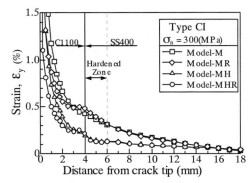

Figure 6 : Distribution of strain ε_y on the prolongation of crack

CONCLUSIONS

The following conclusions may be drawn.

(1) In inhomogeneous specimens, material ahead of the bonded interface of lower strength to the crack tip material was observed to facilitate deformation near the crack tip. Higher strength material ahead of the bonded interface conversely restrained deformation at the crack tip.

(2) Compressive residual stresses near the crack tip restrain the strain near the crack tip and the crack opening displacement. Tensile residual stresses at the bonded interface facilitate strain increase and the crack opening displacement.

(3) The hardened zone produced by explosive bonding exerts a decisive influence on the deformation and restrains the strain near the crack tip and the crack opening displacement.

(4) The crack opening displacement in the inhomogeneous Type CI specimen, which has a higher strength material ahead of the bonded interface than at the crack tip, is remarkably low when compared with that in a homogeneous specimen made of the crack tip material. An explosion clad plate like Type CI therefore exhibits increased fracture strength compared to the homogeneous specimen providing there are no considerable differences in the fracture toughness.

REFERENCES

Kikuchi M. and Kobayashi M. (1994). Two-Dimensional Crack Growth Simulation across the Fusion Line. *Trans. JSME* **60:569**. 25-30 (in Japanese).

Nishioka T., Nishi M. and Fujimoto T. (1993). Measurements of the Near-Tip Deformation in Inhomogeneous Elastic-Plastic Specimens Using the Moire Interferometry. *Trans. JSME* **59:559**, 558-565 (in Japanese).

Vijayakumar S. and Cormac D. E. (1983). Stress Behavior in the Vicinity of a Crack Approaching a Bimaterial Interface. *Engineering Fracture Mechanics* **17:4**, 313-321.

A STUDY ON THE DYNAMIC BRITTLE FRACTURE SIMULATION

D.-T. Chung[1], C. Hwang[2], S. I. Oh[2] and Y. H. Yoo[3]

[1]School of Mechatronics, Korea University of Technology & Education
P.O. box 55, Chonan, Chungnam, 330-600, Korea
[2]School of Mechanical and Aerospace Engineering, Seoul National University,
San 56-1, Shillim, Kwanak, Seoul, 151-742, Korea
[3]Agency for Defense Development
P.O. Box 35-1, Yuseong, Taejon, 305-600, Korea

ABSTRACT

A two-dimensional explicit time integration finite element code of Lagrangian description for analyzing the dynamic brittle fracture was developed. It has several improved features: (1) Robust defense-node contact algorithm is adopted and can now handle multi-body and self contact with friction accurately. (2) Eroding algorithm for effective penetration simulation is included. (3) The macro-crack whose length is larger than that of element is described by creation of the new surfaces using the node separation scheme. (4) Certain amount of plastic deformation for the brittle materials followed by the comminution, whose characteristic crack length is much shorter, is modeled by the granular material behavior. Simulation of the penetration into a ceramic block by long-rod penetrator exhibited the key generic features observed experimentally.

KEYWORDS

Brittle fracture, Node separation, High velocity impact, Long-rod penetration, FEM, Contact treatment

INTRODUCTION

Ceramic materials have good mechanical properties such as high strength in compressive loading, low density and high melting temperature. These properties make ceramic materials employed in advanced armor design. However, their brittle behavior makes it difficult to design ceramic component. For realistic design, robust material modeling of brittle behavior is required. Most material models(Taylor *et al.*, 1986; Rajendran, 1994; Rajendran and Grove, 1995) are based on continuum damage theories in which the net effect of fracture is homogenized as a degradation of the strength of the materials and the fragmentation size and density are represented by damage parameters. However, continuum damage theories for brittle materials suffer from obvious shortcoming: the discrete nature of cracks is lost. The failure of brittle specimen is frequently governed by the growth of single dominant crack. This situation is not consistent with homogenizing damage model. So, in order to describe brittle material behavior explicitly, including crack propagation and fragmentation, node separation scheme was proposed.(Xu and Needlman, 1994; Camacho and Ortiz, 1996)

In this paper, two-dimensional computational modeling for analyzing the dynamic behavior of brittle material is described. Explicit time integration and Lagrangian description were used. Node separation scheme was adopted to describe the dynamic brittle fracture. In order to simulate the penetration process, element eroding scheme(Sewell et al., 1990) was used. Node separation and element eroding usually give rise to complex contact situation. So, the multi-body and self-contact handling capabilities were improved. To verify the validity of the developed code, numerical simulation of penetration into a ceramic block by the long-rod penetrator was carried out.

MATERIAL MODEL

To analyze the behavior of brittle materials under impact loading, material model should be built up including dynamic fracture and failure. In this section, proposed material model, including brittle fracture models for macro and micro crack, is described.

Macro-crack

Node separation scheme was adopted to describe explicitly the initiation and propagation of the macro-cracks, whose length is longer than that of element. The creation of a new crack surface is accomplished by opening the element boundary when the normal force acting on the element segment is greater than the critical spall strength. These cracks can branch and coalesce and lead to the fragments eventually. In node separation scheme, it is assumed that all element boundaries are potential crack paths. The fracture criterion is evaluated at all potential crack paths. As shown in Figure 1, if the fracture occurs at the element side drawn by thick line, new outer surfaces are created.

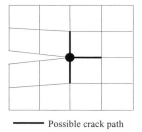

——— Possible crack path

Figure 1: Possible crack path at each node in node separation scheme

As shown in Figure 2, normal and tangential tractions for all element boundaries are calculated. Effective traction σ^{eff} considering mixed-mode fracture was adopted to fracture criterion written in Eqn. 1 where σ, τ and β_τ are normal and tangential tractions and shear stress factor, respectively. The time dependency of fracture could be represented by Eqn. 2, where σ_{fr}, t_c and t are quasi-static spall strength, characteristic time and pulse duration time, respectively(Camacho and Ortiz, 1996).

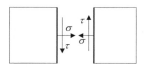

Figure 2: Normal and tangential component of element side traction

$$\sigma^{eff} = \sqrt{\sigma^2 + \beta_\tau \tau^2} \geq \sigma_{in} \qquad (1)$$

$$\sigma_{in} = \sigma_{fr} e^{t/t_c} / \left(e^{t/t_c} - 1\right) \qquad (2)$$

Continuum damage model (Camacho and Ortiz, 1996) was adopted to describe the radial cracks in axisymmetric configuration.

Micro-crack

To describe the inelastic deformation of brittle materials, plasticity model is adopted. Elastic-plastic model is applied to intact material. If the equivalent plastic strain of an element increases beyond a critical value, the element is assumed to be comminuted and full of micro-cracks whose characteristic crack length is much shorter than that of the element. Thus all the boundaries of the element are separated from the body and the strength degrades accordingly. This separation consequently makes the material have no strength in tension and degraded strength in compression.

$$\overline{\sigma} = \begin{cases} Y_c & if \ \overline{\varepsilon}_p \leq \varepsilon_f \\ Y_f & if \ \overline{\varepsilon}_p > \varepsilon_f \end{cases} \qquad (3)$$

$$Y_f = \text{Min}(sY_c, \alpha + \beta P) \qquad (4)$$

The behavior of the comminuted materials follows that of granular materials whose strength is directly proportional to applied pressure. Such behavior modeling may be reasonable for low pressure condition. However, for high velocity impact, the pressure level at the interface of projectile and target is much higher. So, there must be a limiting value and certainly it must be lower than that of the intact material(Rajendran and Grove, 1995). The comminuted material strength Y_f has the form written in Eqn. 4, where α and β are model parameters, ε_f fracture strain, P applied pressure, s degradation factor and Y_c compressive yield strength of intact material. In Eqn. 4, sY_c is the upper limit of the strength. The model parameters α, β, s and ε_f should be determined by parametric studies. However, in certain range of parameter values, overall results are rather insensitive to the values.

Eroding

The assumption behind the element eroding scheme is that the highly strained elements will no longer contribute to the physics and may therefore be eroded. On the computational point of view, severely distorted elements cause excessively small time step size and the difficulties in contact detection. In the implementation, if a condition variable of element, such as damage temperature and plastic strain, reaches a certain critical value, the element is assumed totally failed and made to disappear substantially. And then, the related nodes and outer surfaces are created and updated. Contact surfaces, boundary terms and element connectivity array should be reconstructed subsequently.

CONTACT TREATMENTS

In the process of numerical computation, new surfaces and fragments are created by the results of node separation and element eroding. Careful treatments for multi-body and self-contact situations are required. The contact treatment is composed of three parts: global search, local search and contact forces computation. Global search procedure, based on the bucket sorting algorithm, is to find candidate contact pairs including self-contact situations(Oldenburg and Nilsson, 1994; Zhong, 1993). If geometry of contact surface has sharp corner, special treatment on local contact search are required. Basic assumption of local search is entire area of the elements attached to contacting slave node should be outer side of the master segment. This local search algorithm is robust for handling self-contact and multi-body contact situations involving fragmentations. For more accurate contact force computation, defense node contact algorithm(Zhong, 1993) was adopted and implemented.

NUMERICAL SIMULATION

Numerical experiment of long-rod penetration into ceramic target was performed to validate the developed computational modeling. The experimentally observed mechanisms by which a long-rod penetrates thick ceramic plate were recently reviewed by Curran *et al.*(1993). Upon impact, a radially divergent shock wave is generated. Stress wave interactions with the nearby free surfaces, together with circumferential strains, cause tensile stress generating shallow ring cracks that later extend outward from the initial surface normal to form Hertzian cone cracks. And then fractures and fragmentations around the penetrator form a comminuted zone. The comminuted zone typically consists of a thin layer of finely fragmented particles and of larger fragments. The fragment size increases with distance from the penetrator. The comminuted zone continues to expand and fly-off. From the rear side of the ceramic plate, lateral cracks initiate and propagate.

Simulation configuration

Numerical experiments of the penetration into an AD85 ceramic block by tungsten heavy alloy(WHA) penetrator were carried out. L/D ratio of the long-rod penetrator was 10.7. Material model for intact AD85 was simple elastic-plastic model. The spall and compressive strength of AD85 were 0.2GPa and 4.3GPa respectively. Model parameters ε_f, s, α and β were 0.4, 0.5, 0 and 1, respectively. These post-fracture model parameters were evaluated from the parametric studies. The ceramic block was backed by mild steel plates. Initial velocity of the penetrator was 1.5km/s. Material model for the tungsten heavy alloy and the mild steel was Johnson-Cook model(Johson and Cook, 1985) which includes rate dependent plasticity with isotropic hardening, adiabatic heating, and thermal softening. Element eroding by equivalent plastic strain of 200% was applied to all materials.

Results and discussions

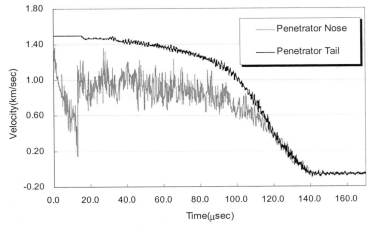

Figure 3: Velocity history of penetrator

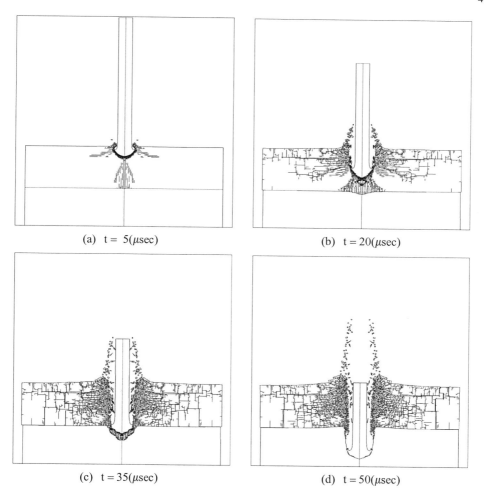

Figure 4: Fracture pattern and deformation shape

Figure 4 (a), shows Hertzian cone cracks, which were generated by the intersection of compressive waves and release at the front free surface of ceramic. Upon contact of penetrator and ceramic block, since the generated compressive stress was beyond the compressive yield strength, ceramic material was subjected to inelastic deformation and then comminuted. The thin layer of comminuted zone around the penetrator nose is consists of comminuted ceramic particle. Initiation of lateral cracks due to compressive wave at the rear side of ceramic block is observed. In Figure 4 (b), the penetrator passes through the triangular fractured region weakened by the wave interactions. This fractured region is separated from the block and independently resists the penetration. As shown in Figure 3, nose velocity of the penetrator decreases rapidly up to 15μsec and then jumps back. At the early stage, high strength of ceramic material resists the penetration effectively. However, the penetration resistance is completely lost, once the ceramic block is weakened by compressive waves and following tensile waves. In Figure 4 (b), (c) and (d), as the penetration progresses, fragmentation and comminution occur with more ejecta forming. The fragment size increases with distance from the penetrator. Ejecta fly opposite to the penetration direction and toward the penetrator axis. The ceramic block is fragmented, but the total energy cannot be absorbed by the deformation alone. The locked energy generates the motion of fragmentations: vibration and rotation. Consequently, the macroscopic behavior of dilation occurs. The shape of crater hole in the first backup plate is different from those of other mild steel plates. Front petaling, occurring only at the free surface, is shown near the crater of first steel plate. The effect of free surface, causing front petaling, was generated by the dilation followed by ejection of ceramic particles.

CONCLUSIONS

A two-dimensional explicit time integration finite element code of Lagrangian description for analyzing the dynamic brittle fracture was developed. Defence node algorithm and contact searching based on bucket sorting algorithm are applied for robust and effective contact treatment including self-contact. Careful local contact searching algorithms were applied considering complex contact situations. Element eroding algorithm was adopted for proper simulation of tasks involving highly distorted elements. The penetration of long-rod into ceramic block backed by metal plates was analyzed and the results of numerical simulation show good agreement with key generic features experimentally observed. It validates the developed brittle material model.

REFERENCES

Camacho, G. T. and Ortiz, M. (1996), Computational modeling of impact damage in brittle materials. *Int. J. Solids Structures* **33**, 2899-2938

Curran, D. R., Seaman, L., Cooper, T. and Shockey, D. A. (1993), Micromechanical model for continuum and granular flow of brittle material under high strain rate application to penetration of ceramic targets, *Int. J. Impact Engng.* **13:1**, 53-83

Johnson, G. R. and Cook, W. H. (1985), Fracture characteristics of three metals subjected to various strains, strain rates, temperatures and pressures, *Engng. Fract. Meh.* **21**, 31-48

Oldenburg, M. and Nilsson, L. (1994), The position code algorithm for contact searching, *Int. J. Numer. Meth. Engng.* **37**, 359-386

Rajendran, A. M. (1994), Modeling the impact behavior of AD85ceramic under multiaxial loading, *Int. J. Impact Engng.* **15:6**, 749-768

Rajendran, A. M. and Grove, D. J. (1995), Effect of pulverized material strength on penetration resistance of ceramic targets. *Proceedings of the 1995 APS Shock Compression Conference*

Sewell, D. A., Ong, A. C. J. and Hallquist, J. O. (1990), Penetration calculation using an erosion algorithm in DYNA, *The 12th International Symposium on Ballistics*

Taylor, L. M., Chen, E. P. and Kuszmaul, J. S. (1986), Microcrack-induced damage accumulation in brittle rock under dynamic loading, *Comp. Meth. App. Mech.* **55**, 301-320

Xu, X. P. and Needleman, A. (1994), Numerical simulation of fast crack growth in brittle solids, *J. Mech. Phys. Solids* **42:9**, 1397-1434

Zhong, Z. H. (1993), Finite Element Procedures for Contact-Impact Problems, Oxford University Press

DYNAMIC FRACTURE TEST OF ROCK UTILIZING UNDERWATER SHOCK WAVE

S. Kubota[1], Y. Ogata[2], R. Takahira[1], H. Shimada[1], K. Matsui[1] and M. Seto[2]

[1] Department of Earth Resources Engineering, Faculty of Engineering,
Kyushu University, Fukuoka, 812-8581, Japan
[2] National Institute of Advanced Industrial Science and Technology,
Ibaraki 305- 8569, Japan

ABSTRACT

This paper presents the experimental technique to estimate the behaviors of dynamic fracture of rock. In this technique, explosive is used as the explosion source, and the pipe filled with water is arranged between the explosive and cylindrical rock specimen. After the explosion of the explosive, the detonation wave interacts with water, and the underwater shock wave generate in the pipe. Since the underwater shock wave attenuates with propagation of it, the strength of the incidence shock wave into the rock specimen can be easily adjusted by changing the length of the pipe. One of the aims of this test is to obtain the dynamic tensile strength of the rock for wide range of the strain rate utilizing Hopkinson's effect. Therefore, the free surface velocity at the end of rock specimen and the position of crack on the rock are observed by using laser vibration meter and high-speed camera, respectively. The results of the fracture test for sandstone and the validity of this test are discussed in this paper.

KEYWORDS

Dynamic fracture, Rock, Explosive, Underwater shock wave, Hopkinson's effect, High-speed camera, Free surface velocity, Dynamic tensile strength

INTRODUCTION

The explosion energy of explosive has been utilized for blasting which is necessary to mining and quarrying operations and some civil engineering applications. Since a structure build in boom year will become too old, the demolition of a large number of it will be require in the near future in Japan (Ma et al. (1998)). Because of low cost and high efficiency of work, the demolition utilized blasting may be one of the powerful methods when the concrete technique is established in Japan. In order to promote blasting efficiency and to establish concrete blasting demolition, it is important to know the mechanism of the dynamic fracture process on rock or construction materials. The fragmentation by blasting is the joint action of gaseous pressure and stress waves (Bhandari (1997)). Since the stress wave greatly depend on the fracture condition near the explosion part, the fragmentation near the free surface by the reflection of the stress wave also depend on it. Therefore, it is necessary to consider the experiments that can be simultaneously estimated both of the fragmentation conditions near the explosion source and free surface.

This paper presents the new technique to estimate the behaviors of dynamic fracture of rock. In this technique, explosive is used as the explosion source, and the pipe filled with water is arranged between the explosive and cylindrical rock specimen. One of the aims of this test is to estimate the dynamic tensile strength of the rock for wide range of the strain rate utilizing Hopkinson's effect. Therefore, the free surface velocity at the end of rock specimen and the position of crack on the rock are observed by using laser vibration meter and high-speed camera, respectively. The results of the fracture test for Kimachi sandstone and the validity of this test are discussed in this paper.

EXPERIMENTAL SET UP

Fig.1 shows the experimental set up for the proposed fracture test. During the fracture process of rock, the free surface velocity and the fracture surface near the free surface were observed by laser vibration meter (OFV-300; made by Polytec) and high speed camera (model 124 framing type camera; made by

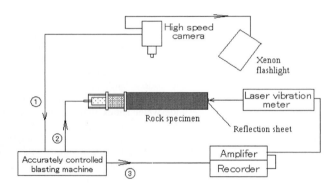

(a) Whole system

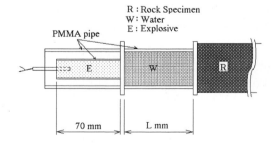

(b) Around the explosion source

Figure 1. Experimental set up for proposed fracture test of rock utilizing underwater shock wave

Cordin). The light source for high speed camera was used Xenon flashlight. The precise detonator was used to control the initiation time of the explosive by using an accurately controlled blasting machine, which was made by Nihon Kayaku Co. Ltd.

Figure 1(b) shows the around the explosive part. For the explosion source, the emulsion explosive was used. Because the detonation product rapidly expands and obstructs the view of the high speed camera, the explosive is set in the double layer pipe. After the explosion of the explosive by the precise detonator, the detonation wave interacts with water, and the underwater shock wave generate in the pipe filled with water. Since the underwater shock wave attenuates with propagation of it, the strength of the incidence shock wave into the rock specimen can be easily adjusted by changing the length of the pipe. In this experiments the length of the water pipe was varied as 30, 50 and 70mm. Kimachi sandstone was used as the rock specimen with 60mm diameter and 300mm length. Arrival time at the free surface of the stress wave from the initiation of emulsion can be roughly estimated in the following for the trigger setting. The transit time of detonation wave in the explosive pipe is about 18 μ s estimated by 4000m/s average detonation velocity and 70mm length of the pipe. The underwater shock wave spends 25 μ s to pass through the water part estimated by the 2000 m/s average velocity of the underwater shock wave and the 50mm length of water pipe. By using the average velocity of 2700m/s longitudinal wave of Kimachi sandstone, it is understand that the stress wave pass through the rock specimen about 110 μ s.

RESULTS AND DISCUSSION

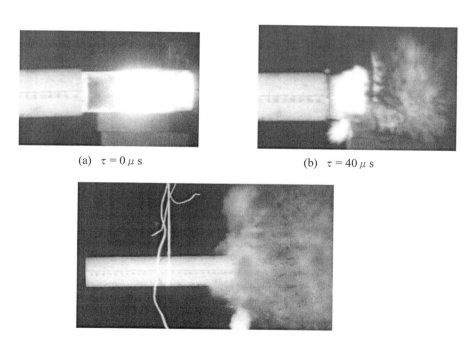

(a) $\tau = 0 \mu$ s (b) $\tau = 40 \mu$ s

(c) $\tau = 200 \mu$ s

Figure 2 Photographs of the expansion process of the detonation product in this fracture test

(a) Water pipe: 50 mm

(b) Water pipe: 70 mm

Figure 3. Recovered rock specimen (Kimachi sandstone) after fracture test

The expansion process of the detonation product in this fracture test is shown in the Figure 2. The time τ is counted from the initiation of the detonator. At time $\tau = 0$, because of the interaction between the detonation wave and the air in the double layer PMMA pipe, the air emits as a flash of lighting. Figure 2(b) corresponds to about the underwater shock wave reaches the interface of the rock specimen and the water. Although the underwater shock wave reaches the interface and interacts with the rock specimen, the view of the rock specimen for high speed camera does not disturb by the expansion gas. The stress wave arrives at the free surface of the rock specimen about 150 μ s after the initiation of explosive. Therefore, Figure 2(c) corresponds the photograph that the time is about 50 μ s passed after the stress wave arrive at the free surface. It can be seen that rapid expansion of the detonation product well prevent by the double layer pipe.

Figure 3 shows the photographs of the rock specimen after the fracture test. Fracture conditions at the explosion side correspond to the right hand side in figure 3. In the case of 50mm water pipe, the end of the specimen at the explosion side was broken into pieces over the 70-80 mm from the end of the specimen, and the fracture part could not be recovered. While in the case of the 70mm water pipe, the fractured part near the explosion source was a few parts in the periphery of specimen. It can be seen that when the water pipe more than 70 mm length is used in this fracture test, the stress wave that arrival at free surface unrelated fracture condition near the explosion source. In such a case when we try to estimate the phenomenon, which occur from the initiation of the explosive to the fracture near the free surface by using numerical simulation, it is unnecessary to consider the fracture mechanism of the rock near the explosion source. Figure 4 indicates the relation of thickness of the fragment vs. length of the water pipe. The thickness of the fragment means the distance from the free surface to position of the cross section as seen left hand side in Figure 3. Since the incident shock pressure to the rock specimen is a high value in the case of shorter water pipe, the strain rate near the free surface is also higher. From Figure 4 it can be seen that as the length of the water pipe increases, the thickness of the fragment is decreased. The dynamic tensile strength depends on the strain rate; however, it is difficult to estimate how the relationship of the dynamic tensile strength and strain rate in the specimen related to the results. It may be necessary to introduce the numerical simulation technique.

Figure 5 shows the profile of the free surface velocity obtained by the laser vibration meter in the case of the 50mm water pipe. In this figure, δ is the thickness of the fragment, V(tp) is the maximum velocity of the free surface and tp is the time needed to become the maximum velocity from the free surface begin moving. Using the profile in Figure 5 and δ, the dynamic tensile strength of the Kimachi sandstone is estimated by one dimensional wave propagation theory. In the case of the 50mm water pipe, the average value of dynamic tensile strength is 5.5 MPa, and 4.2 for the 70mm water pipe, respectively.

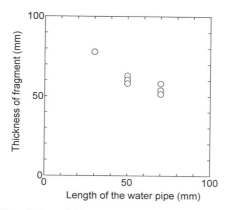

Figure 4. The relation of thickness of the fragment vs. length of the water pipe

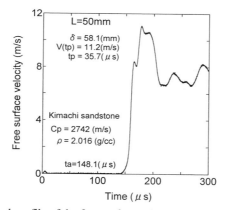

Figure 5. Typical profile of the free surface velocity in the case of the 50mm water pipe

CONCLUSION

The experimental technique to estimate the behaviors of dynamic fracture of rock was proposed, and the behavior of Kimachi sandstone under dynamic loading was investigated. The explosive was used as the explosion source, and the pipe filled with water was arranged between the explosive and cylindrical rock specimen. The length of the water pipe was varied as 30, 50 and 70mm. In the case of the 70mm water pipe, the fractured part near the explosion source was a few parts in the periphery of specimen. The thickness of the fragment is decreased as the length of the water pipe increases in this experiment.

References

Bhandari S. (1997). Engineering rock blasting operations, A.A.Balkema, Rotterdam, Brookfield
Ma G., Miyake A., Ogawa T., Ogata Y., Seto M. and Katsuyama K. (1998). Numerical Investigation on breakage behavior of reinforced concrete by blasting demolition, Journal of the Japan explosives society 59:2, 93-102

IMPACT RESISTANCE AND THERMAL COUPLINGS DURING FAILURE AND FRACTURE OF ENGINEERING MATERIALS

J.R. KLEPACZKO

Laboratory of Physics and Mechanics of Materials, Metz University, 57045 Metz, France

ABSTRACT

Within the framework of fracture or failure mechanics a process zone, or a failure zone, mostly in metallic materials which can undergo plastic deformation, are analyzed as an isothermal process. However, when the loading rate is increased not only the local strain rates may be very high, but adiabatic coupling becomes also very important. Mechanism of failure by Adiabatic Shear Bands (ASB) is a typical example. The so-called rate spectra of failure by ASB, or by Mode II, are a valuable piece of information with regard to fragmentation of impact resistant alloys. Recent experimental studies on high-speed shearing performed on 4340 steel (52 HRC) and titanium alloy Ti-6Al-4V are discussed. They demonstrate that those alloys fail at high strain rate regime by ASB. Combination of extreme conditions (high strain rate and temperature), amplified by appearance of plastic waves, leads to the local limit of material resistance to failure. When stress singularities are neglected, that is the fracture dynamics is not applicable, the failure criteria, discussed in a more detailed way in this contribution, are based on the local strain energy density or on the equivalent plastic strain. If the local shear velocities are of order ~100 m/s and higher, they trigger so-called plastic wave trapping. It has been shown recently that a new material constant can be defined called Critical Impact Velocity (CIV) in shear, Klepaczko (1994). The CIV in shear has been evaluated by FE method for 4340 steel, Klepaczko and Klosak (1999) and for Ti-6Al-4V alloy, Lebouvier and Klepaczko (2001). Those analyses confirm usefulness of the failure criterion based on the equivalent plastic strain. It can be shown that this criterion is rate-dependent and the equivalent plastic strain diminishes rapidly when the nominal strain rate increases, reaching minimum at ~10^4 1/s, Klepaczko (2000).

KEYWORDS

Dynamic Failure Mechanics (DFM), Adiabatic Shear Bands (ASB), Critical Impact Velocity (CIV) in Shear, Thermo-Visco-Plasticity.

INTRODUCTION

Advances in experimental techniques in impact testing of materials and structural components, as well as progress in computer techniques, have allowed for better understanding of dynamic fracture and failure mechanisms. A new research area is emerging which can be called Dynamic Failure Mechanics

(DFM). Typical example is dynamic tension test, where initially there is no a strong stress concentrator, but later the instability occurs (maximum force), next localization in the form of necking takes place, and finally the failure is observed, typically in the form of cup-cone fracture, Klepaczko (1968). It must be noticed the presence of high local strain rates combined with adiabatic heating which leads in turn to thermal softening. Another example of the DFM is dynamic shear localization in the form of Adiabatic Shear Bands (ASB).

Thus, this paper addresses fast and impact shearing and adiabatic shear banding which occurs in: Ti-6AL-4V alloy. Mechanical tests at high strain rates as well as ballistic tests performed on pure titanium and titanium alloys, show its a high sensitivity to formation of the ASBs, Me-Bar and Rosenberg (1997), Meyer at al. (1997). Consequently, studies of ASBs in titanium alloys are the most important part of research for improvement of titanium performance against impact. The adiabatic shear banding in Ti-6Al-4V have already been studied for some time, including experiment and analytic approach. Although some data are available how the ASBs are formed, no systematic study exists in the open literature on the effect of the nominal velocities of shearing, or impact velocities, on the critical conditions for the onset and evolution of catastrophic thermoplastic shear in this alloy. It may be mentioned that ASBs, with or without phase transformation, often act as the sites of fracture initiation in Mode II, Varfolomeyev and Klepaczko (1995).

The main purpose of this study was to clarify, using experiments and numerical methods, the role of short-time local plastic fields with thermal coupling, and the local high strain rates, in development of impact failure in shear. Because the most important part of FE analyses are precise constitutive relations, an updated version of the previous relations used for VAR-4240 steel, Klepaczko and Klosak (1999), Klepaczko (1999), was applied for Ti-6Al-4V. All material constants for the updated version have been identified. Recently, new local failure criteria in shear, based on the adiabatic strain energy density and the local effective adiabatic strain have been implemented in FE numerical code to study the shear failure triggered by ASB. Recently; the local failure criterion for Ti-6Al-4V based on the maximum failure strain in the adiabatic conditions of shear deformation is analyzed and implemented into a FE code, Lebouvier and Klepaczko (2001).

EXERIMENTAL TECHNIQUE OF IMPACT SHEARING

An original experimental technique for fast and impact shearing has been developed in LPMM - Metz, Klepaczko (1994). This experimental technique is based on application of fast hydraulic machine and the direct impact on the Modified Double Shear (MDS) specimen. It combines several positive features not available with other experimental techniques. A very wide strain rate spectrum of shear strain rates can be covered, that is: 10^{-4} 1/s $\leq \dot{\Gamma} \leq 10^5$ 1/s. The scheme of direct impact technique is shown in Fig.1. The direct impact of a long striker is applied to deform plastically up to failure the MDS specimen. Flat-ended strikers of different lengths made of maraging steel and of diameter D_p = 10.0 mm are launched from an air gun with predetermined velocities Vo, 1.0 m/s < Vo < 150 m/s. The impact velocity is measured by the setup with three sources of light L, fiber optic leads 1, 2, 3 and three independent photodiodes F which activate starts and stops of two time counters TC1 and TC2. The time intervals of dark signals from the photodiodes generated during the passage of strikers are transmitted to the start/stop gates of the time counters. Thus, the impact velocity at the instant of striker contact with the specimen can be precisely measured, because with this scheme acceleration/deceleration of a striker can be determined. Axial displacement $U_x(t)$ of the central part of the MDS specimen is measured as a function of time by an optical two channel sensor E, acting as a non-contact displacement gage. The second channel can measure the axial displacement of the striker/specimen interface. The optical gage reacts to the axial movements of a small black and white target cemented to the middle part of the MDS specimen. Axial force which is transmitted by the

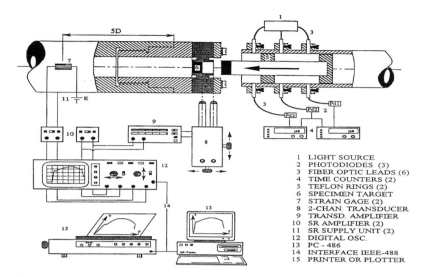

Figure 1: Scheme of experimental setup for impact shearing of the MDS specimen

specimen symmetric supports attached to a long Hopkinson tube can be determined as a function of time from the transmitted longitudinal wave $\varepsilon_t(t)$ measured by SR gages T_1 cemented on the surface of the Hopkinson tube, DC supply unit A_2 and wide band amplifier A_1. All electric signals (from the optical gage and SR amplifier) are recorded by digital oscilloscope DO and next stored in a PC hard disk for further analyses.

The direct impact configuration of experimental setup permits for a wide variation of the nominal strain rate in shear, typically from 10^3 1/s to 10^5 1/s. As it is mentioned above the lower range of nominal strain rates, from 10^{-4} 1/s to $5*10^2$ 1/s, were covered by use of the servo-hydraulic fast machine with special device to support the MDS specimen and to measure precisely the mid-specimen displacement.

SHEAR EXPERIMENTS WITH Ti-6Al-4V

One of the most utilized titanium alloys is Ti-6Al-4V, Grade 5 alloy. This α-β alloy has many industrial applications. Two series of tests have been performed at room temperature on Ti-6Al-4V within a wide range of strain rates in shear, from 10^{-3} 1/s to $\sim 10^5$ 1/s (that is eight decimal orders in strain rate). The standard MDS specimen geometry was used, Klepaczko (1994), Klepaczko (1999), with two shear zones, 2.0 mm each. The specimens were machined out of an electron beam single melt plate of thickness 25 mm. More details on the material composition and processing can be found elsewhere, Wells (1998). The Ti-alloy was delivered by ARL-WM-MC. The total number of standard specimens machined and milled was 60. The specimens were cut in such way as to assure the shear direction perpendicular to the direction of rolling.

The first series of tests was performed at lower strain rates with use of the fast servo-hydraulic universal machine with very precise measurement of force and displacement, Klepaczko (1999). At strain rate range from 10^{-3} 1/s to $2*10^2$ 1/s. For every strain rate at least three good tests were completed. The second series of experiments was performed with the direct impact on the MDS specimens, Fig.1, at impact velocities 10 m/s < V_0 < 110 m/s (nominal strain rates: $5*10^3$ << $5.5*10^4$ 1/s). The complete theory of the direct impact test on MDS specimen has been outlined elsewhere, Klepaczko (1994). The final result in the form of the mean "proof shear stress" obtained at lower strain rates (first series) and all experimental data for high strain rates (second series) are shown as a function of the logarithm of strain rate in Fig.2. The experimental data shown in Fig.2, respectively for the low and high strain rate ranges, indicate as expected, that at lower strain rates the critical shear stress is proportional to the logarithm of strain rate. However, at the strain rate range from $5*10^3$ 1/s to about 10^4 1/s the mean nominal stress increases substantially and at strain rates higher than 10^4 1/s the maximum stress saturates. Such behavior is discussed later on. The mean values of the "proof stress" at the level of shear deformation ~0.02 are shown as black points in Fig.2, where the tendency of stress saturation is well visible. In addition to the stress levels, the following characteristic points have also been determined from all quasi-static and dynamic tests: the maximum shear stress τ_m and the nominal shear strain Γ_m for this maximum, called also the instability strain, the results are given elsewhere, Klepaczko (1999). The failure shear stress τ_c and the nominal strain Γ_c at failure (final localization strain) can also be found by a properly tuned-up constitutive relation under assumption of adiabatic process of deformation. In order to analyze directly the final plastic strain of localization, that is when the uniform deformation is ceased.

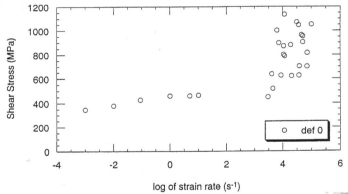

Figure 2: Complete strain rate spectrum at the level of the "proof shear stress", at lower strain rates are the mean values

In order to find directly the instability strain Γ_m the specimen deformed under impact were analyzed with the measuring microscope. The slopes of the deformed sides of the specimen, that is the upper and the lower ones in respect to the impact direction, determine the shear strain of instability. The results are shown in Fig.3.

It is clear after figure 3 that an increase of the nominal strain rate accelerates instability and strain localization. This trend is continued up to strain rate ~10^4 1/s. For the higher range of strain rates the instability strain shows a tendency to increase. A minimum is observed near 10^4 1/s. The decrease of the instability strain is due to combined effect of a positive rate sensitivity and adiabatic heating, Klepaczko (1994). An increase of the nominal strain rate, or impact velocity, leads to an early strain localization and failure by ASBs. A more general discussion of such behavior can be found in the literature, for example Klepaczko (1994, 1995).

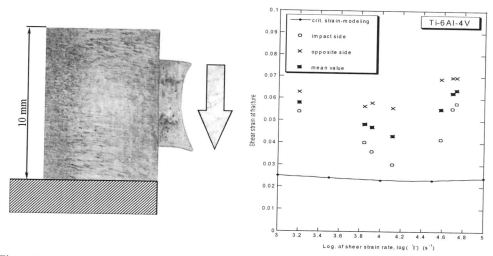

Figure 3: Photograph of the MDS specimen deformed at high strain rate and the nominal shear strain at instability point by microscopy vs. logarithm of the nominal strain rate.

CONSTITUTIVE RELATIONS FOR Ti-6Al-4V

After analyses of experimental data available in the open literature for Ti-6Al-4V alloy and the results obtained with the MDS specimens the following explicit form of the constitutive relation has been worked out

$$\tau = \frac{\mu(T)}{\mu_0}\left[B\left(\frac{T}{T_0}\right)^{-\nu}\left(\Gamma_0 + \Gamma_p\right)^{n(T)} + \tau_0\left(1 + \frac{T}{D}\log\frac{\dot{\Gamma}_0}{\dot{\Gamma}}\right)^{1/m}\right] \quad (1)$$

where B, μ_0, ν, n, m are respectively, the modulus of plasticity, the shear modulus at $T = 300$ K, the temperature index, the strain hardening exponent and the logarithmic rate sensitivity, Γ_0 is the level of initial plastic strain; T_0, $\dot{\Gamma}_0$, and D are the normalization constants. The temperature change of the shear modulus in Eq.(1) is given by

$$\mu(T) = \mu_0\left(1 - AT^* - CT^{*2}\right) \qquad T^* = T - 300 \quad (2)$$

where A and B are constants, and $T^* = T - 300$ K is the modified temperature. In principle, this version of constitutive equations can be applied at RT and at temperatures above 300 K. Since it is well known that the rate of strain hardening in metals and alloys is temperature dependent and it diminishes with an increase of temperature, Klepaczko (1987), the strain hardening exponent n was assumed as a linearly decreasing function of the homologous temperature, Klepaczko and Klosak (1999), Labouvier and Klepaczko (2001).

$$n(T) = n_0\left(1 - \frac{T}{T_m}\right) \quad (3)$$

where n_o is the strain hardening exponent at $T = 300$ K, and T_m is the melting point. This modification was found essential for very high strain rates where adiabatic increase of temperature may be substantial. The total number of constants in Eqs (1, 2, 3) is 12 and they are determined by an optimization procedure. The values of constants for Ti-6Al-4V are:

$B = 1053.8$ MPa; $v = 0113$; $\Gamma_0 = 1.6*10^{-3}$; $\tau_0 = 750$ MPa; $D = 4.347*10^3$; $K = 10^{-6}$ 1/s; $m = 0.291$; $\mu_0 = 43.91$ GPa; $A = 6.808*10^{-4}$ 1/K; $C = 1.036*10^{-7}$ K^{-2}; $n_0 = 0.0336$; $T_m = 1900$ K

Since a large part of the plastic work is converted into heat the temperature of a deformed solid increases when plastic deformation advances. The principle of energy balance with the heat conduction leads to relation (4)

$$\beta\tau = \rho C_v \frac{\partial T}{\partial t} - \lambda \frac{\partial^2 T}{\partial y^2} \qquad (4)$$

where y is the axis of heat conduction, β is coefficient of stored energy, ρ, C_v and λ are respectively the density, the specific heat and the heat conductivity. When the process is entirely adiabatic then λ = 0. The material constants for Ti-6Al-4V are as follows: $\beta = 0.9$, $C_v = 543$ J/(kg K); $\rho = 4.51$ g/cm^3. Recent studies on pure Ti indicate that the conversion coefficient β increases as a function of strain rate, Rosakis (1999). An increase of temperature during adiabatic process of deformation can be calculated after integration of Eq. (4). If the temperature sensitivity of the flow stress $\partial\tau/\partial T$ is known as a function of temperature, it can be found from Eqs (1)-(3), the thermal softening of stress can be found, Klepaczko (1968). Application of this procedure leads to the analytic determination of the nominal shear stress τ_m and Γ_m which correspond to the condition

$$(\partial\tau/\partial \Gamma)_A = 0 \qquad (5)$$

in the adiabatic process of deformation. Such result is shown as the solid line in Fig 3. Theoretical values yield lower values for the instability strain in comparison to the values determined by the direct microscopy observation, but the tendency is correct, a minimum is found. Recently an optimized constitutive relations for Ti-6Al-4V is applied in newer numerical analyses. The constitutive relations together with the thermal characteristics for Ti-6Al-4V, which are discussed above, permit for numerical simulations of all temperature-coupled problems as purely adiabatic as well as with the heat conduction, Klepaczko and Klosak (1999), Labouvier and Klepaczko (2001).

When the ASB is formed during localization the local strain rate in is core is very high, in the theoretical limit the following condition holds

$$\lim \dot{\Gamma}_{core} = \infty \qquad (6)$$

Since the core strain rate in ASB is very high, plastic strain accumulates very rapidly reaching the carrying capacity of the material, and the failure occurs. The critical shear strain in the ASB core is

$$\Gamma_C = \dot{\Gamma}_{core} t_C \qquad (7)$$

The final localization strain Γ_C has been estimated by an inverse method which is outlined elsewhere, Klepaczko (1999). It appears that the critical localization strain Γ_C is much larger than the instability strain Γ_m, that is $\Gamma_c = \alpha \Gamma_m$. Value of α estimated for Ti-6Al-4V is ~50. The maximum localization strain has been used in the FE numerical code to calculate the critical conditions of failure in shear, the results are presented elsewhere, Lebouvier and Klepaczko (2001).

The numerical study of the MDS geometry has confirmed that at high nominal strain rates the stress concentrators trigger ASBs by a local increase of the shear strain rate and temperature. Such scenario explains reduction of the failure energy, for example, in perforation mechanics or in MDS specimen at very high impact velocities (order 100 m/s). At lower rates of deformation the failure energy for Ti-6Al-4V specimens steadily increases up to the moderate high strain rates, say up to 10^4 1/s, and next, at higher impact velocities a substantial drop of the failure energy occurs: the Critical Impact Velocity is reached, Klepaczko (1968), Klepaczko and Klosak (1999), Lebouvier and Klepaczko (2001).

CRITICAL IMPACT VELOCITY IN SHEAR

If the local strain rate in shear is very high it leads to the trapping of plastic waves by adiabatic shearing. This phenomenon is called the Critical Impact Velocity (CIV) in shear, Klepaczko (1995). By analogy to the CIV in tension, Klepaczko (1968), it is possible to construct a model of the plastic wave trapping in shear which is based on the criterion of the adiabatic instability, $(\partial \tau / \partial \Gamma)_A = 0$, Klepaczko (1994). According to the rate-independent theory of elastic-plastic wave propagation, as applied to shear deformation, the wave equation is as follows:

$$\frac{\partial^2 U}{\partial t^2} = C_{2p}(\Gamma) \frac{\partial^2 U}{\partial y^2} \qquad (8)$$

where U is the displacement along the x-axis (shear direction) and the wave propagation is along the y-axis (perpendicular direction). The wave speed in the adiabatic conditions of deformation is defined as follows: (9)

$$C_{2p} = \left(\frac{1}{\rho} \frac{d\tau}{d\Gamma} \right)_A^{1/2} \qquad (9)$$

Equation (9) reduces to $C_2 = \sqrt{\mu/\rho}$ for the elastic wave propagation. Along the characteristics the characteristics $C_{2p} = \pm (dy/dt)$ the following relation holds

$$d \left[\int_0^{\Gamma_m} C_{2p}(\Gamma) d\Gamma \pm v \right] = 0 \qquad (10)$$

where v is the mass velocity, $v = (\partial U / \partial t)$ and Γ_m is the wave amplitude of shear strain at the instability point. Combination of the constitutive relations (1,2,3) with the instability condition for adiabatic shearing (5), that is $d\tau_A = 0$, leads to the result that the wave speed is zero at Γ_m and the wave is trapped. This yields the formula for the CIV in shear v_c

$$v_c = \int_0^{\Gamma_m} C_{2p}(\Gamma)_{\dot{\Gamma}} d\Gamma \qquad (11)$$

Integral (11) can be split into elastic and plastic part, with the shear strain at the limit of elasticity Γ_e,

$$v_c = \int_0^{\Gamma_e} C_2 \, d\Gamma + \int_{\Gamma_e}^{\Gamma_m} C_{2p}(\Gamma) d\Gamma \qquad (12)$$

The procedure for determination of v_c was applied for Ti-6Al-4V using the constitutive relations discussed above. A more detailed discussion of this procedure together with the verification by FE method can be found elsewhere, Klepaczko and Klosak (1999), for VA4340 steel, and in Lebouvier and Klepaczko (2001) for Ti-6Al-4V. In the case of Ti-6Al-4V the first term in Eq.(12) constitutes an important contribution to the final value of v_c (CIV in shear), the second term is relatively small. This is caused by a low rate of strain hardening for this alloy deformed at high strain rate. The nominal strain rate assumed in those calculations was $10^E 4$ 1/s. The values of these two terms are respectively $v_{ce} = 106.4$ m/s and $v_{cp} = 14.0$ m/s, thus the final value is $v_c \approx 121$ m/s. This value is comparable with so far known other values for the CIV in shear, for example ~.110 m/s for VAR4340 - 52 HRC. The CIV in shear can be treated as a new material constant, very important in estimation of fragmentation processes by ASBs.

ACKNOWLEDGEMENTS

The research reported herein was sponsored in part by the US Army trough its European Research Office, Contract N° N68171 -98-M-5829, and in part by CNRS-France.

REFERENCES

Klepaczko J.R. (1968). *Generalized Conditions for Stability in Tension Tests.* Int. J. Mech. Sci. **10**, 207.
Klepaczko J.R. (1987). *A Practical Stress-Strain-Strain Rate-Temperature Constitutive Relation of the Power Form.* J. Mech. Working Technology **15**,143.
Klepaczko J.R.(1994). *An Experimental Technique for Shear testing at High and Very High Strain Rates; The Case of mild Steel.* Int. J. Impact Engng. **15,** 25.
Klepaczko J.R. (1994). *Plastic Shearing at High and Very High Strain Rates.* Proc. Int. Conf. EURODYMAT 94, C8-35.
Klepaczko J.R. (1995). *On the Critical Impact Velocity in Plastic Shearing.* Proc. Int. Conf. EXPLOMET'95, Elsevier Sci. Publ. 413.
Klepaczko J.R. and Rezaig B. (1996). *A Numerical Study of Adiabatic Shear Banding in Mild Steel by Dislocation Based Constitutive relations.* Mech. of Mat. **18**,125.
Klepaczko J.R. (1999).*Effects of Impact Velocity and Stress Concentrators in Titanium on Failure by Adiabatic Shearing*, Final Technical Report for the US Army ERO, Contract DAJA 49-90-C-0052, LPMM, Metz University.
Klepaczko J.R. and Klosak M. (1999). *Numerical Study of the Critical Impact Velocity in Shear.* Eur. J. Mech., A/Solids **18** ,93.
Klepaczko J.R. (2000). *Behavior of Ti-6Al-4V Alloy at High Strain rates, Shear Testing up to $6*10^E 4$ 1/s and Failure Criterion.* Proc. Int. Conf. DYMAT 2000, 191.
Lebouvier A.S. and Klepaczko J.R. (2001). *Numerical Study of Shear Deformation in Ti-6Al-4V at Medium and High Strain Rates, Critical Impact Velocity in Shear.* Int. J. Impact Engng., submitted.
Me-Bar Y. and Rosenberg Z. (1997). *On the Correlation Between the Ballistic Behavior and Dynamic Properties of Titanium-alloy Plates.* Int.J.Impact Engng. **19**, 311.
Meyer L.W., Krueger L. Gooch W.A. and Burkins M.S. *Analysis of Shear Band Effects in Titanium Relative to High Strain-Rate Laboratory/Ballistic Impact Tests.* Proc. Int. Conf. EURODYMAT'97.
Rosakis A.S. (1999). Private communication.
Wells M.G.H. (1998). *The Mechanical and Ballistic Properties of an Electron Beam Single Melt of Ti-6Al-4V Plate*, AMSRL-WM-MC Memo.
Varfolomeyev I.V. and Klepaczko J.R. (1995). *Approximate Analysis on Strain Rate Effects and Behavior of Stress and Strain Fields at Crack Tip in Mode II in Metallic Materials.* Strength of Materials **27** ,138.

Impact Engineering and Application
Akira Chiba, Shinji Tanimura and Kazuyuki Hokamoto (Eds)
©2001 Elsevier Science Ltd. All rights reserved.

THE EFFECT OF PRE-FATIGUE ON DYNAMIC BEHAVIOR OF SOME STEELS AND ALUMINUM ALLOYS

M. Itabashi and H. Fukuda

Department of Materials Science and Technology, Faculty of Industrial Science and Technology,
Science University of Tokyo, 2641, Yamazaki, Noda, Chiba 278-8510, Japan

ABSTRACT

For pre-fatigued structural metals, remaining mechanical properties under dynamic tension are one of the practical themes to be clarified, from the viewpoint of not only prevention of unexpected structural failure but also damage mechanics. Two steels for building structure, SN490B (JIS G 3136-1994) and SM490A (G 3106-1999), were not so degraded by high-cycle pre-fatigue. Two aluminum alloys, 2219-T87 and 6061-T6, were degraded by low-cycle pre-fatigue with respect to ultimate strength under dynamic tension. For the aluminum alloys, the cause of this degradation was explored with an electron probe microanalyzer (EPMA) and a scanning electron microscope (SEM). The number of thin linear damages from opened cracks on the side surface of the fractured specimen might be related with the strength degradation.

KEYWORDS

Pre-Fatigue, Dynamic Behavior, Steels, Aluminum Alloys, Surface Damage, Impact, Strain Rate, Strength Degradation

INTRODUCTION

Kawata et al. (1970) reported the degradation of elongation for low-cycle pre-fatigued 2024-T3 aluminum alloy in quasi-static tension. Basing upon this experimental result, Kawata et al. (1996) carried out dynamic tensile tests, at the strain rate of $\dot{\varepsilon} = 1 \times 10^3 \text{s}^{-1}$, for three aluminum alloys damaged by pre-fatigue. Some alloys were severely degraded in strength and elongation. Although published experimental results of this kind of characterization are not enough, in fact, structural materials are fatigued in service. In an accident, the materials may be exposed to impact loading possibly. Hence, in design of cars, aircraft and building structure in earthquake zones, the above-mentioned characteristics should be introduced in structural design.

Formerly, the present authors reported dynamic tensile mechanical properties for low-cycle pre-fatigued steels, SN490B and SM490A (Itabashi & Fukuda (1999)). In this paper, data for the steels damaged by high-cycle pre-fatigue are presented. And some trials with EPMA and SEM to detect what is the cause of the dynamic strength degradation for low-cycle pre-fatigued aluminum alloys, 2219-T87 and 6061-T6, are introduced.

EXPERIMENTAL

The chemical compositions for SN490B steel were Fe-0.18C-0.46Si-1.41Mn-0.019P-0.008S-0.01Cu-0.01Ni-0.02Cr in wt%, for SM490A steel Fe-0.10C-0.21Si-1.52Mn-0.016P-0.004S-0.04Nb-0.02V, for 2219-T87 alloy Al-0.07Si-0.16Fe-5.90Cu-0.26Mn-0.03Zn-0.03Ti-0.12Zr-0.12V-0.01others and for 6061-T6 alloy Al-0.7Si-0.25Fe-0.3Cu-0.12Mn-1.1Mg-0.19Cr-0.01Zn-0.01Ti-0.01others. These investigated materials were supplied in the form of plate. In case of the aluminum alloys, specimens were machined, as shown in Figure 1, in two directions. One was parallel to the rolling direction (L) and another one was perpendicular to it (T). The steel specimens were only L direction. After machining, a parallel part with round fillets of the specimen was ground by #800 emery paper.

First of all, tensile tests of virgin specimens were carried out at two strain rates. For a quasi-static strain rate, $1 \times 10^{-3} \mathrm{s}^{-1}$, a universal material testing machine (Shimadzu Autograph AG-10TA with specially designed specimen attachments, crosshead speed: 0.5mm/min) was used. For a dynamic strain rate, $1 \times 10^{3} \mathrm{s}^{-1}$, the horizontal slingshot machine (Itabashi & Kawata (2000)) adopting the one bar method (Kawata et al. (1979)) was used. Next, in order to set pre-fatigue conditions, S-N curves were obtained under sinusoidal pulsating tensile loading (stress ratio: 0, frequency: 20Hz) with a hydraulic fatigue testing machine (Shimadzu Servopulser EHF-FB1 type 1111 with special attachments). The obtained S-N curves with set pre-fatigue conditions are shown in Figures 2 and 3. The high-cycle pre-fatigue conditions for the steels were set at the maximum (pre-fatigue) stresses of 100MPa and

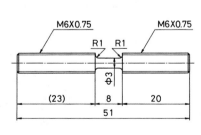

Figure 1: Specimen dimensions (in mm)

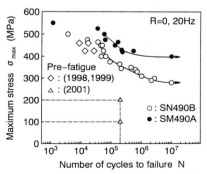

Figure 2: S-N curves and pre-fatigue conditions for SN490B and SM490A steels

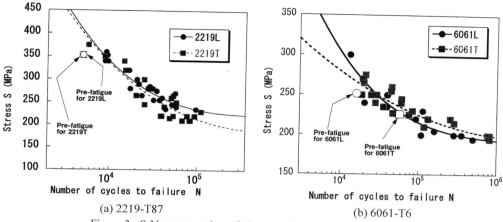

Figure 3 : S-N curves and pre-fatigue conditions for aluminum alloys

200MPa and at the number of cycles of 2×10^5. On the other hand, for the low-cycle pre-fatigued alloys, according to the former results (Kawata et al. (1996)), the condition was decided as follows. A maximum stress was 78% of the quasi-static (virgin) tensile strength of each kind of the specimens and the number of repeated (pre-fatigue) loading cycles was 60% of the number of cycles to failure. After preparation of the pre-fatigued specimens with the same hydraulic fatigue testing machine, tensile tests at $\dot{\varepsilon}=1\times10^{-3}\mathrm{s}^{-1}$ and $1\times10^{3}\mathrm{s}^{-1}$ were executed. For the aluminum alloys only, fractured specimen surface was observed microscopically with EPMA and SEM, in order to detect the cause of the strength degradation under dynamic tension.

RESULTS AND DISCUSSION

High-cycle Pre-fatigued steels

Typical stress-strain curves of the virgin and high-cycle pre-fatigued specimens for SN490B steel are shown in Figure 4. In case of these steels, extra dynamic tensile tests were performed at a low temperature of 0℃. There is no significant difference between the virgin and pre-fatigued specimens in each tensile test condition. In Figure 5, maximum (pre-fatigue) stress dependence of tensile strength is indicated. The maximum stress of 0MPa means the virgin condition. In tensile strength of SN490B, only strain rate dependence was appeared. Dynamic strength was not so varied with temperature, since the temperature difference was only 19℃. However, the dynamic behavior at 0℃ plays the important role in the standard, because Charpy impact test at 0℃ is the duty of steel companies. Anyway, SN490B steel did not need to pay any special attention to the possibility of the dynamic strength degradation even damaged by high-cycle fatigue. SM490A steel also showed the same behavior.

Low-cycle Pre-fatigued Aluminum Alloys

Kawata et al. (1996) reported that, as a pre-fatigue condition, the maximum stress at 78% of quasi-

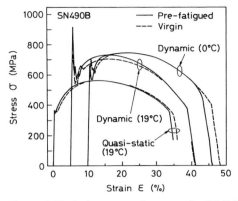
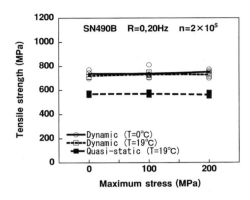

Figure 4: Typical stress-strain curves for SN490B steel (Dynamic: $\dot{\varepsilon} = 1 \times 10^3 \text{s}^{-1}$, Quasi-static: 1×10^{-3} s^{-1})

Figure 5: Maximum (pre-fatigue) stress dependence of tensile strength for SN490B steel

static strength and the number of repeated cycles at 60% of the number of repeated cycles to failure made some aluminum alloys weak. Basing upon the results, the authors of that article proposed that dynamic stress concentration around inclusions (precipitations) was occurred and micro-fractures concentrated in the neighborhood of them. The stress concentration could be introduced by the difference in mechanical and acoustic properties between inclusions and matrix. If this assumption was correct, the dynamically fractured surface of the pre-fatigued specimen should be covered by spots of elements composing the precipitations, such as $Al_{20}Cu_2Mn_3$ and Al_3Zr for 2219-T87 alloy and Mg_2Si for 6061-T6.

In the present series of experiments, the specimens were machined from the same plates of the former work in 1996. Typical stress-strain curves for 2219L (2219-T87 alloy, L direction) is indicated in Figure 6. As shown also in Figure 7, dynamic tensile strength for the pre-fatigued 2219L and 2219T specimens was weakened to the level of the quasi-static strength. The other kinds of the specimens, 6061L and 6061T, showed such degradation in strength, too.

The fractured surfaces were observed with EPMA and qualitative analyses failed to detect distinctive maps of the elements included in the precipitations. The above-mentioned assumption was not realized. Auzanneau (1999) pointed out that some opened cracks existed on the side surface of the pre-fatigued and dynamically-fractured 2017-T3 alloy. So, the present authors observed the side surface of the fractured specimens and found many opened cracks. Some cracks were accompanied with thin linear damage(s) as shown in Figure 8. The number of the thin linear damages was counted for each experimental condition, as indicated in Figure 9. The counted area was limited within the distance of approximately 200μm from the fractured surface. And for each experimental condition, only one specimen was observed in this manner. Strictly speaking, only a half of one specimen was observed, because the specimen was not rotated during the observation. However the data are not enough to conclude that these linear damages are the cause of the dynamic strength degradation, Figure 9 tells that no linear damage is formed for the dynamically fractured specimen without the pre-fatigue.

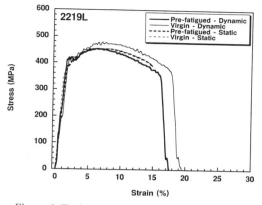
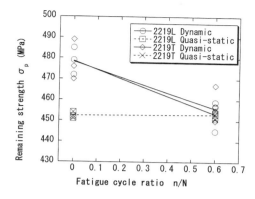

Figure 6: Typical stress-strain curves for 2219L (Dynamic: $\dot{\varepsilon}=1\times10^{3}\text{s}^{-1}$, Quasi-static: $1\times10^{-3}\text{s}^{-1}$)

Figure 7: Fatigue cycle ratio dependence of tensile strength for 2219-T87 alloy

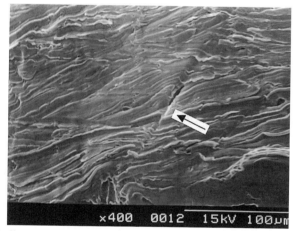

Figure 8: Thin linear damage from opened crack (2219L, Pre-fatigued, Dynamic tension)

CONCLUSIONS

Pre-fatigued SN490B and SM490A steels were mechanically characterized in tension at two strain rates of $1\times10^{-3}\text{s}^{-1}$ and $1\times10^{3}\text{s}^{-1}$. Tensile strength of the steels was unchanged by the high-cycle pre-fatigue. According to the former work (Itabashi & Fukuda (1999)), the low-cycle pre-fatigue did not give its serious effect to dynamic strength of the steels, also. On the other hand, low-cycle pre-fatigued 2219-T87 and 6061-T6 aluminum alloys were degraded in strength under dynamic tension only. This phenomenon might be related to the number of thin linear damages from opened cracks on the side surface of the fractured specimens. This micro-observation needs additional data to verify the fact obviously.

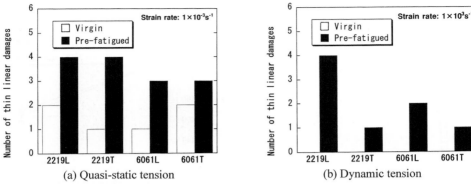

Figure 9: The number of thin linear damages on the side surface for aluminum alloys

Acknowledgment

This work was supported by a grant of KAWASAKI STEEL 21st Century Foundation and a Grant-in-Aid for Encouragement of Young Scientists, The Ministry of Education, Science, Sports and Culture (Project No.10750077). The authors wish to express thanks to Mr. Tomoharu Yasukawa and Mr. Kenji Tokuno for their eager assistance in carrying out the experiments.

References

Auzanneau T. (1999). *Influence d'un pré-endommagement par fatigue sur la tenue au choc. Application à un alliage d'aluminium 2017A T3*. Thèse de Doctrat ENSAM, Bordeaux, France.

Itabashi M. and Fukuda H. (1999). Dynamic Tensile Properties of Pre-Fatigued Steel for New Seismic Proof Structural Design Method. In: Shim V.P.W., Tanimura S. and Lim C.T., editors. *Impact Response of Materials & Structures*, Oxford University Press, Oxford, UK, 117-122.

Itabashi M. and Kawata K. (2000). Carbon Content Effect on High-Strain-Rate Tensile Properties for Carbon Steels. *Int. J. Impact Engng* **24:2**, 117-131.

Kawata K., Hashimoro S., Kurokawa K. and Kanayama N. (1979). A New Testing Method for the Characterization of Materials in High Velocity Tension. In: Harding J., editor. Mechanical Properties at High Rates of Strain 1979, Inst. of Physics, Bristol and London, UK, 71-80.

Kawata K., Hashimoto S. and Hondo A. (1970). Mechanical Degradation of Aeroplane Materials by Their Fatigue and Its Detection (1st Report). *Report of Institute of Space and Aeronautical Science, Univ. of Tokyo* **6:3(B)**, 716-728 (in Japanese).

Kawata K., Itabashi M. and Kusaka S. (1996). Behaviour Analysis of Pre-Fatigue Damaged Aluminum Alloys under High-Velocity and Quasi-Static Tension. In: Pineau A. and Zaoui A., editors. *IUTAM Symp. on Micromechanics of Plasticity and Damage of Multiphase Materials*, Kluwer Academic Publishers, Dordrecht, The Netherlands, 397-404.

COMPRESSION FLOW STRESS OF ULTRA LOW CARBON MILD STEEL AT HIGH STRAIN RATE

N.Kojima, Y.Nakazawa and N.Mizui

Corporate R&D Labs., Sumitomo Metal Industries Ltd.,
1-8 Fuso-cho Amagasaki 660-0891, Japan

ABSTRACT

Precise stress-strain curves of sheet steels at a wide range of strain-rate are important for automotive crash simulation and usually measured by tensile test. However it is often difficult to measure the true stress in large strain region. Because mild steel tends to start necking at a very small strain when strain-rate increases. For avoiding necking, compression test was applied in this research and the flow stress of mild steel at large strain was investigated. The ultra low carbon Ti-added steel was laboratory processed and hot-rolled to 7mm thickness. The amount of Ti was enough to stabilize solute carbon for eliminating yield point elongation and strain aging. Cylindrical specimens with 5mm diameter and 7mm length were subjected to compression test at various strain-rate ranging from 0.01 to 570/s. With the increase in strain-rate, the flow stress increased and the work-hardening exponent (n-value) decreased. The n-value at 570/s is 0.05 and much smaller than at 0.01/s. The small n-value would lead to the small uniform elongation in tensile test. It was verified by tensile test experiment with the same steel and the numerical simulation.

KEYWORDS

High strain-rate, Compression test, Tensile test, Mild steel, Work-hardening, Crash safety

INTRODUCTION

The numerical simulation technology is widely used in the automotive industry. Regarding crash safety, it is a useful tool, which can reduce the number of crash trails and contribute toward saving a lot of cost and time. Reliability of the numerical simulation depends on both the simulation technique and the material data used in it. As the computer hardware and software make progress, more precise material data is required. The body structure of automobile is made of thin sheet steels but those dynamic mechanical properties are not sufficiently investigated, as compared to structural steels and armors. Although high strength sheet steels became recently popular for automobile, studying mild steels intensely would be helpful in considering strain-rate effect in crash simulation. Because the lower strength steel generally exhibits more significant strain-rate sensitivity.

In the tensile test, the true stress at a large strain beyond necking cannot be measured directly. Especially the mild steel shows a very small uniform elongation at higher strain-rate, in spite of quite large uniform elongation at static test. To solve this problem, the compression test using the sandwich specimen by Zhao and the shear test using the double shear specimen by Rusinek have been reported for thin sheet steels. However tensile test is an easier method and widely used for thin sheet steels. Thus it is important to consider necking behavior in tensile test deeply.

In this research, the material with a similar chemistry to the commercial sheet steel was laboratory made so that the thickness was enough for the conventional compression test. The compression test was conducted for mild steel in addition to tensile test and they were compared with each other.

EXPERIMENTAL PROCEDURE

The ultra low carbon interstitial free steel was molted and cast in laboratory. The chemical composition is listed in Table 1. As the excess titanium (Ti*) is much more than the carbon atomic percent, the solute carbon can be completely stabilized as TiC. The slab was heated at 1523K, hot rolled to 7mm in thickness above 1173K. After rolling, the hot band was cooled to 923K and kept at the temperature for 1800sec for precipitating TiC, then furnace cooled to the room temperature. The cylindrical compression specimen in Fig.1(a) and the flat plate tensile specimen in Fig.1(b) for various strain-rate test were machined from the hot band. The JIS No.5 type tensile specimen with 3.0mm thickness was also machined. Testing direction of all specimens was along rolling direction of the hot band.

Table 1 : Chemical composition (wt%)

C	Si	Mn	P	S	N	Sol Al	Ti	Ti*	Ti*/C
0.0022	<0.01	0.13	0.009	0.002	0.0025	0.022	0.041	0.029	13

Ti*=Ti-48*(N/14+S/32)

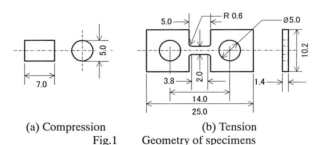

(a) Compression (b) Tension
Fig.1 Geometry of specimens

Static tensile test with JIS No.5 type specimen was conducted by Instron type test machine and the strain for 50mm gage length was measured by extensometer. Aging index (AI) was also measured as following. Specimen was prestrained by 10% at first test and aged at 373K for 3600sec, then second test was conducted. AI is defined as the difference between the yield strength of second test and 10% flow stress of first test. The compressive and the tensile test at various strain-rates from 0.01 to 570/s were carried out by means of the sensing block type high speed testing machine, which was developed by Tanimura. Its loading device is a hydraulic servo system with the 10kN capability at static.

RESULTS

Conventional tensile properties
Conventional tensile test result is listed in Table 2. AI shows almost zero and it implies that the

interstitial solute atoms are fully stabilized. Thus results of this research never involve the effect of discontinuous yielding and dynamic strain aging. The properties except the r-value are almost same as the commercial cold rolled thin sheet steel. As the steel was not subjected to cold rolling and recrystallization annealing, the r-value is quite low in comparison to that of cold rolled steel.

Table 2 : Static tensile test properties by JIS No.5

YS (MPa)	TS (MPa)	EL (%)	n-value	r-value	AI (MPa)
126	291	54.3	0.203	0.75	-3

Dynamic compression test

The experimental data of compression test are shown in Fig. 2. When strain-rate is less than 10/s, strain-rate decreases with strain increasing because the hydraulic capability of the machine limits the maximum stress to about 550MPa. For more than 100/s, the inertia of the machine lets the stress reach higher and keep the strain-rate constant during test. The true stress - true strain relations in Fig. 3 are derived for the strain region where the strain-rate is between the maximum ((a) 0.009, (b) 3.0, (c) 350, (d) 570/s) and half of it.

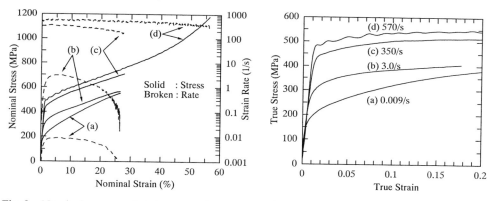

Fig. 2 Nominal compressive stress - strain curves

Fig. 3 True compressive stress - strain curves

Yield strength (defined as 0.2% offset stress) and true stresses at 0.05, 0.10 and 0.15 strains are plotted in Fig. 4. The work-hardenable exponent (n-value) defined by two points at 0.05 and 0.15 strain is also plotted. The flow stress at any strain increases as increasing strain-rate. The flow stress at smaller strain has more significant strain-rate sensitivity. The n-value decreases with increasing strain-rate. The n-value at 570/s becomes only one fifth of static.

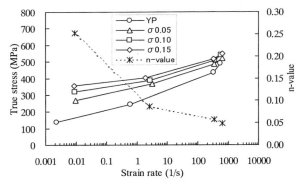

Fig. 4   Effect of strain-rate on flow stress and n-value

Dynamic tensile test

The nominal stress - strain curves of dynamic tensile test are shown in Fig.5. The nominal strain was measured as the change of distance between the pins at both ends of the specimen. Thus it includes deformation around the pins besides on the parallel portion. As increasing strain-rate, the maximum stress increases and the strain, which gives the maximum stress, becomes smaller.

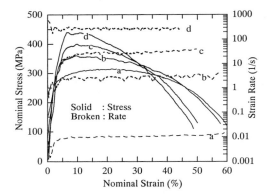

Fig. 5   Nominal stress - strain curves of tensile test

DISCUSSION

Work-hardening behavior

The compressive true stress - true strain curves and work-hardenability curves ($d\sigma/d\varepsilon$ - ε) are superimposed in Fig.6. The work-hardenability decreases with increasing strain and this trend at high strain-rate is more rapidly than static. The strain, where two curves crosses ($\sigma = d\sigma/d\varepsilon$), decreases with increasing strain-rate. In case of tensile test, this strain point would be the onset of instability (necking) and give the uniform elongation. The uniform elongation estimated from compressive test and that obtained by tensile experiment are shown in Fig.7. They decreases suddenly as increasing strain-rate from 0.009/s to 3.0/s and gradually decrease beyond 3.0/s. The estimation agrees well with the experimental result.

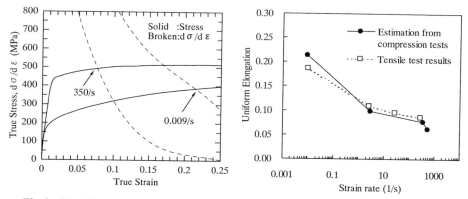

Fig.6 Work-hardenability curves in compression test

Fig.7 Effect of strain rate on uniform elongation

Numerical analysis

The aim of the numerical analysis is to verify the early necking phenomena discussed above. Numerical simulation for tensile test was performed by means of the static implicit code MARC Ver.7.3. For static tensile test, only compressive stress strain curve at 0.009/s was used as material data. For dynamic, that at 350/s was used and any other strain-rate dependency was not taken into account. Nominal stress - strain curves derived from the simulation are compared with the experimental result at corresponding strain-rate in Fig.8. The nominal stresses of them agree with each other before reaching the maximum nominal stress. But after the maximum stress, calculation shows lower stress than experiment. Because strain-rate of necking part increases quickly and the true stress becomes higher but the same stress - strain curve is used in this calculation. Figure 9 shows the strain distribution over the parallel portion of specimen. In case of 0.009/s, the strain distributes relatively uniform over parallel portion as shown Fig.9(a) when the nominal strain reached about 10%. The strain starts localizing around the center of the parallel portion beyond the nominal strain of 20% as shown in Fig.9(b). On the other hand, in case of 350/s, the strain shown in Fig.9(c) localizes even at the nominal strain as small as Fig.9(a). Thus it was confirmed that the necking occurs at smaller nominal strain at higher strain-rate.

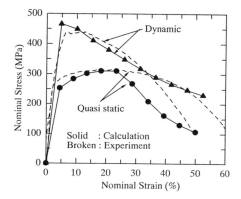

Fig.8 Comparison of tensile test calculation with experimental results

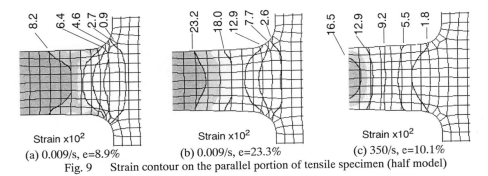

Fig. 9 Strain contour on the parallel portion of tensile specimen (half model)

CONCLUSION

Compression test at various strain-rates has been performed on the interstitial free mild steel. As increasing strain-rate, the flow stress increased and the work-hardenability decreased. Tensile test also has been performed on the same steel. As increasing strain-rate, the necking of the parallel portion occurred at smaller nominal strain. This necking behavior corresponds to the work-hardenability measured by compression test. The numerical simulation for tensile test demonstrated that the strain distribution localized easier at higher strain-rate.

REFERENCES

Mizui N., Fukui K., Kojima N., Yamamoto M., Kawaguchi Y., Okamoto A. and Nakazawa Y. (1997). Fundamental Study on Improvement in Frontal Crashworthiness by Application of High-Strength Sheet Steels. SAE Technical Paper 970156

Tanimura S., Mimura K. and Takada S. (1998). Development of Dynamic Testing Apparatus with a Sensing Block and Its Applications. Proc.Annual Meeting of JSME/MMD Vol.B **98-5**, 303-304.

Rusinek A. and Klepaczko J.R. (2001). Shear testing of a sheet steel at wide range of strain rates and a constitutive relation with strain-rate and temperature dependence of the flow stress. Int. J. Plasticity **17**, 87-115.

Zhao H. and Gray G. (1995). Strain-rate sensitivity of the behaviour of steel sheets in uniaxial compression. Proc. fifth Int. Sym. on Plasticity and its current applications, 215-218.

SPALL DAMAGE GROWTH UNDER REPEATED PLATE IMPACT TESTS

N. Nishimura, K. Kawashima, T. Yamakawa and M. Kondo

Department of Mechanical Engineering, Nagoya Institute of Technology,
Gokisocho, Showaku, Nagoya 466-8555, JAPAN

ABSTRACT

Evolution of spall damage under repeated impacts was evaluated nondestructively with an acoustic image as well as ultrasonic wave velocity, amplitude ratio (B2/B1) and backscattering intensity. The change in acoustic images under repeated impact was well correlated with the change in these ultrasonic variables. A simulation of stress wave propagation made clear that the initial spall damage is nucleated in the periphery of a circular plate, because the maximum tensile stress appear in this position by the superposition of axial and radial stress waves. In the stress wave measurement under the second impact, the fast arrival of the spall signal meant the damage growth in the back surface side. These methods give us an advanced means to evaluate nondestructively spall damage and to make clear the mechanisms of the spall damage growth under repeated impacts.

KEYWORDS

Spall damage, Repeated impacts, Plate impact test, Stress wave, Distinct element method, Ultrasonic inspection, Nondestructive evaluation, PVDF pressure gauge.

INTRODUCTION

Spall damage, namely microvoids or microcracks, is nucleated by an intensive tensile stress pulse within a solid which has been impacted at high velocity. In order to understand the spall damage nucleation mechanism, the stress wave propagation should be calculated for the whole area of specimen. Traditionally the extent of spall damage has been examined by destructive means (see Davison & Graham [1979] and Curran et al. [1987]). However these destructive methods are applied to neither actual structural parts nor investigation of the spall damage growth under repeated impacts on the identical sample.

For nondestructive evaluations of the spall damage, we should know the size and the quantity of the minute damage as well as the location. Among many nondestructive methods, an ultrasonic method is versatile and suitable for quantitative evaluation. Namely, we can visualize minute defects within solids by C- and B-scan images, which give us the defect distribution at an arbitrary depth. In addition, with the conventional pulse-echo measurement, we can measure ultrasonic wave velocity, attenuation (amplitude ratio of B2 to B1 echo), backscattering intensity scattered at the defect. Thus, we can evaluate nondestructively the spall damage growth under repeated impacts.

The aim of the present paper is to reveal the spall damage growth process of the identical target plate under repeated plate impacts with a computer simulation of stress wave propagation, ultrasonic technique and measurements of stress history using PVDF pressure gauge.

SPALLATION UNDER PLATE IMPACT TEST

Spall damage, microvoids or microcracks, is nucleated within a target plate by nearly triaxial tensile stress pulse which is generated by the plate impact test. When a target plate is impacted by a flyer plate at a high velocity, compressive stress waves travel toward the free surfaces of both plates. The waves reflected at the free surface with the reversed phase, namely tensile nature, travel back toward the impacted surface with canceling compressive wave. Thus it is called as rarefied wave. When both rarefied waves from the free surfaces are superposed, the real tensile stress pulse is generated. When the stress is higher than the spall threshold stress of the solid, the spall damage is nucleated there. The extent of the damage depends on the magnitude and duration of the tensile stress pulse.

SIMULATION OF STRESS WAVE PROPAGATION

Stress wave propagation in a target plate was analyzed by the distinct element method (see Wolf et al. [1978]). Flyer and target plates are assumed to be elastic. The equation of motion was replaced by the spring-lumped mass model, as shown in Figure 1. When one of principal stresses exceeded the spall threshold stress, spall damage are introduced by cutting off the springs. A flyer and target plates are discretized into N elements in the radial direction and into M1, 2M1 elements in the axial direction respectively, as shown in Figure 1. Each lumped mass $M(i,j)$ is coupled by the elastic springs. Axial, shear and circumferential forces affect lumped mass $M(i,j)$, as shown in Figure 2. Stress wave propagation is simulated by numerically calculating the equation of motion at each lumped mass in each time step (see Fukatsu & Hori [1993]).

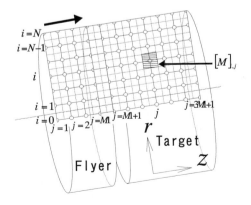

Figure 1 : Computational model

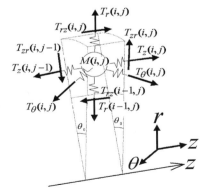

Figure 2 : Internal forces on lumped mass

EXPERIMENTAL METHOD

Plate Impact Tests

Target plates of commercially pure aluminum and medium carbon steel were impacted repeatedly by flyer plates of the same material. Their dimensions are shown in Figure 3. The flyer plate bonded to the plastic sabot was accelerated by a single stage gas gun. The target plate was also bonded to the plastic holder which was fastened to the target holder and velocity measurement unit, as shown in Figure 3. The velocity of the flyer was measured by the optical fiber switches away 20 and 30mm from the target plate. After each impact test, the front and rear surfaces of the target plate were ground. At maximum, three cumulative impacts were given to the identical target plate with different impact velocities.

The compressive stress at the impacted surface is estimated by

$$\sigma = \frac{1}{2}\rho CV \quad , \quad C = \sqrt{\frac{K}{\rho}} \tag{1}$$

, where ρ, K and V are the density, bulk modulus of the material and the velocity of the flyer.

Stress history of the target plate were measured by a PVDF pressure gauge at the interface between the target plate and thick PMMA backing plate. Stress is measured after transmission into the backing plate. In order to account for the acoustic mismatch between the target and the PMMA backing, the following relation is used,

$$\sigma_{Target} = \frac{Z_{Target} + Z_{PMMA}}{2Z_{PMMA}} \sigma_{PMMA} \tag{2}$$

where Z stands for shock impedance and σ_{PMMA} is the stress measured by the gauge.

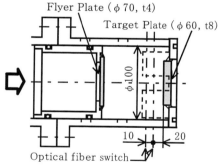

Figure 3 : Target, flyer and velocity measurement unit

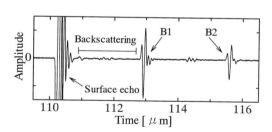

Figure 4 : Received waveform (medium carbon steel, No.S22)

Ultrasonic Evaluation of Spall Damage

Spall damage distribution in the target plate was examined by ultrasonic C- or B-scan image which visualized the void or crack distribution in a plane parallel or perpendicular to the impact plane. The image is constructed with the amplitude of the longitudinal wave reflected at voids or cracks by scanning the focused ultrasonic beam. We have used a PVDF transducer of which central frequency, diameter and focal length are 30MHz, 6.3 and 25.4mm.

For quantitative evaluation of the spall damage evolution, the time-of-flight (velocity), the backwall echo (B1 and B2) height of the target plate and the backscattering intensity scattered at the spall damage were measured with a digital ultrasonic measurement system (see Kawashima & Fujii [1995]). A longitudinal transducer of 10MHz frequency and 6.4mm in diameter was used. The waveforms of the backwall echos and the backscattering echo were recorded at three locations on the target plate. As an example, Figure 4 shows a received waveform on the impacted specimen (carbon steel). The backscattering intensity depending on frequency is evaluated by a non-dimensional value,

$$N = \int_{f_1}^{f_2} B(f) df \bigg/ \int_{f_1}^{f_2} U(f) df \tag{3}$$

,where $B(f)$ and $U(f)$ are the amplitude spectrum of the backscattering wave and surface echo, and f_1 and f_2 are lower and upper limits of frequency range, namely 3 and 14 MHz. After the final ultrasonic measurement, the target plate was sectioned and the damage was observed by an optical microscope.

RESULTS

Simulation of Stress Wave Propagation

Impact speed of the flyer plate was assumed to be V=113m/s (σ =2.6GPa) as an initial condition. Flyer and target plates are discretized by 135 elements in radial direction and 30, 60 elements in axial direction respectively.

Figure 5 shows the C-scan image of spall damage within the target plate of medium carbon steel subjected to impact velocity 113m/s. The initial spall damage is nucleated near the circumference of the specimen, not at the center. The dimensionless principal stress distribution is shown in Figure 6. The stress have a maximum value in the position near the circumference of specimen by the superposition of the axial and radial stress waves. Therefore, the initial spall damage is generated near the circumference.

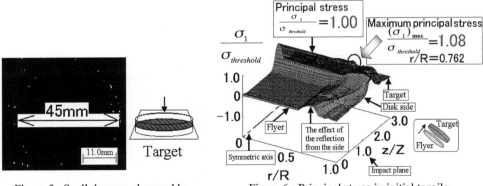

Figure 5 : Spall damage observed by acoustic microscope

Figure 6 : Principal stress in initial tensile stress generation

Nondestructive Evaluation of Spall Damage under Pepeated Impacts

For aluminum, the damage density after the second impact is much greater than that in the first, when the second impact stress is higher than the first. In contrast, when the second impact stress is lower than the first, the ductile voids nucleated by the first impact are squashed and voids surfaces are bonded by a kind of explosive welding. For carbon steel, however, the spall damage generated by the first impact have increased in its density and sizes by the second impact irrespective of higher or lower stress than the first impact stress. Small cracks once generated have been extended by lower impact stress in successive impacts (see Nishimura et al. [1998]).

The C-scan images of the spall damage of medium carbon steel are shown in Figure 7. These two images are similar, where the second impact stress is much lower than the threshold spall stress. To confirm whether the cracks grow or not, the other more sensitive measurement is required.

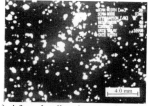   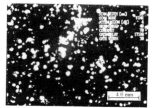

(a) After the first impact (3.0GPa)   (b) After the second impact (1.8GPa)
Figure 7 : Change in cracks distribution under repeated impacts (Carbon steel, No.S18, depth 3.5mm)

When a solid includes a lot of voids or cracks, the wave velocity decrease, but the attenuation and the backscattering intensity increase due to the ultrasonic wave scattering at voids or cracks. Thus the spall damage extent is correlated with ultrasonic wave velocity, attenuation (amplitude ratio, B2/B1) and backscattering intensity. These ultrasonic variables which measured on medium carbon steel are shown in Figure 8 for the first impact stress. When the impact stress is lower than the spall threshold stress (approximately 2.6GPa), the wave velocity is equal to that of non-impacted plates, and the amplitude ratio and the backscattering intensity are also unchanged. It is clear that the higher the impact stress, the lower the velocity and the amplitude ratio. The backscattering intensity increases with the impact stress. Thus, these ultrasonic variables characterize the spall damage quantitatively.

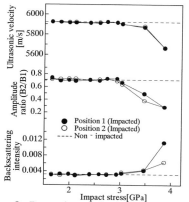   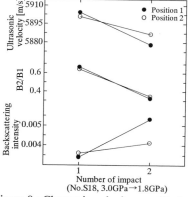

Figure 8 : Dependence of velocity, amplitude ratio, backscattering intensity on impact stress

Figure 9 : Change in velocity, amplitude ratio, backscattering intensity under repeated impacts (No.S18, 3.0GPa→1.8GPa)

These ultrasonic variables were measured for the sample of C-scan images shown in Figure 7. In this sample, the change in spall damage after the second impact was not detected by acoustic images. The velocity, amplitude ratio and backscattering intensity after second impact are shown in Figure 9. The change of these variables clearly shows the growth of the spall damage under repeated impacts. Thus, these ultrasonic methods are useful to evaluate nondestructively the spall damage growth under repeated impact.

Measurement of Stress Wave

Stress history on mild carbon steel measured by a PVDF pressure gauge under repeated impacts are shown in Figure 10. The points A and B represents the arrival of the elastic and plastic wave from the impact plane. The elastic release wave from the back surface of the flyer plate arrives at point C. When the spall damage occurs inside the target, the stresses is reduced to zero and the unloading wave is generated. This wave arrives at point D. The signal beyond point D is called as spall signal. In the stress history after the second impact (dashed line) in Figure 10, the elastic wave velocity lowers, namely rise time after the arrival of elastic wave is long. This is because of lowering of the elastic limit of the specimen and attenuation of the stress wave by the spall damage which generated by the first impact. Also, spall signal (D) arrives earlier than that in the first impact. It confirmed that the spall damage has been increased by repeated impacts from the acoustic images shown in Figure 11. Therefore, the arrival of the spall signal in the early time is due to the spall damage growth in the back surface side.

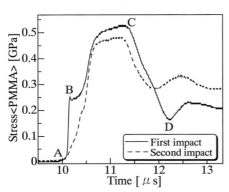
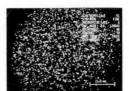

After the first impact (3.7GPa)

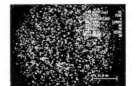

After the second impact (4.4GPa)

Figure 10 : Stress history measured by PVDF gauge under repeated impacts (No.S46)

Figure 11 : Change in spall damage under repeated impacts (No.S46, depth 3.0mm)

CONCLUSIONS

The stress wave propagation in the target plate was simulated by using the distinct element method. The change in the spall damage of commercially pure aluminum and medium carbon steel under repeated impacts were nondestructively evaluated with ultrasonic C- and B-scan images as well as ultrasonic velocity, attenuation and backscattering intensity. For commercially pure aluminum, impact bonding of void surfaces were confirmed when the second impact stress was lower than the first. On the contrary, the cracks in medium carbon steel extend under the same condition just mentioned. The evolution or repair of the spall damage under repeated impacts was able to be monitored well by ultrasonic methods. In the stress history under repeated impacts, the spall signal arrived in the early time by the damage growth in the back surface side.

References

Davison L. and Graham R. A. (1979). Shock Compression of Solids. *Physical Report* **55**, 255-379.
Curran D. R., Seaman L. and Shockey D. A. (1987). Dynamic Failure of Solids. *Physical Report* **147**, 253-388.
Wolf J. P., Bucher K. M. and Skrikerud P. E. (1978). Response of equipment to aircraft impact.
 Nuclear Engineering and Design **47**,169-193.
Fukatsu K. and Hori T. (1993). Analysis of Fracturing Generated by High-Velocity Impact of Dry Ice
 Cylindrical Flier. *Proc. Plasticity and Impact Mechanics*, 440-451.
Kawashima K. and Fujii I. (1995). Digital Measurement of Ultrasonic Velocity.
 Review of Progress in QNDE **14**, 203-209.
Nishimura N., Kawashima K. and Nakayama O. (1998). Ultrasonic Evaluation of Spall Damage Accumulation
 of Aluminum and Steel under Repeated Plate Impact Tests. *Proc. of Int. Sympo. on Strength Theory*, 133−138.

IMPACT BEHAVIOR OF MERCURY DROPLET

Hidefumi Date[1], Masatoshi Futakawa[2] and Shuichi Ishikura[2]

[1] Department of Mechanical Engineering, Tohoku Gakuin University,
1-13-1, Chuo, Tagajo 985-8537, Japan
[2] Japan Atomic Energy Research Institute,
Tokai-mura, Naka-gun, Ibaraki-ken 319-1195, Japan

ABSTRACT

In order to examine the impact behavior of mercury, which is an important key-issue in a facility for high intensity neutron sources, the falling and colliding profiles of mercury droplets were recorded by high-speed video recorder. The impact force was also measured using an elastic bar glued to a strain gage. The falling mercury droplet oscillated and the profile changed from a longitudinal ellipsoid to a transverse one, repeatedly. The regathering and jumping of mercury at the collision point on the impact face of the target were observed after impact because of the strong surface tension of mercury. The impact force of mercury was in proportion to the impact velocities and the square root of the potential energy as a solid steel ball. The non-dimensional-duration-time K that obtained experimentally is independent of the impact velocity and the size of the droplet.

KEYWORDS

Mercury droplet, Impact behavior, Surface tension, Erosion

INTRODUCTION

It has been deduced that the quantity of supply of neutrons will decrease to a quarter of current level by the beginning of the 21st century. Therefore, it is planned to establish the facility for a high intensity neutron source which can produce lots of neutrons by the injection of protons against a heavy liquid metal like mercury. The proton beam is injected pulsatively into a container made of stainless steel filled with mercury. The duration of the pulse will be about 1 µs. The pulse of the injected protons induces to heat locally up in the mercury and causes a rapid temperature elevation. The pressure waves are generated due to the thermal expansion in it, and impose dynamically to the container (Futakawa et al [2000a]). A negative pressure might be caused at the interface between the solid container and the liquid mercury by the pulsative pressure waves. The cavitations could be induced by the negative pressure, and then the following cavitation collapse forms cavities on the inner wall of the container (Futakawa et al [2000b]). The safety of the facility depends on the structural integrity of the container. It is important, therefore, to examine the impact behavior of mercury from the viewpoint of the structural integrity.

Here, the impact behavior of a mercury droplet was examined by the free fall method to clarify the erosion of material due to the liquid metal. The remarkable behavior of a mercury droplet falling, colliding and spreading were recorded with the high-speed video recorder and it was clarified that the impact force is estimated uniformly from potential energy and the impact erosion stress is deduced the equation derived using the non-dimensional impact duration time.

EXPERIMENTAL METHOD

The experimental apparatus for the free falling of a mercury droplet is shown in Fig. 1. The falling test was carried out in PMMA pipe with a thickness of 3 mm and an inner diameter of 100 mm. The pipe was held vertically on a thick PMMA plate using three screw pins. The elastic bar, made of aluminum alloy for measuring the impact load, was supported vertically at the center of the pipe. The total length and diameter of the elastic bar were 1000 mm and 10 mm, respectively. The pipette for keeping the falling mass of the mercury droplet constant was supported vertically at the center of the pipe.

The impact force was measured by a strain gage that was glued at a distance of 300 mm from the impact end. The output of the strain gage was amplified and recorded by a storage oscilloscope. The stainless steel target, with a diameter of 30 mm, was also used for the observation of the colliding profile. In this experimental condition, the tip of pipette with the inner diameter of 4 mm is the most suitable to form enough size of droplet to make observation of it. The mass of the mercury droplets obtained using the pipette was 0.37 ± 0.02 g. The falling and colliding profiles of the mercury droplets were recorded by high-speed video recorder. A recording rate was 2000 ~ 4000 frames /s. The mercury droplets were dropped from a height of 70 to 900 mm, and the equivalent impact velocities were 1.2 to 4.2 m/s. Three types of steel balls (S,M,L) with diameters of 2.4, 4.8 and 11.0 mm were also used to compare with the results of the mercury droplet.

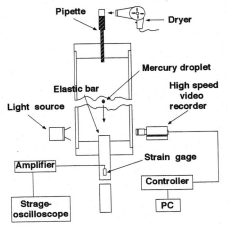

Fig.1   Experimental apparatus

PROFILES OF FALLING AND COLLIDING

The profile of the mercury droplet leaving the pipette was spindle shaped like a water droplet. The droplet begins to oscillate after leaving because of the reaction against the cut off of the spindle profile of the mercury droplet. The period of oscillation was about 40ms. The profile of the falling droplet changed from a longitudinal to a transverse ellipsoid.

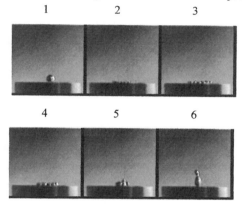

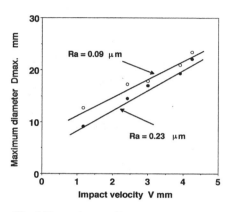

Fig. 2 Collision of a droplet
Recording time = 28 ms

Fig. 3 Dependence of impact velocity on maximum diameter after colliding

The collision at an impact velocity of 1.2 m/s are shown in Fig. 2. The droplet with an almost sphere profile before the collision, as illustrated in the frame1, changed into a hemispheric one during the collision. Then, the mercury droplet spreads to take the form of a circular flat plate. The diameter of circular plate got to be the largest around 2 ms after collision. It has been reported that the impact duration time of a melting metal droplet, including solidification, is about 2 ms (Hirata et al [1997]). Accordingly, it was deduced that the collision of a droplet of liquid metal finished within 2 ms regardless of solidification. The maximum diameter of the spreading plate of mercury plotted against the impact velocities are given in Fig. 3. The maximum diameter increases with an increase in the impact velocity, but the diameter depends on the surface roughness of the target face described later. Finally, after spreading to the maximum diameter, mercury regathered at the collision point due to the strong surface tension and jumped up and into the air as shown in the frame of 5 and 6 of Fig. 2. The surface roughness of the target face was 0.09 µm. When the inertia force in mercury produced by spreading behavior is larger than the surface tension of the mercury, the mercury around the circumference of the circular plate scattered as the particles. The behavior after spreading on the target with a surface roughness of 0.23 µm is shown in Fig. 4. The impact velocity was 4.43 m/s. The thickness of the spreading mercury plate became thinner with the increase of the impact velocity. When the thickness was comparable to the maximum surface roughness of the target face, the mercury plate was rent at the point with the maximum surface roughness. In the case as shown in Fig.4, a mercury ring like a doughnut was formed on the target. The number of the rending points increased with the increasing surface roughness and the mercury circular plate contracted violently because of the strong surface tension.

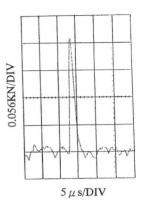

Fig. 4 Spreading behavior on the target with a surface roughness of 0.23 μm
Recording time = 9 ms

Fig. 5 Typical impact force at an impact velocity of 3.6 m/s

IMPACT FORCE AND DURATION TIME

The typical impact force at an impact velocity of 3.6 m/s is shown in Fig. 5. The maximum force was about 250 N and the impact duration time was about 3~5 μs. It was not easy for all the droplets of mercury and steel balls to collide with the center of the target face because the diameter of the target face against the falling distance was very small. The impact forces plotted against impact velocity are shown in Fig. 6. The impact velocity was calculated from the falling distance using the equation of free falling. The data on the steel ball hardly scattered. However, the data on the mercury droplets scattered relatively. Though the colliding position of the steel ball on the target face could be confirmed by visual observation, the position of the mercury droplet was difficult to be confirmed, because the mercury droplet deformed largely. Finally the only results of the steel balls that collided at the center of the elastic bar and all of the results of mercury are plotted in Fig. 6. Since the scattering of the mass of mercury is small and the falling distance is constant, it was deduced that the data on the scattering of mercury depends on the collision position on the impact face of the elastic bar. Then, the maximum value at every falling height was deduced to be the accurate impact force in the experiment.

The accuracy of the impact force of the steel ball shown in Fig.6 is confirmed by the following equation (Okada et al [1995]);

$$F = m(1+e)v_0/t \qquad (1)$$

where m is the mass of the steel ball, e the coefficient of restitution, v_0 the impact velocity and t the duration of impact. Since there was good agreement between the calculated and experimental results of the steel ball, it proved that the impact force was measured accurately using the elastic bar. The maximum force of mercury at the various impact velocities as shown in Fig. 6 is approximately expressed by a linear equation. However, the gradient of mercury is much smaller than that of a steel ball because the response of the mercury droplet is not elastic. The square root of the potential energy plotted against the impact force is shown in Fig. 7. In the case of mercury, the few experimental results that are close to the maximum impact force in Fig. 6 were potted in Fig. 7. The relation between the square root of the potential energy and the impact force is also approximately expressed by a linear equation, regardless of materials.

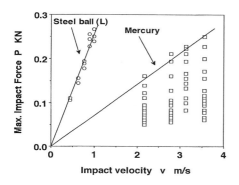
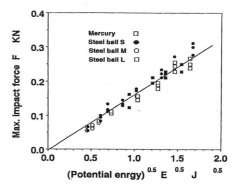

Fig. 6 Impact force plotted against impact velocity

Fig. 7 Potential energy plotted against impact velocity

Huang et al. (1973) numerically analyzed the impact behavior of a spherical liquid droplet. They gave the following relationship about the time required to nearly reach the steady-state stagnation force;

$$K = Ct/D = 1.5, \qquad (2)$$

where c is the sound velocity of liquid, D the diameter of the liquid and K the non-dimensional time. A time of about 4 μs is calculated by Eqn. (2) for the mercury droplet. The experimental results of the duration time for the mercury droplet, as measured by the strain gage, almost agreed with the numerical ones. The Ks calculated by Eqn. 2 are plotted against the impact velocities in Fig. 8. The Ks for steel or mercury appears hardly dependent on the impact velocity and the size of steel balls or droplet.

The cavitations could be induced by the negative pressure, and then the following cavitation collapse forms cavities on the inner wall of the container (Futakawa et al [2000b]) as described before. If it is deduced that the impact force indicated by Eqn. (1) causes the cavitations in the mercury droplet to collapse and the collapse forms cavities on the inner wall of the container, the approximate force F_e due to the impact force is given using Eqn. (1) and (2) as follows.

$$F_e = \frac{\pi}{6} D^2 \rho \left(\frac{1+e}{K}\right) v_0 C. \qquad (3)$$

Since the Ks are almost constant regardless of the impact velocity and the size of the droplet as shown in Fig. 8, and the e is determined by the combination of the droplet and target, the following approximate erosion stress σ_e due to the impact force is derived easily from Eqn.(3),

$$\sigma_e = \frac{2}{3}\rho\left(\frac{1+e}{K}\right)v_0 C. \qquad (4)$$

The cross-sectional area of the mercury droplet as the contact area is used to obtain the erosion stress indicated by Eqn. (4).The K of the mercury droplet is about 2 as shown in Fig. 8. The e estimated roughly from Fig. 2 is about 0.01. Since the yield stress of the container made of stainless steel is about 200 MPa, the critical impact velocity for the erosion obtained by Eqn. (4) is about 3 m/s. It is important to pay an attention to the erosion on the inner wall of the target container because the collision of the mercury droplet at an impact velocity around 3 m/s is possible to occur at the inner wall /mercury interface.

CONCLUSIONS

The mercury droplet with a diameter of 4 mm collided on the target face for measuring the impact force and observing the spreading profile. The following results were obtained.
1) The impact behaviors of mercury droplets were influenced significantly by the surface tension.
2) The relationship of the square root of the potential energy and the impact velocity of the mercury droplet is linear, as with the steel balls.
3) The non-dimensional–duration-time K is independent of the impact velocity and the size of droplet. The stress due to droplet impact is predictable by the equation derived using the K value if the impact velocity is known.

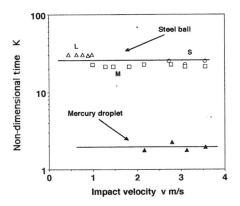

Fig.8 Non-dimensional duration time

REFERENCES

Futakawa, M., Kikuchi, K., Conrad, H., and Stechemesser, H.., (2000a), "Pressure and stress waves in a spallation neutron source mercury target generated by high-power proton pulses," Nuc. Instruments and Methods in Physics Research, A 439,1-7.

Futakawa, M., H. Kogawa and R. Hino, (2000b), "Measurement of dynamic response of liquid metal subjected to uniaxial strain wave," J. Phys. IV France 10, Prq-237-242.

Hirata, Y., Matsumoto, Y., Maruo, H., Ozawa, T. and Ohji, T., (1997), " Deformation and Solidification Process of a Molten Droplet," J. Soc. High Temp., 25, 222-229.

Huang, T., Hammitt, F. G. and Yang, W-J. (1973), " Hydrodynamic Phenomena During High-Speed Collision Between Liquid Drop and Rigid Plane," ASME J. of Fluids Engineering, 95, 276-292.

Okada, T., Iwai, Y., Hattori, S. and Tanimura, N. (1995), " Relation between impact load and the damage produced by cavitation bubble collapse," Wear, 184, 231-239.

STRESS-STRAIN RESPONSE OF S15C CARBON STEEL FRICTION WELDED BUTT JOINTS UNDER IMPACT TENSILE LOADING

T. Yokoyama

Department of Mechanical Engineering, Okayama University of Science
Okayama 700-0005, Japan

ABSTRACT

Stress-strain response of S15C carbon steel friction welded butt joints under impact tensile loading is determined using the split Hopkinson bar. Round tensile specimens machined from as-welded butt joints of 13-mm diameter are used in both quasi-static and impact tension tests. Friction welding is carried out using a brake type friction welding machine under fixed welding conditions. The effect of strain rate up to $\dot{\varepsilon}$ =700/s on the tensile stress-strain characteristics of the joint specimens and a base material (S15C carbon steel) is studied. It is shown that the flow stress for friction welded butt joints is always higher than that for the base material at low and high rates of strain, but the former are less sensitive to strain rate than the latter. Microhardness measurements are conducted to examine the extent of the heat-affected zone (HAZ) across the weld interface. The enhanced strength of joints is found to be attributed to the presence of a locally-hardened region within the HAZ.

KEYWORDS

Impact tension test, S15C carbon steel, Friction-welded butt joint, Hopkinson bar, Stress-strain characteristics, Strain rate, Microhardness

INTRODUCTION

Friction welding has been widely used as a solid phase welding technique to bond similar or dissimilar metals. Using friction welding, it is possible to reduce the number of processes, the amount of materials used and the welding time, and hence improve the operational efficiency. Therefore, wider applications of this technique can be expected if the strength of the friction welded butt joints are fully assured. Most mechanical tests (such as tension, torsion and three-point-bend tests) for determining strength properties have been performed under quasi-static loading. In general, friction welded butt joints in carbon steel are known to have higher tensile and fatigue strengths than the base material, except for a relatively low Charpy impact value (Shioya, Yamada and Kuzuya, 1965) . However, basic stress-strain characteristics of friction welded butt joints in carbon steel under impact tensile loading have not been well investigated yet, due to the experimental difficulties involved.

The present paper is concerned with the determination of the tensile stress-strain response of S15C carbon steel friction welded butt joints at high rates of strain. Impact tension tests on joint speicmens and a S15C base material were conducted using the split Hopkinson bar. The effect of strain rate on the tensile stress-strain response of the joint specimens and the base material was investigated. Microhardness tests were performed to examine the heat-affected zone near the weld interface of joints.

PREPARATION OF FRICTION-WELDED BUTT JOINTS

The base material used was a 13 mm diameter bar of S15C low carbon steel. The chemical composition and nominal tensile properties of the base material are given in Tables 1 and 2. Joints were produced using a brake type friction welding machine under fixed welding conditions as shown in Figure 1. The influence of the welding conditions on the strength and quality of mild steel friction welded butt joints was discussed by Duffin and Bahrani (1973). Joints and the base material were machined to the specimen geometry as depicted in Figure 2. Note that the weld interface is located in the middle of the gage length of joint specimens.

TABLE 1
CHEMICAL COMPOSITION OF BASE MATERIAL

Material	Chemical composition (mass %)								
	C	Si	Mn	P	S	Cu	Ni	Cr	Fe
JIS S15C	0.15	0.27	0.44	0.009	0.013	0.02	0.03	0.07	bal.

TABLE 2
MECHANICAL PROPERTIES OF BASE MATERIAL

Material	Upper yield point σ_{uy} (MPa)	Tensile strength σ_B (MPa)	Elongation δ (%)	Hardness HV (0.3)
JIS S15C	350	512	39	140

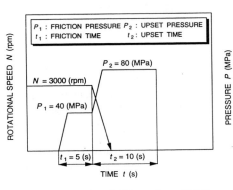

Figure 1: Schematic illustration of friction welding conditions

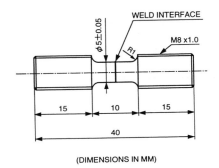

Figure 2: Form and dimensions of tensile specimen

TEST PROCEDURE

Tensile Split Hopkinson Bar

A schematic diagram of the tensile split Hopkinson bar apparatus is given in Figure 3. The apparatus consists principally of a striker tube, two Hopkinson bars (input and output bars) and associated recording system. The striker tube (650 mm long) made of steel has an inside diameter of 16.2 mm and an outside diameter of 22.5 mm. The specimen is tightly attached to the two Hopkinson bars through a screw connection (see the inset in Figure 3). When the loading block is impacted with the striker tube fired through the barrel by compressed air released from the pressure tank, a compressive strain pulse is generated and reflected at its free end as a tensile strain pulse, which is propagated along the input bar to the specimen. When the tensile strain pulse arrives at the specimen, part of the pulse is reflected into the input bar at the bar/specimen interface because of the impedance mismatch, and part of the pulse is transmitted through the specimen to the output bar. The strain pulses are measured by two sets of strain gages with a gage length of 2 mm mounted on the Hopkinson bars. The strain gage signals are recorded using a 10-bit digital storage oscilloscope, where the signals are digitized and stored at a sampling rate of 1μs/word. The digitized data are then transferred to a 32-bit microcomputer for data processing.

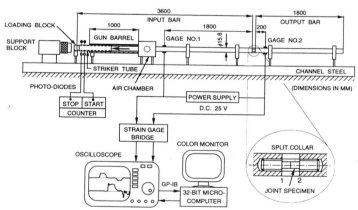

Figure 3: Schematic of tensile split Hopkinson bar apparatus and recording system

Method of Data Reduction

The theory and analysis of the tensile split Hopkinson bar test is identical to that of the compressive one (e.g. Lindholm, 1964), except for the change in sign of the strain pulses. By application of elementary one-dimensional elastic wave theory (e.g. Graff, 1975), the average nominal tensile strain, strain rate and stress in the specimen are determined from the Hopkinson bar records as

$$\varepsilon(t) = \frac{c_0}{l_0} \int_0^t \{\varepsilon_i(t') - \varepsilon_r(t') - \varepsilon_t(t')\} dt' \qquad (1)$$

$$\dot{\varepsilon}(t) = \frac{c_0}{l_0} \{\varepsilon_i(t) - \varepsilon_r(t) - \varepsilon_t(t)\} \qquad (2)$$

$$\sigma(t) = \frac{AE}{2A_s} \{\varepsilon_i(t) + \varepsilon_r(t) + \varepsilon_t(t)\} \qquad (3)$$

Here A is the cross-sectional area, E is Young's modulus, c_o is the longitudinal elastic wave velocity in the Hopkinson bars; l_o and A_s are the initial gage length and the cross-sectional area of the specimen, ε_i, ε_r and ε_t are the incident, reflected and transmitted strain pulses, t is the time from the start of the pulse. When the forces are equal at both ends of the specimen, the average tensile stress in the specimen reduces to

$$\sigma(t) = \frac{A E}{A_s} \varepsilon_t(t) \qquad (4)$$

Eliminating time t yields the stress and strain rate vs. strain relations for the specimen.

RESULTS AND DISCUSSION

Quasi-Static Stress-Strain Response

Tension tests at quasi-static and intermediate rates were conducted on the same designs of specimen (see Figure 2) in an Instron testing machine at two different crosshead velocities of 1mm/min and 100 mm/min. The nominal tensile properties of the joint specimens are listed in Table 3. The joint efficiency η (= ratio of the tensile strengths of the joint specimen and the base material) was estimated to be 117 %.

TABLE 3
TENSILE PROPERTIES OF FRICTION WELDED BUTT JOINTS

Material	Upper yield point σ_{uy} (MPa)	Tensile strength σ_B (MPa)	Elongation δ (%)
S15C friction welded butt joint	445	598	27

Impact Stress-Strain Response

A number of impact tension tests on joint specimens and the base material were performed at room temperature. Figure 4 shows typical oscilloscope records from the impact tension test on the joint specimen. The upper trace from gage No. 1 gives the incident and reflected strain pulses, and the lower trace from gage No. 2 gives the strain pulse transmitted through the specimen. The resulting dynamic stress-strain curve together with the stress-strain curves at quasi-static and intermediate rates are shown in Figure 5. The specimen exhibits a very sharp upper yield point at high strain

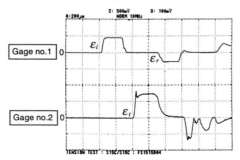

Sweep rate: 200μs/div
Vertical sensitivity
 Upper trace: 500mV/div
  (1176 με/div)
 Lower trace: 100mV/div
  (233 με/div)

Figure 4: Oscilloscope records from the impact tension test on joint specimen

rates, and the drop in the flow stress on the dynamic stress-strain curve corresponds to the unloading process, but does not indicate the onset of necking in the joint specimen. Figure 6 presents typical tensile stress-strain curves for the base material at different strain rates. As in the joint speimen, the base material shows a very high upper yield point at high strain rates. Similar results on mild steel from impact tension tests were reported by Langseth, Lindholm, Larsen and Lian (1991). Note that the flow stress level for joint specimens is always higher than that for the base material at strain rates of the same order.

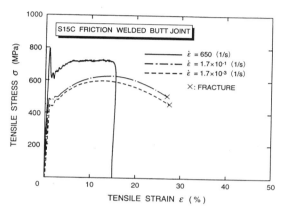

Figure 5: Tensile stress-strain curves for joint specimens at different strain rates.

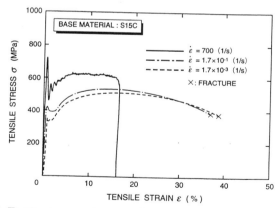

Figure 6: Tensile stress-strain curves for base material at different strain rates.

Microhardness Tests

In order to discuss the enhanced strength of joints, microhardness tests were carried out across the weld interface of an original joint. Figure 6 shows the microhardness distributions across the weld interface of the joint. It is observed that the higher strength of joints arises from the presence of a thermally hardened region close to the weld interface. In addition, it is found that the heat-affected zone (HAZ) due to friction welding is confined to a region about 6mm wide across the weld interface.

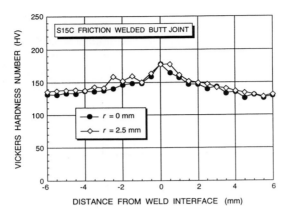

Figure 7: Microhardness distributions across weld interface of joint

CONCLUSIONS

Tensile stress-strain characteristics of S15C carbon steel friction welded butt joints and the base material at low and high rates of strain have been investigated. From the experimental results, it can be concluded that:
(1) The flow stress for joints is always higher than that for the base material due to a microstructural change near the weld interface, causing an increase in the hardness within the heat-affected zone.
(2) Both the joints and the base material exhibit a very sharp upper yield point at high rates of strain.
(3) The joints show higher strain rate-sensitivity than the base material.
(4) Microhardness tests indicate that the heat-affected zone is limited to a region about 6mm wide across the weld interface.

References

Duffin, F.D. and Bahrani, A.S. (1973). Friction Welding of Mild Steel: The Effects of Varying the Value of Deceleration. *Metal Constr. Brit. Weld. J.* **5:4**, 125-132.
Duffin, F.D. and Bahrani, A.S. (1973). Frictional Behaviour of Mild Steel in Friction Welding. *Wear,* **26**, 53-74.
Graff, K. F. (1975). *Wave Motion in Elastic Solids.* Oxford University Press, UK
Langseth, M., Lindholm, U.S., Larsen, P.K. and Lian, B. (1991). Strain-Rate Sensitivity of Mild Steel Grade St52-3N. *J. Eng. Mech.* **117:4**, 719-732.
Lindholm, U.S. (1964). Some Experiments with the Split Hopkinson Bar. *J. Mech. Phys. Solids,* **12**, 317-335.
Shioya, T., Yamada, S. and Kuzuya, Y. (1965). Study on Mechanical Properties of Friction Welded S20C and S45C. *Journal of Japanese Welding Society,* **34:11**, 1197-1203.

EVALUATION OF IMPACT TENSILE STRENGTH FOR PMMA/Al BUTT ADHESIVE JOINTS

H. Wada[1], K. Suzuki[1], K. Murase[2] and T. C. Kennedy[3]

[1] Department of Mechanical Engineering, Daido Institute of Technology,
10-3 Takiharu-cho, Minami-ku, Nagoya, 457-8530, Japan
[2] Department of Transportation Eng., Meijo University,
1-501 Shiogamaguchi, Tenpaku-ku, Nagoya, 468-8502, Japan
[3] Department of Mechanical Engineering, Oregon State University,
Corvallis, OR 9733 1-5001, USA

ABSTRACT

The impact tensile strength was evaluated for a butt adhesive joint on Al alloy and PMMA round bars. The materials were bonded together by cold-setting epoxy, commercial base adhesive. An impact tension test was carried out using an air-gun with a special jig. A time-history of the load applied to the specimen was measured by an electrical resistance strain gage. The fracture initiation time in the adhesive interface was determined from the measured strain gage signal. The time-history of the measured load was used as an external force for a simulation of the dynamic stress by the finite element method. The impact tensile strength of the adhesive joint was evaluated by the stress singularity field parameter method. Fracture toughness in the dynamic stress field was determined from the stress distribution in the vicinity of the edge of the adhesive interface calculated by the finite element method at the fracture initiation time. The static fracture toughness was also determined by a similar method using the same type of specimen used in the dynamic test. The fracture toughness measured by this method was displayed in a Weibull distribution. It was found that the dynamic strengths exhibited considerably larger values than the static strengths.

KEYWORDS

Impact tensile strength, Butt adhesive joints, Dissimilar adhesively joints, Loading rate, Stress singularity, Strain gage, Finite element method, Stress wave, Fracture

INTRODUCTION

Adhesive joints have been widely used in mechanical structures for the purpose of lightening, cost reduction and the ease joining dissimilar materials. However, the variability of strength is larger for adhesive joints. Hence, the establishment of a fracture criterion for an adhesive interface is required for design. So far, a number of studies (Sawa et al. (1995), Tong (1998), Hattori et al. (1988), Sato et al.

(1997), Yokoyama and Shimizu (1997)) on stress analysis and strength evaluation of an adhesive interface have been carried out mainly for the static problem. In particular, a stress singularity exists in the vicinity of the edge in adhesive joints between dissimilar materials. Therefore, a strength evaluation method based on the stress singularity field parameter (Hattori et al. (1988)) for the adhesive joint between dissimilar materials has been proposed for the static problem. On the other hand, the establishment of an impact strength evaluation method for an adhesive joint is needed for the case of severe loading of the structure. In this study, the impact tensile strengths are investigated for dissimilar butt adhesive shaft joints using fracture toughness. The interface between poly-methyl-methacrylate (PMMA) and Al alloy was analyzed to study the joint strength using a finite element analysis at the fracture initiation time. An impact tensile fracture test was carried out for the joint specimens using a special jig with an air gun. The results of impact strengths were compared with static results.

EXPERIMENTAL METHOD

Materials and Specimen

The joint specimen configuration used in the present impact fracture experiment is a solid cylindrical bar of 20 mm diameter and 600 mm total length with 300 mm length for each material as shown in Figure 1(a). Figure 1(b) shows a joint specimen for a static test. The total length of the joint specimen is 200 mm as shown in this figure. The ends of specimen are threaded for applying a tensile load.

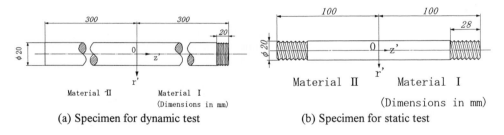

(a) Specimen for dynamic test   (b) Specimen for static test

Figure 1: Specimen configuration and dimensions

TABLE 1
MECHANICAL PROPERTIES OF MATERIALS

	Young's modulus E(GPa)	Density ρ(kg/m^3)	Poisson's ratio ν
Material I (PMMA)	3.50 (static) 4.87 (dynamic)	1.18×10^3	0.388
Material II (Al)	71.78 (static and dynamic)	2.70×10^3	0.331

Figure 2: Photomicrograph of adhesive layer

PMMA was used for material I and Al alloy for material II. The end of material I was threaded in order to apply the impact tensile load. The specimen without a clamp breaks at the adhesive interface from its own inertial force. The physical properties of both adherents are shown in TABLE 1. Planes to adhere the two materials were polished in random directions by sandpaper. The surface roughness of the plane was measured with a surface roughness meter. As a result, the center-line average roughness was Ra=0.12 µm for the PMMA material and Ra=0.14 µm for the Al alloy. The two materials were bonded together using a special jig developed in this study in order to avoid slippage of the butt and non-uniformity of the thickness of the adhesive layer. The materials were bonded together by a cold-setting epoxy type adhesive (Epoxy resin 100%, Polyamide resin 100%). The specimen was used in the experiment after allowing the adhesive to set for 24 hours. Figure 2 shows a photomicrograph around the adhesive interface of the specimen. We can see the thickness, about 0.1 mm, and uniformity of the thickness of the adhesive layer from this figure. Strain gages were bonded at the z'=200 mm position for measuring the time-history of the incident applied load and at the z'=1 mm position for measuring a fracture initiation time, respectively.

Impact loading equipment

Figure 3 shows an impact tension jig with supports for the dissimilar round bar joint specimen. This jig is used to convert the compressive load from an air-gun into a tensile load. Namely, a PMMA pipe provides the impact compressive load from the incident bar as shown in Figure 3, and the impact tensile load is applied to the specimen through a nut fitting on the threaded PMMA side end of the specimen. The specimen fractures at the end of the adhesive interface without any support in the direction of loading, while a collar at the end of Al side is used to hold the vertical position of specimen.

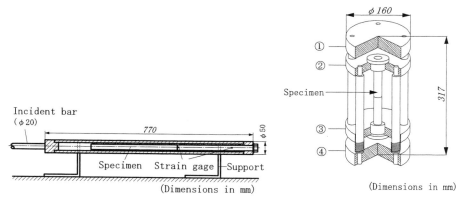

Figure 3:Impact tension jig   Figure 4:Static tension jig

Static tension jig

Figure 4 shows a tension jig for a static experiment. It is constructed using 4 steel disks (ϕ 160 x 30) and 8 steel pins (ϕ 20 x 260), as shown in the figure. Steel disks ① and ③, and ② and ④ of the jig are respectively moved together by the pins. The specimen is installed in the center hole of steel disks ② and ③. Disks ① and ③ move downward as load is applied to the disk ① using the universal testing machine, and the specimen installed between disks ② and ④ is pulled.

FINITE ELEMENT STRESS ANALYSIS

Analysis of stress singularity field

Static and dynamic stresses in the adhesive joint between the dissimilar materials used in this experiment were analyzed using the general-purpose program ANSYS5.6. A second-order quadrilateral element with 8 nodes was used in the calculation, and the stress analysis was carried out assuming axisymmetric conditions. As an example, Figure 5 shows the finite element mesh pattern for the static specimen. One half of specimen was divided into 1673 elements and 5290 nodal points. The smallest element size was e=0.002R in order to analyze the stress distribution in the vicinity of the edge of interface accurately.

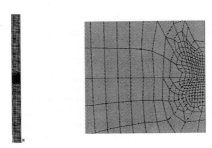

Figure 5:Meshing condition of joint for FEM analysis

Figure 6 shows an example of the static stress distribution at the interface for a fracture load. The figure includes the results of the FEM analysis and calculation results for plane strain conditions by Bogy's equation. We can confirm the accuracy of the FEM results from the figure. Figure 7 shows also an example of the dynamic stress distribution at the interface. The stress and stress singularity show the same tendency as the static results.

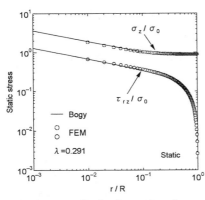
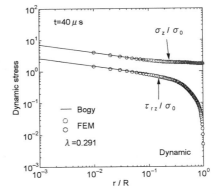

Figure 6:Static stress distribution on interface   Figure 7:Dynamic stress distribution on interface

Determination of stress singularity field parameter

The static stress distribution in the vicinity of the edge of an interface between dissimilar materials has been determined by Bogy (1971). Namely, the stress distribution is given in equation (1),

$$\sigma(r) = K/r^\lambda \tag{1}$$

where $\sigma(r)$ is the stress component, K is the intensity of the stress singularity field, and λ (Dunders

(1967)) is the exponent in the singular term which is determined by a combination of the properties of the dissimilar materials, and r is the distance along the interface from the singular point. A convenient technique (Wada et al. (2001)) shown next was used to determine the intensity K of the singular stress field. Namely, it is possible to obtain the theoretical resultant force in a small region (a~c, a<b<c) in the vicinity of the singular point from equation (1). On the other hand, the stress in the 3 nodes (a,b,c) of the 2 elements in the vicinity of the interface edge gives the resultant force from the finite element analysis. The resultant force from equation (1) becomes equal to the resultant force from the FEM analysis, if the stress in the vicinity of the edge of the adhesive interface is obtained accurately by the finite element method. Therefore, we obtained equation (2).

$$K = \frac{(\sigma_a + \sigma_b)(D-a-b)e_1 + (\sigma_b + \sigma_c)(D-b-c)e_2}{4\left\{\frac{D(c^{1-\lambda}-a^{1-\lambda})}{2(1-\lambda)} - \frac{(c^{2-\lambda}-a^{2-\lambda})}{(2-\lambda)}\right\}} \quad (2)$$

In equation (2), D is the diameter of the specimen bar, and e_1=(b-a) and e_2=(c-b)

RESULTS AND DISCUSSIONS

FEM simulation of dynamic stress

Figure 8 shows measured results of the time-history of stress in the impact fracture test and the finite element simulation results for the dynamic test specimen shown in Figure 1(a). In this figure, the title of Input stress indicates an incident stress wave in the finite element simulation. The title of G1 indicates the time-histories of stress at the z'=200 mm point and G2 indicates the time-histories of stress at the z'=1 mm point. The results measured with the strain gage and the results calculated by FEM are in excellent agreement as shown in Figure 8. Strain gage signal G1 exhibits a rapid decrease by the release of the stress when the fracture initiates at the end of the interface. Therefore, we obtain the fracture time t_f from this gage signal. We also obtain the stress distribution at the adhesive interface at the fracture initiation time with the finite element method.

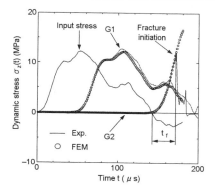

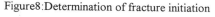

Figure8:Determination of fracture initiation

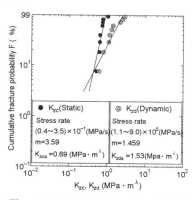

Figure 9:Weibull distribution of K_{zc}, K_{zd}

Fracture strength characteristics

Figure 9 shows results of a Weibull distribution of the critical strengths K_{zc} and K_{zd} of the stress singularity field, namely fracture toughness, obtained from the normal stress component σ_z which is perpendicular to the adhesive interface. Cumulative fracture probability F(%) was determined by the median-rank method (Daimaruya et al. (1997)). The exponent λ in the singular term is $\lambda=0.2910$ in the static stress field. In the dynamic stress field it is $\lambda=0.2735$ since the dynamic longitudinal modulus of PMMA becomes larger than the static one. The increase in λ becomes a factor that causes the strength to be lower. Therefore, the difference in the two strengths is increased slightly because of the difference in λ. However, there is considerable difference in strengths under static and dynamic load conditions as shown in Figure 9. In both the static and dynamic tests, fracture was observed to occur in the PMMA/epoxy adhesion interface.

CONCLUSIONS

In this study the impact and static strengths of a PMMA/Al butt adhesive round bar joint were evaluated with the stress singularity field parameter. From this the following conclusions are derived.
(1) It was confirmed that the stress singularity for the static stress field and the dynamic one are similar in the vicinity of the interface edge for an adhesive joint between dissimilar materials.
(2) The stress distribution from the finite element analysis was in good agreement with calculated results by Bogy's equation at the interface of an adhesive joint between dissimilar materials.
(3) The evaluation technique using the stress singularity field parameter is an effective one for determining the strength of an adhesive joint between dissimilar materials.
(4) The dynamic tensile strengths K_{zda} of the PMMA/Al butt adhesion round bar joint were about 2.2 times of the static values K_{zca}.

REFERENCES

Bogy D.B. (1971). Two Edge-Bonded Elastic Wedges of Different Materials and Wedge Angles Under Surface Tractions. *J. Appl. Mech.* **38**, 377-386.
Daimaruya M., Kobayashi H., Chiba M. and Maeda H. (1997). Measurement of Impact Tensile Strength of Concretes. *Trans. of the JSME(A)* **63**, 2592-2597.
Dunders J. (1967). Effect of Elastic Constants on Stress in a Composite under Plane Deformations. *J. Composite Materials* **1**, 310.
Hattori T., Sakata S., Hatsuda T. and Murakami G. (1988). A Stress Singularity Parameters Approach for Evaluating Adhesive Strength. *Trans. of the JSME(A)* **54**, 597-602.
Sato C., Iwata H. and Ikegami K. (1997). Dynamic Strength of Adhesive Layer under Combined Impact Loading using Clamped Hopkinson Bar Method. *Trans. of the JSME(A)* **63**, 341-346.
Sawa T., Nakano K., Toratani H. and Horiuchi M. (1995). Two-Dimensional Stress Analysis of Single-Lap Adhesive Joints Subjected to Tensile Shear Loads. *Trans. of the JSME(A)* **61**, 1994-2002.
Tong L. (1998). Strength of Adhesively Bonded Single-Lap and Lap-Shear Joints. *Int. J. Solids Structures* **35**, 2601-2616.
Wada H., Kubo S. and Murase K. (2001). Dynamic Tensile Strength of PMMA/Al Plate Butt Adhesive Joints. *J. Soc. Mat. Sci., Japan*, **50**, 223-228.
Yokoyama T. and Shimizu H. (1997). Determination of Impact Shear Strength of Adhesive Bonds with the Split Hopkinson Bar. *Trans. of the JSME(A)* **63**, 2604-2609.

FINITE ELEMENT STRESS RESPONSE ANALYSIS OF BUTT ADHESIVE JOINTS OF HOLLOW CYLINDERS SUBJECTED TO IMPACT TENSILE LOADS

Toshiyuki SAWA [1], Yoshihito SUZUKI [2] and Izumi HIGUCHI [3]

[1] Department of Mechanical Engineering, Yamanashi University
4-3-11, Takeda, Kofu, Yamanashi 400-8511, Japan
[2] Department of Mechanical Engineering, Yamanashi University
[3] Kofu Jyousai High School 1-9-1, Shimoiida, Kofu, Yamanashi 400-0064, Japan

ABSTRACT

The stress wave propagation in butt adhesive joints of similar hollow cylinders subjected to impact tensile loadings are analyzed in elastic and elasto-plastic deformation ranges using a three-dimensional finite-element method. The impact load is applied to the joint by dropping a weight. The upper end of the upper adherend is fixed and the lower adherend of which the lower end is connected to a guide bar is subjected to the impact load. The FEM code employed is DYNA3D. The effect of the adhesive thickness, Young's modulus of the adhesive, and the inside diameter of hollow cylinders on the stress wave propagation at the interfaces are examined. In addition, the characteristics of the joints subjected to the impact loadings are compared with those of the joints under static loadings. It is found that the maximum value of the maximum principal stress σ_1 occurs at the outside edge of the interface of the lower adherend to which the impact loading is applied. The maximum value of the maximum principal stress σ_1 increases as Young's modulus of the adhesive increases when the joints are subjected to impact loadings. It is found that the characteristics of the joints subjected to impact loadings are opposite to those subjected to static loadings. In addition, experiments were carried out to measure the strain response of the butt adhesive joints subjected to impact and static tensile loadings using strain gauges. Fairy good agreements are observed between the numerical and the measured results.

KEYWORDS

Stress response, Impact tensile load, Hollow cylinders, Butt adhesive joint, FEM, Maximum principal stress, Interface, strength

INTRODUCTION

Adhesive joints have been widely used in mechanical structures, automobile, aerospace, electrical industry and so on. A lot of researchers have been carried out on interface stress distributions and strength in butt, scarf and lap adhesive joints subjected to static or thermal loadings (Chen & Cheng (1990), Nakagawa & Sawa (2001), Goland & Reissner (1944), Adams & Wake (1984), Adams et al. (1973), Sawa & Suga (1996), Sawa et al. (1996), Wah (1976), Wah (1976), Bascom & Oroshnik (1978)). However, only a few investigations have been carried out on stress wave propagation and the stress distributions in adhesive joints subjected to impact loads (Higuchi et al. (1999), Zachary & Burger (1980), Sawa et al. (1996)). Adhesive joints used in mechanical structures are subjected to impact as well as static loadings in partice. An important issue is how to design adhesive joint, that is, how to determine the material property of the adhesive. In partice, butt adhesive joints are commonly used. However, the characteristics of the joints of hollow (solid) cylinders under static and impact

impact loading must be investigated from a design standpoint. A few studies on the subject have been carried out. Some investigations using finite-element method, boundary-element method, theory of elasticity and photoelasticity have been performed on the stress analysis and the strength evaluation in single-lap adhesive joints, butt adhesive joints, scarfed adhesive joints and steped-lap adhesive joints subjected to tensions, bending moments, and cleavage loadings under static load. The butt adhesive joints were primarily studied for impact loading. This paper presents the results of investigations on lap adhesive joints of hollow cylinders. The stress wave propagation and stress distribution in those subjected to impact and static loads were analyzed using three-dimensional finite-element method considering the elastic deformation. The code employed is DYNA3D for impact loading and MARC for static loading. The effects of Young's modulus of the adherends and the adhesives, the overlap length and the diameter of outer adherend cylinders on the stress distribution at the adhesive interface in the joints are examined. For verification of the calculations, experiments of the joints were conducted to measure the strain. Comparisons of the analytical predictions and experimental results were carried out.

FEM ANALYSIS

Figure 1 shows a model for analysis of the adhesive joint. The cylindrical coordinate system (r, θ, z) used is as shown in Figure 1, where the outer hollow cylinder is fixed. Static tensile load is applied to the free end of an inner hollow cylinder. A weight is dropped on a circular plate connected to an inner hollow cylinder and the edge of the inner hollow cylinder is then impacted. The hollow cylinder must be long enough to neglect the reflection of stress wave which occurred at a fixed edge of the outer hollow cylinder, and the analyses were conducted it. Young's modulus and Poisson's ratio of the adherend cylinders are designated as E_1 and ν_1, and those of adhesive as E_2 and ν_2, respectively. The length and thickness of the adherend cylinder are designated as l_1 and t_1, and thickness of the adhesive as t_2, respectively. The overlap length was denoted by l_2. The inside diameter of the outer hollow cylinder was denoted by d_1. Quarter part of the joint is analyzed because the cylinder is symmetric to the circumference. As for the boundary conditions, the displacement is constrained in the all directions at the fixed end of the outer cylinder and in the circumference directions at each cross section of the cylinder. In addition, impact loads are applied to the circular plate such that the weight has an initial velocity V, the initial velocity of the weight is chosen as $V = -1400$mm as shown in Figure 1 for impact loading, and static loads are applied to the free end of the inner cylinder for static loading. In the FEM analysis, hexahedron elements were used, and the numbers of finite elements and nodes employed were 2576 and 1704 respectively. The FEM code employed is DYNA3D for impact loading and MARC for static loading, and the three-dimensional elastic analysis was carried out. Mild steel SS400 was chosen as the material of the adherends. The adhesive chosen epoxy resin (Scotch-Weld/SUMITOMO 3M). Table 1 lists the material properties for the adherends and the adhesion used in this study. Young's modulus of the adhesive is so small than adhesion that we should pay attention to the adhesion on stress distribusion.

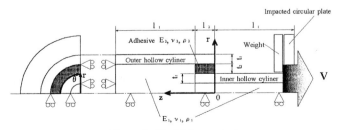

Figure 1: Model for analysis

Table 1: Mechanical properties of the adherend and the adhesive

	Adherend (SS400)	Adhesive (Scotch-Weld)
Young's modulus, E (GPa)	206	33.3
Poisson's ratio, ν	0.30	0.38

EXPERIMENTAL METHOD

Figure 2 shows the dimensions and schematic diagram of an experimental set-up for impact and static tests. Their dimensions were corresponded to Figure 1. The bonded surfaces of the adhrends were ground by sandpaper and any oil present was removed from the bonding surfaces with 2-butanone ($C_2H_5COOCH_3$). After the adherends had been bonded with the epoxy adhesive, they were heated at 20℃ for 24 h and at 80℃ for 8 h. Adhesive thickness t_2 was held constant at 0.1 mm. Two couples of strain gauges which were paired in symmetry with an axis z ($\theta = \pm 90$) are attached to the position z=2.5 and 20.0mm from the edge of the adhesive interface on the outside of the outer hollow cylinder (r=11.1mm). Under impact loading, the outer cylinder was fixed on the experimental setup and the weight of 0.22kgf was dropped from the height of H=100mm. Strain responses of cylinders subjected to the impact load were recorded by an analyzing recorder. Under static load, the specimen was subjected to static tensile load W of 300kgf.

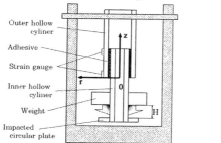

t_1=4mm
t_2=0.1mm
l_1=200mm
l_2=20mm
d_i=14.1mm
H=100mm

Figure 2: Schematic diagram of the experimental set-up and dimensions

NUMERICAL RESULTS

Stress Wave Propagation under Impact loading

Figure 3(a), and (b) show stress wave propagations at the interface between the outer cylinder and the adhesive (r=7.0, z=0.0, 1.9, 4.7, 11.2 and 20.0), and the interface between the outer cylinder and the adhesive (r=7.0, z=0.0, 1.9, 4.7, 11.2 and 20.0) under impact loadings, respectively. The ordinate is the maximum principal stress σ_1 at the interface and the abscissa indicates the time initiation from the impact loading. From Figure 3, it is found that maximum principal stress σ_1 is highest near the point (z=1.9, r=7.1). The failure will occur from it so that we should pay attention to it.

Stress distributions under static loading

Figure 4(a) shows the stress distribution in the adhesive (r=7.00, 7.05, 7.10) subjected to static load of 300kgf, and Figure 4(b) is a magnified figure of Figure 4(a) in nearby the free edge of adhesives (z=0, 1.9). The ordinate is the maximum principal stress σ_1 and the abscissa indicates the distance from the edge of the adhesive joint. In the adhesives, the maximum principal stress σ_1 shows the highest at the inside adhesive interface (z=0, r=7.00).

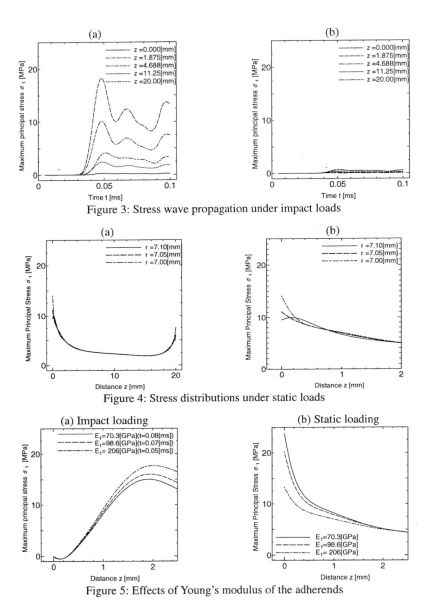

Figure 3: Stress wave propagation under impact loads

Figure 4: Stress distributions under static loads

Figure 5: Effects of Young's modulus of the adherends

Comparison of adhesive joints under impact and static loads as changing the properties of the adherend and the adhesive

Figure 5, 6,7 and 8 show the stress distribution under impact and static loads. In these figures, Young's modulus of adherend, adhesive, overlap length and the inside diameter of outer hollow cylinder was changed. The ordinate is the maximum principal stress σ_1 and the abscissa indicates the distance of the adhesive joint. Figure 5 shows the effects of Young's modulus of the adherend on the stress in the joints under impact and static loads. the maximum principal stress σ_1 increased as the Young's moduli of the adherend increased under impact loads. The other hand, maximum principal stress σ_1 decreased as it increased under static loads. Figure 6 shows the effects of Young's modulus of the adhesive on the joint strength under impact and static loads. Young's modulus of adherend cylinder is constant (E_1=206GPa), and that of the adhesive was changed. The maximum principal stress σ_1 decreased under impact loads as the ratio of Young's moduli of adherends to adhesives increase.

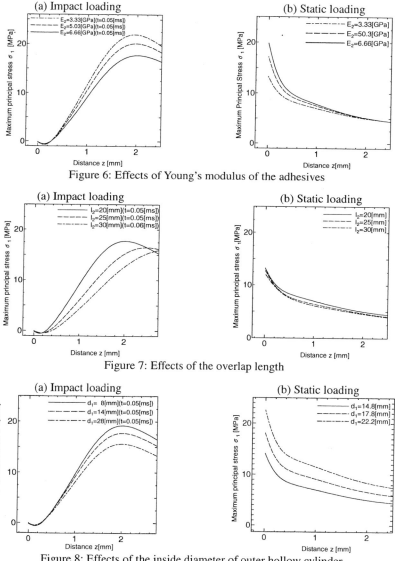

Figure 6: Effects of Young's modulus of the adhesives

Figure 7: Effects of the overlap length

Figure 8: Effects of the inside diameter of outer hollow cylinder

The other hand, maximum principal stress σ_1 increased under static loads as the ratio of Young's moduli of the adherends to that the adhesives increased. Figure 7 shows the effects of the over lap length on the stress in the joints under impact and static loads. Figure 8 shows the effects of the inside diameter of outer hollow cylinder on the stress in the joints under impact and static loads. The maximum principal stress σ_1 decreased under impact loads as the overlap length and the diameter of outer hollow cylinder increased. Under the static loads, the characteristics of the joints were as the same as those under the impact loads on changing the overlap length and the diameter of outer hollow cylinder.

NUMERICAL RESULTS AND COMPARISON WITH EXPERIMENTS

The experiments were carried out using the model of Figure 1. Figure 9(a) shows the results of the experiments concerning the stress wave propagation under impact load on the outside of the outer hollow cylinder (r=11.1, z=2.5, 20.0mm). Figure 9(b) shows the results of the experiments concerning stress distribution under static load on the outside of the outer hollow cylinder (r=11.1, z=2.5, 20.0mm).

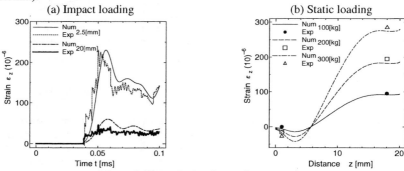

Figure 9: Numerical and experiment results

CONCLUSIONS

In the paper, it was dealt with the stress wave propagation and stress distributions in lap adhesive joints of hollow cylinders under impact load and static load. The following results were obtained;
(1) The maximum principal stress σ_1 occured at the end of the outer interface of which the outer adherend was impacted between the outer hollow cylinder and the adhesive under impact load, and at the end of the inner interface of which the inner adherend was free between the inner hollow cylinder and the adhesive under static load.
(2) The maximum principal stress σ_1 increased under impact loads as the ratio of Young's moduli of the adherends to those of the adhesives increased. The other hand, the maximum principal stress σ_1 decreased under static loads as the ratio of Young's moduli of the adherends to those of the adhesives increased.
(3) The maximum principal stress σ_1 decreased under impact loads as the characteristic of the joints under static loads was the same as the joints under impact loads.
(4) The maximum principal stress σ_1 decreased under impact loads as the diameter of outer hollow cylinder increased. Under the static loads, it was as good as impact loads.
(5) Experiments were carried out to measure the strain response of under impact and static loads. The numerical predictions were in fairly good agreements with the experimental results concerning the joints.

IMPACT ENERGY ABSORPTION OF CFRP-ALUMINIUM ALLOY HYBRID MEMBERS BONDED ADHESIVELY

*Masanori WASAKI, *Takaomi SUWA, **Takuya KARAKI and *Chiaki SATO

*Precision and Intelligence Laboratory, Tokyo Institute of Technology,
4259 Nagatsuta, Midori-ku, Yokohama 226-8503, JAPAN
**Composite material Research laboratories, TORAY corporation
1515 Tsutsui, Ohaza, Matsumae-cho, Iyo-gun, Ehime, 791-3193, JAPAN

ABSTRACT

The paper presents how to reinforce aluminium extrusions by adhering CFRP plate on it. Using the method, not only the static strength, but also the impact strength was improved. Furthermore, the absorbed energy in case of crash is also improved drastically with very small weight increase. Effects of bonding position and used adhesive were investigated analytically and experimentally.

KEYWORDS

Impact strength, Absorbed energy, CFRP, Aluminium extrusions, Adhesion, Crash.

INTRODUCTION

Lightweight bodies of transport such as automobiles and trains have become important in terms of saving energy consumption and reducing CO_2 emission. Aluminium alloy is the most promising material for the lightweight bodies because the cost is reasonable rather than the other sophisticated materials such as composites. Good recyclability of the material is also a big advantage.
Recent aluminium cars have space frame structures consisting of aluminium alloy extrusions, which mainly provide the stiffness and the strength of the bodies. However the crash worthiness of the aluminium space frame structures is not still enough. Broughton et al showed an efficient method to reinforce aluminium alloy extrusions by bonding thin CFRP plates adhesively on their surfaces (Broughton J.G. et al. (1997)). The reinforcement seemed to be available also under impact loading which might occur in case of car crash. Thus the aim of this work is to investigate the absorbed strain energy of the members in the process of crash deformation.

EXPERIMENTAL

Figure 1 shows the configurations and the dimensions of beam specimens used in the work. The specimens consisted of aluminum alloy extrusion having box cross-section and CFRP plates bonded together adhesively. There were three types of specimens such as Type A, B and C. Type A specimen was a specimen of only an extrusion without CFRP. Type B specimen had the CFRP plate adhered on the supporting side. Type C specimen had two CFRP plates on the both loading and supporting sides. The aluminium alloy was 6063-T3 and the CFRP was unidirectional type including carbon fibers (T700) and a matrix (#2500) produced by TORAY Corporation. The thickness of the CFRP was 0.4mm. We used four types of adhesives. One of them is ductile epoxy resin modified with rubber and including electro-isolating particles to prevent erosion caused by the difference of ionization tendencies between aluminium alloy and CFRP. The other was typical epoxy adhesive (3M,DP-460). Second generation acrylic adhesive and cyanoacrylate adhesive were also used for the experiments. The CFRP plates were bonded on the upper surface or the lower surface of the extrusions as the fibers of the CFRP aligned parallel to the length of the beams. In advance of the bonding, the surfaces of the extrusions were milled using grid papers (#500) and degreased with acetone. The ductile epoxy adhesive and DP-460 were cured at 70 ° C for 2 hours after the bonding. The other adhesives were cured in room temperature.

Impact tests of the beam specimens in three-point flexure were carried out using an instrumented Charpy tester which could measure the variations of the impact load and displacement together with respect to time. The specimens were simply supported by two wedges in the Charpy tester. The distance between the wedges was 90mm. The impact velocity of the hammer was kept about 3.5m/s in every experiment. Figure 2 shows the profiles of the specimens bonded by the ductile epoxy adhesive after the impact tests. An extrusion without CFRP was cracked. However the others reinforced by CFRP were not flawed, and they were only plastically deformed.

Static tests of the specimens in three-point flexure were also carried out using Instron-type mechanical tester. Both the radiuses of the loading and supporting rollers were 4mm, which was similar to the radius of the Charpy hammer.

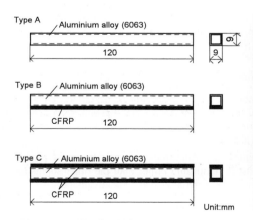

Figure 1: Configuration and dimensions of specimens consisting of aluminium extrusions and CFRP bonded adhesively

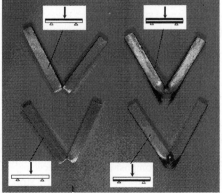

Figure 2: Photograph of the deformed profiles of the specimens after impact tests.

NEUMERICAL ANALYSIS

The deformation and the stress distribution of the specimens were calculated using the finite element

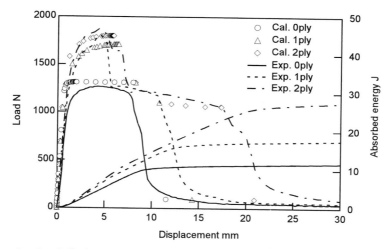

Figure 3: Load-displacement curves and absorbed energies of the specimens in the static tests and their analytical predictions.

method with ABAQUS ver.5.7. Three-dimensional elasto-plastic analysis was carried out using cubic solid elements with 8 nodes. In the case of the impact tests, dynamic analysis was also conducted considering the inertia effects of the materials.

EXPERIMENTAL AND ANALYTICAL RESULTS

Figure 3 shows the load-displacement curves obtained by the static tests. Absorbed energies, which were calculated from the load-displacement curves, were also shown in the figure. An extrusion without CFRP (Type A specimen) was deformed elastically in the initial stage and showed plastic deformation in the next stage. After the displacement of the specimen reached about 8mm, a crack caused at the opposite side of the loading point and propagated to the loading side. The load decreased in this process and finally the specimen collapsed. Maximum loads of Type B specimens were greater than that of Type A specimen and the transitions from the elastic deformation to the plastic deformation occurred more gradually. In the adhesive of Type B specimens, delaminations occurred prior to the crack generation in the extrusions and their collapse by the crack propagation delayed as the thickness of the CFRP increased. Therefore, the presence and the thickness of the CFRP influenced the load-displacement relations and the absorbed energy.

Analytical predictions of the load-displacement curves are also shown in figure 3 comparing from the experimental results. In the analysis, the yielding condition of bulk adhesive was used as the failure criterion of the adhesive layer. The generation of the initial cracks and their propagation were predicted based on a maximum strain in the direction of the length of the beams. It is shown that the analytical predictions agree with the experimental results very well.

Figure 4 shows the experimental results of the impact tests. The load-displacement curve of Type A specimen was similar to that of the static test although the maximum stress of the impact test was higher. The adhesive layers of Type B and Type C specimens didn't delaminate in any case of the impact tests and the CFRP plates didn't separate from the extrusions. The CFRP plates prevented the extrusion surface bonded to them from too much elongation causing the cracks. The other side of the extrusion was subject to compressive load and deformed much more with buckling. Therefore the absorbed energies by impact was greater than those of the static tests. In terms of energy absorption, Type B specimen is similar to Type C specimen. However both of them are superior to Type A

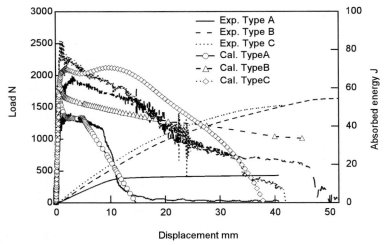

Figure 4: Load-displacement curves and variations of absorbed energy of the specimens subjected to impact loads and their analytical predictions.

specimen about three times although weight increase added by bonding the CFRP was few percents of the total weight. In other words, we can improve the performance of energy absorption with small weight increase.

Analytical results of the impact tests are shown in Figure 4. In this analysis, adhesive failure was simulated using a criterion based on stress parameters obtained impact experiments. However the failure was not predicted in the analysis and didn't occur also in the actual tests. The analytical predictions of load-displacement curves can describe the characteristics of the experimental results.

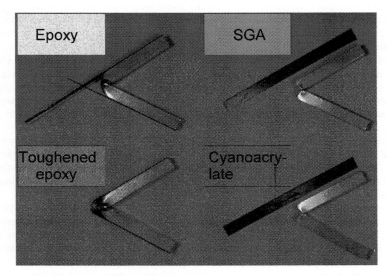

Figure 5: Profiles of the specimens bonded by different types of adhesives after impact tests.

STRENGTH OF ADHESIVES

Figure 5 shows the profiles of the specimens which have each different adhesives after the impact tests. It is seen in the figure that only the ductile adhesive can not be broken under impact loading. The other adhesives were failed and impact loads also separated the CFRP plates. Therefore, the ductile adhesive is most suitable to join the CFRP plates to the aluminium extrusions. Absorbed energies of the specimens by the impact tests are shown in Figure 6. The specimen connected with the ductile adhesive had the greatest absorbed energy of course. DP-460 showed good performance of energy absorption, but it was less than the ductile one. SGA and cyanoacrylate adhesive was not so effective to join the CFRP to aluminium because their absorbed energies were not improved from the aluminium extrusions without CFRP.

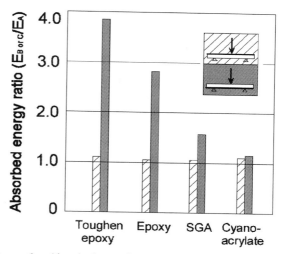

Figure 6: Absorbed energies of the specimens by bonded with different adhesives

CONCLUSIONS

Combining aluminum alloy extrusions with CFRP by adhesion is very effective method to improve the stiffness and the strength of the material under not only static loading but also impact loading. Furthermore, the method also makes the absorption of strain energy increase drastically (about four times) with few percents of weight increase. Fractures in the adhesive layers are vital to the overall strength of the material under static loading. However, the impact strength and the absorbed energy of the material under impact loading doesn't depend on the adhesive fractures so much because the adhesive layers are strong enough under the condition. The ductile adhesive used in the experiment was most effective to join the CFRP to the aluminium extrusions. The other adhesives were not suitable rather than that.

REFERENCE

Broughton J.G., Beevers A. and Hutchinson A.R. (1997). Carbon-fiber-reinforced plastic (CFRP) strengthening of aluminium extrusions, *Int. J. Adhesion and Adhesives*, **3:17**, 271-278.

HIGH PRESSURE GENERATION SYSTEMS FOR MATERIALS PROCESSING AND EVALUATION

T. Aizawa

Research Center for Advanced Science and Technology, University of Tokyo,
4-6-1 Komaba, Meguro-ku, Tokyo153-8903, Tokyo, Japan

ABSTRACT

Intense shock loading devices have been developed for materials processing and materials evaluation. Mini-laser driven shock-loading device is favored mainly for fundamental studies on the dynamic response of materials subjected to stress/pressure pulses. Time-resolved measurement of particle velocity profiles and stress histories enables us to construct various types of new mechanical testing instead of conventional static materials evaluation. Ion-beam driven shock-loading device is adaptive to shock recovery testing and dynamic impact experiments for relatively wide range of impact velocity. Since there is no limitation to acceleration of velocity in principle, higher velocity can be obtained even for solid flyers in the order of 100 mili-gram to grams. Magnetic pulse loading device is used for relatively large-scaled powder compaction in cold or warm. Compared to the first two approaches, it has longer time duration so that power compaction can be made for the dynamic range of pressure up to several GPa. Several characteristic features of each device are discussed with some comments on the developed prototype system and its demonstration.

KEYWORDS

Intense shock loading, Mini-laser-driven shock loading, Ion-beam driven shock loading, Magnetic pulse loading, Dynamic behavior of materials, High pressure material science

INTRODUCTION

A couple of decades ago, high-speed camera drastically changed our mechanical point of view in describing the dynamic behavior of materials. Looking through a series of images in the order of μs, the dynamic cracking behavior (Dally (1979)) or dynamic deformation of structures (Meyers (1994)) can be described quantitatively. At present, time-resolved measurement of material kinematic change in the order of femto-second is changing our physical and chemical senses in description of various processes in the solid state. Intermolecular interaction (Lee et al. (1995)), fast rate reaction (Hambir et al. (1998)), shock loading behavior (Nakamura et al. (1997)) or fundamental steps in solid synthesis of matters (MaCluney (1994)) can be quantitatively described by these new types of time resolved measurements. High-strain rate behavior, which is an essential fundamental for materials science and engineering, is never exceptional. In order to analyze the time response of materials by using these time-resolved measurements, the intense high-strain-rate, loading device might well be customized to include measurement units. Especially, precise control of mode locking for optical drivers becomes indispensable to attain high reproducibility of signals. Author has been developing three types of shock loading devices driven by the mini/micro lasers and the ion beam driven shock loading devices (Ito et al. (1998), Ito et al. (1999), Tsuchida et al. (1998), Tsuchida et al. (1999)) together with the magnetic-pulse loading device (Aizawa et al. (1999)). Each device has its own dynamic range as shown in Fig. 1. In order to deal with various phenomena and mechanical behavior of materials having different dynamic range, adequate high-strain-rate, loading device to each research field must be selected.

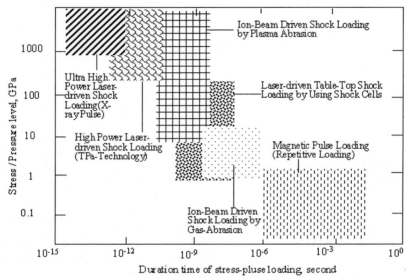

Figure 1: Effective intense high-strain-rate loading devices in the diagram of the applied stress versus the duration time.

In the present paper, three typical shock-loading devices are introduced to have their own characteristics understood: laser-driven, ion-beam-driven shock loading and magnetic pulse loading devices. The first one is mainly used for time-resolved measurement with the time duration from sub-nano-second to 100 ns and the pressure up to 100 GPa. Through focussing the laser beams, microscopic shock loading can be realized. The second is mainly for shock recovery and high-velocity-physics/mechanics with the time duration up to micro-second order and the velocity up to 20 km/s. Different from the laser-driven loading device, relatively larger solid flyer can be launched. The third is for dynamic consolidation of powder compact. Since the applied load is proportional to the magnetic pulse density and the self-pinching current density, relatively longer pulse duration can be utilized in mechanical testing. Although the developed systems are still in the prototype version, the fundamentals of each device as well as its feasibility in applications are discussed for further development in the materials science and technology.

MINI-LASER DRIVEN SHOCK LOADING DEVICE

High power laser enables us to broaden the experimental flame of research and development not only in high-pressure physics and chemistries but also in materials science and engineering. When using the conventional powder guns, gas guns or the explosive methods, large amount of samples are required for preparation. In order to obtain the Hugoniot of the targeting matter, for an example, several to ten samples, having the same properties and geometric configuration, must be prepared before shock experiments. The attached sensors to the sample are all destroyed during measurement. Hence, the conventional experiments for Hugoniot measurement are always costly. In addition, local modification of materials by application of high pressure is nearly impossible.

In the laser driven shock loading, the solid flyer is launched and shot on the sample, as shown in Fig. 2. Through direct measurement of the velocity history of a flyer by the optical interferometry, both the elasto-plastic behavior and the Hugoniot of any sample materials can be obtained by a series of launching. Since only one sample can be used for the repetitive measurement with different flyer velocities, relatively accurate data can be obtained. Figure 3 illustrates the developed laser-driven shock-loading device with VISAR for optical interferrometry. The Joule-class power laser (Nd:YAG laser with 0.86 J), the reference laser (Ar: die laser) and the sample holder can be placed on the same table so that no special rooms are necessary for this type of experiment. As had been discussed in Ito et al. (1998), Ito et al. (1999), the shock cell was designed to make launching of a solid flyer. The simple configuration of a shock cell consists of the fuel and flyer layers.

The optical energy by a laser is transformed into an explosive power of fuel material with a plasma pressure. Usually, aluminum coating by PVD (physical vapor deposition) is utilized as a fuel layer. The thickness of fuel layer must be designed to preserve the solid state of a flyer or to keep the flyer cold. If no solid parts were left or the flyer became hot, the flyer velocity could be difficult to be determined. The velocity of cold flyer can be controlled by the input energy. In order to accelerate this flyer velocity, three-layered shock cell is recommended to use: (fuel) – (accelerator) – (flyer). Of much interest is an optical fiber cell. Once the small planar cell is fixed at one end of an optical fiber, the local high-pressure pulse can be ejected by transmitting the laser pulse from its other end. When using these shock cells in the shock loading device, the targeting sample or the window materials must be equipped into a sample cell including the shock cell. In the normal operation of the laser-driven shock loading devices, time history of launched flyer trajectory can be traced by the time-resolved measurement as shown in Fig. 4. More advanced measurements of Hugoniot and elasto-plastic materials response can be also performed by using nearly the same configuration.

In the present configuration of a unit cell, the sapphire was used as a back plate to transfer the plasma pressure to the kinetic energy of a cold flyer. The laser has maximum power of 570 mJ/shot with the pulse duration of 12 ns: maximum energy density reaches 1 GW/cm^2. Most metallic flyer with the maximum thickness of 25 μm can be launched up to about 800 m/s. In the velocity range up to 800 m/s, the energy conversion efficiency, which is defined by (kinetic energy of a flyer) / (Driving laser energy), became about 40 %: this high efficiency is favored to a table-top experiment.

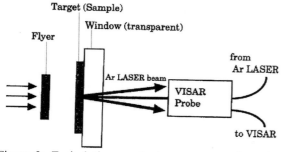

Figure 2: Typical time-resolved measurement by using a shock cell to be working for energy conversion.

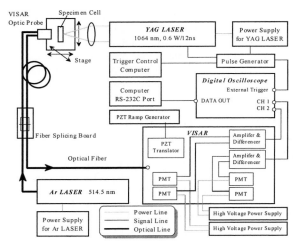

Figure 3: A schematic illustration of the developed multi-purposed table-top laser-driven shock loading device.

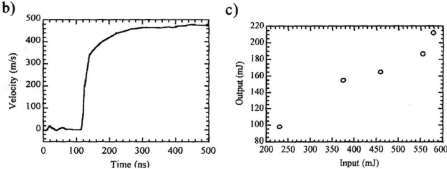

Figure 4: Time solved measurement of a solid flyer launched by the laser-driven shock-loading device: b) Time history of flyer trajectory, and c) relationship between the input energy calculated from the condenser bank and the kinetic energy of a flyer.

ION-BEAM DRIVEN SHOCK LOADING DEVICE

In dealing with the high-velocity impact problems, relatively large mass of matters is often necessary. In the space debris problem, its size or mass distribution must be considered for safety gourd design of space crafts and structures. The developed ion-beam-driven shock loading device might be suitable to research and development in the fields of high-velocity physics and mechanics. As shown in Fig. 5, the large condenser bank is utilized to generate the plasmoid and to accelerate it for launching the sold flyer. Different from the laser-driven shock loading device, the particle velocity of plasmoid can be directly controlled and accelerated by the power of applied pulse. Then, there is no limitation of accelerated velocity of plasmoid. Due to interaction between the accelerated plasmoid and the solid flyer surface, the actual velocity of a flyer is also affected by abrasion of flyer matter.

Both the plasma abrasion and the gas abrasion modes are available to control the flyer velocity in the

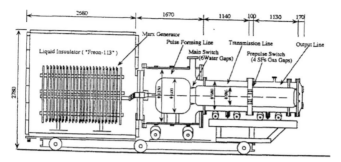

a) Pulse transmission by using LAIMEI-I at TIT.

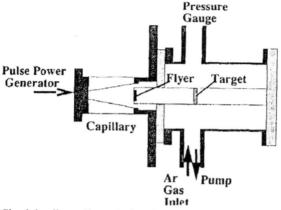

b) Shock loading cell attached to the RHS of a).

Figure 5: A schematic illustration of ion-beam-driven shock loading device.

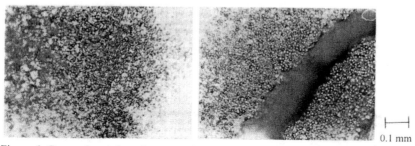

Figure 6: Comparison of starting amorphous powder compact and consolidated specimen.

relatively wide range from 100 m/s to 15 km/s. The energy conversion efficiency is defined by (kinetic energy of cold Flyer) / (Condenser bank stored energy). This efficiency also varies with increasing the flyer velocity. In the range from 1 km/s to 3 km/s, the efficiency became about 6 %. Although this value is lower than that reported in the case of laser-driven shock loading device, the conventional plasma guns cannot attain this high efficiency.

This ion-beam shock lading was applied to the powder compaction. As well-known, most of amorphous or non-equilibrium-phase material powders have high strength and low thermal stability. Hence, both cold and hot pressing are usually in vain for powder compaction. In this experiment, aluminum-mesh metal powders were employed for dynamic compaction. This amorphous phase has amorphous-to-crystalline transformation temperature at 648 K. Comparison of XRD profiles before and after ion-beam shock loading showed that most of phase is still amorphous with crystalline peaks of intermetallics. Figure 6 depicted the change of morphology before and after dynamic compaction. Hence, aluminum base amorphous powder compact was successfully consolidated by using the present ion-beam-driven shock loading device.

MAGNETIC-PULSE DRIVEN IMPACT LOADING DEVICE

In this approach, the short-duration magnetic pulse B(t) is generated by induced high current density J(t) and the pulsed stress by J x B. The most interesting features of this magnetic pulse loading are: 1) relative longer duration time in the order of 100 s to 1 ms and 2) applied stress or pressure range from sub-GPa to a few GPa. In the developed prototype system, the loading capacity is summarized in the following.

TABLE 1
Magnetic pulse loading capacity of the developed prototype system.

Duration Time	Peak Magnetic Density	Maximum Applied Stress
1-3 ms	27. 5 T	300 MPa

In addition, repetitive loading can be done with the same configuration preserved although the successive pulse loading. In fact, the powder compaction testing was preformed to demonstrate the successive powder compaction by this repetitive shock loading. Figure 7 shows the variation of the relative density of aluminum powder compacts with increasing the number (N) of pulse loading. Initial relative density of cold-pressed compact was limited to 60 % T.D. The relative density continues to increase with N. The compaction behavior is equivalent of the powder pressing up to 300 MPa.

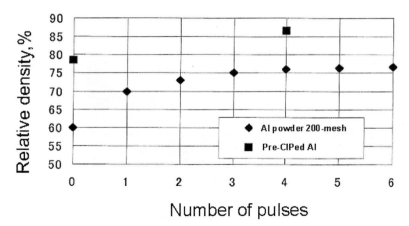

Figure 7: Variation of the relative density of powder compact with increasing the number of pulses in the magnetic pulse loading device.

To be noticed here is the difference of compaction behavior and relative density distribution in the powder compact. In the case of mechanical pressing, the vicinity of moving upper punches becomes high dense while the vicinity near the lower punch has lower density. On the contrary, since the uniform stress is applied in the axi-symmetric manner to the powder mixture, more uniform relative density can be attained in the whole samples.

In order to investigate the effect of initial density on the compaction behavior, the original powder compact was subjected to cold isostatic pressing (CIP) to increase the relative density up to 78 % T.D. by applying the static pressure of 300 MPa. Without any effect of dynamic pressure on the densification, there could be no increase of the relative density in the present method. In fact, as shown in Figure 8, the relative density can be increased up to nearly 90 % T.D. by the present method.

CONCLUSION

Due to advancement of time-resolved measurement and short-pulse laser/beam technologies, intense shock loading has become a powerful tool to describe the mechanical response of advanced materials subjected to stress pulses for relatively wide dynamic range. Instead of the conventional mechanical testing, these tools might play a significant role to cultivate new fields in materials synthesis and evaluation. As had been seen in the history of mechanical testing and evaluation, various types of shock-loading and dynamic loading devices has better be developed to cultivate the fundamentals in the materials Science and engineering and to explore new ways and new waves in the application of manufacturing and materials processing.

ACKNOWLEDGMENTS

Author would like to express his gratitude to previous graduate students, Mr. K. Ito (IHI, Co. Ltd.) and M. (Isuzu. Co. Ltd.), and undergraduate student, Mr. Haneda (Tokyo Rope, Co. Ltd.) for their help in experiments. This study is financially supported in part by the Grand-in-Aid from the Ministry of Education, Science and Culture with the contract number of # 11225205.

REFERENCES

Aizawa T., Haneda Y. and Kido G. (1999) *Report on NRIM (National Research Institute of Metals)* 121-126.
Dally J.W. (1979). Experimental Mechanics **19**, 349-357.
Hambir A., Franken J., Hill J.R. and Dlott D.D. (1998). *Shock Compression of Condensed Matter – 1997*, 823-826.
Ito K., Aizawa T. and Paisley D.L. (1998). *Rev. High Pressure Sci. Technol.* **7**, 876-878.
Ito K. and Aizawa T. (1999). *Materials Processing Technology.* **85**, 91-96.
Lee I-Y.S., Hill J.R., Suzuki H., Baer B.J., Chronister E.L. and Dlott D.D. (1995). *J. Chem. Phys.* **103**, 8313-8321.
MaCluney R. (1994). *Introduction to Radiometry and Photonometry*, Artech House, Boston.
Meyers M.A. (1994). *Dynamic Behavior of Materials*, JohnWiley.
Nakamura M., Uchino M. and Mashimo T. (1997). *SPIE*, 4669-4672.
Tsuchida M., Aizawa T. and Horioka K. (1998). *Rev. High Pressure Sci. Technol.* **7**, 948-950.
Tsuchida M., Aizawa T. and Horioka K. (1999). *Materials Processing Technology.* **85**, 148-152.

AUTHOR INDEX

Volumes I and II

Abe, A. 857, 863
Adachi, T. 815
Aizawa, T. 487, 821, 826
Akahoshi, Y. 213, 219, 225, 231, 845
Akiyama, M. 779
Amaki, H. 395
Anteby, I. 625
Arai, H. 503, 509
Arimitsu, Y. 699, 767
Asada, K. 243
Asakawa, M. 117
Ayabe, K. 261

Bae, Y.Z. 945
Bakhrakh, S.M. 893
Balakrishna Bhat, T. 939
Bartnicki, E. 839
Baruchel, J. 633
Belov, G.V. 893
Ben-Dor, G. 625
Beppu, M. 737
Berthe, L. 839
Bin Li 963
Bivin, Yu.K. 189
Blanc, R. 71
Boller, E. 633
Bontaz-Carion, J. 633, 839
Botvinkin, A.K. 951
Bryukhanov, N.V. 951
Buchar, J. 195
Bussac, M.N. 49, 71

Carton, E.P. 495
Chang, S.N. 945
Chatani, A. 345, 799
Chen Yuze 663
Chen, X.W. 183
Cheng, W. 565
Chengli Yan 963
Chenguang Huang 803
Chengwei Sun. 963
Cheong, C.H. 291, 297
Chiba, A. 267, 273, 899, 905, 939
Chuan Yu 963
Chung, D.-T. 413
Collet, P. 49, 71

Daimaruya, M. 29
Dandekar, D.P. 785
Date, H. 339, 451
Dejin, H. 743
Deribas, A. 527
Dong Qindong 963
Duan Zhuping 803

Egroshin, S.P. 893
Erheng, Z. 663
Ermenko, A.S. 951

Fabo Li 963
Fan, X. 267
Fedorova, Ju.G. 893
Ferranti, L. 321
Fujimoto, T. 381, 389
Fujita, K. 683
Fujita, M. 911, 917, 923, 927, 939, 957
Fujiwara, K. 117, 851, 869
Fukuda, H. 433
Fukuda, K. 773
Futakawa, M. 339, 451, 779

Gao, L. 589
Gary, G. 23, 49, 71
Gonda, T. 791
Gotoh, M. 809
Greening, D. 201
Grolleau, V. 651
Gui Yulin 963
Gunawan, F. 705

Haham, O. 625
Halle, T. 91
Hamate, Y. 157
Hanada, T. 145
Harada, K. 737
Harrigan, J.J. 15
Hata, H. 225
Hayashi, Y. 869
He Yingbo 663
Held, M. 207, 279
Henmi, N. 821, 826
Higashi, K. 237, 547, 577
Higuchi, I. 469
Hikosaka, H. 619, 731

Hino, R. 339, 779
Hirai, K. 821
Hiroe, T. 117, 851, 869
Hode, S. 243
Hojo, A. 345, 799
Hokamoto, K. 911, 923, 927, 939, 957
Homma, H. 705
Honma, H. 151
Horie, Y. 201
Horikawa, N. 43
Hu, Q.H. 381
Hu, S.S. 309
Hu, X.Z. 309
Huang Chenguang 803
Huang Dejin 743
Huaqiang Zhang 803.

Hwang, C. 413
Hyunmo, Y. 571

Iida, M. 171
Ikai, S. 395
Im, K.H. 815, 875
Inamura, E. 333
Inou, K. 515
Inoue, A. 577
Isbell, W.M. 131
Ishii, S. 619
Ishikawa, K. 547
Ishikawa, N. 737, 761
Ishikawa, S. 345, 451, 779
Isuzugawa, K. 773
Itabashi, M. 433
Ito, C. 583
Ito, M. 375
Itoh, M. 737
Itoh, S. 607, 917, 923, 927, 933, 939, 957
Iwamoto, T. 713
Iyama, H. 957
Izuma, T. 401

Jamolkin, E.L. 893
Jeon, B.S. 945
Jiang, D. 749
Jingrun, L. 663
Jinyan Yang 963
Jones, N. 1, 105

Jordan J.L. 255
Jung, Y.H. 969

Kadono, K. 395
Kainou, H. 249
Karagiozova, D. 105
Karaki, T. 481
Kasai, T. 395
Katayama, M. 737
Kato, A. 249, 927
Kato, I. 139, 157
Kato, Y. 933
Kawakita, S. 225
Kawakubo, H. 880
Kawamura, Y. 273, 577
Kawasaki, T. 333
Kawashima, K. 445
Keller, A. 321
Kennedy, G. 321
Kennedy, T.C. 463
Khvorostin, V.N. 951
Kikuchi, W. 832
Kim, M.S. 725
Kim, S.K. 815
Kim, Y.N. 875
Kimura, S. 905
Kira, A. 911, 917, 923, 957
Kishida, K. 395
Kitaura, K. 315
Klepaczko, J.R. 425, 541
Kobayashi, H. 29
Kobayashi, T. 65, 363
Kogawa, H. 339, 779
Kohara, K. 863
Kojima, N. 439
Kondo, M. 445
Koshi, M. 175
Kouda, K. 219
Krüger, L. 91
Kubota, S. 419, 521
Kuri, S. 243
Kurokawa, T. 559
Kusaka, T. 43
Kyoungjoon, P. 571

Lee, H.C. 725
Lee, J.S. 939
Lee, N.K. 327
Lee, W.Y. 565
Li Bin 963
Li Fabo 963
Li, J.R. 309, 535
Li, Q.M. 183
Li, Y.C. 309, 357
Liew, Y.K. 755

Li-Lih Wang 743
Lim, C.L. 553
Lim, C.T. 291, 297, 553, 565
Lim, H.C. 755
Liu, K. 589, 595, 601
Liu, Y. 595
Low, K.H. 975, 981
Lu, G. 719
Luo Jingrun 663

Ma, J. 565
Maekawa, I. 669
Maeno, K. 151
Maillard, S. 651
Manczur, P. 633, 839
Mashimo, T. 267
Masuda, M. 43
Masuda, N. 761
Masuda, T. 363
Masuki, A. 401
Matsui, K. 419, 521
Matsunaga, T. 171
Matsuo, H. 117, 851, 869
Matsuoka, T. 243
Meyer, L.W. 91
Meyers, M.A. 123
Midorikawa, Y. 339
Mimura, K. 57, 77, 111, 657, 821
Minamoto, H. 639
Mitake, S. 737
Mitsuyama, M. 639
Miyake, A. 163, 171, 503,509
Miyamoto, K. 171
Miyazaki, M. 887
Miyoshi, T. 237
Mizui, N. 439
Mochihara, M. 911
Moiseev, E.B. 951
Mori, K. 607
Mori, M. 761
Mori, S. 509
Morioka, T. 151
Morita, S. 65
Morizono, Y. 899, 905
Muaki, T. 237, 547, 577
Murakimi, Y. 607
Murase, K. 463
Murata, S. 880

Nagahiro, J. 731
Nagai, M. 639
Nagata, S. 773

Nagayama, K. 515, 521
Nakagawa, H. 475
Nakahara, M. 515
Nakai, K. 699
Nakamura, Y. 933
Nakano, M. 395, 767
Nakano, S. 237
Nakazawa, Y. 439
Nakiyama, Y. 911
Negishi, H. 285, 887
Nesterenko, V.F. 123
Nicollet, M. 633
Nieh, T.G. 577
Nishida, M. 375
Nishigaki, K. 395
Nishimura, N. 445
Nishioka, T. 381, 389
Nizri, E. 625
Novikov, S.A. 951
Nowacki, W.K. 83

Oakkey, M. 571
Oda, I. 401, 407
Ogata, Y. 419, 509
Ogawa, K. 99
Ogawa, N. 657
Ogawa, T. 171, 503, 509
Oh, S.I. 413
Oh, S.Y. 945
Ohba, M. 779
Ohsono, K. 243
Okabe, T. 261
Okada, H. 315
Okagawa, K. 821, 826
Othman, R. 49, 71
Ozaki, N. 395

Pan, Y. 35
Panhai Xie 963
Park, J.W. 875
Pellegrini, Y.-P. 633
Petrushin, A.V. 893
Peyre, P. 839
Protat, J.C. 839

Qindong Dong 963
Qiu, J. 691

Raghukandan, K. 939
Ramjaun, D.H. 139
Reid, S.R. 15
Rio, G. 651
Rolc, S. 195
Ruan, D. 719
Rusinek, A. 541

Sadot, O. 625
Saito, T. 657
Sano, R. 857
Sano, Y. 857, 863
Sasatani, Y. 395
Sassa, M. 917
Sato, C. 481
Sato, T. 657
Sato, Y. 832, 845
Satoh, K. 683
Sawa, T. 469, 475
Sawairi, Y. 809
Schenker, A. 625
Seto, M. 419
Shaoqiu, S. 743
Shaverdov, S.A. 893
Shen, J. 799
Shen, W. 749
Shen, W.Q. 755
Shi Shaoqiu 743
Shibuya, H. 151, 880
Shim, V.P.W. 327, 369
Shimada, H. 419, 521
Shin, H.S. 725, 945
Shinohara, M. 139, 157
Shioya, T. 683
Shirai, K. 583
Shiraishi, K. 407
Shiramoto, K. 923
Sim, J.K. 875
Simonov, I.V. 189
Siva Kumar, K. 939
Sogabe, Y. 699, 767
Song, K.N. 969
Sonoda, Y. 619, 731
Spagnoli, A. 607
Stronge, W.J. 675
Stuivinga, M. 495
Sun Chengwei 963
Suwa, T. 481
Suzuki, K. 463, 761
Suzuki, Y. 469

Tachiya, H. 345, 799
Taira, M. 559
Takahara, K. 503, 509
Takahira, R. 419
Takayama, K. 139, 157
Takeishi, H. 339

Takezono, S. 333, 639, 791
Tamura, S. 737
Tan, G.E.B. 565
Tan, P.J. 15
Tan, V.B.C. 291, 297, 327, 553
Tan, Y. 243
Tanaka, K. 29, 375
Tanaka, K.A. 395
Tanaka, M. 231
Tanaka, Y. 401
Tang, Z.P. 357
Tani, J. 691
Tanimura, S. 57, 77, 111, 589, 657, 821
Tao, K. 333, 791
Tashiro, T. 880
Thadhani, N.N. 255, 321, 851
Toda, H. 363
Tomoshige, R. 249, 927
Tong Yanjin 963
Tonokura, K. 175
Torigoe, I. 607
Tsunetomi, T. 225
Tsuta, T. 713

Uda, K. 669
Ueguri, H. 669
Ujimoto, Y. 923, 927
Umeda, T. 77, 111, 657
Unosson, M. 613
Urushiyama, Y. 691
Usui, T. 351

Voldrich, J. 195

Wada, H. 463
Wada, Y. 503
Wakamori, T. 111
Wang, B. 719
Wang, L. 743
Wang, X. 35
Wang, Y. 799
Wasaki, M. 481
Washio, T. 705
Watanabe, D. 213
Watanabe, T. 899

Wei, Z.G. 35, 309, 357, 535
Widijaja, J. 175
Wong, P.S. 755
Wu, Z. 699, 767

Xie Panhai 963
Xu, Y. 123
Xue, Q. 123

Yagishita, T. 583
Yamakawa, T. 445
Yamamoto, M. 407
Yamamoto, N. 345
Yamashita, K. 111
Yamashita, M. 809
Yamauchi, Y. 395
Yan Chengli 963
Yang Jinyan 963
Yang, I.Y. 815, 875
Yang, L.M. 369
Yanjin Tong 963
Yano, K. 201
Yasaka, T. 145
Ye, S. 175
Yin, Y. 713
Yingbo, H. 663
Yokoyama, T. 303
Yokoyama, T. 457, 767
Yoo, Y.H. 413
Yoon, K.H. 969
Yoshie, S. 351
Yoshizawa, M. 821, 826
Yu Chuan 963
Yu, J.-L. 35, 535, 645
Yu, T.X. 645
Yulin Gui 963
Yuze, C. 663

Zeng, W. 589
Zhang Erheng 663
Zhang Huaqiang 803
Zhang, J.-Y. 645
Zhang, S.B. 285
Zhang, X. 975, 981
Zhang, Y. 267
Zhao, H. 23
Zhou, M. 321
Zhu, W. 77
Zhuping Duan 803

KEYWORD INDEX

Volumes I and II

Ablation of metal film, 521
Absorbed energy, 29, 65, 481
Absorbing capacity, 815
Accuracy of measurement, 77
Activation energy, 255
Adhesion, 481
Adiabatic heating, 541
Adiabatic Shear Bands (ASB), 309, 425, 535
Adiabatic shear failure, 91
Al-Mg series alloys, 363
Al-Ni alloy, 857
Alumina, 905
Alumina AD 995, 785
Alumina plates, 945
Aluminium, 71
Aluminium extrusions, 481
Aluminium foam, 15, 35
Aluminium honeycomb, 49
Aluminized explosives, 207
Aluminum, 285, 839
Aluminum alloys, 99, 433
Aluminum foams, 237, 625
Aluminum jointing, 827
Aluminum sheet, 827
Aluminum tube, 375
Ammonium nitrate, 163
Amorphous alloy, 273
ANFO, 503
Anisotropic, 369
Anisotropic damage, 645
Anisotropic plastic softening, 645
Anticollision, 111
AO, 683
Application of explosive, 171
Aramid fibre, 327
Artificial viscous stress, 863
Art object, 917
Atomic oxygen, 683
AUTODYN, 736
AUTODYN-2D, 509
Automobile reinforcing members, 809
Average collapse stress, 875
Axial impact, 105
Axial load, 887

Back plate material, 724
Back ply, 297
Ball-impact, 724
Ballistic, 327
Ballistic limit, 291, 375

Ballistic range, 139, 157
Balsa wood, 243
Basis vector method, 699
Biaxial, 91
Bicharacteristics, 601
Bimaterial, 381, 905
Bonded dissimilar plate, 401, 407
Boron, 261
Boundary condition, 969
Bowing, 291
Brittle fracture, 413
Brittle material, 724
Buckling mode, 111
Build-up distances, 279
Bulk metallic glass, 577
Bumper shield, 213
Butt adhesive joints, 463, 469

CAD, 699
CAI System, 65
Calibration, 773
Calibration curve, 509
Carbon, 683
Carbon steel, 345
Card gap test, 509
Cash-II code, 243
Cask, 243
Casting direction, 15
Caustics, 381
Cavitation, 779
Cell Size, 15, 237
Ceramic composites, 249, 321
CFRP, 395, 481
CFRP (Carbon-Fiber Reinforced Plastics) tubes, 875
Characteristic impedance, 767
CHEETAH code, 503
Chopped strand mat glass, 705
Circular tubes, 105
Cladding, 273
Closed cell, 15, 237
Cobalt, 267
Coefficient of Restitution, 639
Coil spring, 797
Collapse Characteristics, 875
Collision, 675
Combustion synthesis, 261
Comparison table with weighing factors, 207
Complex Young's modulus, 71
Compliant contacts, 675

Composite armour, 803
Composite laminated plates, 749
Composite materials, 43, 939
Compression, 91
Compression-shear, 91
Compression test, 237, 439
Compressive strength, 583
Concrete, 583
Concrete structure, 607
Concrete target, 183
Confinement, 945
Consolidation condition, 111
Constitutive equation, 91, 111
Constitutive law, 23, 803
Contact, 651
Contact/non-contact condition, 381
Contact pressure, 945
Contact treatment, 413
Continuum damage mechanics, 749
Control, 869
Controller of energy absorption, 815
Copper, 267
Covered high explosives, 279
Crack, 559
Cracked beam, 663
Crack opening displacement, 407
Crack-tip stress field, 401
Crash, 481
Crash safety, 439
Crashworthiness, 1
Critical impact strength, 969
Critical impact velocity, 969
Critical Impact Velocity (CIV) in shear, 425
Cross-ply CFRP, 303
Crush strength, 809
Crushable foam, 369
Curling, 285
Curved beam, 797
Cyclic creep, 315

Damage, 633, 839
Damage-ability, 809
Damage evolution, 743
Damage mechanics, 731
Damage mechanism, 705
Damage model, 851
Damage thresholds, 749
De-bonding, 705
Deceleration, 243
Deconvolution, 15
Deep penetration, 183
Defect, 607
Deformation and failure mechanism, 357
Deformation process under impact load, 957
Deformed shape, 969

Degradation, 683
Delamination, 43, 395, 645, 705
Delay times, 279
Density, 249
Desktop two-stage light gas gun, 845
Detonation, 157, 869
Detonation pressure, 163, 503
Detonation properties, 207
Detonation velocity, 163, 503
Dimple, 363
Direct absorption, 521
Direct central impact, 639
Direct electric heating, 691
Discrete element method, 589
Displacement rate, 303
Dissimilar adhesively joints, 463
Dissipation, 675
Distinct element method, 445, 619
Drop hammer test, 175
Drop/impact analysis, 975
Drop/impact simulation, 981
Dry powder, 117
Ductile metal, 839
Ductility, 547
Dynamic behavior, 433
Dynamic behavior of materials, 487
Dynamic behaviour, 719
Dynamic buckling, 111
Dynamic compression test, 29
Dynamic crushing strength, 15
Dynamic deformation, 887
Dynamic Failure Mechanics (DFM), 425
Dynamic failure resistance, 321
Dynamic fracture, 419, 657, 663
Dynamic J integral, 381
Dynamic loading, 547
Dynamic local buckling, 761
Dynamic material behavior, 565
Dynamic material property, 351
Dynamic nonlinear analysis, 969
Dynamic plastic buckling, 105
Dynamic plasticity, 195
Dynamic powder consolidation, 905
Dynamic progressive buckling, 105
Dynamic properties, 767
Dynamic recrystallization, 123
Dynamic response, 35, 755, 791, 981
Dynamics, 1, 71
Dynamic simple shear, 83
Dynamic strength, 57
Dynamic stress intensity factor, 663
Dynamic tensile loading, 577
Dynamic tensile strength, 419
Dynamic tension, 821
Dynamic testing, 49

Dynamic thermal stress, 333
Dynamic yield stress, 91

EFP, 963
Elastic bar, 773
Elasticity, 333
Elastic material, 857
Elastic region, 475
Elastic wave, 857
Elasto visco-plastic fracture, 389
Elasto-plastic impact, 639
Electric resistivity, 249
Electromagnetic forming, 285, 827
Electronic product, 981
Energetic coefficient of restitution, 675
Energetic material, 175
Energy absorber, 559
Energy absorption, 111, 237, 291, 815, 887
Energy absorption characteristics, 875
Energy transfer, 175
Equilibrium (Thermo-elastic) stress, 863
Erosion, 339, 451
Experimental tests, 755
Experiments, 23, 57, 625, 651, 683
Explosion, 117
Explosion products, 171
Explosion welding, 401, 407
Explosive, 171, 419, 869, 899, 905, 917
Explosive cladding, 495
Explosive cleansing, 495
Explosive compaction, 495, 899
Explosive forming, 495, 917
Explosive generator of pressure (EGP), 951
Explosive loading, 195, 851
Explosive materials, 201
Explosive materials processing, 495
Explosive removal of slag, 495
Explosive transformation, 893
Explosive welding, 273, 495, 923
Extremely high shock pressure, 927

Fabric armour, 327
Failure, 1, 755
Failure criteria, 195, 736
Failure mechanisms, 297
Falling rock speed, 881
Fast crack propagation, 389
Fatigue strength, 345
FDM, 791, 863
FEM, 65, 333, 413, 469, 475, 639
FEM simulation and experiments, 713
Fibrillation, 291
Finite difference scheme, 857
Finite element analysis, 613

Finite Element Method (FEM), 981, 407, 463, 657, 975, 981
Finite element simulation, 535, 553
Firply-wood, 243
Flexible rock-fall fence, 619
Fluid pressure, 791
Flying projectile, 713
Foam core sandwich beam, 35
Foam model, 369
Food, 117
Fracture, 463
Fragment, 893
Fracture behavior, 315
Fracture criterion, 43
Fracture mechanics parameter, 389
Fracture mechanics, 401, 407
Fracture mechanism, 547
Fracture shock waves, 395
Fracture strength, 401
Fracture surface, 363
Fracture toughness, 43
Fragment acceleration, 207
Fragment creation, 145
Fragmentation, 951
Free surface velocity, 419
Friction, 291
Friction-welded butt joint, 457
Front ply, 297
FRP, 705
Functionally graded material, 333, 905
Fuse network model, 743

Gauge length, 57
Geometrical initial imperfection, 887
GFRP, 705
Global-local modeling, 975
Golf club, 699
Grain size, 547
Granite, 951

Hardened zone, 407
Hardness, 249
Hazard assessments, 1
Heat affected zone, 821
Heat wave, 857
Hertzian cone fracture, 945
High cycle fatigue failure and impact load, 731
High energy atom, 683
High-energy forming, 827
High-energy-rate forming, 911
High explosive, 521, 893, 933
High performance concrete, 613
High pressure, 933
High pressure material science, 487

High-speed camera, 419
High speed diagnostic, 279
High-speed response, 691
High strain rate, 83, 99, 351, 633, 803, 833, 839, 99, 439, 571
High strain rate testing, 565
High strength fabric, 291, 297
High strength fiber, 803
High strength steel, 315
High-temperature air and nitrogen, 151
High-temperature plasma, 515
High-velocity deep penetration, 189
High velocity impact, 413
Hollow cylinders, 469, 475
Hopkinson bar, 49, 71, 303, 321, 363, 457
Hopkinson's effect, 419
Hot-spot ignition, 201
Hugoniot, 863
Hugoniot elastic limit, 785
Hydro code, 736, 851
Hydrodynamic code, 521
Hydrostatic pressure-shear, 91
Hydroxyapatite, 905
Hypervelocity, 261
Hypervelocity impact, 139, 219, 225, 395, 845

ICCD, 151
Ignition and growth model, 521
Imaging spectroscopy, 151
Impact, 1, 23, 291, 309, 321, 339, 351, 369, 433, 553, 559, 601, 625, 651, 683, 755, 767, 893, 663
Impact analysis, 699
Impact behavior, 451
Impact behaviour, 541
Impact consolidation, 261
Impact-echo method, 607
Impact energy absorption, 15
Impact-energy absorption capacity, 809
Impact experiments, 669
Impact failure, 35
Impact fatigue, 315, 345
Impact flash, 139
Impact fracture, 381
Impact load, 475, 797, 887
Impact loading, 613, 815
Impact of two balls, 639
Impact punch shear test, 303
Impact response analysis, 619
Impact strength, 43, 375, 481
Impact stress-strain curve, 363
Impact tensile load, 469
Impact tensile strength, 463
Impact tensile test, 363

Impact tension test, 457
Impact test, 779, 969
Impact velocity, 669
Improved method, 863
Impulsive pressure, 285, 911
Impulsive push-up test, 761
Inelastic, 1
Inertia, 15
Infrared thermography, 401
Initial elastic coefficient, 583
Initial shape irregularity, 809
Initiation, 279, 515, 869, 893
Initiation times, 279
Inside temperature method, 863
Instrumented Charpy impact test, 65
Instrumented long bar impact, 945
Instrumented striker, 65
Intact and damaged honeycombs, 719
Intense shock loading, 487
Interface, 469
Interface stress, 475
Interfacial fracture mechanics, 381
Interference effect, 669
Intermetallic compound, 261
Intermetallics, 939
Intersonic crack propagation, 381
Ion-beam driven shock loading, 487
Iron, 863

Johnson-Cook model, 553
Joining, 273

Kolsky bar, 23

Lap jointing, 827
Large deformation, 389
Laser, 395
Laser-driven flyer plate, 521
Laser heating, 357
Laser impact, 839
Laser vibrometer, 607
Laser welding method, 821
Lead cover, 279
Length/radius ratio, 963
LEO, 683
Liner, 963
Line-spring model, 663
Liquid metal, 339
Loading rate, 463
Local melting, 577
Long-rod penetration, 413
Low-and medium-cycle, 315
Low-velocity impact, 145

Mach detonation, 933

Magnesium alloys, 547
Magnetic pressure, 827
Magnetic property, 267
Magnetic pulse loading, 487
Manganin gauge, 509
Marble, 951
Mass conversion ratio, 963
Mass distributions, 219
Material, 57
Material inhomogeneity, 407
Material model, 613
Materials synthesis, 255
Material testing, 77, 833
Maximum principal stress, 469, 475
Mechanical alloying, 261, 267
Mechanical property, 911
Mercury droplet, 451
Mercury target, 779
Meshless, 213
Metal jets, 927
Metallic glass, 273
Metastable bulk alloy, 267
Mg particle, 899
Micro void evolution process, 713
Microhardness, 457
Microstructures, 249
Middle-size slope model, 881
Mild steel, 439
Mini-laser-driven shock loading, 487
Mixed mode, 43
Morphology, 15
Moving finite element method, 381, 389
Multi-body dynamics, 675
Multi-layered shells, 791
Multi layer explosive welding, 957
Multi-layer structure, 231
Multiple regression analysis, 881
Multi-point initiation, 963

Near-source earthquake, 657
Necking, 821
Negative pressure, 779
Neutron scattering facility, 779
Nickel aluminide, 261
Node separation, 413
Non-coaxial Hopkinson bar method, 821
Non-destructive evaluation, 445, 607
Non-ideal detonation, 163, 503
Nose factor, 183
Notched specimen, 315
Numerical analysis, 77, 601, 669, 809, 887
Numerical inverse Laplace transformation, 797
Numerical simulation,, 57, 83, 195, 589, 736, 869, 933, 957

Oblique impact, 157, 327
Optimal design, 699
Orbital debris, 145, 395
Orthotropic, 589
Orthotropic media, 601
Oxidation, 683

Paper, 917
Paper foam board, 29
Particle velocity, 863
Path independent integral, 381
PCB, 171
Pendulum, 625
Penetration, 351
Perfect cone crack, 724
Perforation, 291, 297, 351, 375
Perforation energy, 303
Perforation of plate, 713
Permanent strain rate, 315
Personal protection, 803
Phase transformation, 691
Phonon, 175
Pit, 339
Plain-weave CFRP, 303
Plain-weave GFRP, 303
Plastic buckling, 887
Plastic deformation, 639, 375, 559
Plastic instability, 541
Plastic target, 189
Plastic wave propagation, 541
Plate, 651
Plateau stress, 237
Plate impact, 633, 851
Plate impact experiment, 833
Plate impact test, 445
Plugging, 375
Plugging fracture, 351
PMMA, 509, 669, 724, 833
Polymeric foams, 23
Polystyrene foam board, 29
Poroelastic material, 791
Porosity, 633, 839
Porous material, 111
Porous media, 595
Porous surface, 899
Powder explosive, 515
Powder metallurgy aluminum, 565
Pre-fatigue, 433
Pressure transducer, 773
Pressure wave, 339, 779
Pressure welding, 827
Projectile impact, 375
Projectile motion, 131
Projectile shapes, 291, 297

Pulse laser, 515
Pulsed proton beam, 779
Punching processing, 724
Punching, 911
PVDF, 773
PVDF gauge, 833
PVDF pressure gauge, 445

Radial breakthroughs, 279
Radial crack, 945
Radiation, 151
Raindrop, 705
Raman spectra, 175
Random choice method, 157
Rate dependence, 43
Rate-dependent constitutive relation, 743
Rayleigh wave, 595
Recovery force, 691
Recrystallization, 123
Rectangular plate, 755
Rectangular tube, 969
Redwood, 243
Reformed bamboo/aluminum laminate, 645
Reinforced concrete, 736
Relaxation time, 857
Repeated impact tension, 345
Repeated impacts, 445, 705
Repeated shock, 785
Repeated tensile load, 315
Residual stress, 407
Resistance welding, 827
Rigid body impact, 675
Rigid plastic constitutive model, 713
Rock, 419
Rockfall prevention wall, 881
Rock massive, 951
Rotating disk method, 669
Rupture, 1, 291

S15C carbon steel, 457
Safety, 1
Safety engineering, 809
Seam welding, 827
Second debris clouds, 219
Self organization, 123
Self-shielding bumper, 225
SEM, 363
Sensing block method, 77, 821
Sensitivity, 207
Separated dynamic J integral, 381
Separation of slabs, 951
Shape memory alloy, 691
Shape of edge, 911
Shape of falling rock, 881
Shape optimization, 699

Shaped charge jet initiation, 279
Shear bands, 123, 577
Shearing deformation, 911
Shear strength, 785
Shearing tensile strength, 827
Sheet metal, 83
Sheet-metal forming, 827
Sheet metal plasticity, 541
Shock-absorber, 243
Shock analysis, 981
Shock compaction, 249, 255, 495, 905
Shock compression, 267
Shock consolidation of powders, 939
Shock Hugoniot, 509, 785
Shock pressure, 509
Shock sensitivity, 175
Shock synthesis, 495, 939
Shock tube, 607
Shock wave instrumentation, 131
Shock waves, 117, 151, 607, 633, 839, 851, 869, 927, 939
SHPB, 571
SHS, 249
Silicon, 683
Simulation, 285, 625, 651
Single-edge notched specimen, 663
Slip bands, 345
Slope gradient, 881
Slope length, 881
Smooth specimen, 315
Smoothed particle hydrodynamics, 213
Solid metal, 339
Space, 683
Spacecraft, 145
Space debris, 139, 213, 219, 225, 231, 845
Spall, 195
Spallation, 395, 851
Spall damage, 445
Spall strength, 321
Specimen, 57
Specimen geometry, 535
Specimen length, 669
Spectroscopy, 139
Split Hopkinson bar, 99, 553, 803
Split Hopkinson pressure bar, 339, 565, 767
Splitting, 559
Square tubes, 105, 285, 887
Stacking sequence, 875
Star shaped tail, 963
Static and dynamic tests, 875
Static or dynamic perturbing stresses, 189
Steady shock wave, 863
Steel plate, 195
Steel tube, 761
Steels, 433, 651

Sterilization, 117
Strain gage, 463
Strain gage position, 65
Strain rate, 15, 91, 433, 457, 785
Strain rate-dependence, 29
Strain rate dependency, 77, 363
Strain rate effect, 99, 583, 743
Strain rate sensitivity, 1, 111, 821
Strain rates, 23, 57
Strain response, 749
Strain waves, 651
Strength, 91, 321, 469
Strength degradation, 433
Strength of the weld, 821
Stress distribution in HAZ, 821
Stress intensity factor, 401
Stress pulse, 773
Stress response, 469
Stress singularity, 463
Stress-strain characteristics, 457
Stress-strain curve, 833
Stress-strain relation, 99
Stress wave propagation, 105, 475
Stress wave propagation test, 29
Stress waveform, 345
Stress waves, 1, 445, 463, 589, 601, 797
Structural analysis, 77, 333, 657, 791
Structure instability, 357
Structures, 1
Structure steels, 315
Surface damage, 433
Surface roughness, 515
Surface tension, 451
Surge wave, 797
Survivability, 207
Suspending structure, 981
Synthesis behaviour, 249

T* integral, 389
Tall building, 657
Tantalum, 839
Target beam window, 779
Taylor rod test, 553
Taylor test, 541
Temperature, 91
Temperature dependence, 99, 571
Temperature-dependent properties, 333
Temperature-strain rate parameter, 99
Tensile strength, 57, 547, 577, 583
Tensile, 91
Tensile test, 439
Tension test, 99
Testing method, 57
The adiabatic exponent, 163
Thermal activation, 99

Thermal shock, 857
Thermal softening, 553
Thermal viscoplastic material, 357
Thermo-plastic behavior, 83
Thermo-visco-plasticity, 425
Thermo-viscoplastic modelling, 541
Thick shells, 333
Thickness effect, 303
Thin metal layer, 515
Thin walled-tube, 111
Three-dimensional, 601
Three-dimensional spatial distributions, 219
TIGER code, 163
Ti particle, 899
Tiny structure, 975
Titanium, 905
Titanium-aluminum nitride, 255
Titanium implant, 899
Titanium-silicon carbide, 255
Tomography X, 633
Toxic waste, 171
Transient response, 663
Transient stage, 657
Transmissivity, 773
Transonic crack propagation, 381
Transverse impact, 749
Transversely isotropic, 595
Trinitrotoluol, 893
Truncated-ogive-nose projectile, 183
Tube, 559
Tungsten heavy alloy penetrator, 309
Tungsten heavy alloys, 535
Tunnel, 607
Two notches, 669
Two-stage light gas gun, 157, 219, 225

Ultrasonic inspection, 445
Underwater discharge, 773
Underwater explosion, 923
Underwater shock pressure, 917
Underwater shock wave, 419, 923, 939, 957

Unifying framework, 201
Unilateral constraints, 675
Urethane foam filler, 111

Variable thickness, 761
Vectran and polyerethane, 231
Vein pattern, 577
Velocity interferometry, 131
Vibration, 175
VISAR, 131, 851
Viscoelastic bars, 71
Viscoelasticity, 767
Viscoelastic material, 49

Visco-plastic, 369
Viscoplasticity, 633, 839
Viscosity, 595
Viscous stress, 863

Warhead tests, 207
Waste treatment, 171
Water-filled tube, 375
Wave dispersion, 71
Wave profiles, 131
Wave propagation, 675

Wave separation, 49
Wedge test, 175
Weight reduction, 809
Work-hardening, 439

YNA3D, 475

¼ scale model, 243
2-ply system, 297
6061-T6 aluminum, 863
9m drop test, 24